ABOUT THE AUTHORS

REED WICANDER

Reed Wicander is a geology professor at Central Michigan University where he teaches physical geology, historical geology, prehistoric life, and invertebrate paleontology. He has co-authored several geology textbooks with James S. Monroe. His main research interests involve various aspects of Paleozoic palynology, specifically the study of acritarchs, on which he has published many papers. He is currently the President of the American Association of Stratigraphic Palynologists.

JAMES S. MONROE

James S. Monroe is a professor of geology at Central Michigan University where he teaches physical geology, historical geology, prehistoric life, and stratigraphy and sedimentology. He has co-authored several textbooks with Reed Wicander for West Publishing Company, and has interests in Cenozoic geology and geologic education.

Mount Aetna from Taormina, Thomas Cole, Oil on Canvas, 78⅝ × 120⅝ inches. Wadsworth Atheneum, Hartford, CT, dated 1843.

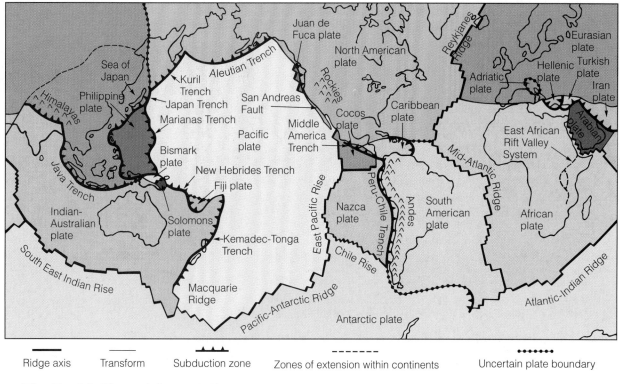

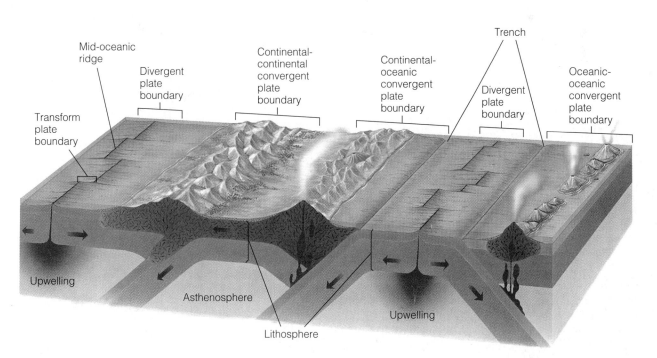

➤ The Earth's Plates (Figure 1-9)

➤ Three Principal Types of Plate Boundaries (Figure 1-10)

Essentials of
GEOLOGY

Reed Wicander
James S. Monroe
Central Michigan University

WEST PUBLISHING COMPANY

Minneapolis/St. Paul   New York   Los Angeles   San Francisco

PRODUCTION CREDITS

Copyediting and Indexing   Patricia Lewis
Interior Design and Cover Design   Diane Beasley
Artwork   Darwen and Vally Hennings, Carlyn Iverson, Precision Graphics, and Victor Royer. Individual credits follow index.
Composition   Carlisle Communications, Ltd.
Cover Image   The cover painting, *Mount Aetna from Taormina,* is by Thomas Cole (dated 1843), and represents the Hudson River School of Painting. Cole, the most noted of this group of artists, was known for his inspiring views of the natural world. A number of geologic phenomena are represented in this painting, as well as in the other paintings in this book. Courtesy of the Wadsworth Atheneum.

WEST'S COMMITMENT TO THE ENVIRONMENT

In 1906, West Publishing Company began recycling materials left over from the production of books. This began a tradition of efficient and responsible use of resources. Today, up to 95 percent of our legal books and 70 percent of our college and school texts are printed on recycled, acid-free stock. West also recycles nearly 22 million pounds of scrap paper annually-the equivalent of 181,717 trees. Since the 1960s, West has devised ways to capture and recycle waste inks, solvents, oils, and vapors created in the printing process. We also recycle plastics of all kinds, wood, glass, corrugated cardboard, and batteries, and have eliminated the use of Styrofoam book packaging. We at West are proud of the longevity and the scope of our commitment to the environment.

Production, Prepress, Printing and Binding by West Publishing Company.

COPYRIGHT ©1995   By WEST PUBLISHING COMPANY
610 Opperman Drive
P.O. Box 64526
St. Paul, MN 55164–0526

All rights reserved

Printed in the United States of America

02 01 00 99 98 97 96   8 7 6 5 4 3 2 1

Library of Congress Cataloging-in-Publication Data

Wicander, Reed.
    Essentials of geology / Reed Wicander, James S. Monroe.
        p.   cm.
    Includes index.
    ISBN 0-314-04562-7
    1. Geology.   I. Monroe, James S. (James Stewart), 1938-
II. Title.
    QE26.2.W53   1995
    550--dc20   94-44477
                                                          CIP

British Library Cataloguing-in-Publication Data. A catalogue record for this book is available from the British Library.

TEXT IS PRINTED ON 10% POST CONSUMER RECYCLED PAPER

Printed with **Printwise**
Environmentally Advanced Water Washable Ink

∞

BRIEF
CONTENTS

CONTENTS

 Chapter 1

UNDERSTANDING THE EARTH: AN INTRODUCTION TO PHYSICAL GEOLOGY

 Chapter 2

PLATE TECTONICS: A UNIFYING THEORY

 *Chapter 3*

MINERALS

Chapter 4:

IGNEOUS ROCKS AND INTRUSIVE IGNEOUS ACTIVITY

 *Chapter 5*

VOLCANISM

 *Chapter 6*

WEATHERING, EROSION, AND SOIL

 *Chapter 7*

SEDIMENT AND SEDIMENTARY ROCKS

 *Chapter 8*

METAMORPHISM AND METAMORPHIC ROCKS

 *Chapter 11*
MASS WASTING

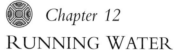 *Chapter 12*
RUNNING WATER

 *Chapter 13*

GROUNDWATER

Chapter 14

GLACIERS AND GLACIATION

 Chapter 17

GEOLOGIC TIME: CONCEPTS AND PRINCIPLES

 Chapter 18

EARTH HISTORY

Chapter 19
LIFE HISTORY

PREFACE

The Earth is a dynamic planet that has changed continuously during its 4.6 billion years of existence. The size, shape, and geographic distribution of the continents and ocean basins have changed through time, as have the atmosphere and the organisms that inhabit the Earth. Over the past 20 years, bold new theories and discoveries concerning the Earth's origin and how it works have sparked a renewed interest in geology. We have become increasingly aware of how fragile our planet is and, more importantly, how interdependent all of its various systems are. We have learned that we cannot continually pollute our environment and that our natural resources are limited and, in most cases, nonrenewable. Furthermore, we are coming to realize how central geology is to our everyday lives. For these and other reasons, geology is one of the most important college or university courses a student can take.

Essentials of Geology is designed for a one-semester introductory course and is written with the student in mind. One of the problems with any introductory science course is that students are overwhelmed by the amount of material they must learn. Furthermore, much of the material may not seem to be linked by any unifying theme or appear to be relevant to their lives.

One of the goals of this book is to provide students with a basic understanding of geology and its processes and, more importantly, with an understanding of how geology relates to the human experience; that is, how geology affects not only individuals, but society in general. It is also our intention to provide students with an overview of the geologic and biologic history of the Earth, not as a set of encyclopedic facts to memorize, but rather as a continuum of interrelated events that reflect the underlying geologic and biologic principles and processes that have shaped our planet and life upon it. With these goals in mind, we introduce the major themes of the book in the first chapter to provide students with an overview of the subject and enable them to see how the various systems of the Earth are interrelated. We also discuss the economic and environmental aspects of geology throughout the book rather than treating these topics in separate chapters. In this way students can see, through relevant and interesting examples, how geology impacts our lives.

TEXT ORGANIZATION

Plate tectonic theory is the unifying theme of geology and this book. This theory has revolutionized geology because it provides a global perspective of the Earth and allows geologists to treat many seemingly unrelated geologic phenomena as part of a total planetary system. Because plate tectonic theory is so important, it is covered in Chapter 2 and is discussed in most subsequent chapters as it relates to the subject matter of that chapter.

We have organized *Essentials of Geology* into several informal categories. Chapter 1 provides an introduction to geology and discusses its relevance to the human experience, plate tectonic theory, the rock cycle, geologic time and uniformitarianism, and the origin of the solar system and Earth. Chapter 2 deals with plate tectonics, while Chapters 3–8 examine the Earth's materials (minerals and igneous, sedimentary, and metamorphic rocks) and the geologic processes associated with them including the role of plate tectonics in their origin and distribution. Chapters 9 and 10 deal with the related topics of the Earth's interior, earthquakes, and deformation and mountain building. Chapters 11–16 cover the Earth's surface processes, and Chapter 17 discusses geologic time, introduces several dating methods, and explains how geologists correlate rocks. Chapter 18 and 19 provide an overview of the geologic history of the Earth and its biota.

We have found that presenting the material in this order works well for most students. We know, however, that many instructors prefer an entirely different order of topics depending on the emphasis in their course. We have therefore written this book so that instructors can present the chapters in any order that suits the needs of their course.

CHAPTER ORGANIZATION

All chapters have the same organizational format. Each chapter opens with a photograph relating to the chapter material, a detailed outline, and a Prologue, which is designed to stimulate interest in the chapter by discussing relevant aspects of the material.

The text is written in a clear informal style, making it easy for students to comprehend. Numerous color diagrams and photographs complement the text, providing a visual representation of the concepts and information presented. Each chapter contains a Perspective that briefly discusses an interesting aspect of geology or geological research. Mineral and energy resources are discussed in the final sections of several chapters.

The end-of-chapter materials begin with a concise review of important concepts and ideas in the Chapter Summary. The Important Terms, which are printed in boldface type in the chapter text, are listed at the end of each chapter for easy review, and a full glossary of important terms appears at the end of the text. The Review Questions are another important feature of this book; they include multiple-choice questions with answers as well as short essay questions. More challenging questions appear under the Points to Ponder heading.

SPECIAL FEATURES

This book contains a number of special features that set it apart from other geology textbooks. Among them are a critical thinking and study skills section, the chapter Prologues, the integration of economic and environmental geologic issues throughout the book, and a set of multiple-choice questions with answers for each chapter. A separate section entitled "Points to Ponder" contains thought-provoking and quantitative questions.

Study Skills

Immediately following the Preface is a section devoted to developing critical thinking and study skills. This section offers hints to help students improve their study habits, prepare for exams, and generally get the most out of every course they take. While these tips can be helpful in any course, many of them are particularly relevant to geology. Whether you are just beginning college or about to graduate, take a few minutes to read over this section as these suggestions can help you in your studies and later in life.

Prologues

Many of the introductory Prologues focus on the human aspects of geology such as the eruption of Krakatau (Chapter 1), the Northridge earthquake (Chapter 9), and the Flood of '93 (Chapter 12).

Economic and Environmental Geology

The topics of environmental and economic geology are discussed throughout the text. Integrating economic and environmental geology with the chapter material helps students see the importance and relevance of geology to their lives. In addition, several chapters close with a section on resources, further emphasizing the importance of geology in today's world.

Figures

Many of the illustrations depicting geologic processes or events are block diagrams rather than cross sections so that students can more easily visualize the salient features of these processes and events. Our color paleogeographic maps in Chapter 18 are designed to illustrate clearly and accurately the geography during the various geologic periods. Full-color scenes showing associations of plants and animals in Chapter 19 are based on the most current interpretations. Great care has been taken to ensure that the art and captions provide an attractive, informative, and accurate illustration program.

Figure and Table Reference System

A color cue (➤) will be found in the text next to the first reference for each figure, and a (◉) appears beside the first reference for each table. This system is designed to help students quickly return to their place in the text when they interrupt their reading to examine an illustration or table.

Perspectives

The chapter Perspectives generally focus on aspects of environmental and economic geology such as asbestos (Chapter 8), radioactive waste disposal (Chapter 13), and radon (Chapter 17). The topics for the Perspectives were chosen to provide students with an overview of the many fascinating aspects of geology. The Perspectives can be assigned as part of the chapter reading, used as the basis for lecture or discussion topics, or even used as the starting point for student papers.

ANCILLARY MATERIALS

To assist you in teaching this course and supplying your students with the best in teaching aids, West Publishing Company has prepared a complete package available to all adopters:

■ The comprehensive instructor's manual and test bank includes teaching ideas, learning objectives, lecture outlines in the form of a point-by-point summary, discussions of common student misconceptions, a list of media sources, Consider This lecture questions, enrichment topics, lists of acetates and slides that accompany the text, and a test bank. The test bank contains approximately two thousand multiple-choice, true/false, fill-in-the-blank, matching, and short-answer questions.

■ We can provide the entire text bank on diskette along with WESTEST, a computerized testing package. Using WESTEST 3.0, it is possible to generate examinations using either questions selected by the instructor or those randomly generated by the computer. The WESTEST 3.0 edit function makes it possible to modify these questions, add new questions, or delete existing questions. Additionally, West's Classroom Management Software allows student data to be recorded, stored, and used for various reports.

■ The new CD-ROM disk, *In-TERRA-Active*, developed through West by Phil Brown (University of Wisconsin—Madison) and Jeremy Dunning (University of Indiana—

Bloomington), provides instructors and students with meaningful new ways to enhance the textbook. The illustrations, animations, photos, and video make this a teaching/learning tool of great value. Additionally, the interactive modules provide students with a new mode for mastering the material.

- West's Geology Videodisc allows instructors to display photographic images, illustrations, video segments, and computer animations in lecture or lab settings. Developed specifically for our larger text *Physical Geology: Exploring the Earth,* 2d ed., the disk contains more than 1,500 still photographs of geologic features organized by region. These can be used to show students examples of the formations discussed that are from the local area, from other regions of the country, or from around the world. The videodisc also includes illustrations and diagrams from the text.

- West's Geology Videotape Library includes the entire Planet Earth film series as well as additional programs that discuss earthquakes, mineral resources, and environmental geology topics. For a complete, up-to-date listing of the titles available, please contact your West representative.

- Three slide sets are available to qualified adopters. The first set includes approximately 150 of the most important and attractive illustrations and photographs from this text and our larger *Physical Geology,* 2d ed., and the second set contains over 450 slides illustrating important geologic features. The majority of these photographs are from North America, but the set also includes examples from around the world and the solar system. The third set includes images from the In-TERRA-Active CD, plus more photos of the 1994 Northridge earthquake.

- A new set of 225 full-color transparency acetates provides clear and effective illustrations of important artwork and maps from the text.

- A new ancillary containing all acetates printed and bound with perforated, three-hole-punched pages allows students to take notes as the acetates are shown in lecture. It also contains *Study Skills for Science Students* by Daniel Chiras, which is described below.

- *Current Issues in Geology: Selected Readings,* 2d ed., prepared by Michael L. McKinney and Robert L. Tolliver of the University of Tennessee—Knoxville, is a collection of approximately 65 very current articles that supplement material students will encounter in their coursework. The articles have been selected from a number of general interest and science magazines. West can make this supplement available with the text as a set, or it can be purchased separately.

- A copy of *Great Ideas for Teaching Geology* is available free to all adopters of the textbook. This 100-page book is a collection of lecture suggestions, demonstrations, analogies, and other ideas contributed by geology teachers from across the country. These ideas are intended to provide instructors with a variety of approaches to teaching some of the difficult concepts in geology.

- *The Changing Earth Update,* West's biannual geology newsletter, is provided to adopters twice each year to update the book with recent and relevant research news. This will ensure that your students have the most current information available.

- West Publishing also offers tutorial software for students' review and software for lecture presentations. For example, GeoTutor by Vicki Harder is a hypercard stack designed as a review program in the format of a question and answer game. Interactive Geoscience Tutorials take topics and figures in geology and illustrate them using full-color animation. Modules include the Rock Cycle, Igneous Rocks, Sedimentary Rocks, Plate Tectonics, Weathering and Erosion, Earthquakes, Minerals, and Mass Wasting, among others. Each tutorial begins with an introduction file that illustrates a concept related to the title for the tutorial. Three software programs are available from Micro-Innovations, Inc. *Quake* helps students understand the distribution of earthquakes. *Groundwater* allows students to manipulate several hydrologic variables simultaneously, then rapidly solve the modified program, and display the results graphically, and *Coastal* is an instructional program to simulate the effects of wave action on beach shape.

- *Perspectives in Canadian Geology,* prepared by I. Peter Martini and Ward Chesworth of the University of Guelph, is a collection of essays that expands upon topics in the text by highlighting the geologic features of Canada.

- Lastly, West has available *Study Skills for Science Students* by Daniel Chiras. This supplement emphasizes critical thinking and developing a positive lifestyle and provides students simple ways to improve memory, learn more quickly, get the most out of lectures, prepare for tests, produce top-notch term papers, and improve critical-thinking skills. West can make this supplement available with the text as a set, or it can be purchased separately.

ACKNOWLEDGMENTS

As the authors, we are, of course, responsible for the organization, style, and accuracy of the text, and any mistakes, omissions, or errors are our responsibility. During the preparation of our previous geology textbooks, we received numerous comments and advice from many geologists who reviewed parts of the various texts. We wish to express our sincere appreciation to all of the many reviewers whose contributions were invaluable during the writing of those books. Because *Essentials of Geology* combines material from our previous books, the reviewers of those books thus contributed to this book.

More specifically, we would like to thank the following reviewers who advised us on developing *Essentials*. Their thoughts and suggestions were instrumental in organizing the text.

Lawrence Balthaser
California Polytechnic State University—San Luis Obispo
Fredric R. Goldstein
Trenton State College

Brian Grant
Brock University

William F. Kean
University of Wisconsin—Milwaukee

Susan Morgan
Utah State University

Louis Pinto
Monroe Community College

Robert J. Smith
Seattle University

James L. Talbot
Western Washington University

We also wish to thank Richard V. Dietrich, Eric L. Johnson, David J. Matty, Jane M. Matty, Wayne E. Moore, and Stephen D. Stahl of the Geology Department of Central Michigan University, and Bruce M. C. Pape of the Geography Department for providing us with photographs. We also thank Pam Iacco and Kathleen Butzier for their assistance in the preparation of this manuscript. In addition, we are also grateful for the generosity of the various agencies and individuals from many countries who provided photographs.

Special thanks must go to Jerry Westby, college editorial manager for West Publishing Company, who made many valuable suggestions and patiently guided us through this project. His continued encouragement provided constant inspiration and helped us produce the best possible book. Thanks to West Publishing developmental editors Dean DeChambeau and Betsy Friedman, who have overseen the extensive ancillary package. We are equally indebted to our production editor, Matt Thurber, whose attention to detail and consistency is greatly appreciated. We would also like to thank Patricia Lewis for her excellent copyediting and indexing skills. We appreciate her help in improving our manuscript. Because geology is such a visual science, we extend special thanks to Carlyn Iverson who rendered the reflective art and to the artists at Precision Graphics who were responsible for much of the rest of the art program. They did an excellent job, and we enjoyed working with them. We would also like to acknowledge our promotion manager, Stephanie Buss, for her help in the development of the promotional material for this edition, Maureen Rosener, media editor at West, who developed the excellent videodisc that accompanies this book, and Maureen, Lucinda Gatch, Kent Baird, and Robyn Thorson for their hard work developing the CD-ROM project.

Our families were, as always, patient and encouraging when most of our spare time and energy were devoted to this book. We thank them for their continuing support and understanding.

DEVELOPING
CRITICAL THINKING
AND STUDY SKILLS

INTRODUCTION

College is a demanding and important time, a time when your values will be challenged, and you will try out new ideas and philosophies. You will make personal and career decisions that will affect your entire life. One of the most important lessons you can learn in college is how to balance your time among work, study, and recreation. If you develop good time management and study skills early in your college career, you will find that your college years will be successful and rewarding.

This section offers some suggestions to help you maximize your study time and develop critical thinking and study skills that will benefit you, not only in college, but throughout your life. While mastering the content of a course is obviously important, learning how to study and to think critically is, in many ways, far more important. Like most things in life, learning to think critically and study efficiently will initially require additional time and effort, but once mastered, these skills will save you time in the long run.

You may already be familiar with many of the suggestions and may find that others do not directly apply to you. Nevertheless, if you take the time to read this section and apply the appropriate suggestions to your own situation, we are confident that you will become a better and more efficient student, find your classes more rewarding, have more time for yourself, and get better grades. We have found that the better students are usually also the busiest. Because these students are busy with work or extracurricular activities, they have had to learn to study efficiently and manage their time effectively.

One of the keys to success in college is avoiding procrastination. While procrastination provides temporary satisfaction because you have avoided doing something you did not want to do, in the long run it leads to stress. While a small amount of stress can be beneficial, waiting until the last minute usually leads to mistakes and a subpar performance. By setting clear, specific goals and working toward them on a regular basis, you can greatly reduce the temptation to procrastinate. It is better to work efficiently for short periods of time than to put in long, unproductive hours on a task, which is usually what happens when you procrastinate.

Another key to success in college is staying physically fit. It is easy to fall into the habit of eating junk food and never exercising. To be mentally alert, you must be physically fit. Try to develop a program of regular exercise. You will find that you have more energy, feel better, and study more efficiently.

GENERAL STUDY SKILLS

Most courses, and geology in particular, build upon previous material, so it is extremely important to keep up with the coursework and set aside regular time for study in each of your courses. Try to follow these hints, and you will find you do better in school and have more time for yourself:

- Develop the habit of studying on a daily basis.
- Set aside a specific time each day to study. Some people are day people, and others are night people. Determine when you are most alert and use that time for study.
- Have an area dedicated for study. It should include a well-lighted space with a desk and the study materials you need, such as a dictionary, thesaurus, paper, pens and pencils, and a computer if you have one.
- Study for short periods and take frequent breaks, usually after an hour of study. Get up and move around and do something completely different. This will help you stay alert, and you'll return to your studies with renewed vigor.
- Try to review each subject every day or at least the day of the class. Develop the habit of reviewing lecture material from a class the same day.
- Become familiar with the vocabulary of the course. Look up any unfamiliar words in the glossary of your textbook or in a dictionary. Learning the language of the discipline will help you learn the material.

GETTING THE MOST FROM YOUR NOTES

If you are to get the most out of a course and do well on exams, you must learn to take good notes. This does not mean you should try to take down every word your professor says. Part of being a good note taker is knowing what is important and what you can safely leave out.

Early in the semester, try to determine whether the lecture will follow the textbook or be predominantly new material. If much of the material is covered in the textbook, your notes do not have to be as extensive or detailed as when the material is new. In any case, the following suggestions should make you a better note taker and enable you to derive the maximum amount of information from a lecture:

- Regardless of whether the lecture discusses the same material as the textbook or supplements the reading assignment, read or scan the chapter the lecture will cover before class. This way you will be somewhat familiar with the concepts and can listen critically to what is being said rather than trying to write down everything. Later a few key words or phrases will jog your memory as to what was said.
- Before each lecture, briefly review your notes from the previous lecture. Doing this will refresh your memory and provide a context for the new material.
- Develop your own style of note taking. Do not try to write down every word. These are notes you're taking, not a transcript. Learn to abbreviate and develop your own set of abbreviations and symbols for common words and phrases: for example, w/o (without), w (with), = (equals), ∧ (above or increases), ∨ (below or decreases), < (less than), > (greater than), & (and), u (you).
- Geology lends itself to many abbreviations that can increase your note-taking capability: for example, pt (plate tectonics), ig (igneous), meta (metamorphic), sed (sedimentary), rx (rock or rocks), ss (sandstone), my (million years), and gts (geologic time scale).
- Rewrite your notes soon after the lecture. Rewriting your notes helps reinforce what you heard and gives you an opportunity to determine whether you understand the material.
- By learning the vocabulary of the discipline before the lecture, you can cut down on the amount you have to write—you won't have to write down a definition if you already know the word.
- Learn the mannerisms of the professor. If he or she says something is important or repeats a point, be sure to write it down and highlight it in some way. Students have told me (RW) that when I stated something twice during a lecture, they knew it was important and probably would appear on a test. (They were usually right!)
- Check any unclear points in your notes with a classmate or look them up in your textbook. Pay particular attention to the professor's examples. These usually elucidate and clarify an important point and are easier to remember than an abstract concept.
- Go to class regularly, and sit near the front of the class if possible. It is easier to hear and see what is written on the board or projected onto the screen, and there are fewer distractions.
- If the professor allows it, tape record the lecture, but don't use the recording as a substitute for notes. Listen carefully to the lecture and write down the important points; then fill in any gaps when you replay the tape.

- If your school allows it, and they are available, buy class lecture notes. These are usually taken by a graduate student who is familiar with the material; typically they are quite comprehensive. Again use these notes to supplement your own.
- Ask questions. If you don't understand something, ask the professor. Many students are reluctant to do this, especially in a large lecture hall, but if you don't understand a point, other people are probably confused as well. If you can't ask questions during a lecture, talk to the professor after the lecture or during office hours.

GETTING THE MOST OUT OF WHAT YOU READ

The old adage that "you get out of something what you put into it" is very true when it comes to reading textbooks. By carefully reading your text and following these suggestions, you can greatly increase your understanding of the subject:

- Look over the chapter outline to see what the material is about and how it flows from topic to topic. If you have time, skim through the chapter before you start to read in depth.
- Pay particular attention to the tables, charts, and figures. They contain a wealth of information in abbreviated form and illustrate important concepts and ideas. Geology, in particular, is a visual science, and the figures and photographs will help you visualize what is being discussed in the text and provide actual examples of features such as faults or unconformities.
- As you read your textbook, highlight or underline key concepts or sentences, but make sure you don't highlight everything. Make notes in the margins. If you don't understand a term or concept, look it up in the glossary.
- Read the chapter summary carefully. Be sure you understand all of the key terms, especially those in boldface or italic type. Because geology builds on previous material, it is imperative that you understand the terminology.
- Go over the end-of-chapter questions. Write out your answers as if you were taking a test. Only when you see your answer in writing will you know if you really understood the material.

DEVELOPING CRITICAL THINKING SKILLS

Few things in life are black and white, and it is important to be able to examine an issue from all sides and come to a logical conclusion. One of the most important things you will learn in college is to think critically and not accept everything you read and hear at face value. Thinking critically is particularly important in learning new material and relating it to what you already know. Although you can't know everything, you can learn to question effectively and arrive at conclusions consistent with the facts. Thus, these suggestions for critical thinking can help you in all your courses:

- Whenever you encounter new facts, ideas, or concepts, be sure you understand and can define all of the terms used in the discussion.
- Determine how the facts or information was derived. If the facts were derived from experiments, were the experiments well executed and free of bias? Can they be repeated? The controversy over cold fusion is an excellent example. Two scientists claim to have produced cold fusion reactions using simple experimental laboratory apparatus, yet other scientists have as yet been unable to achieve the same reaction by repeating the experiments.
- Do not accept any statement at face value. What is the source of the information? How reliable is the source?
- Consider whether the conclusions follow from the facts. If the facts do not appear to support the conclusions, ask questions and try to determine why they don't. Is the argument logical or is it somehow flawed?
- Be open to new ideas. After all, the underlying principles of plate tectonic theory were known early in this century, yet were not accepted until the 1970s in spite of overwhelming evidence.
- Look at the big picture to determine how various elements are related. For example, how will constructing a dam across a river that flows to the sea affect the stream's profile? What will be the consequences to the beaches that will be deprived of sediment from the river? One of the most important lessons you can learn from your geology course is how interrelated the various systems of the Earth are. When you alter one feature, you affect numerous other features as well.

IMPROVING YOUR MEMORY

Why do you remember some things and not others? The reason is that the brain stores information in different ways and forms, making it easy to remember some things and difficult to remember others. Because college requires that you learn a vast amount of information, any suggestions that can help you retain more material will help you in your studies:

- Pay attention to what you read or hear. Focus on the task at hand, and avoid daydreaming. Repetition of any sort will help you remember material. Review the previous lecture before going to class, or look over the last chapter before beginning the next. Ask yourself questions as you read.
- Use mnemonic devices to help you learn unfamiliar material. For example, the order of the Paleozoic periods (Cambrian, Ordovician, Silurian, Devonian, Mississippian, Pennsylvanian, and Permian) of the geologic time scale can be remembered by the phrase, *Campbell's Onion Soup Does Make Peter Pale*, or the order of the Cenozoic epochs (Paleocene, Eocene, Oligocene, Miocene, Pliocene, and Pleistocene) can be remembered by the phrase, *Put Eggs On My Plate Please*. Using rhymes can also be helpful.

- Look up the roots of important terms. If you understand where a word comes from, its meaning will be easier to remember. For example, *pyroclastic* comes from *pyro* meaning fire and *clastic* meaning broken pieces. Hence a pyroclastic rock is one formed by volcanism and composed of pieces of other rocks. We have provided the roots of many important terms throughout this text to help you remember their definitions.
- Outline the material you are studying. This will help you see how the various components are interrelated. Learning a body of related material is much easier than learning unconnected and discrete facts. Looking for relationships is particularly helpful in geology because so many things are interrelated. For example, plate tectonics explains how mountain building, volcanism, and earthquakes are all related. The rock cycle relates the three major groups of rocks to each other and to subsurface and surface processes (Chapter 1).
- Use deductive reasoning to tie concepts together. Remember that geology builds on what you learned previously. Use that material as your foundation and see how the new material relates to it.
- Draw a picture. If you can draw a picture and label its parts, you probably understand the material. Geology lends itself very well to this type of memory device because so much is visual. For example, instead of memorizing a long list of glacial terms, draw a picture of a glacier and label its parts and the type of topography it forms.
- Focus on what is important. You can't remember everything, so focus on the important points of the lecture or the chapter. Try to visualize the big picture, and use the facts to fill in the details.

PREPARING FOR EXAMS

For most students, tests are the critical part of a course. To do well on an exam, you must be prepared. These suggestions will help you focus on preparing for examinations:

- The most important advice is to study regularly rather than try to cram everything into one massive study session. Get plenty of rest the night before an exam, and stay physically fit to avoid becoming susceptible to minor illnesses that sap your strength and lessen your ability to concentrate on the subject at hand.
- Set up a schedule so that you cover small parts of the material on a regular basis. Learning some concrete examples will help you understand and remember the material.
- Review the chapter summaries. Construct an outline to make sure you understand how everything fits together. Drawing diagrams will help you remember key points. Make up flash cards to help you remember terms and concepts.
- Form a study group, but make sure your group focuses on the task at hand, not on socializing. Quiz each other and compare notes to be sure you have covered all the

material. We have found that students dramatically improved their grades after forming or joining a study group.

■ Write out the answers to all of the Review Questions. Before doing so, however, become thoroughly familiar with the subject matter by reviewing your lecture notes and reading the chapter. Otherwise, you will spend an inordinate amount of time looking up answers.

■ If you have any questions, visit the professor or teaching assistant. If review sessions are offered, be sure to attend. If you are having problems with the material, ask for help as soon as you have difficulty. Don't wait until the end of the semester.

■ If old exams are available, look at them to see what is emphasized and what type of questions are asked. Find out whether the exam will be all objective or all essay or a combination. If you have trouble with a particular type of question (such as multiple choice or essay), practice answering questions of that type—your study group or a classmate may be able to help.

⊛ TAKING EXAMS

The most important thing to remember when taking an exam is not to panic. This, of course, is easier said than done. Almost everyone suffers from test anxiety to some degree. Usually, it passes as soon as the exam begins, but in some cases, it is so debilitating that an individual does not perform as well as he or she could. If you are one of those people, get help as soon as possible. Most colleges and universities have a program to help students overcome test anxiety or at least keep it in check. Don't be afraid to seek help if you suffer test anxiety. Your success in college depends to a large extent on how well you perform on exams, so by not seeking help, you are only hurting yourself. In addition, the following suggestions may be helpful:

■ First of all, relax. Then look over the exam briefly to see its format and determine which questions are worth the most points. If it helps, quickly jot down any information you are afraid you might forget or particularly want to remember for a question.

■ Answer the questions that you know the best first. Make sure, however, that you don't spend too much time on any one question or on one that is worth only a few points.

■ If the exam is a combination of multiple choice and essay, answer the multiple-choice questions first. If you are not sure of an answer, go on to the next one. Sometimes the answer to one question can be found in another question. Furthermore, the multiple-choice questions may contain many of the facts needed to answer some of the essay questions.

■ Read the question carefully and answer only what it asks. Save time by not repeating the question as your opening sentence to the answer. Get right to the point. Jot down a quick outline for longer essay questions to make sure you cover everything.

■ If you don't understand a question, ask the examiner. Don't assume anything. After all, it is your grade that will suffer if you misinterpret the question.

■ If you have time, review your exam to make sure you covered all the important points and answered all the questions.

■ If you have followed our suggestions, by the time you finish the exam, you should feel confident that you did well and will have cause for celebration.

⊛ CONCLUDING COMMENTS

We hope that the suggestions we have offered will be of benefit to you not only in this course, but throughout your college career. Though it is difficult to break old habits and change a familiar routine, we are confident that following these suggestions will make you a better student. Furthermore, many of the suggestions will help you work more efficiently, not only in college, but also throughout your career. Learning is a lifelong process that does not end when you graduate. The critical thinking skills that you learn now will be invaluable throughout your life, both in your career and as an informed citizen.

Essentials of
GEOLOGY

Chapter 1

UNDERSTANDING THE EARTH: AN INTRODUCTION TO PHYSICAL GEOLOGY

OUTLINE

As a result of numerous eruptions like the one shown here, Anak Krakatau emerged above sea level in 1928 from the 275 meter deep caldera formed by the 1883 eruption of Krakatau.

On August 26, 1883, Krakatau, a small, uninhabited volcanic island in the Sunda Straits between Java and Sumatra, exploded (▷ Figure 1-1a). In less than one day, 18 cubic kilometers (km³) of rock were erupted in an ash cloud 80 kilometers (km) high. The explosion was heard as far away as Australia and Rodriguez Island, 4,653 km to the west in the Indian Ocean. Where the 450 meter (m) high peak of Danan once stood, the water was now 275 m deep, and only one-third of the 5 × 9 km island remained above sea level (Figure 1-1b). The explosions and the collapse of the chamber that held the magma (molten rock) beneath the volcano produced giant sea waves, some as high as 40 m. On nearby islands, at least 36,000 people were killed and 165 coastal villages destroyed by the sea waves that hurled ashore coral blocks weighing more than 540 metric tons.

So much ash was blown into the stratosphere that the Sunda Straits were completely dark from 10 A.M., August 27, until dawn the next day. Ash was reported falling on ships as far away as 6,076 km. For the next three years, vivid red sunsets were common around the world due to these airborne products. The volcanic dust also reflected incoming solar radiation back into space; the average global temperature dropped as much as 1/2°C during the following year and did not return to normal until 1888.

Of course, all animal life was destroyed on Krakatau. A year after the eruption, however, a few shoots of grass appeared, and three years later 26 species of plants had colonized the island, thus providing a suitable habitat for the animals that soon followed.

Why have we chosen the eruption of Krakatau as an introduction to geology? The reason is that it illustrates several of the aspects of geology that we will examine, including the way the Earth's interior, surface, and atmosphere are interrelated.

Sumatra, Java, Krakatau, and the Lesser Sunda Islands are part of a 3,000 km long chain of volcanic islands comprising the nation of Indonesia. Their location is a result of a collision between two pieces of the Earth's outer layer, generally called the lithosphere. The theory that the Earth's crust is divided into rigid plates that move over a plastic zone is known as *plate tectonics* (see Chapter 2). This unifying theory explains and ties together such apparently unrelated geologic phenomena as volcanic eruptions, earthquakes, and the origin of mountain ranges.

▷ FIGURE 1-1 (*a*) Krakatau, part of the island nation of Indonesia, is located in the Sunda Straits between Java and Sumatra. (*b*) Krakatau before and after the 1883 eruption.

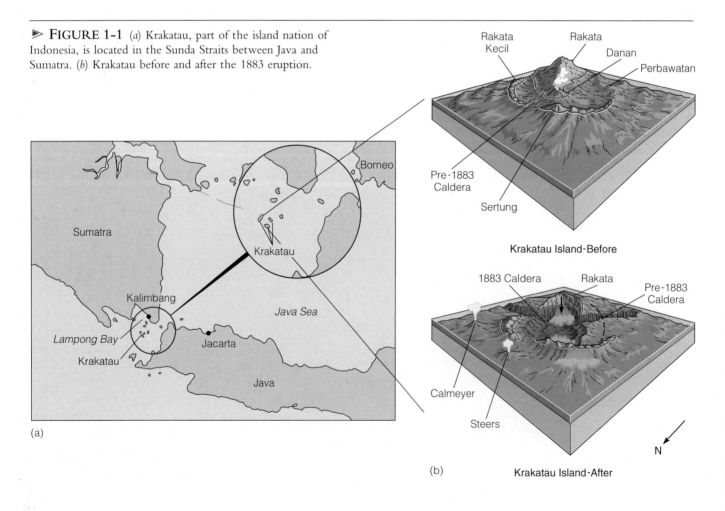

(a)

(b)

Rakata Kecil Rakata Danan Perbawatan

Pre-1883 Caldera

Sertung

Krakatau Island-Before

1883 Caldera Rakata Pre-1883 Caldera

Calmeyer

Steers

Krakatau Island-After

In tropical areas such as Indonesia, physical and chemical processes rapidly break down ash falls and lava flows, converting them into rich soils that are agriculturally productive and can support large populations (see Chapter 6). In spite of the dangers of living in a region of active volcanism, a strong correlation exists between volcanic activity and population density. Indonesia has experienced 972 eruptions during historic time, 83 of which have caused fatalities. Yet these same eruptions are also ultimately responsible for the high food production that can support large numbers of people.

Volcanic eruptions also affect weather patterns; recall that the eruption of Krakatau caused a global cooling of 1/2°C. More recently, the 1991 eruption of Mount Pinatubo in the Philippines resulted in lower global temperatures and abnormal weather patterns the following summer (see Chapter 5).

As you read this book, keep in mind that the different topics you are studying are parts of dynamic interrelated systems, not isolated pieces of information. Volcanic eruptions such as Krakatau are the result of complex interactions involving the Earth's interior and surface. These eruptions not only have an immediate effect on the surrounding area, but also contribute to climatic changes that affect the entire planet.

INTRODUCTION

The Earth is unique among the planets of our solar system in that it supports life and has oceans of water, a hospitable atmosphere, and a variety of climates. It is ideally suited for life as we know it because of a combination of factors, including its distance from the Sun and the evolution of its interior, crust, oceans, and atmosphere. Over time, changes in the Earth's atmosphere, oceans, and, to some extent, its crust have been influenced by life processes. In turn, these physical changes have affected the evolution of life.

The Earth is not a simple, unchanging planet. Rather, it is a complex dynamic body in which innumerable interactions are occurring among its many components. The continuous evolution of the Earth and its life makes geology an exciting and ever changing science in which new discoveries are continually being made.

WHAT IS GEOLOGY?

Just what is geology and what is it that geologists do? **Geology**, from the Greek *geo* and *logos*, is defined as the study of the Earth. It is generally divided into two broad areas—physical geology and historical geology. *Physical geology* is the study of Earth materials, such as minerals and rocks, as well as the processes operating within the Earth and upon its surface. *Historical geology* examines the origin and evolution of the Earth, its continents, oceans, atmosphere, and life.

The discipline of geology is so broad that it is subdivided into many different fields or specialties. ◉ Table 1-1 shows many of the diverse fields of geology and their relationship to the sciences of astronomy, physics, chemistry, and biology.

Nearly every aspect of geology has some economic or environmental relevance. Many geologists are involved in exploration for mineral and energy resources, using their specialized knowledge to locate the natural resources on which our industrialized society is based. As the demand for

these nonrenewable resources increases, geologists are intensifying their search and applying the basic principles of geology in increasingly sophisticated ways (▷ Figure 1-2).

Although locating mineral and energy resources is extremely important, geologists are also being asked to use their expertise to help solve many of our environmental problems. Some geologists are involved in finding groundwater for the ever burgeoning needs of communities and industries or in monitoring surface and underground water pollution and suggesting ways to clean it up. Geological engineers help find safe locations for dams, waste disposal sites, and power plants, and design earthquake-resistant buildings.

Geologists are also involved in making short- and long-range predictions about earthquakes and volcanic eruptions and the potential destruction that may result. In addition, they are working with civil defense planners to help draw up contingency plans should such natural disasters occur.

As this brief survey illustrates, geologists are employed in a wide variety of pursuits. As the world's population increases and greater demands are made on the Earth's limited resources, the need for geologists and their expertise will become even greater.

GEOLOGY AND THE HUMAN EXPERIENCE

Many people are surprised at the extent to which we depend on geology in our everyday lives and also at the numerous references to geology in the arts, music, and literature. Rocks and landscapes are realistically represented in many sketches and paintings. Examples by famous artists include Leonardo da Vinci's *Virgin of the Rocks* and *Virgin and Child with Saint Anne*, Giovanni Bellini's *Saint Francis in Ecstasy* and *Saint Jerome*, and Asher Brown Durand's *Kindred Spirits* (▷ Figure 1-3).

In the field of music, Ferde Grofé's *Grand Canyon Suite* was, no doubt, inspired by the grandeur and timelessness of Arizona's Grand Canyon and its vast rock exposures. The

TABLE 1-1 Specialties of Geology and Their Broad Relationship to the Other Sciences		
Specialty	**Area of Study**	**Related Science**
Geochronology	Time and history of the Earth	Astronomy
Planetary geology	Geology of the planets	
Paleontology	Fossils and life history	Biology
Economic geology	Mineral and energy resources	Chemistry
Environmental geology	Environment	
Geochemistry	Geology of chemical change	
Hydrogeology	Water resources	
Mineralogy	Minerals	
Petrology	Rocks	
Geophysics	Earth's interior	Physics
Structural geology	Rock deformation	
Seismology	Earthquakes	
Geomorphology	Landforms	
Oceanography	Oceans	
Paleogeography	Ancient geographic features and locations	
Stratigraphy/sedimentology	Layered rocks and sediments	

▷ **FIGURE 1-2** Geologists increasingly use computers in their search for petroleum and other natural resources.

rocks on the Island of Staffa in the Inner Hebrides provided the inspiration for Felix Mendelssohn's famous *Hebrides Overture.*

References to geology abound in *The German Legends of the Brothers Grimm,* and Jules Verne's *Journey to the Center of the Earth* describes an expedition into the Earth's interior. On one level, the poem "Ozymandias" by Percy B. Shelley deals with the fact that nothing lasts forever and even solid rock eventually disintegrates under the ravages of time and weathering. References to geology can even be found in comics, two of the best known being *B.C.* by Johnny Hart and *The Far Side* by Gary Larson.

Geology has also played an important role in history. Wars have been fought for the control of such natural resources as oil, gas, gold, silver, diamonds, and other valuable minerals. Empires throughout history have risen and fallen on the distribution and exploitation of natural resources. The configuration of the Earth's surface, or its topography, which is shaped by geologic agents, plays a critical role in military tactics. Natural barriers such as mountain ranges and rivers have frequently served as political boundaries.

◉ HOW GEOLOGY AFFECTS OUR EVERYDAY LIVES

Destructive volcanic eruptions, devastating earthquakes, disastrous landslides, large sea waves, floods, and droughts are headline-making events that affect many people (▷ Figure 1-4). Although we are unable to prevent most of these natural disasters, the more we know about them, the better we are able to predict, and possibly control, the severity of their impact. The environmental movement has forced everyone to take a closer look at our planet and the delicate balance between its various systems.

> ► **FIGURE 1-3** *Kindred Spirits* by Asher Brown Durand (1849) realistically depicts the layered rocks occurring along gorges in the Catskill Mountains of New York State. Asher Brown Durand was one of numerous artists of the nineteenth-century Hudson River School, who were known for their realistic landscapes.

As society becomes increasingly complex and technologically oriented, we, as citizens, need an understanding of science so that we can make informed choices about those things that affect our lives. We are already aware of some of the negative aspects of an industrialized society, such as problems relating to solid waste disposal, contaminated groundwater, and acid rain. We are also learning the impact that humans, in increasing numbers, have on the environment and that we can no longer ignore the role that we play in the dynamics of the global ecosystem.

Most people are unaware of the extent to which geology affects their lives. For many people, the connection between geology and such well-publicized problems as nonrenewable energy and mineral resources, let alone waste disposal and pollution, is simply too far removed or too complex to be fully appreciated. But consider for a moment just how dependent we are on geology in our daily routines.

Much of the electricity for our appliances comes from the burning of coal, oil, or natural gas or from uranium consumed in nuclear-generating plants. It is geologists who locate the coal, petroleum, and uranium. The copper or other metal wires through which electricity travels are manufactured from materials found as the result of mineral

> ► **FIGURE 1-4** As these headlines from various newspapers indicate, geology affects our everyday lives.

exploration. The buildings we live and work in owe their very existence to geological resources. A few examples are the concrete foundation (concrete is a mixture of clay, sand, or gravel, and limestone), the drywall (made largely from the mineral gypsum), the windows (the mineral quartz is the principal ingredient in the manufacture of glass), and the metal or plastic plumbing fixtures inside the building (the metals are from ore deposits, and the plastics are most likely manufactured from petroleum distillates of crude oil).

Furthermore, when we go to work, the car or public transportation we use is powered and lubricated by some type of petroleum by-product and is constructed of metal alloys and plastics. And the roads or rails we ride over come from geologic materials, such as gravel, asphalt, concrete, or steel. All of these items are the result of processing geologic resources.

It is quite apparent that as individuals and societies, the standard of living we enjoy is directly dependent on the consumption of geologic materials. Therefore, we need to be aware of geology and of how our use and misuse of geologic resources may affect the delicate balance of nature and irrevocably alter our culture as well as our environment.

THE ORIGIN OF THE SOLAR SYSTEM AND THE DIFFERENTIATION OF THE EARLY EARTH

According to the currently accepted theory for the origin of the solar system (➤ Figure 1-5), interstellar material in a spiral arm of the Milky Way Galaxy condensed and began collapsing. As this cloud gradually collapsed under the influence of gravity, it flattened and began rotating counter-clockwise, with about 90% of its mass concentrated in the central part of the cloud. As the rotation and concentration of material continued, an embryonic sun formed, surrounded by a turbulent, rotating cloud of material called a *solar nebula*.

The turbulence in this solar nebula formed localized eddies where gas and solid particles condensed. As condensation proceeded, gaseous, liquid, and solid particles began accreting into ever larger masses called *planetesimals* that eventually became true planetary bodies. While the planets were accreting, material that had been pulled into the center of the nebula also condensed, collapsed, and was heated to

➤ **FIGURE 1-5** The currently accepted theory for the origin of our solar system involves (*a*) a huge nebula condensing under its own gravitational attraction, then (*b*) contracting, rotating, and (*c*) flattening into a disk, with the Sun forming in the center and eddies gathering up material to form planets. As the Sun contracted and began to visibly shine, (*d*) intense solar radiation blew away unaccreted gas and dust until finally, (*e*) the Sun began burning hydrogen and the planets completed their formation.

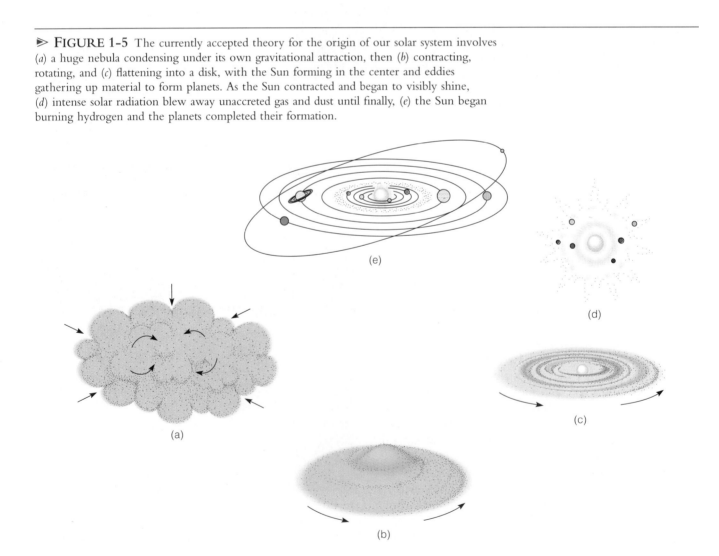

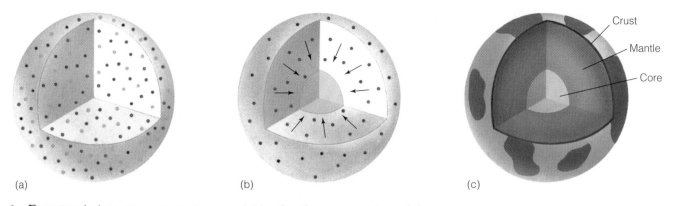

(a)　　　　　　　　　　　　(b)　　　　　　　　　　　　(c)

➤ FIGURE 1-6 (*a*) The early Earth was probably of uniform composition and density throughout. (*b*) Heating of the early Earth reached the melting point of iron and nickel, which, being denser than silicate minerals, settled to the Earth's center. At the same time, the lighter silicates flowed upward to form the mantle and the crust. (*c*) In this way, a differentiated Earth formed, consisting of a dense iron-nickel core, an iron-rich silicate mantle, and a silicate crust with continents and ocean basins.

several million degrees by gravitational compression. The result was the birth of a star, our Sun.

Some 4.6 billion years ago, enough material eventually gathered together in one of the turbulent eddies that swirled around the early Sun to form the planet Earth. Scientists think that this early Earth was rather cool, so the accreting elements and nebular rock fragments were solids rather than gases or liquids. This early Earth is also thought to have been of generally uniform composition and density throughout (➤ Figure 1-6a). It was composed mostly of compounds of silicon, iron, magnesium, oxygen, aluminum, and smaller amounts of all the other chemical elements. Subsequently, when the Earth underwent heating, this homogeneous composition disappeared (Figure 1-6b), and the result was a differentiated planet, consisting of a series of concentric layers of differing composition and density (Figure 1-6c). This differentiation into a layered planet is probably the most significant event in the history of the Earth. Not only did it lead to the formation of a crust and eventually to continents, but it was also probably responsible for the emission of gases from the interior that eventually led to the formation of the oceans and the atmosphere.

⬣ THE EARTH AS A DYNAMIC PLANET

The Earth is a dynamic planet that has continuously changed during its 4.6-billion-year existence. The size, shape, and geographic distribution of continents and ocean basins have changed through time, the composition of the atmosphere has evolved, and life-forms existing today differ from those that lived during the past. We can easily visualize how mountains and hills are worn down by erosion and how landscapes are changed by the forces of wind, water, and ice. Volcanic eruptions and earthquakes reveal an active interior, and folded and fractured rocks indicate the tremendous power of the Earth's internal forces.

The Earth consists of three concentric layers: the core, the mantle, and the crust (➤ Figure 1-7). This orderly

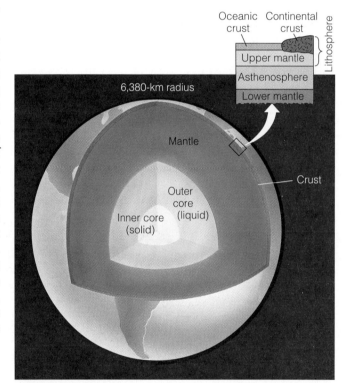

➤ FIGURE 1-7 A cross section of the Earth illustrating the core, mantle, and crust. The enlarged portion shows the relationship between the lithosphere, composed of the continental crust, oceanic crust, and upper mantle, and the underlying asthenosphere and lower mantle.

division results from density differences between the layers as a function of variations in composition, temperature, and pressure.

The **core** has a calculated density of 10 to 13 grams per cubic centimeter (g/cm^3) and occupies about 16% of the Earth's total volume. Seismic (earthquake) data indicate that the core consists of a small, solid inner core and a larger, apparently liquid, outer core. Both are believed to consist largely of iron and a small amount of nickel.

The **mantle** surrounds the core and comprises about 83% of the Earth's volume. It is less dense than the core (3.3−5.7 g/cm^3) and is thought to be composed largely of *peridotite,* a dark, dense igneous rock containing abundant iron and magnesium. The mantle can be divided into three distinct zones based on physical characteristics. The lower mantle is solid and forms most of the volume of the Earth's interior. The **asthenosphere** surrounds the mantle. It has the same composition as the lower mantle but behaves plastically and slowly flows. Partial melting within the asthenosphere generates magma, some of which rises to the Earth's surface because it is less dense than the rock from which it was derived. The upper mantle surrounds the asthenosphere. The solid upper mantle and the overlying crust constitute the **lithosphere,** which is broken into numerous individual pieces called **plates** that move over the asthenosphere as a result of underlying *convection cells* (➤ Figure 1-8). Interactions of these plates are responsible for such phenomena as earthquakes, volcanic eruptions, and the formation of mountain ranges and ocean basins.

The **crust,** the outermost layer of the Earth, consists of two types. The *continental crust* is thick (20−90 km), has an average density of 2.7 g/cm^3, and contains considerable silicon and aluminum. The *oceanic crust* is thin (5−10 km), denser than continental crust (3.0 g/cm^3), and is composed of the dark igneous rock *basalt.*

Since the widespread acceptance of plate tectonic theory about 25 years ago, geologists have viewed the Earth from a global perspective in which all of its systems are intercon-nected. Thus, the distribution of mountain chains, major fault systems, volcanoes and earthquakes, the origin of new ocean basins, the movement of continents, and several other geological processes and features are perceived to be interre-lated.

GEOLOGY AND THE FORMULATION OF THEORIES

The term **theory** has various meanings. In colloquial usage, it means a speculative or conjectural view of something—hence the widespread belief that scientific theories are little more than unsubstantiated wild guesses. In scientific usage, however, a theory is a coherent explanation for one or several related natural phenomena that is supported by a large body of objective evidence. From a theory are derived predictive statements that can be tested by observation and/or experiment so that their validity can be assessed. The law of universal gravitation is an example of a theory describing the attraction between masses (an apple and the Earth in the popularized account of Newton and his discov-ery).

Theories are formulated through the process known as the **scientific method.** This method is an orderly, logical approach that involves gathering and analyzing the facts or data about the problem under consideration. Tentative explanations or **hypotheses** are then formulated to explain the observed phenomena. Next, the hypotheses are tested to see if what they predicted actually occurs in a given situation (see Perspective 1-1). Finally, if one of the hypotheses is found, after repeated tests, to explain the phenomena, then that hypothesis is proposed as a theory. One should remem-

---

➤ **FIGURE 1-8** The Earth's plates are thought to move as a result of underlying mantle convection cells in which warm material from deep within the Earth rises toward the surface, cools, and then, upon losing heat, descends back downward into the interior. The movement of these convection cells is believed to be the mechanism responsible for the movement of the Earth's plates, as shown in this diagrammatic cross section.

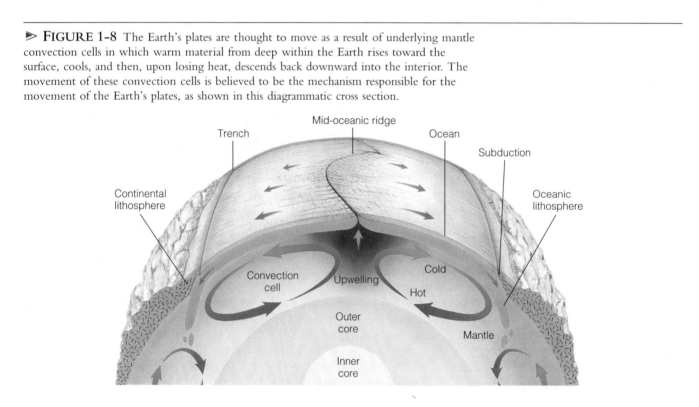

THE GAIA HYPOTHESIS

In 1785, James Hutton said, "I think the Earth is a super organism." In 1972, James Lovelock, a British scientist-inventor, proposed his controversial Gaia hypothesis, which he updated in 1988. According to Lovelock and other proponents of this hypothesis, "the dynamic forces of life so dominate our planet that life has a controlling influence on the oceans and atmosphere." In other words, life has shaped the environment in order to keep it within a comfortable range rather than adapting to an otherwise benign environment.

There is little doubt that interrelationships exist between life-forms and the environment. No one would deny that microorganisms aid in such processes as soil formation and deposition of some rocks, or that green plants obtain carbon dioxide from the atmosphere and release oxygen as a waste product.

Although most scientists accept such relationships as those just mentioned, the Gaia hypothesis, named for the Greek Earth goddess, proposes an even more fundamental relationship between organisms and the environment. According to Lovelock, rather than adapting to an evolving environment determined solely by physical and chemical processes, organisms have the capacity to control the environment, especially the atmosphere and oceans, in such ways as to make continued life possible.

Life exists within a narrow range of physical and chemical conditions, and Lovelock proposes that feedback mechanisms exist that control the environment to suit the needs of organisms. The Earth's temperature, for example, has stayed within the narrow limits suitable for the existence of life for at least the last 3.5 billion years, even though the Sun now produces about two-thirds more heat and is brighter than it was when the first organisms existed on Earth.

To demonstrate how organisms maintain environmental parameters within narrow limits, Lovelock proposed a mathematical model that he called Daisyworld, in which an imaginary planet is populated only by white and black daisies. If the temperature on Daisyworld rises, the black daisies absorb too much heat and die, thus leaving mostly white daisies that reflect more heat and cool the planet down. When Daisyworld cools sufficiently, black daisies thrive again and absorb more heat. In short, there is a feedback mechanism for temperature control.

As one would expect, there are strong objections to the Gaia hypothesis. Many biologists dismiss it because it is teleological; that is, it appeals to design or purpose in nature and thus cannot be tested. Some geologists point out that plate tectonics alone can control the Earth's temperature through the recycling of carbon dioxide.

While the Gaia hypothesis is, to say the least, controversial, it remains to be seen whether it will eventually become an acceptable theory. As in any scientific endeavor, new and radical ideas must demonstrate their worth in the competitive field of hypothesis, evidence testing, and prediction. Perhaps Gaia will be supported as scientists investigate its theoretical postulates, or it may be rejected or modified, depending on future discoveries. In any case, Gaia has forced scientists to critically evaluate the relationship between life and the global environment.

---

ber that in science, even a theory is still subject to further testing and refinement as new data become available.

The fact that a scientific theory can be tested and is subject to such testing separates science from other forms of human inquiry. Because scientific theories can be tested, they have the potential of being supported or even proved wrong. Accordingly, science must proceed without any appeal to beliefs or supernatural explanations, not because such beliefs or explanations are necessarily untrue, but because we have no way to investigate them. For this reason, science makes no claim about the existence or nonexistence of a supernatural or spiritual realm.

Each scientific discipline has certain theories that are of particular importance for that discipline. In geology, the formulation of plate tectonic theory has changed the way geologists view the Earth. Geologists now view Earth history in terms of interrelated events that are part of a global pattern of change.

PLATE TECTONIC THEORY

The acceptance of **plate tectonic theory** is recognized as a major milestone in the geological sciences. It is comparable to the revolution caused by Darwin's theory of evolution in biology. Plate tectonics has provided a framework for interpreting the composition, structure, and internal processes of the Earth on a global scale. It has led to the realization that the continents and ocean basins are part of a lithosphere-atmosphere-hydrosphere (water portion of the planet) system that evolved together with the Earth's interior.

According to plate tectonic theory, the lithosphere is divided into plates that move over the asthenosphere (▶ Figure 1-9). Zones of volcanic activity, earthquake activity, or both mark most plate boundaries. Along these boundaries, plates diverge, converge, or slide sideways past each other.

At **divergent plate boundaries**, plates move apart as

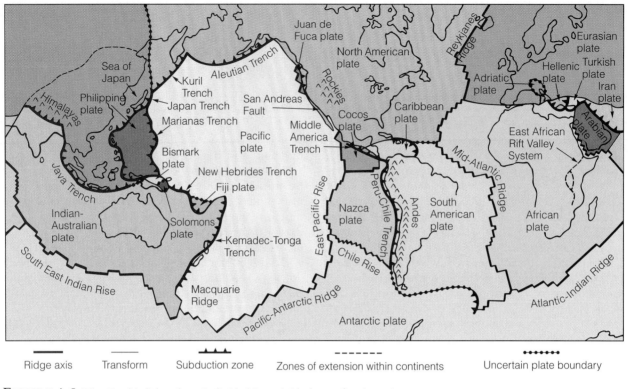

| Ridge axis | Transform | Subduction zone | Zones of extension within continents | Uncertain plate boundary |

▶ **FIGURE 1-9** The Earth's lithosphere is divided into rigid plates of various sizes that move over the asthenosphere.

magma rises to the surface from the asthenosphere (▶ Figure 1-10). The margins of divergent plate boundaries are marked by mid-oceanic ridges in oceanic crust, and are recognized by linear rift valleys, such as the Mid-Atlantic Ridge, where newly forming divergent boundaries occur beneath continental crust.

Plates move toward one another along **convergent plate boundaries** where one plate sinks beneath another plate along what is known as a **subduction zone** (Figure 1-10). As the plate descends into the Earth, it becomes hotter until it melts, or partially melts, thus generating a magma. As this magma rises, it may erupt at the Earth's surface, forming a chain of volcanoes. The Andes Mountains on the west coast of South America are a good example of a volcanic mountain range formed as a result of subduction along a convergent plate boundary (Figure 1-9).

Transform plate boundaries are sites where plates slide sideways past each other (Figure 1-10). The San Andreas fault in California is a transform plate boundary separating the Pacific plate from the North American plate (Figure 1-9). The earthquake activity along the San Andreas fault results from the Pacific plate moving northward relative to the North American plate.

A revolutionary concept when it was proposed in the 1960s, plate tectonic theory has had significant and far-reaching consequences in all fields of geology because it provides the basis for relating many seemingly unrelated geologic phenomena. For example, the Appalachian Mountains in eastern North America and the mountain ranges of Greenland, Scotland, Norway, and Sweden are not the result of unrelated mountain-building episodes, but rather are part of a larger mountain-building event that involved the closing of an ancient "Atlantic Ocean" and the formation of the supercontinent Pangaea about 245 million years ago.

THE ROCK CYCLE

A **rock** is an aggregate of minerals. **Minerals** are naturally occurring, inorganic, crystalline solids that have definite physical and chemical properties. Minerals are composed of elements such as oxygen, silicon, and aluminum, and elements are made up of atoms, the smallest particles of matter that still retain the characteristics of an element. More than 3,500 minerals have been identified and described, but only about a dozen make up the bulk of the rocks in the Earth's crust.

Geologists recognize three major groups of rocks—*igneous, sedimentary,* and *metamorphic*—each of which is characterized by its mode of formation. Each group contains a variety of individual rock types that differ from one another on the basis of composition or texture (the size, shape, and arrangement of mineral grains).

The **rock cycle** is a way of viewing the interrelationships between the Earth's internal and external processes (▶ Figure 1-11). It relates the three rock groups to each other; to surficial processes such as weathering, transportation, and deposition; and to internal processes such as magma generation and metamorphism. Plate movement is the

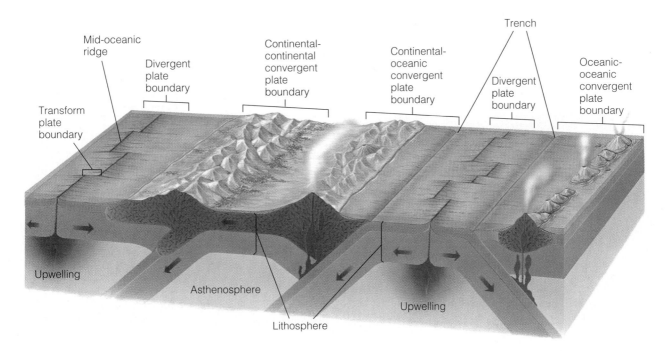

➤ **FIGURE 1-10** An idealized cross section illustrating the relationship between the lithosphere and the underlying asthenosphere and the three principal types of plate boundaries: divergent, convergent, and transform.

➤ **FIGURE 1-11** The rock cycle showing the interrelationships between the Earth's internal and external processes and how each of the three major rock groups is related to the others.

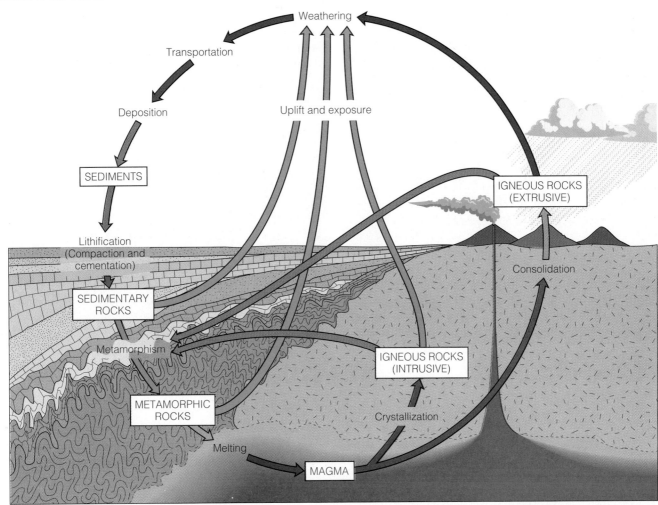

mechanism responsible for recycling rock materials and therefore drives the rock cycle.

Igneous rocks result from the crystallization of magma or the accumulation and consolidation of volcanic ejecta such as ash. As a magma cools, minerals crystallize, and the resulting rock is characterized by interlocking mineral grains. Magma that cools slowly beneath the Earth's surface produces *intrusive igneous rocks* (➤ Figure 1-12a), while magma

(a)

(b)

(c)

(d)

(e)

(f)

➤ **FIGURE 1-12** Hand specimens of common igneous (*a,b*), sedimentary (*c, d*), and metamorphic (*e, f*) rocks. (*a*) Granite, an intrusive igneous rock. (*b*) Basalt, an extrusive igneous rock. (*c*) Conglomerate, a sedimentary rock formed by the consolidation of rock fragments. (*d*) Marine limestone, a sedimentary rock formed by the extraction of mineral matter from seawater by organisms or by the inorganic precipitation of the mineral calcite from seawater. (*e*) Gneiss, a foliated metamorphic rock. (*f*) Quartzite, a nonfoliated metamorphic rock. (Photos courtesy of Sue Monroe.)

that cools at the Earth's surface produces *extrusive igneous rocks* (Figure 1-12b).

Rocks exposed at the Earth's surface are broken into particles and dissolved by various weathering processes. The particles and dissolved material may be transported by wind, water, or ice and eventually deposited as *sediment.* This sediment may then be compacted or cemented into sedimentary rock.

Sedimentary rocks originate by consolidation of rock fragments, precipitation of mineral matter from solution, or compaction of plant or animal remains (Figure 1-12c and d). Because sedimentary rocks form at or near the Earth's surface, geologists can make inferences about the environment in which they were deposited, the type of transporting agent, and perhaps even something about the source from which the sediments were derived (see Chapter 7). Accordingly, sedimentary rocks are very useful for interpreting Earth history.

Metamorphic rocks result from the alteration of other rocks, usually beneath the Earth's surface, by heat, pressure, and the chemical activity of fluids. For example, marble, a rock preferred by many sculptors and builders, is a metamorphic rock produced when the agents of metamorphism are applied to the sedimentary rocks limestone or dolostone. Metamorphic rocks are either *foliated* (Figure 1-12e) or *nonfoliated* (Figure 1-12f). Foliation, the parallel alignment of minerals due to pressure, gives the rock a layered or banded appearance.

The Rock Cycle and Plate Tectonics

Interactions among plates determine, to a certain extent, which of the three rock groups will form (➤ Figure 1-13).

For example, weathering and erosion produce sediments that are transported by agents such as running water from the continents to the oceans, where they are deposited and accumulate. These sediments, some of which may be lithified and become sedimentary rock, become part of a moving plate along with the underlying oceanic crust. When plates converge, heat and pressure generated along the plate boundary may lead to igneous activity and metamorphism within the descending oceanic plate, thus producing various igneous and metamorphic rocks.

Some of the sediment and sedimentary rock is subducted and melts, while other sediments and sedimentary rocks along the boundary of the nonsubducted plate are metamorphosed by the heat and pressure generated along the converging plate boundary. Later, the mountain range or chain of volcanic islands formed along the convergent plate boundary will once again be weathered and eroded, and the new sediments will be transported to the ocean to begin yet another cycle.

GEOLOGIC TIME AND UNIFORMITARIANISM

An appreciation of the immensity of geologic time is central to understanding the evolution of the Earth and its biota. Indeed, time is one of the main aspects that sets geology apart from the other sciences. Most people have difficulty comprehending geologic time because they tend to think in terms of the human perspective—seconds, hours, days, and years. Ancient history is what occurred hundreds or even thousands of years ago. When geologists talk of ancient geologic history, they are referring to events that happened hundreds of millions or even billions of years ago. To a

➤ **FIGURE 1-13** Plate tectonics and the rock cycle. The cross section shows how the three major rock groups, igneous, metamorphic, and sedimentary, are recycled through both the continental and oceanic regions.

geologist, recent geologic events are those that occurred within the last million years or so.

The **geologic time scale** resulted from the work of many nineteenth-century geologists who pieced together information from numerous rock exposures and constructed a sequential chronology based on changes in the Earth's biota through time. Subsequently, with the discovery of radioactivity in 1895 and the development of various radiometric dating techniques, geologists have been able to assign absolute age dates in years to the subdivisions of the geologic time scale (▶ Figure 1-14).

One of the cornerstones of geology is the **principle of uniformitarianism**. It is based on the premise that present-day processes have operated throughout geologic time. Therefore, in order to understand and interpret the geologic record, we must first understand present-day processes and their results.

Uniformitarianism is a powerful principle that allows us to use present-day processes as the basis for interpreting the past and for predicting potential future events. We should keep in mind that uniformitarianism does not exclude such sudden or catastrophic events as volcanic eruptions, earthquakes, landslides, or flooding. These are processes that shape our modern world, and, in fact, some geologists view the history of the Earth as a series of such short-term or punctuated events. Such a view is certainly in keeping with the modern principle of uniformitarianism.

Furthermore, uniformitarianism does not require that the rates and intensities of geological processes be constant through time. We know that volcanic activity was more intense in North America 5 to 10 million years ago than it is today, and that glaciation has been more prevalent during the last several million years than in the previous 300 million years.

What uniformitarianism means is that even though the rates and intensities of geological processes have varied during the past, the physical and chemical laws of nature have remained the same. Although the Earth is in a dynamic state of change and has been ever since it was formed, the processes that shaped it during the past are the same ones that are in operation today.

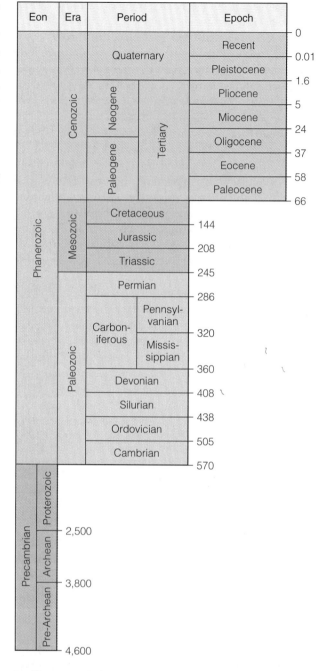

Eon	Era	Period		Epoch	
Phanerozoic	Cenozoic	Quaternary		Recent	0
					0.01
				Pleistocene	
					1.6
		Neogene	Tertiary	Pliocene	
					5
				Miocene	
					24
		Paleogene		Oligocene	
					37
				Eocene	
					58
				Paleocene	
					66
	Mesozoic	Cretaceous			
					144
		Jurassic			
					208
		Triassic			
					245
	Paleozoic	Permian			
					286
		Carboniferous	Pennsylvanian		
					320
			Mississippian		
					360
		Devonian			
					408
		Silurian			
					438
		Ordovician			
					505
		Cambrian			
					570
Precambrian	Proterozoic				2,500
	Archean				3,800
	Pre-Archean				4,600

▶ FIGURE 1-14 The geologic time scale. Numbers to the right of the columns are ages in millions of years before the present.

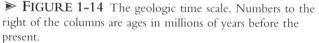

CHAPTER SUMMARY

1. Geology is the study of the Earth. It is divided into two broad areas: physical geology is the study of the composition of Earth materials as well as the processes that operate within the Earth and upon its surface; historical geology examines the origin and evolution of the Earth, its continents, oceans, atmosphere, and life.

2. Geology is part of the human experience. We can find examples of it in the arts, music, and literature. A basic understanding of geology is also important for dealing with the many environmental problems and issues facing society.

3. Geologists engage in a variety of occupations, the main one being exploration for mineral and energy resources.

They are also becoming increasingly involved in environmental issues and making short- and long-range predictions of the potential dangers from such natural disasters as volcanic eruptions and earthquakes.

4. About 4.6 billion years ago, the solar system formed from a rotating cloud of interstellar matter. Eventually, as this cloud condensed, it collapsed under the influence of gravity and flattened into a rotating disk. Within this rotating disk, the Sun, planets, and moons formed from the turbulent eddies of nebular gases and solids.

5. The Earth is differentiated into layers. The outermost layer, is the crust, which is divided into continental and oceanic portions. Below the crust is the upper mantle. The crust and upper mantle, or lithosphere, overlie the asthenosphere, a zone that behaves plastically. The asthenosphere is underlain by the solid lower mantle. The Earth's core consists of an outer liquid portion and an inner solid portion.

6. The lithosphere is broken into a series of plates that diverge, converge, and slide sideways past one other.

7. The scientific method is an orderly, logical approach that involves gathering and analyzing facts about a particular phenomenon, formulating hypotheses to explain the phenomenon, testing the hypotheses, and finally proposing a theory. A theory is a testable explanation for some natural phenomenon that has a large body of supporting evidence.

8. Plate tectonic theory provides a unifying explanation for many geological features and events. The interaction between plates is responsible for volcanic eruptions, earthquakes, the formation of mountain ranges and ocean basins, and the recyling of rock material.

9. Igneous, sedimentary, and metamorphic rocks comprise the three major groups of rocks. Igneous rocks result from the crystallization of magma or the consolidation of volcanic ejecta. Sedimentary rocks are formed mostly by the consolidation of rock fragments, precipitation of mineral matter from solution. Metamorphic rocks are produced from other rocks, generally beneath the Earth's surface, by heat, pressure, and chemically active fluids.

10. The rock cycle illustrates the interactions between internal and external processes and shows how the three rock groups of the Earth are interrelated.

11. Time sets geology apart from the other sciences, except astronomy. The geologic time scale is the calendar geologists use to date past events; it is divided into eras, periods, and epochs.

12. The principle of uniformitarianism is basic to the interpretation of Earth history. This principle holds that the laws of nature have been constant through time and that the same processes operating today have operated in the past, although at different rates.

IMPORTANT TERMS

asthenosphere	geology	mineral	rock cycle
convergent plate boundary	hypothesis	plate	scientific method
core	igneous rock	plate tectonic theory	sedimentary rock
crust	lithosphere	principle of	subduction zone
divergent plate boundary	mantle	uniformitarianism	theory
geologic time scale	metamorphic rock	rock	transform plate boundary

REVIEW QUESTIONS

1. The study of Earth materials is:
   a. ____ paleontology;
   b. ____ stratigraphy;
   c. ____ physical geology;
   d. ____ historical geology;
   e. ____ environmental geology.
2. Which of the following statements about a mineral is not true?
   a. ____ it is organic;
   b. ____ it has definite physical and chemical properties;
   c. ____ it is naturally occurring;
   d. ____ it is a crystalline solid;
   e. ____ none of these.
3. The Earth's core is inferred to be:
   a. ____ hollow;
   b. ____ composed of rock with a high silica content;
   c. ____ completely molten;
   d. ____ composed mostly of iron;
   e. ____ completely solid.
4. The layer between the core and the crust is the:
   a. ____ mantle;
   b. ____ lithosphere;
   c. ____ hydrosphere;
   d. ____ biosphere;
   e. ____ asthenosphere.
5. What fundamental process is thought to be responsible for plate motion?
   a. ____ hot spot activity;
   b. ____ subduction;
   c. ____ spreading ridges;
   d. ____ convection cells;
   e. ____ density differences.

6. Which of the following statements about a scientific theory is not true?
   a. ____ it is an explanation for some natural phenomenon;
   b. ____ it has a large body of supporting evidence;
   c. ____ it is a conjecture or guess;
   d. ____ it is testable;
   e. ____ none of these.

7. Mid-oceanic ridges are examples of what type of boundary?
   a. ____ divergent;
   b. ____ convergent;
   c. ____ transform;
   d. ____ subduction;
   e. ____ answers (b) and (d).

8. The San Andreas fault separating the Pacific plate from the North American plate is an example of what type of boundary?
   a. ____ divergent;
   b. ____ convergent;
   c. ____ transform;
   d. ____ subduction;
   e. ____ answers (b) and (d).

9. Which rocks form from the cooling of a magma?
   a. ____ igneous;
   b. ____ metamorphic;
   c. ____ sedimentary;
   d. ____ all of these;
   e. ____ none of these.

10. The premise that present-day processes have operated throughout geologic time is known as the principle of:
    a. ____ plate tectonics;
    b. ____ sea-floor spreading;
    c. ____ continental drift;
    d. ____ volcanism;
    e. ____ uniformitarianism.

11. The rock cycle implies that:
    a. ____ metamorphic rocks are derived from magma;
    b. ____ any rock type can be derived from any other rock type;
    c. ____ igneous rocks only form beneath the Earth's surface;

    d. ____ sedimentary rocks only form from the weathering of igneous rocks;
    e. ____ all of these.

12. Why is it important for people to have a basic understanding of geology?

13. Describe some of the ways in which geology affects our everyday lives.

14. Name the major layers of the Earth, and describe their general composition.

15. Describe the scientific method, and explain how it may lead to a scientific theory.

16. Briefly describe plate tectonic theory, and explain why it is a unifying theory in geology.

17. Describe the rock cycle, and explain how it may be related to plate tectonics.

18. Does the principle of uniformitarianism allow for catastrophic events? Explain.

POINTS TO PONDER

1. Propose a pre–plate tectonic hypothesis explaining the formation and distribution of mountain ranges.

2. Provide several examples of how a knowledge of geology would be useful in planning a military campaign against another country.

PLATE TECTONICS: A UNIFYING THEORY

OUTLINE

View looking down the Great Rift Valley of Africa. Little Magadi, seen in the background, is one of numerous soda lakes forming in the valley. Because of high evaporation rates and lack of any drainage outlets, these lakes are very saline. The Great Rift Valley is part of the system of rift valleys resulting from stretching of the crust as plates move away from each other in eastern Africa.

A fundamental question in planetary geology has been the nature and extent of volcanic and mountain-building activity within the solar system. All of the terrestrial planets—Mercury, Venus, Earth, and Mars—experienced a similar early history that was marked by widespread volcanism and meteorite impacts, both of which helped modify their surfaces. The volcanic and mountain-building activity and resultant surface features (other than meteorite craters) of these planets are clearly related to the way they transport heat from their interiors to their surfaces.

The Earth appears to be unique in that its surface is broken into a series of individual moving plates. The other three terrestrial planets and the moon all have a single, globally continuous plate and thus exhibit fewer types of volcanic and mountain-building activities than the Earth.

Mercury's surface is heavily cratered and shows little in the way of primary volcanic structures. Its largest impact basins are filled with what appear to be lava flows similar to the lava plains on the Moon.

Mars has numerous features that indicate an extensive early period of volcanism. These include *Olympus Mons,* the solar system's largest volcano, lava flows, and uplifted regions thought to have resulted from mantle convection. In addition to volcanic features, Mars also displays abundant evidence of tensional forces, including numerous faults (fractures along which movement has occurred) and large fault-produced valley structures. There is, however, no evidence that plate tectonics comparable to that on Earth has ever occurred on Mars.

Radar images from the *Magellan* spacecraft, which has been mapping the surface of Venus, reveal a wide variety of spectacular, fascinating, and, in many cases, enigmatic landscapes (➤ Figure 2-1). Among the features discovered are volcanoes; extensive lava flows and channels; pancake-shaped volcanic features, many of which are surrounded by complex networks of cracks and fractures; intricate networks of fractures and faults, some producing elongated valleys and ridges; and areas of crumpled crust that superficially resemble mountain ranges formed by compression. Though there is no evidence of plate tectonics on Venus, the volcanic and associated features certainly attest to an internally active planet.

While all of the terrestrial planets and the Earth's moon underwent a period of active volcanism during their early history, only three bodies in the solar system appear to show any signs of present-day volcanism: the Earth, the Jovian moon Io, and perhaps the Neptunian moon Triton.

Io, the innermost of the four large moons of Jupiter, is probably the most volcanically active body yet observed in the solar system. It is brilliantly colored in red, oranges, and yellows resulting from the various sulfur compounds spewed forth by geysers and volcanoes. To date, at least 10 active volcanoes have been discovered on Io.

The Neptunian moon Triton also appears to be volcanically active and to have experienced numerous episodes of deformation. Evidence from images returned by *Voyager 2* indicates that Triton has geysers that are erupting frozen nitrogen crystals and organic compounds.

➤ FIGURE 2-1 A nearly complete map of the northern hemisphere of Venus based on radar images beamed back to Earth from the *Magellan* space probe. (Photo courtesy of NASA.)

INTRODUCTION

The recognition that the Earth's geography has changed continuously through time has led to a revolution in the geological sciences, forcing geologists to greatly modify the way they view the Earth. Although many people have only a vague notion of what plate tectonic theory is, plate tectonics has a profound effect on all of our lives. It is now realized that most earthquakes and volcanic eruptions occur near plate margins and are not merely random occurrences. Furthermore, the formation and distribution of many important natural resources, such as metallic ores, are related to plate boundaries, and geologists are now incorporating plate tectonic theory into their prospecting efforts.

The interaction of plates determines the location of continents, ocean basins, and mountain systems, which in turn affects the atmospheric and oceanic circulation patterns that ultimately determine global climates. Plate movements have also profoundly influenced the geographic distribution, evolution, and extinction of plants and animals.

Plate tectonic theory is now almost universally accepted among geologists, and its application has led to a greater understanding of how the Earth has evolved and continues to do so. This powerful, unifying theory accounts for many apparently unrelated geologic events, allowing geologists to view such phenomena as part of a continuing story rather than as a series of isolated incidents.

◉ EARLY IDEAS ABOUT CONTINENTAL DRIFT

The idea that the Earth's geography was different during the past is not new. During the late nineteenth century, the Austrian geologist Edward Suess noted the similarities between the Late Paleozoic plant fossils of India, Australia, Africa, Antarctica, and South America as well as evidence of glaciation in the rock sequences of these southern continents. In 1885 he proposed the name Gondwanaland (or **Gondwana** as we will use here) for a supercontinent composed of these southern landmasses. Gondwana is a province in east-central India where evidence exists for extensive glaciation as well as abundant fossils of the ***Glossopteris*** flora (▷ Figure 2-2), an association of Late Paleozoic plants found only in India and the Southern Hemisphere continents. Suess thought that extensive land bridges once connected the continents and later sank beneath the ocean.

▷ **FIGURE 2-2** *Glossopteris* leaves from the Upper Permian Dunedoo Formation, Australia. Fossils of the *Glossopteris* flora are found on all five of the Gondwana continents. (Photo courtesy of Patricia G. Gensel, University of North Carolina.)

Alfred Wegener, a German meteorologist (▷ Figure 2-3), is generally credited with developing the hypothesis of **continental drift.** In his monumental book, *The Origin of Continents and Oceans,* first published in 1915, Wegener proposed that all of the landmasses were originally united into a single supercontinent that he named **Pangaea,** from the Greek meaning "all land." Wegener portrayed his grand concept of continental movement in a series of maps showing the breakup of Pangaea and the movement of the various continents to their present-day locations. Wegener had amassed a tremendous amount of geological, paleontological, and climatological evidence in support of continental

▷ **FIGURE 2-3** Alfred Wegener, a German meteorologist, proposed the continental drift hypothesis in 1912 based on a tremendous amount of geological, paleontological, and climatological evidence. He is shown here waiting out the Arctic winter in an expedition hut.

drift, but the initial reaction of scientists to his then-heretical ideas can best be described as mixed.

Nevertheless, the eminent South African geologist Alexander du Toit further developed Wegener's arguments and gathered more geological and paleontological evidence in support of continental drift. In 1937, du Toit published *Our Wandering Continents,* in which he contrasted the glacial deposits of Gondwana with coal deposits of the same age found in the continents of the Northern Hemisphere. To resolve this apparent climatological paradox, du Toit moved the Gondwana continents to the South Pole and brought the northern continents together such that the coal deposits were located at the equator. He named this northern landmass **Laurasia.** It consisted of present-day North America, Greenland, Europe, and Asia (except for India).

In spite of what seemed to be overwhelming evidence, most geologists still refused to accept the idea that continents moved. Not until the 1960s, when oceanographic research provided convincing evidence that the continents had once been joined together and subsequently separated, did the hypothesis of continental drift finally become widely accepted.

THE EVIDENCE FOR CONTINENTAL DRIFT

The evidence used by Wegener, du Toit, and others to support the hypothesis of continental drift includes the fit of the shorelines of continents; the appearance of the same rock sequences and mountain ranges of the same age on continents now widely separated; the matching of glacial deposits and paleoclimatic zones; and the similarities of many extinct plant and animal groups whose fossil remains are found today on widely separated continents.

Continental Fit

Wegener, like some before him, was impressed by the close resemblance between the coastlines of continents on oppo-

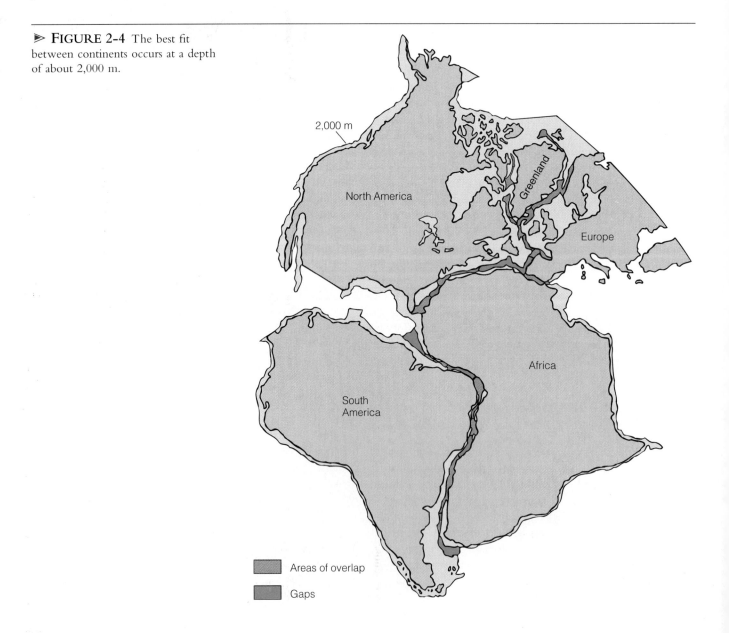

➤ FIGURE 2-4 The best fit between continents occurs at a depth of about 2,000 m.

2,000 m

North America

Greenland

Europe

Africa

South America

Areas of overlap

Gaps

site sides of the Atlantic Ocean, particularly between South America and Africa. He cited these similarities as partial evidence that the continents were at one time joined together as a supercontinent that subsequently split apart. As his critics pointed out, though, the configuration of coastlines results from erosional and depositional processes and therefore is continually being modified. Even if the continents had separated during the Mesozoic Era, as Wegener proposed, it is not likely that the coastlines would fit exactly.

A more realistic approach is to fit the continents together below sea level where erosion would be minimal. In 1965 Sir Edward Bullard, an English geophysicist, and two associates showed that the best fit between the continents occurs at a depth of about 2,000 m (▷ Figure 2-4). Since then, other reconstructions using the latest ocean basin data have confirmed the close fit between continents when they are reassembled to form Pangaea.

Similarity of Rock Sequences and Mountain Ranges

If the continents were at one time joined together, then the rocks and mountain ranges of the same age in adjoining locations on the opposite continents should closely match. Such is the case for the Gondwana continents. Marine, nonmarine, and glacial rock sequences of Pennsylvanian to Jurassic age are almost identical for all five Gondwana continents, strongly indicating that they were at one time joined together.

The trends of several major mountain ranges also support the hypothesis of continental drift. These mountain ranges seemingly end at the coastline of one continent only to apparently continue on another continent across the ocean. The folded Appalachian Mountains of North America, for example, trend northeastward through the eastern United States and Canada and terminate abruptly at the Newfoundland coastline. Mountain ranges of the same age and structure occur in eastern Greenland, Ireland, Great Britain, and Norway. Even though these mountain ranges are currently separated by the Atlantic Ocean, they form an essentially continuous mountain range when the continents are positioned next to each other (▷ Figure 2-5).

Glacial Evidence

During the Late Paleozoic Era, massive glaciers covered large continental areas of the Southern Hemisphere. Evidence for this glaciation includes layers of till (sediments deposited by glaciers) and striations (scratch marks) in the bedrock beneath the till. Fossils and sedimentary rocks of the same age from the Northern Hemisphere, however, give no indication of glaciation. Fossil plants found in coals indicate that the Northern Hemisphere had a tropical climate during the time that the Southern Hemisphere was glaciated.

All of the Gondwana continents except Antarctica are currently located near the equator in subtropical to tropical climates. Mapping of glacial striations in bedrock in Austra-

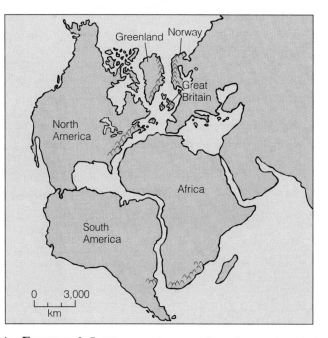

▷ **FIGURE 2-5** When continents are brought together, their mountain ranges form a single continuous range of the same age and style of deformation throughout. Such evidence indicates the continents were at one time joined together and were subsequently separated.

lia, India, and South America indicates that the glaciers moved from the areas of the present-day oceans onto land. This would be highly unlikely because large glaciers on land flow outward from their central area of accumulation toward the sea.

If the continents did not move during the past, one would have to explain how glaciers moved from the oceans onto land and how large glaciers formed near the equator. But if the continents are reassembled as a single landmass with South Africa located at the South Pole, the direction of movement of Late Paleozoic glaciers makes sense. Furthermore, this geographic arrangement places the northern continents nearer the tropics, which is consistent with the fossil and climatological evidence from Laurasia (▷ Figure 2-6).

Fossil Evidence

Some of the most compelling evidence for continental drift comes from the fossil record (▷ Figure 2-7). Fossils of the *Glossopteris* flora are found in equivalent Pennsylvanian- and Permian-aged coal deposits on all five Gondwana continents. The *Glossopteris* flora is characterized by the seed fern *Glossopteris* (Figure 2-2) as well as by many other distinctive and easily identifiable plants. Plant pollen and spores can be dispersed over great distances by wind, but *Glossopteris*-type plants produced seeds that are too large to have been carried by winds. It is also unlikely that the seeds floated across an ocean, because they probably would not have remained viable for any length of time in salt water.

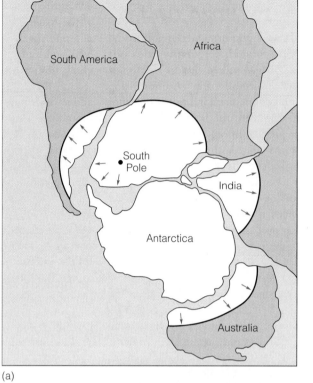

Glaciated area
Arrows indicate the direction of glacial movement based on striations preserved in bedrock.

South America

Africa

South Pole

India

Antarctica

Australia

(a)

➤ FIGURE 2-6 (*a*) If the continents are brought together so that South Africa is located at the South Pole, then the glacial movements indicated by the striations make sense. In this situation, the glacier, located in a polar climate, moved radially outward from a thick central area toward its periphery. (*b*) Permian-aged glacial striations in bedrock exposed at Hallet's Cove, Australia, indicate the direction of glacial movement more than 200 million years ago.

(b)

The present-day climates of South America, Africa, India, Australia, and Antarctica range from tropical to polar and are much too diverse to support the type of plants that compose the *Glossopteris* flora. Wegener reasoned therefore that these continents must once have been joined such that these widely separated localities were all in the same latitudinal climatic belt.

The fossil remains of animals also provide strong evidence for continental drift. One of the best examples is *Mesosaurus*, a freshwater reptile whose fossils are found in Permian-aged rocks in certain regions of Brazil and South Africa and

nowhere else in the world. Because the physiology of freshwater and marine animals is completely different, it is hard to imagine how a freshwater reptile could have swum across the Atlantic Ocean and found a freshwater environment nearly identical to its former habitat. Moreover, if *Mesosaurus* could have swum across the ocean, its fossil remains should be widely dispersed. It is more logical to assume that *Mesosaurus* lived in lakes in what are now adjacent areas of South America and Africa, but were then united into a single continent.

Lystrosaurus and *Cynognathus* are both land-dwelling rep-

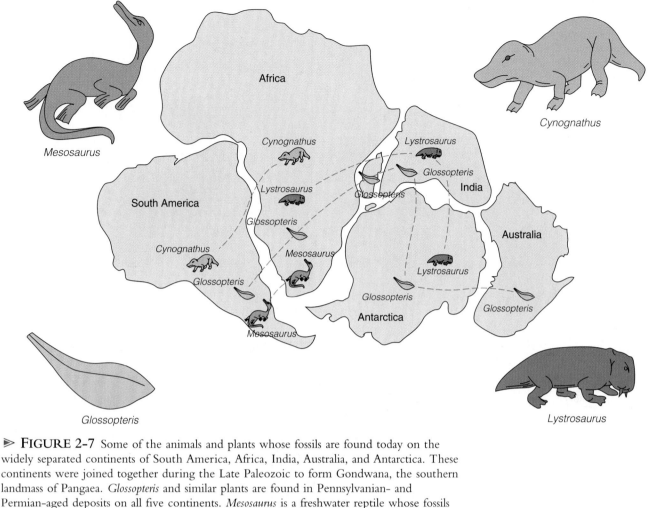

▷ FIGURE 2-7 Some of the animals and plants whose fossils are found today on the widely separated continents of South America, Africa, India, Australia, and Antarctica. These continents were joined together during the Late Paleozoic to form Gondwana, the southern landmass of Pangaea. *Glossopteris* and similar plants are found in Pennsylvanian- and Permian-aged deposits on all five continents. *Mesosaurus* is a freshwater reptile whose fossils are found in Permian-aged rocks in Brazil and South Africa. *Cynognathus* and *Lystrosaurus* are land reptiles that lived during the Early Triassic Period. Fossils of *Cynognathus* are found in South America and Africa, while fossils of *Lystrosaurus* have been recovered from Africa, India, and Antarctica.

tiles that lived during the Triassic Period; their fossils are found only on the present-day continental fragments of Gondwana. Because they are both land animals, they certainly could not have swum across the oceans currently separating the Gondwana continents. Therefore, the continents must once have been connected.

Paleomagnetism and Polar Wandering

Interest in continental drift revived in the 1950s as a result of new evidence from studies of the Earth's ancient magnetic field. **Paleomagnetism** is the remanent magnetism in ancient rocks recording the direction of the Earth's magnetic poles at the time of the rock's formation. The Earth can be thought of as a giant dipole magnet in which the magnetic poles correspond closely to the location of the geographic poles (▷ Figure 2-8). This arrangement means that the strength of the magnetic field is not constant, but varies, being weakest at the equator and strongest at the poles. The

Earth's magnetic field is thought to result from convection within the liquid outer core.

When a magma cools, the magnetic iron-bearing minerals align themselves with the Earth's magnetic field, recording both its direction and strength. The temperature at which iron-bearing minerals gain their magnetization is called the **Curie point**. As long as a rock is not subsequently heated above the Curie point, it will preserve that remanent magnetism. Thus, an ancient lava flow provides a record of the orientation and strength of the Earth's magnetic field at the time the lava flow cooled.

As paleomagnetic research progressed in the 1950s, some unexpected results emerged. When geologists measured the magnetism of recent rocks, they found it was generally consistent with the Earth's current magnetic field. The paleomagnetism of ancient rocks, however, showed different orientations. For example, studies of Silurian lava flows in North America indicated that the north magnetic pole was located in the western Pacific Ocean at that time, while the

► **FIGURE 2-8** (*a*) The magnetic field of the Earth has lines of force just like those of a bar magnet. (*b*) The strength of the magnetic field changes uniformly from the magnetic equator to the magnetic poles. This change in strength causes a dip needle to parallel the Earth's surface only at the magnetic equator, whereas its inclination with respect to the surface increases to 90° at the magnetic poles.

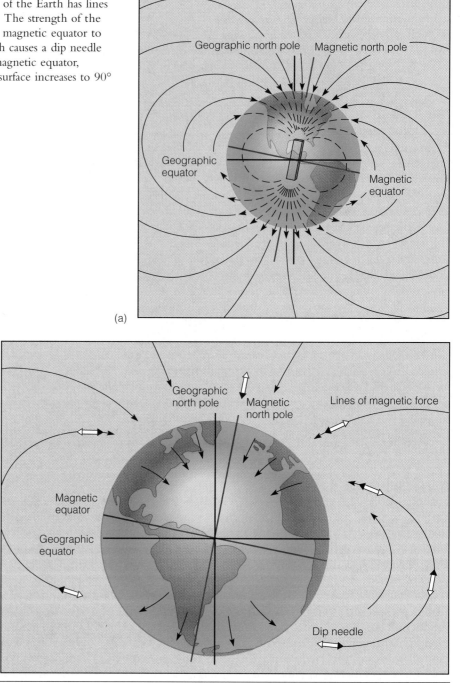

(a)

(b)

paleomagnetic evidence from Permian lava flows indicated a pole in Asia, and that of Cretaceous lava flows pointed to yet another location in northern Asia. When plotted on a map, the paleomagnetic readings of numerous lava flows from all ages in North America trace the apparent movement of the magnetic pole through time (► Figure 2-9). This paleomagnetic evidence from a single continent could be interpreted in three ways: the continent remained fixed and the north magnetic pole moved; the north magnetic pole stood still and the continent moved; or both the continent and the north magnetic pole moved.

Upon analysis, magnetic minerals from European Silurian and Permian lava flows pointed to a different magnetic pole location than those of the same age from North America (Figure 2-9). Furthermore, analysis of lava flows from all continents indicated each continent had its own series of magnetic poles. Does this mean there were different north magnetic poles for each continent? That would be highly unlikely and difficult to reconcile with the theory accounting for the Earth's magnetic field. The best explanation for this data is that the magnetic poles have remained at their present locations near the geographic north and south poles and the continents have moved. When the continental margins are fitted together so that the paleomagnetic data point to only one magnetic pole, we find, just as Wegener did, that the rock sequences and glacial deposits match, and

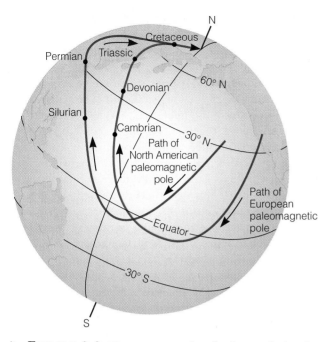

> FIGURE 2-9 The apparent paths of polar wandering for North America and Europe. The apparent location of the north magnetic pole is shown for different periods on each continent's polar wandering path.

that the fossil evidence is consistent with the reconstructed paleogeography.

MAGNETIC REVERSALS AND SEA-FLOOR SPREADING

Geologists refer to the Earth's present magnetic field as being normal, that is, with the north and south magnetic poles located approximately at the north and south geographic poles. At numerous times in the geologic past, the Earth's magnetic field has completely reversed. The existence of such **magnetic reversals** was discovered by dating and determining the orientation of the remanent magnetism in lava flows on land (▷ Figure 2-10). Once their existence was well established, magnetic reversals were also discovered in igneous rocks of the oceanic crust as part of the extensive mapping of the ocean basins that took place during the 1960s. Although the cause of magnetic reversals is still uncertain, their occurrence in the geologic record is well documented.

In addition to the discovery of magnetic reversals, mapping of the ocean basins also revealed a ridge system 65,000 km long, constituting the most extensive mountain range in the world. Perhaps the best-known part of the ridge system

> FIGURE 2-10 The sequence of magnetic anomalies preserved within the oceanic crust on both sides of an oceanic ridge is identical to the sequence of magnetic reversals already known from continental lava flows. Magnetic anomalies are formed when magma intrudes into oceanic ridges; when the magma cools below the Curie point, it records the Earth's magnetic polarity at the time. Subsequent sea-floor spreading splits the previously formed crust in half, so that it moves laterally away from the oceanic ridge. Repeated intrusions record a symmetrical series of magnetic anomalies that reflect periods of normal and reversed polarity. The magnetic anomalies are recorded by a magnetometer, which measures the strength of the magnetic field.

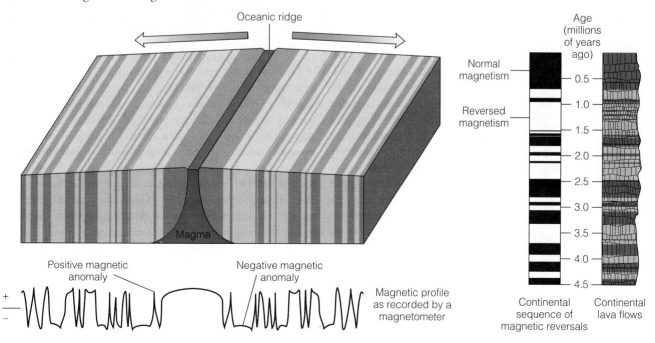

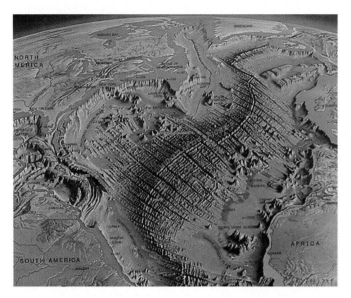

> **FIGURE 2-11** Artistic view of what the Atlantic Ocean basin would look like without water. The major feature is the Mid-Atlantic Ridge. (Photo courtesy of ALCOA.)

is the Mid-Atlantic Ridge, which divides the Atlantic Ocean basin into two nearly equal parts (▷ Figure 2-11).

In 1962, as a result of the oceanographic research conducted in the 1950s, Harry Hess of Princeton University proposed the theory of **sea-floor spreading** to account for continental movement. Hess suggested that continents do not move across oceanic crust, but rather that the continents and oceanic crust move together. He suggested that sea floor separates at oceanic ridges where new crust is formed by upwelling magma. As the magma cools, the newly formed oceanic crust moves laterally away from the ridge. As a mechanism to drive this system, Hess revived the idea of **thermal convection cells** in the mantle; that is, hot magma rises from the mantle, intrudes along fractures defining oceanic ridges, and thus forms new crust. Cold crust is subducted back into the mantle at oceanic trenches (long, narrow, deep features, along which subduction occurs), where it is heated and recycled, thus completing a thermal convection cell.

How could Hess's hypothesis be confirmed? Magnetic surveys of the oceanic crust revealed striped **magnetic**

▷ **FIGURE 2-12** The age of the world's ocean basins established from magnetic anomalies demonstrates that the youngest oceanic crust is adjacent to the oceanic ridges and that its age increases away from the ridge axis.

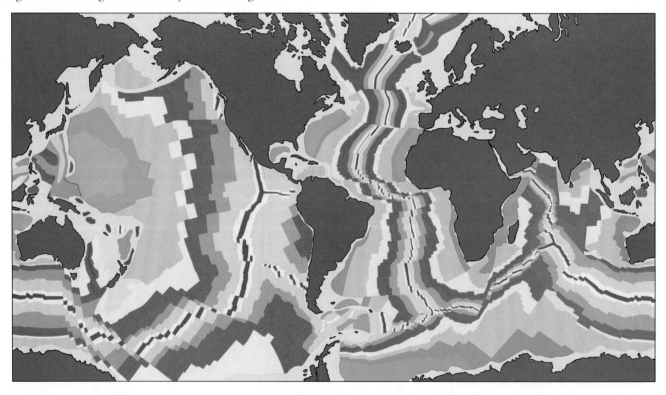

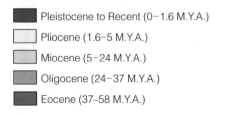

■ Pleistocene to Recent (0–1.6 M.Y.A.)	■ Paleocene (58–66 M.Y.A.)
□ Pliocene (1.6–5 M.Y.A.)	□ Late Cretaceous (66–88 M.Y.A.)
▨ Miocene (5–24 M.Y.A.)	▨ Middle Cretaceous (88–118 M.Y.A.)
▨ Oligocene (24–37 M.Y.A.)	▨ Early Cretaceous (118–144 M.Y.A.)
■ Eocene (37–58 M.Y.A.)	■ Late Jurassic (144–161 M.Y.A.)

anomalies (deviations from the average strength of the Earth's magnetic field) in the rocks that were both parallel to and symmetrical with the oceanic ridges (Figure 2-10b). Furthermore, the pattern of oceanic magnetic anomalies matched the pattern of magnetic reversals already known from studies of continental lava flows. When magma wells up and cools along a ridge summit, it records the Earth's magnetic field at that time as either normal or reversed. As new crust forms at the summit, the previously formed crust moves laterally away from the ridge. These magnetic stripes, representing times of normal or reversed polarity, are parallel to and symmetrical around oceanic ridges (where upwelling magma forms new oceanic crust), conclusively confirming Hess's theory of sea-floor spreading.

One of the consequences of the sea-floor spreading theory is its confirmation that ocean basins are geologically young features whose openings and closings are partially responsible for continental movement (Figure 2-11). Radiometric dating reveals that the oldest oceanic crust is less than 180 million years old, whereas the oldest continental crust is 3.96 billion years old (➢ Figure 2-12).

PLATE TECTONIC THEORY

Plate tectonic theory is based on a simple model of the Earth. The rigid lithosphere, consisting of both oceanic and continental crust, as well as the underlying upper mantle, consists of numerous variable-sized pieces called **plates**

(➢ Figure 2-13). The plates vary in thickness; those composed of upper mantle and continental crust are as much as 250 km thick, whereas those composed of upper mantle and oceanic crust are up to 100 km thick.

Most geologists accept plate tectonic theory, in part because the evidence for it is overwhelming, and also because it is a unifying theory that accounts for a variety of apparently unrelated geologic features and events. Consequently, geologists now view many geologic processes, such as mountain building, earthquake activity, and volcanism, from the perspective of plate tectonics (see Perspective 2-1). Furthermore, because all of the terrestrial planets have had a similar origin and early history (see the Prologue to this chapter), geologists are interested in determining whether plate tectonics is unique to Earth or whether it operates in the same way on the other terrestrial planets.

The lithosphere overlies the hotter and weaker semiplastic asthenosphere. It is thought that movement resulting from some type of heat transfer system within the asthenosphere causes the overlying plates to move. As plates move over the asthenosphere, they separate, mostly at oceanic ridges, while in other areas such as at oceanic trenches, they collide and are subducted back into the mantle.

PLATE BOUNDARIES

Plates move relative to one another such that their boundaries can be characterized as *divergent, convergent,* and

➢ **FIGURE 2-13** A map of the world showing the plates, their boundaries, relative motion and rates of movement in centimeters per year, and hot spots.

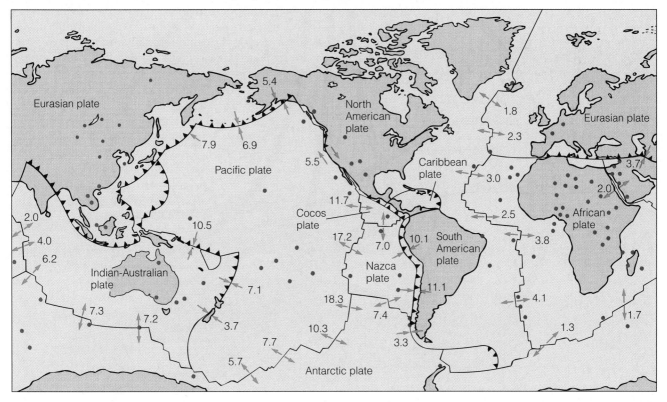

• Hot spot   → Direction of movement

THE SUPERCONTINENT CYCLE

At the end of the Paleozoic Era, all continents were amalgamated into the supercontinent Pangaea. Pangaea began fragmenting during the Triassic Period and continues to do so, thus accounting for the present distribution of continents and oceans. It now appears that another supercontinent existed at the end of the Proterozoic Eon, and there is some evidence for even earlier supercontinents. Recently, it has been proposed that supercontinents consisting of all or most of the Earth's landmasses form, break up, and reform in a cycle spanning about 500 million years.

The supercontinent cycle hypothesis is an expansion on the ideas of the Canadian geologist J. Tuzo Wilson. During the early 1970s, Wilson proposed a cycle (now known as the Wilson cycle) that includes continental fragmentation, the opening and closing of an ocean basin, and finally reassembly of the continent. According to the supercontinent cycle hypothesis, heat accumulates beneath a supercontinent because rocks of continents are poor conductors of heat. As a result of the heat accumulation, the supercontinent domes upward and fractures. Magma rising from below fills the fractures. As a fracture widens, it begins subsiding and forms a long narrow ocean such as the present-day Red Sea. Continued rifting eventually forms an expansive ocean basin such as the Atlantic.

According to proponents of the supercontinent cycle, one of the most convincing arguments for their hypothesis is the "surprising regularity" of mountain building caused by compression during continental collisions. Such mountain-building episodes occur about every 400 to 500 million years and are followed by an episode of rifting about 100 million years later. In other words, a supercontinent fragments and its individual plates disperse following a rifting episode, an interior ocean forms, and then the dispersed fragments reassemble to form another supercontinent (➤ Figure 1).

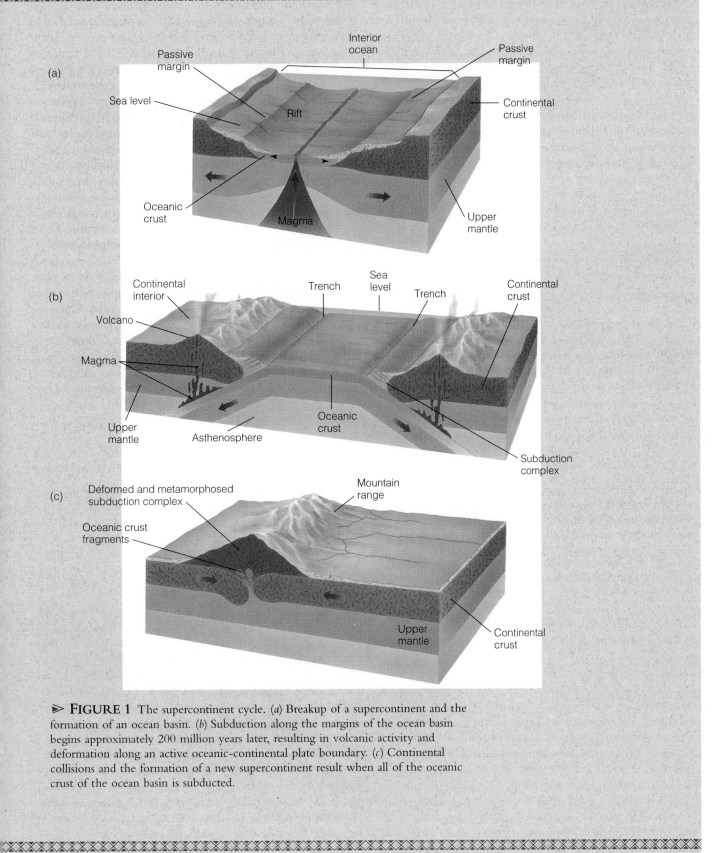

(a)

Passive margin

Interior ocean

Passive margin

Sea level

Rift

Continental crust

Oceanic crust

Magma

Upper mantle

(b)

Continental interior

Trench

Sea level

Trench

Continental crust

Volcano

Magma

Upper mantle

Asthenosphere

Oceanic crust

Subduction complex

(c)

Deformed and metamorphosed subduction complex

Mountain range

Oceanic crust fragments

Upper mantle

Continental crust

➤ **FIGURE 1** The supercontinent cycle. (*a*) Breakup of a supercontinent and the formation of an ocean basin. (*b*) Subduction along the margins of the ocean basin begins approximately 200 million years later, resulting in volcanic activity and deformation along an active oceanic-continental plate boundary. (*c*) Continental collisions and the formation of a new supercontinent result when all of the oceanic crust of the ocean basin is subducted.

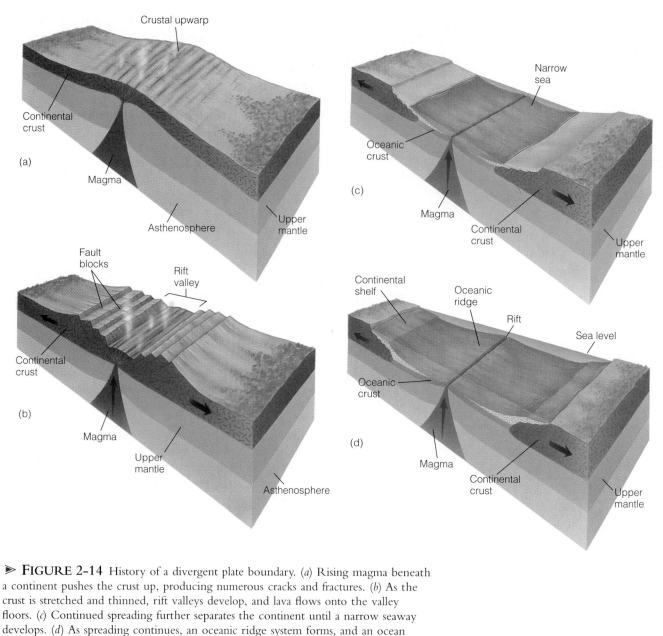

▶ **FIGURE 2-14** History of a divergent plate boundary. (*a*) Rising magma beneath a continent pushes the crust up, producing numerous cracks and fractures. (*b*) As the crust is stretched and thinned, rift valleys develop, and lava flows onto the valley floors. (*c*) Continued spreading further separates the continent until a narrow seaway develops. (*d*) As spreading continues, an oceanic ridge system forms, and an ocean basin develops and grows.

transform. Interaction of plates at their boundaries accounts for most of the Earth's earthquake and volcanic activity and, as will be apparent in Chapter 10, the origin of mountain systems.

Divergent Boundaries

Divergent plate boundaries or *spreading ridges* occur where plates are separating and new oceanic lithosphere is forming. Divergent boundaries are places where the crust is being extended, thinned, and fractured as magma, derived from the partial melting of the mantle, rises to the surface, intrudes into vertical fractures, and flows out onto the sea floor forming pillow lavas (see Figure 5-5). As successive injections of magma cool and solidify, they form new oceanic crust and record the intensity and orientation of the Earth's magnetic field (Figure 2-10). Divergent boundaries most commonly occur along the crests of oceanic ridges, for example, the Mid-Atlantic Ridge. Oceanic ridges are thus characterized by rugged topography with high relief resulting from displacement of rocks along large fractures, shallow earthquakes, high heat flow, and pillow lavas.

Divergent plate boundaries also occur under continents during the early stages of continental breakup (▶ Figure 2-14). When magma wells up beneath a continent, the crust is initially elevated, stretched, and thinned, producing fractures and rift valleys. During this stage, magma typically intrudes into the fractures and flows onto the valley floor. The East African rift valleys are an excellent example of this stage of continental breakup (▶ Figure 2-15). As rifting proceeds, the continental crust eventually breaks. If magma continues welling up, the two parts of the continent will

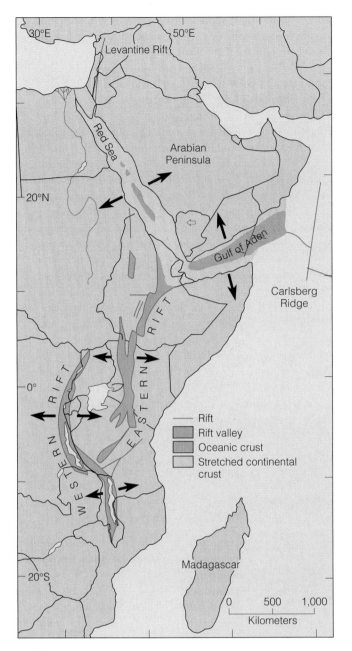

> ► **FIGURE 2-15** The East African rift valley is being formed by the separation of eastern Africa from the rest of the continent along a divergent plate boundary. The Red Sea and Gulf of Aden represent more advanced stages of rifting in which two continental blocks are separated by a narrow sea.

move away from each other, as is happening today beneath the Red Sea. As this newly formed narrow ocean basin continues to enlarge, it will eventually become an expansive ocean basin such as the Atlantic Ocean basin is today.

Convergent Boundaries

Whereas new crust forms at divergent plate boundaries, old crust is destroyed at many **convergent plate boundaries.** One plate is subducted under another and eventually is resorbed in the asthenosphere. When we talk about conver-

gent plate boundaries, we are really talking about three different types of boundaries: *oceanic-oceanic, oceanic-continental,* and *continental-continental.* The basic processes are the same for all three types of boundaries, but because different types of crust are involved, the results are different.

Oceanic-Oceanic Boundaries. When two oceanic plates converge, one of them is subducted under the other along an **oceanic-oceanic plate boundary** (► Figure 2-16). The subducting plate bends downward as it descends into the mantle, where it is heated and partially melted, thus generating magma. This magma is less dense than the surrounding mantle rocks and rises to the surface of the nonsubducted plate, forming a curved chain of volcanic islands called a **volcanic island arc.** This arc is nearly parallel to the oceanic trench and is separated from it by a distance of up to several hundred kilometers—the distance depends on the angle of dip of the subducting plate.

In those areas where the rate of subduction is faster than the forward movement of the overriding plate, the lithosphere on the landward side of the volcanic island arc may be subjected to tensional stress and stretched and thinned, resulting in the formation of a *back-arc basin.* This back-arc basin may grow by spreading if magma breaks through the thin crust and forms new oceanic crust (Figure 2-16). A good example of a back-arc basin associated with an oceanic-oceanic plate boundary is the Sea of Japan between the Asian continent and the islands of Japan.

Most present-day active volcanic island arcs are in the Pacific Ocean basin and include the Aleutian Islands, the Kermadec-Tonga arc, and the Japanese and Philippine Islands. The Scotia and Antillean (Caribbean) island arcs are present in the Atlantic Ocean basin.

Oceanic-Continental Boundaries. An **oceanic-continental plate boundary,** occurs when oceanic crust is subducted under continental crust (► Figure 2-17). The magma generated by subduction rises beneath the continent and either crystallizes as a large igneous body before reaching the surface or erupts at the surface, producing a chain of volcanoes (also called a volcanic arc). An excellent example of an oceanic-continental plate boundary is the Pacific coast of South America where the oceanic Nazca plate is currently being subducted under South America (Figure 2-13). The Peru-Chile Trench marks the site of subduction, and the Andes Mountains are the resulting volcanic mountain chain on the nonsubducting plate.

Continental-Continental Boundaries. Two continents approaching each other will initially be separated by an ocean floor that is being subducted under one continent. The edge of that continent will display the features characteristic of oceanic-continental convergence. As the ocean floor continues to be subducted, the two continents will come closer together until they eventually collide. Because continental lithosphere, which consists of continental crust and the upper mantle, is less dense than oceanic lithosphere (oceanic

> FIGURE 2-16 Oceanic-oceanic plate boundary. An oceanic trench forms where one oceanic plate is subducted beneath another. On the nonsubducted oceanic plate, a volcanic island arc forms from the rising magma generated from the subducting plate.

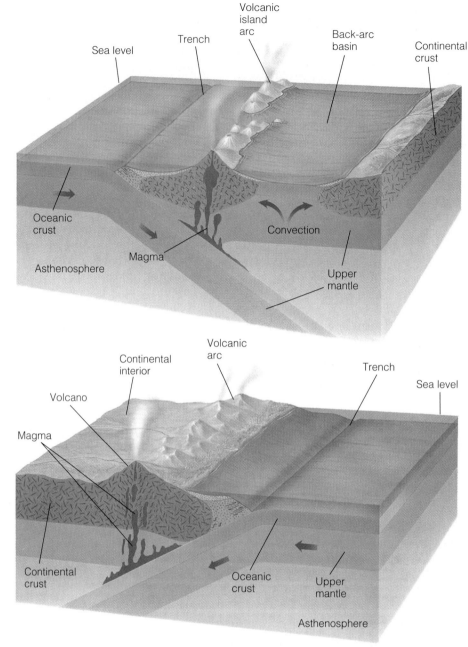

> FIGURE 2-17 Oceanic-continental plate boundary. When an oceanic plate is subducted beneath a continental plate, a volcanic mountain range is formed on the continental plate as a result of rising magma.

crust and the upper mantle), it cannot sink into the asthenosphere. Although one continent may partly slide under the other, it cannot be pulled or pushed down into a subduction zone (> Figure 2-18).

When two continents collide, they are welded together along a zone marking the former site of subduction. At this **continental-continental plate boundary,** an interior mountain belt is formed consisting of deformed sediments, igneous intrusions, metamorphic rocks, and fragments of oceanic crust. In addition, the entire region is subjected to numerous earthquakes. The Himalayas in central Asia are the result of a continental-continental collision between India and Asia that began about 40 to 50 million years ago and is still continuing.

Transform Boundaries

The third type of plate boundary is a **transform plate boundary.** These occur along fractures in the sea floor known as *transform faults* where plates slide laterally past one another roughly parallel to the direction of plate movement. Although lithosphere is neither created nor destroyed along a transform boundary, the movement between plates results in a zone of intensely shattered rock and numerous shallow earthquakes.

Transform faults are particular types of faults that "transform" or change one type of motion between plates into another type of motion. The majority of transform faults connect two oceanic ridge segments, but they can also

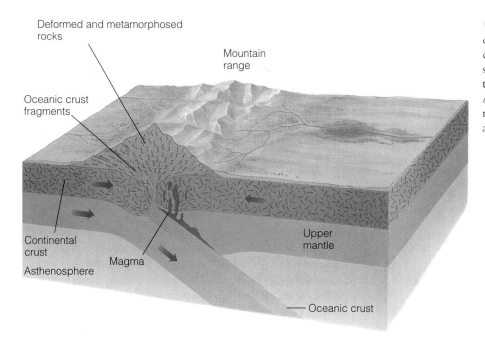

Deformed and metamorphosed rocks

Mountain range

Oceanic crust fragments

Continental crust

Asthenosphere

Magma

Upper mantle

Oceanic crust

⊳ **FIGURE 2-18** Continental-continental plate boundary. When two continental plates converge, neither is subducted because of their great thickness and low and equal densities. As the two plates collide, a mountain range is formed in the interior of a new and larger continent.

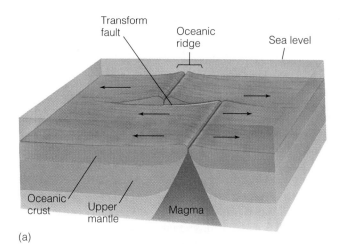

Transform fault

Oceanic ridge

Sea level

Oceanic crust

Upper mantle

Magma

(a)

⊳ **FIGURE 2-19** Horizontal movement between plates occurs along a transform fault. (*a*) The majority of transform faults connect two oceanic ridge segments. Note that relative motion between the plates only occurs between the two ridges. (*b*) A transform fault connecting two trenches. (*c*) A transform fault connecting a ridge and a trench.

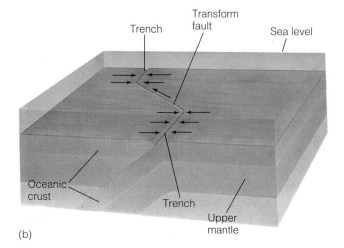

Trench

Transform fault

Sea level

Oceanic crust

Trench

Upper mantle

(b)

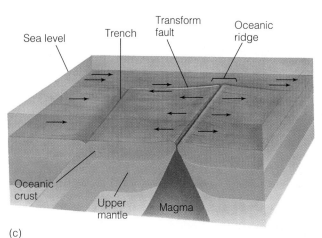

Sea level

Trench

Transform fault

Oceanic ridge

Oceanic crust

Upper mantle

Magma

(c)

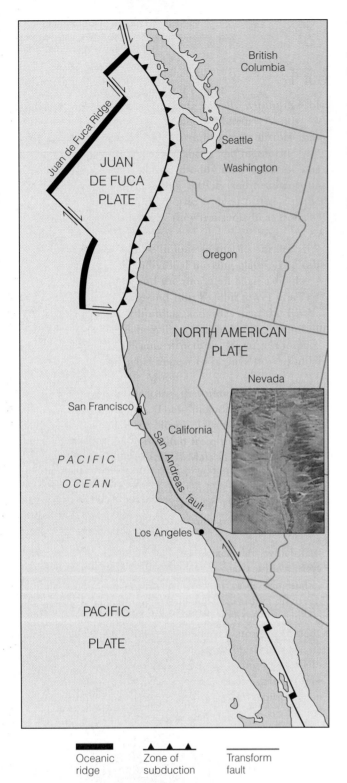

> **FIGURE 2-20** Transform plate boundary. The San Andreas fault is a transform fault separating the Pacific plate from the North American plate. Movement along this fault has caused numerous earthquakes. The photograph shows a segment of the San Andreas fault as it cuts through the Leona Valley, California. (Photo courtesy of Eleanora I. Robbins, U.S. Geological Survey.)

connect ridges to trenches and trenches to trenches (➤ Figure 2-19). Though the majority of transform faults occur in oceanic crust and are marked by distinct fracture zones, they may also extend into continents.

One of the best-known transform faults is the San Andreas fault in California. It separates the Pacific plate from the North American plate and connects spreading ridges in the Gulf of California and off the coast of northern California (➤ Figure 2-20). Many of the earthquakes affecting California are the result of movement along this fault.

PLATE MOVEMENT AND MOTION

How fast and in what direction are the Earth's various plates moving, and do they all move at the same rate? Rates of plate movement can be calculated in several ways. The least accurate method is to determine the age of the sediments immediately above any portion of the oceanic crust and divide that age by the distance from the spreading ridge. Such calculations give an average rate of movement.

A more accurate method of determining both the average rate of movement and relative motion is by dating the magnetic reversals in the crust of the sea floor. The distance from an oceanic ridge axis to any magnetic reversal indicates the width of new sea floor that formed during that time interval. Thus, for a given interval of time, the wider the strip of sea floor, the faster the plate has moved. In this way not only can the present average rate of movement and relative motion be determined (Figure 2-13), but the average rate of movement during the past can also be calculated by dividing the distance between reversals by the amount of time elapsed between reversals.

The average rate of movement as well as the relative motion between any two plates can also be determined by satellite laser ranging techniques. Laser beams from a station on one plate are bounced off a satellite (in geosynchronous orbit) and returned to a station on a different plate. As the plates move away from each other, there is an increase in the length of time that the laser beam takes to go from the sending station to the stationary satellite and back to the receiving station. This difference in elapsed time is used to calculate the rate of movement and relative motion between plates.

Hot Spots and Absolute Motion

Plate motions determined from magnetic reversals and satellite lasers give only the relative motion of one plate with respect to another. To determine absolute motion, we must have a fixed reference from which the rate and direction of plate movement can be determined. **Hot spots,** which may provide reference points, are locations where stationary columns of magma, originating deep within the mantle (mantle plumes), slowly rise to the Earth's surface and form volcanoes or flood basalts (Figure 2-13).

One of the best examples of hot spot activity is that over which the Emperor Seamount–Hawaiian Island chain

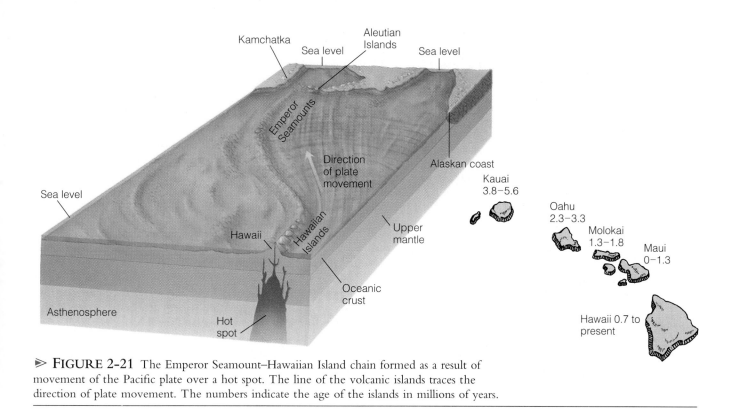

> **FIGURE 2-21** The Emperor Seamount–Hawaiian Island chain formed as a result of movement of the Pacific plate over a hot spot. The line of the volcanic islands traces the direction of plate movement. The numbers indicate the age of the islands in millions of years.

formed (▷ Figure 2-21). Currently, the only active volcanoes in this island chain are on the island of Hawaii and the Loihi Seamount. The rest of the islands and seamounts (structures of volcanic origin rising more than 1 km above the sea floor) of the chain are also of volcanic origin and are progressively older west-northwestward along the Hawaiian chain and north-northwestward along the Emperor Seamount chain.

The reason these islands and seamounts are progressively older as one moves toward the north and northwest is that the Pacific plate has moved over an apparently stationary mantle plume. Thus, a line of volcanoes was formed near the middle of the Pacific plate, marking the direction of the plate's movement. In the case of the Emperor Seamount–Hawaiian Island chain, the Pacific plate moved first north-northwesterly and then west-northwesterly over a single mantle plume.

Mantle plumes and hot spots are useful to geologists in helping to explain some of the geologic activity occurring within plates as opposed to that occurring at or near plate boundaries. In addition, they may prove useful as reference points for determining paleolatitude.

THE DRIVING MECHANISM OF PLATE TECTONICS

A major obstacle to the acceptance of continental drift was the lack of a driving mechanism to explain continental movement. When it was shown that continents and ocean floors moved together and not separately and that new crust formed at spreading ridges by rising magma, most geologists accepted some type of convective heat system as the basic process responsible for plate motion. The question of exactly what drives the plates, however, still remains.

Two models involving thermal convection cells have been proposed to explain plate movement (▷ Figure 2-22). In one model, thermal convection cells are restricted to the asthenosphere, whereas in the second model the entire mantle is involved. In both models spreading ridges mark the ascending limbs of adjacent convection cells, while trenches occur where the convection cells descend back into the Earth's interior. The locations of spreading ridges and trenches are therefore determined by the convection cells themselves and the lithosphere is considered to be the top of the thermal convection cell. Each plate thus corresponds to a single convection cell.

Although most geologists agree that the Earth's internal heat plays an important role in plate movement, problems are inherent in both models. The major problem associated with the first model is the difficulty in explaining the source of heat for the convection cells and why they are restricted to the asthenosphere. In the second model, the source of heat comes from the Earth's outer core, but it is still not known how heat is transferred from the outer core to the mantle. Nor is it clear how convection can involve both the lower mantle and the asthenosphere.

Some geologists think that in addition to thermal convection within the Earth, plate movement also occurs, in part, because of a mechanism involving "slab-pull" or "ridge-push." Both of these mechanisms are gravity driven, but still depend on thermal differences within the Earth. In "slab-pull," the subducting cold slab of lithosphere being

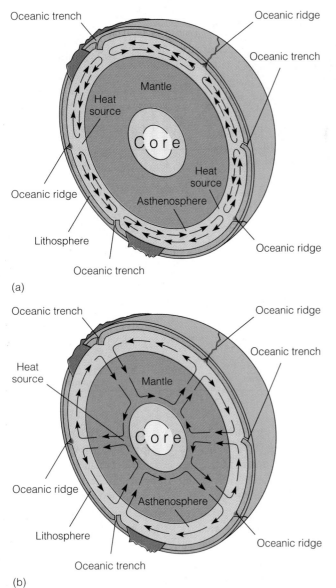

(a)

(b)

► **FIGURE 2-22** Two models involving thermal convection cells have been proposed to explain plate movement. (*a*) In one model, thermal convection cells are restricted to the asthenosphere. (*b*) In the other model, thermal convection cells involve the entire mantle.

ridges are higher than the surrounding oceanic crust. It is thought that gravity pushes the oceanic lithosphere away from the higher spreading ridges and toward the trenches.

Currently, geologists are fairly certain that some type of convective system is involved in plate movement, and the extent to which other mechanisms such as "slab-pull" and "ridge-push" are involved is still unresolved. Consequently, a comprehensive theory of plate movement has not as yet been developed, and much still remains to be learned about the Earth's interior.

⊕ PLATE TECTONICS AND THE DISTRIBUTION OF NATURAL RESOURCES

In addition to being responsible for the major features of the Earth's crust, plate movements also affect the formation and distribution of some natural resources. Consequently, geologists are using plate tectonic theory in their search for new mineral deposits and to explain the occurrence of known deposits.

Many metallic mineral deposits such as copper, gold, lead, silver, tin, and zinc are related to igneous activity, so it is not surprising that a close relationship exists between plate boundaries and the occurrence of these valuable deposits.

The magma generated by partial melting of a subducting plate rises toward the Earth's surface, and as it cools, it precipitates and concentrates various metallic ores. Many of the world's major metallic ore deposits, such as the copper deposits of western North and South America, are excellent examples of the relationship between convergent plate boundaries and the distribution, concentration, and exploitation of metallic ores.

Divergent plate boundaries also yield valuable resources. The island of Cyprus in the Mediterranean is rich in copper and has been supplying all or part of the world's needs for the last 3,000 years. The concentration of copper on Cyprus formed as a result of precipitation adjacent to hydrothermal vents along a divergent plate boundary. This deposit was brought to the surface when the copper-rich sea floor collided with the European plate, warping the sea floor and forming Cyprus.

Studies indicate that minerals containing such metals as copper, gold, iron, lead, silver, and zinc are currently forming in the Red Sea. The Red Sea is opening as a result of plate divergence and represents the earliest stage in the growth of an ocean basin.

It is becoming increasingly clear that if we are to keep up with the continuing demands of a global industrialized society, the application of plate tectonic theory to the origin and distribution of mineral resources is essential.

denser than the surrounding warmer asthenosphere, pulls the rest of the plate along with it as it descends into the asthenosphere. As the lithosphere moves downward, there is a corresponding upward flow back into the spreading ridge.

Operating in conjunction with "slab-pull" is the "ridge-push" mechanism. As a result of rising magma, the oceanic

1. The concept of continental movement is not new. Alfred Wegener is generally credited with developing the hypothesis of continental drift. He provided abundant geological and paleontological evidence to show that the continents were once united into one supercontinent he named Pangaea. Unfortunately, Wegener could not explain how the continents moved, and most geologists ignored his ideas.

2. The hypothesis of continental drift was revived during the 1950s when paleomagnetic studies indicated the presence of multiple magnetic north poles instead of just one as there is today. This paradox was resolved by moving the continents into different positions, making the paleomagnetic data consistent with a single magnetic north pole.

3. Magnetic surveys of the oceanic crust reveal magnetic anomalies in the rocks indicating that the Earth's magnetic field has reversed itself in the past. Because the anomalies are parallel and form symmetric belts adjacent to the oceanic ridges, new oceanic crust must have formed as the sea floor was spreading.

4. Sea-floor spreading has been confirmed by radiometric dating of rocks on oceanic islands. Such dating reveals that the oceanic crust becomes older with increasing distance from spreading ridges.

5. Plate tectonic theory became widely accepted by the 1970s because of the overwhelming evidence supporting it and because it provides geologists with a powerful theory for explaining such phenomena as volcanism, earthquake activity, mountain building, global climatic changes, past and present animal and plant distribution, and the distribution of some mineral resources.

6. Three types of plate boundaries are recognized: divergent boundaries, where plates move away from each other; convergent boundaries, where two plates collide; and transform boundaries, where two plates slide past each other.

7. The average rate of movement and relative motion of plates can be calculated in several ways. The results of these different methods all agree and indicate that the plates move at different average velocities.

8. Absolute motion of plates can be determined by the movement of plates over mantle plumes. A mantle plume is an apparently stationary column of magma that rises to the Earth's surface where it becomes a hot spot and forms a volcano.

9. Although a comprehensive theory of plate movement has yet to be developed, geologists think that some type of convective heat system is involved.

10. A close relationship exists between the formation of some mineral deposits and plate boundaries. Furthermore, the formation and distribution of some natural resources are related to plate movements.

IMPORTANT TERMS

continental-continental
  plate boundary
continental drift
convergent plate boundary
Curie point
divergent plate boundary
Glossopteris flora

Gondwana
hot spot
Laurasia
magnetic anomaly
magnetic reversal
oceanic-continental plate
  boundary

oceanic-oceanic plate
  boundary
paleomagnetism
Pangaea
plate
plate tectonic theory

sea-floor spreading
thermal convection cell
transform fault
transform plate boundary
volcanic island arc

REVIEW QUESTIONS

1. The man who is credited with developing the continental drift hypothesis is:
   a. _____ Wilson;
   b. _____ Hess;
   c. _____ Vine;
   d. _____ Wegener;
   e. _____ du Toit.

2. The southern part of Pangaea, consisting of South America, Africa, India, Australia, and Antarctica, is called:
   a. _____ Gondwana;
   b. _____ Laurasia;
   c. _____ Atlantis;
   d. _____ Laurentia;
   e. _____ Pacifica.

3. Which of the following has been used as evidence for continental drift?
   a. _____ continental fit;
   b. _____ fossil plants and animals;
   c. _____ similarity of rock sequences;
   d. _____ paleomagnetism;
   e. _____ all of these.

4. Plates:
    a. ____ are the same thickness everywhere;
    b. ____ vary in thickness;
    c. ____ include the crust and upper mantle;
    d. ____ answers (a) and (c);
    e. ____ answers (b) and (c).
5. Divergent boundaries are areas where:
    a. ____ new continental lithosphere is forming;
    b. ____ new oceanic lithosphere is forming;
    c. ____ two plates come together;
    d. ____ two plates slide past each other;
    e. ____ answers (b) and (d).
6. Along what type of plate boundary does subduction occur?
    a. ____ divergent;
    b. ____ transform;
    c. ____ convergent;
    d. ____ answers (a) and (b);
    e. ____ answers (b) and (c).
7. The west coast of South America is an example of a(n) _____ plate boundary.
    a. ____ divergent;
    b. ____ continental-continental;
    c. ____ oceanic-oceanic;
    d. ____ oceanic-continental;
    e. ____ transform.

8. Back-arc basins are associated with _____ plate boundaries.
    a. ____ divergent;
    b. ____ convergent;
    c. ____ transform;
    d. ____ answers (a) and (b);
    e. ____ answers (b) and (c).
9. The San Andreas fault is an example of a(n) _____ boundary.
    a. ____ divergent;
    b. ____ convergent;
    c. ____ transform;
    d. ____ oceanic-continental;
    e. ____ continental-continental.
10. Which of the following will allow you to determine the absolute motion of plates?
    a. ____ hot spots;
    b. ____ the age of the sediment directly above any portion of the ocean crust;
    c. ____ magnetic reversals in the sea-floor crust;
    d. ____ satellite laser ranging techniques;
    e. ____ all of these.
11. The formation of the island of Hawaii and the Loihi Seamount are the result of:
    a. ____ oceanic-oceanic plate boundaries;
    b. ____ hot spots;
    c. ____ divergent plate boundaries;

    d. ____ transform boundaries;
    e. ____ oceanic-continental plate boundaries.
12. The driving mechanism of plate movement is thought to be:
    a. ____ composition;
    b. ____ magnetism;
    c. ____ thermal convection cells;
    d. ____ rotation of the Earth;
    e. ____ none of these.
13. What evidence convinced Wegener that the continents were once joined together and subsequently broke apart?
14. What is the significance of polar wandering in relation to continental drift?
15. How can magnetic anomalies be used to show that the sea floor has been spreading?
16. Summarize the geologic features characterizing the three different types of plate boundaries.
17. What are some of the positive and negative features of the various models proposed to explain plate movement?
18. What features would an astronaut look for on the Moon or another planet to find out if plate tectonics is currently active or if it was active during the past?

POINTS TO PONDER

1. If movement along the San Andreas fault, which separates the Pacific plate from the North American plate, averages 5.5 cm per year, how long will it take before Los Angeles is opposite San Francisco?
2. The average rate of movement away from the Mid-Atlantic Ridge between North America and Africa is 3.0 cm per year (Figure 2-13) and the average sedimentation rate in the open ocean is 0.275 cm per thousand years. Calculate the age of the oceanic crust adjacent to Norfolk, Virginia, and the thickness of the sediment overlying the oceanic crust at that location. Does your calculation agree with the age of the oceanic crust at that location as shown in Figure 2-12?

MINERALS

OUTLINE

Two copper minerals, azurite (blue) and malachite (green), in the Flagg Collection, Arizona Mining and Mineral Museum.

Among the hundreds of minerals used by humans none is so highly prized and eagerly sought as gold (➢ Figure 3-1). This deep yellow mineral has been the cause of feuds and wars and was one of the incentives for the exploration of the Americas. Gold has been mined for at least 6,000 years, and archaeological evidence indicates that people in Spain possessed small quantities of gold 40,000 years ago. Probably no other substance has caused so much misery, but at the same time provided so many benefits for those who possessed it.

Why is gold so highly prized? Certainly not for use in tools or weapons, for it is too soft and pliable to hold a cutting edge. Furthermore, it is too heavy to be practical for most utilitarian purposes (it weighs about twice as much as lead). During most of historic time, gold has been used for jewelry, ornaments, and ritual objects and has served as a symbol of wealth and as a monetary standard. Gold is desired for several reasons: (1) its pleasing appearance, (2) the ease with which it can be worked, (3) its durability, and (4) its scarcity (it is much rarer than silver).

Central and South American natives used gold extensively long before the arrival of Europeans. In fact, the

➢ FIGURE 3-1 Specimen of gold from Grass Valley, California—National Museum of Natural History (NMNH) specimen #R121297. (Photo by D. Penland, courtesy of Smithsonian Institution.)

Europeans' lust for gold was responsible for the ruthless conquest of the natives in those areas. In the United States, gold was first profitably mined in North Carolina in 1801 and in Georgia in 1829, but the truly spectacular finds occurred in California in 1848. This latter discovery culminated in the great gold rush of 1849 when tens of thousands of people flocked to California to find riches. Unfortunately, only a few found what they sought. Nevertheless, during the five years from 1848 to 1853, which constituted the gold rush proper, more than $200 million in gold was recovered.

Another gold rush occurred in 1876 following the report by Lieutenant Colonel George Armstrong Custer that "gold in satisfactory quantities can be obtained in the Black Hills [South Dakota]." The flood of miners into the Black Hills, the Holy Wilderness of the Sioux Indians, resulted in the Indian War during which Custer and some 260 of his men were annihilated at the Battle of the Little Bighorn in Montana in June 1876. Despite this stunning victory, the Sioux could not sustain a war against the U.S. Army, and in September 1876, they were forced to relinquish the Black Hills.

Canada, too, has had its gold rushes. The first discovery came in 1850 in the Queen Charlotte islands on the Pacific coast, and by 1858 about 10,000 people were panning for gold there. The greatest Canadian gold rush occurred between 1897 and 1899 when as many as 35,000 men and women traveled to the remote, hostile Klondike region in the Yukon Territory. In fact, Dawson City grew so rapidly that hundreds of people had to be evacuated during the winter of 1897 because of food shortages. As in other gold rushes, local merchants made out better than the miners, most of whom barely eked out a living.

For 50 years following the California gold rush, the United States led the world in gold production, and it still produces a considerable amount, mostly from mines in Nevada, California, and South Dakota. Currently, the leading producer is South Africa with the United States a distant second, followed by Russia, Australia, and Canada.

Although much gold still goes into jewelry, its uses have expanded beyond those of earlier times; now gold is used in the chemical industry and for gold plating, electrical circuitry, and glass making. Consequently, the quest for gold has not ceased or even abated. In many industrialized nations, including the United States, domestic production cannot meet the demand, and much of the gold used must be imported.

INTRODUCTION

The term "mineral" commonly brings to mind dietary components essential for good nutrition such as calcium, iron, potassium, and magnesium. These substances are actually chemical elements, not minerals in the geologic sense. Mineral is also sometimes used to refer to any material that is neither animal nor vegetable. Such usage implies that minerals are inorganic substances, which is correct, but not all inorganic substances are minerals. Water, for example, is not a mineral even though it is inorganic and is composed of the same chemical elements as ice, which is a mineral. Ice, of course, is a solid whereas water is a liquid; minerals are solids rather than liquids or gases. In fact, geologists have a very specific definition of the term **mineral**: a naturally occurring, inorganic, crystalline solid. Crystalline means it has a regular internal structure. Furthermore, a mineral has a narrowly defined chemical composition and characteristic physical properties such as density, color, and hardness. Most rocks are solid aggregates of one or more minerals, so minerals are the building blocks of rocks.

Obviously, minerals are important to geologists as the constituents of rocks, but they are important for other reasons as well. Gemstones such as diamond and topaz are minerals, and rubies are simply red-colored varieties of the mineral corundum. The sand used in the manufacture of glass is composed of the mineral quartz, and ore deposits are natural concentrations of economically valuable minerals. Indeed, industrialized societies depend directly upon finding and using mineral resources such as iron, copper, gold, and many others.

MATTER AND ITS COMPOSITION

Anything that has mass and occupies space is *matter*. The atmosphere, water, plants and animals, and minerals and rocks are all composed of matter. Matter occurs in one of three states or phases, all of which are important in geology: *solids, liquids,* and *gases.* Atmospheric gases and liquids such as surface water and groundwater will be discussed later in this book, but here we are concerned chiefly with solids because all minerals are solids.

Elements and Atoms

All matter is made up of chemical **elements**, each of which is composed of incredibly small particles called **atoms**. Atoms are the smallest units of matter that retain the characteristics of an element. Ninety-two naturally occurring elements have been discovered, some of which are listed in ◉ Table 3-1, and more than a dozen additional elements have been made in laboratories. Each naturally occurring element and

TABLE 3-1 Symbols, Atomic Numbers, and Electron Configurations for Some of the Naturally Occurring Elements

Element	Symbol	Atomic Number	Number of Electrons in Each Shell			
			1	2	3	4
Hydrogen	H	1	1			
Helium	He	2	2			
Lithium	Li	3	2	1		
Beryllium	Be	4	2	2		
Boron	B	5	2	3		
Carbon	C	6	2	4		
Nitrogen	N	7	2	5		
Oxygen	O	8	2	6		
Fluorine	F	9	2	7		
Neon	Ne	10	2	8		
Sodium	Na	11	2	8	1	
Magnesium	Mg	12	2	8	2	
Aluminum	Al	13	2	8	3	
Silicon	Si	14	2	8	4	
Phosphorus	P	15	2	8	5	
Sulfur	S	16	2	8	6	
Chlorine	Cl	17	2	8	7	
Argon	Ar	18	2	8	8	
Potassium	K	19	2	8	8	1
Calcium	Ca	20	2	8	8	2

most artificially produced ones have a name and a chemical symbol (Table 3-1).

Atoms consist of a compact **nucleus** composed of one or more **protons**, which are particles with a positive electrical charge, and **neutrons**, which are electrically neutral (➤ Figure 3-2). The nucleus of an atom makes up most of its mass. Encircling the nucleus are negatively charged particles called **electrons**. Electrons orbit rapidly around the nucleus at specific distances in one or more **electron shells** (Figure 3-2). The number of protons in the nucleus of an atom determines what the element is and determines the **atomic number** for that element. For example, each atom of the element hydrogen (H) has one proton in its nucleus and thus has an atomic number of 1. Helium (He) possesses 2 protons, carbon (C) has 6, and uranium (U) has 92, so their atomic numbers are 2, 6, and 92, respectively (Table 3-1).

Atoms are also characterized by their **atomic mass number**, which is determined by adding together the number of protons and neutrons in the nucleus (electrons contribute negligible mass to an atom). Not all atoms of the same element have the same number of neutrons in their nuclei. In other words, atoms of the same element may have different atomic mass numbers. For instance, different carbon (C) atoms have atomic mass numbers of 12, 13, and 14. All of these atoms possess 6 protons—otherwise they would not be carbon—but the number of neutrons varies. Forms of the same element with different atomic mass numbers are **isotopes** (➤ Figure 3-3).

A number of elements have a single isotope but many, such as uranium and carbon, have several (Figure 3-3). All isotopes of an element behave the same chemically. For example, both carbon 12 and carbon 14 are present in carbon dioxide (CO_2).

Bonding and Compounds

Atoms are joined to other atoms through the process of **bonding**. When atoms of two or more different elements are bonded, the resulting substance is a **compound**. A chemical substance such as gaseous oxygen, which consists entirely of oxygen atoms, is an element, whereas ice, consisting of hydrogen and oxygen, is a compound. Most minerals are compounds although there are several important exceptions, such as gold and silver.

To understand bonding, it is necessary to delve deeper into the structure of atoms. Recall that negatively charged electrons in electron shells orbit the nuclei of atoms. With the exception of hydrogen, which has only one proton and one electron, the innermost electron shell of an atom contains only two electrons. The other shells contain various numbers of electrons, but the outermost shell never contains more than eight (Table 3-1). The electrons in the outermost shell are the ones that are usually involved in chemical bonding.

Two types of chemical bonds, *ionic* and *covalent,* are particularly important in minerals, and many minerals contain both types of bonds. Two other types of chemical bonds, *metallic* and *van der Waals,* are much less common, but are extremely important in determining the properties of some very useful minerals.

➤ **FIGURE 3-2** The structure of an atom. The dense nucleus consisting of protons and neutrons is surrounded by a cloud of orbiting electrons.

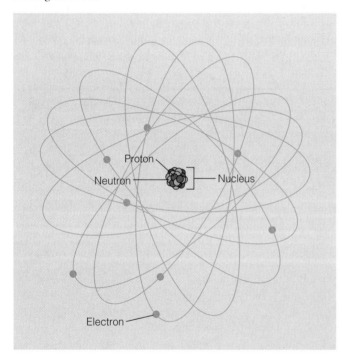

➤ **FIGURE 3-3** Schematic representation of isotopes of carbon. A carbon atom has an atomic number of 6 and an atomic mass number of 12, 13, or 14 depending on the number of neutrons in its nucleus.

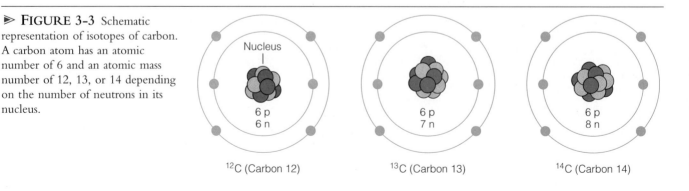

^{12}C (Carbon 12) ^{13}C (Carbon 13) ^{14}C (Carbon 14)

Ionic Bonding. Notice in Table 3-1 that most atoms have fewer than eight electrons in their outermost electron shell. Some elements, such as neon and argon, have complete outer shells containing eight electrons; they are known as the *noble gases*. The noble gases do not react readily with other elements to form compounds because of this electron configuration. Interactions among atoms tend to produce electron configurations similar to those of the noble gases. That is, atoms interact such that their outermost electron shell is filled with eight electrons, unless the first shell (with two electrons) is also the outermost electron shell as in helium.

One way that the noble gas configuration can be attained is by the transfer of one or more electrons from one atom to another. Common salt is composed of the elements sodium (Na) and chlorine (Cl), each of which is poisonous, but when combined chemically, forms the compound sodium chloride (NaCl), the mineral halite. Notice in ▷ Figure 3-4a that sodium has 11 protons and 11 electrons; thus, the positive electrical charges of the protons are exactly balanced by the negative charges of the electrons, and the atom is electrically neutral. Likewise, chlorine with 17 protons and 17 electrons is electrically neutral (Figure 3-4a). However, neither sodium nor chlorine has eight electrons in its outermost electron shell; sodium has only one whereas chlorine has seven. In order to attain a stable configuration, sodium loses the electron in its outermost electron shell, leaving its next shell with eight electrons as the outermost one (Figure 3-4a). Sodium now has one fewer electron (negative charge) than it has protons (positive charge) so it is an electrically charged particle. Such a particle is an **ion** and, in the case of sodium, is symbolized Na^{+1}.

The electron lost by sodium is transferred to the outermost electron shell of chlorine, which originally had seven electrons. This addition of one more electron gives chlorine an outermost electron shell of eight electrons, the configuration of a noble gas. But now its total number of electrons is 18, which exceeds by one the number of protons. Accordingly, chlorine also becomes an ion, but it is negatively charged (Cl^{-1}). An **ionic bond** forms between sodium and chlorine because of the attractive force between the positively charged sodium ion and the negatively charged chlorine ion (Figure 3-4a).

In ionic compounds, such as sodium chloride (the mineral halite), the ions are arranged in a three-dimensional framework that results in overall electrical neutrality. In halite, sodium ions are bonded to chlorine ions on all sides, and chlorine ions are surrounded by sodium ions (Figure 3-4b).

Covalent Bonding. **Covalent bonds** form between atoms when their electron shells overlap and electrons are shared. Atoms of the same element, such as oxygen in oxygen gas, cannot bond by transferring electrons from one atom to another. Carbon (C), which forms the minerals graphite and diamond, has four electrons in its outermost electron shell (▷ Figure 3-5a). If these four electrons were transferred to another carbon atom, the atom receiving the electrons would have the noble gas configuration of eight electrons in

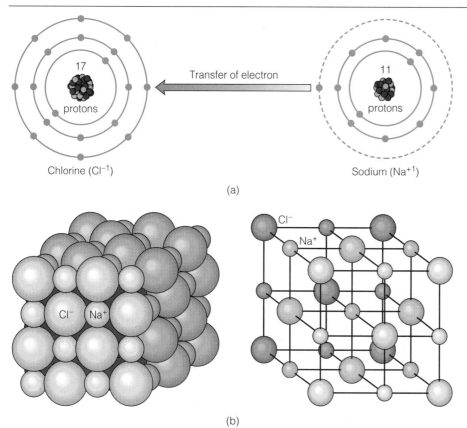

Chlorine (Cl^{-1})   Transfer of electron   Sodium (Na^{+1})

(a)

(b)

▷ **FIGURE 3-4** (*a*) Ionic bonding. The electron in the outermost shell of sodium is transferred to the outermost electron shell of chlorine. Once the transfer has occurred, sodium and chlorine are positively and negatively charged ions, respectively. (*b*) The crystal structure of sodium chloride, the mineral halite. The diagram on the left shows the relative sizes of the sodium and chlorine ions, and the diagram shows the locations of the ions in the crystal structure.

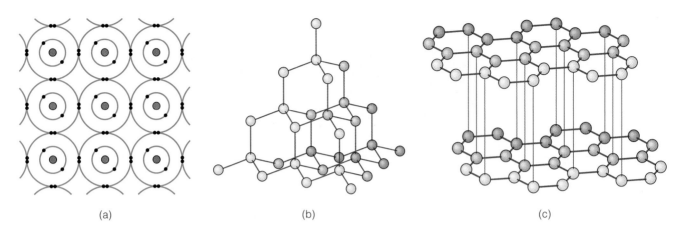

(a)   (b)   (c)

➤ **FIGURE 3-5** (*a*) Covalent bonds formed by adjacent atoms sharing electrons in diamond. (*b*) The three-dimensional framework of carbon atoms in diamond. (*c*) Covalent bonding also occurs in graphite, but here the carbon atoms are bonded together to form sheets that are held to one another by van der Waals bonds. The sheets themselves are strong, but the bonds between sheets are weak.

---

its outermost electron shell, but the atom contributing the electrons would not.

In such situations, adjacent atoms share electrons by overlapping their electron shells. A carbon atom in diamond, for instance, shares all four of its outermost electrons with a neighbor to produce a stable noble gas configuration (Figure 3-5a).

Covalent bonds are not restricted to substances composed of atoms of a single kind. Among the most common minerals, the silicates (discussed later in this chapter), the element silicon forms partly covalent and partly ionic bonds with oxygen.

Metallic and van der Waals Bonds. *Metallic bonding* results from an extreme type of electron sharing. The electrons of the outermost electron shell of such metals as gold, silver, and copper are readily lost and move about from one atom to another. This electron mobility accounts for the fact that metals have a metallic luster (their appearance in reflected light), provide good electrical and thermal conductivity, and can be easily reshaped. Only a few minerals possess metallic bonds, but those that do are very useful; copper, for example, is used for electrical wiring because of its high electrical conductivity.

Some electrically neutral atoms and molecules★ have no electrons available for ionic, covalent, or metallic bonding. Nevertheless, a weak attractive force exists between them when they are in proximity. This weak attractive force is a *van der Waals* or *residual bond.* The carbon atoms in the mineral graphite are covalently bonded to form sheets, but the sheets are weakly held together by van der Waals bonds (Figure 3-5b). Graphite is used for pencil leads because when pencil lead is moved across a piece of paper, small pieces of

graphite flake off along the planes held together by van der Waals bonds and adhere to the paper.

 **MINERALS**

Before we discuss minerals in more detail, let us recall our formal definition: a mineral is a naturally occurring, inorganic, crystalline solid, with a narrowly defined chemical composition and characteristic physical properties. The next sections will examine each part of this definition.

Naturally Occurring, Inorganic Substances

"Naturally occurring" excludes from minerals all substances that are manufactured by humans. Accordingly, synthetic diamonds and rubies and a number of other artificially synthesized substances are not regarded as minerals by most geologists.

Some geologists think the term "inorganic" in the mineral definition is superfluous. It does, however, remind us that animal matter and vegetable matter are not minerals. Nevertheless, some organisms such as corals and clams construct their shells of the compound calcium carbonate ($CaCO_3$), which is either aragonite or calcite, both of which are minerals.

The Nature of Crystals

By definition a mineral is a **crystalline solid,** that is a solid in which the constituent atoms are arranged in a regular, three-dimensional framework (Figure 3-4b). Under ideal conditions, such as in a cavity, mineral crystals can grow and form perfect crystals that possess planar surfaces (crystal faces), sharp corners, and straight edges (➤ Figure 3-6). In other words, the regular geometric shape of a well-formed mineral crystal is the exterior manifestation of an ordered internal atomic arrangement. Not all rigid substances are crystalline solids, however; natural and manufactured glass

---

★A molecule is the smallest unit of a substance having the properties of that substance. A water molecule (H_2O), for example, possesses two hydrogen atoms and one oxygen atom.

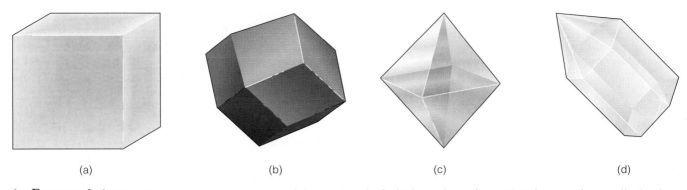

▷ **FIGURE 3-6** Mineral crystals occur in a variety of shapes, several of which are shown here. (*a*) Cubic crystals typically develop in the minerals halite, galena, and pyrite. (*b*) Dodecahedron crystals such as those of garnet have 12 sides. (*c*) Diamond has octahedral or 8-sided crystals. (*d*) A prism terminated by a pyramid is found in quartz.

lack the ordered arrangement of atoms and are said to be *amorphous,* meaning without form.

As early as 1669, a well-known Danish scientist, Nicholas Steno, determined that the angles of intersection of equivalent crystal faces on different specimens of quartz are identical. Since then the *constancy of interfacial angles* has been demonstrated for many other minerals, regardless of their size, shape, or geographic occurrence (▷ Figure 3-7). Steno postulated that mineral crystals are composed of very small, identical building blocks and that the arrangement of these blocks determines the external form of the crystals. Such regularity of the external form of minerals must surely mean that external crystal form is controlled by internal structure.

Crystalline structure can be demonstrated even in minerals lacking obvious crystals. For example, many minerals possess a property known as *cleavage,* meaning that they break or split along closely spaced, smooth planes. The fact that these minerals can be split along such smooth planar surfaces indicates that the mineral's internal structure controls such breakage. The behavior of light and X-ray beams

transmitted through minerals also provides compelling evidence for an orderly arrangement of atoms within minerals.

Chemical Composition

Mineral composition is generally shown by a chemical formula, which is a shorthand way of indicating the number of atoms of different elements composing the mineral. The mineral quartz consists of one silicon (Si) atom for every two oxygen (O) atoms, and thus has the formula SiO_2; the subscript number indicates the number of atoms. Orthoclase is composed of one potassium, one aluminum, three silicon, and eight oxygen atoms so its formula is $KAlSi_3O_8$. A few minerals are composed of a single element. Known as **native elements**, they include such minerals as platinum (Pt), graphite and diamond, both composed of carbon (C), silver (Ag), and gold (Au) (see the Prologue).

The definition of a mineral contains the phrase "a narrowly defined chemical composition," because some minerals actually have a range of compositions. For many

▷ **FIGURE 3-7** Side views and cross sections of three quartz crystals showing the constancy of interfacial angles: (*a*) a well-shaped crystal; (*b*) a larger crystal; and (*c*) a poorly-shaped crystal. The angles formed between equivalent crystal faces on different specimens of the same mineral are the same regardless of the size or shape of the specimens.

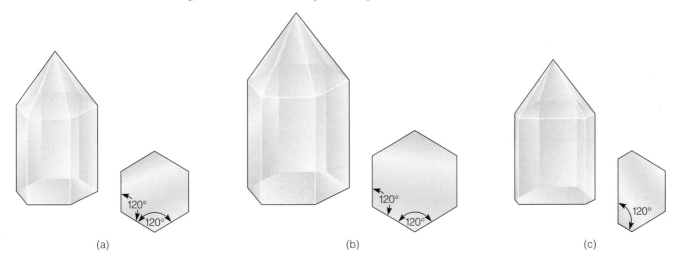

minerals the chemical composition is constant as in quartz (SiO_2), and halite ($NaCl$). Other minerals have a range of compositions because one element may substitute for another if the atoms of two or more elements are nearly the same size and the same charge. Notice in ➤ Figure 3-8 that iron and magnesium atoms are about the same size; therefore they can substitute for one another. The chemical formula for the mineral olivine is $(Mg,Fe)_2SiO_4$, meaning that, in addition to silicon and oxygen, it may contain only magnesium, only iron, or a combination of both. A number of other minerals also have ranges of compositions, so these are actually mineral groups with several members.

Physical Properties

The last criterion in our definition of a mineral, "characteristic physical properties," refers to such properties as hardness, color, and crystal form. These properties are controlled by composition and structure. We shall have more to say about physical properties of minerals later in this chapter.

◉ MINERAL DIVERSITY

More than 3,500 minerals have been identified and described, but only a very few—perhaps two dozen—are particularly common. Considering that 92 naturally occurring elements have been discovered, one might think that an extremely large number of minerals could be formed, but several factors limit the number of possible minerals. For one thing, many combinations of elements are chemically impossible; no compounds are composed of only potassium and sodium or of silicon and iron, for example. Another important factor restricting the number of common minerals is that only eight chemical elements make up the bulk of the Earth's crust (◉ Table 3-2). As Table 3-2 shows, oxygen and silicon constitute more than 74% (by weight) of the Earth's crust and nearly 84% of the atoms available to form compounds. By far the most common minerals in the Earth's crust consist of silicon and oxygen, combined with one or more of the other elements listed in Table 3-2.

◉ MINERAL GROUPS

Geologists recognize mineral classes or groups, each with members that share the same negatively charged ion or ion group (◉ Table 3-3).

Silicate Minerals

Because silicon and oxygen are the two most abundant elements in the Earth's crust (Table 3-2), it is not surprising that many minerals contain these elements. A combination of silicon and oxygen is called **silica**, and the minerals containing silica are **silicates**. Quartz (SiO_2) is composed entirely of silicon and oxygen so it is pure silica. Most silicates, however, have one or more additional elements, as in orthoclase ($KAlSi_3O_8$) and olivine [$(Mg,Fe)_2SiO_4$]. Silicate minerals include about one-third of all known minerals,

---

➤ **FIGURE 3-8** Electrical charges and relative sizes of ions common in minerals. The numbers within the ions are the radii shown in Ångstrom units (1 Å = 10^{-10} m).

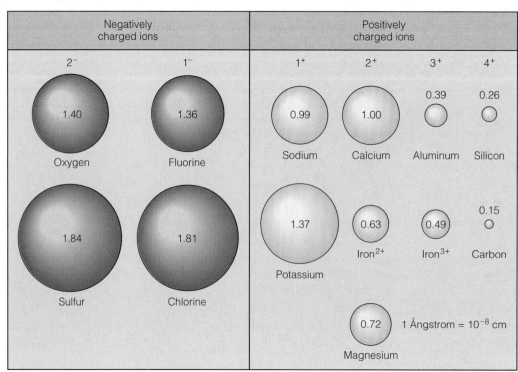

TABLE 3-2 Common Elements in the Earth's Crust

Element	Symbol	Percentage of Crust (by Weight)	Percentage of Crust (by Atoms)
Oxygen	O	46.6%	62.6%
Silicon	Si	27.7	21.2
Aluminum	Al	8.1	6.5
Iron	Fe	5.0	1.9
Calcium	Ca	3.6	1.9
Sodium	Na	2.8	2.6
Potassium	K	2.6	1.4
Magnesium	Mg	2.1	1.8
All others		1.5	0.1

TABLE 3-3 Some of the Mineral Groups Recognized by Geologists

Mineral Group	Negatively Charged Ion or Ion Group	Examples	Composition
Carbonate	$(CO_3)^{-2}$	Calcite	$CaCO_3$
		Dolomite	$CaMg(CO_3)_2$
Halide	Cl^{-1}, F^{-1}	Halite	$NaCl$
Native element	—	Gold	Au
		Diamond	C
Oxide	O^{-2}	Hematite	Fe_2O_3
Silicate	$(SiO_4)^{-4}$	Quartz	SiO_2
		Olivine	$(Mg,Fe)_2SiO_4$
Sulfate	$(SO_4)^{-2}$	Gypsum	$CaSO_4 \cdot 2H_2O$
Sulfide	S^{-2}	Galena	PbS

but their abundance is even more impressive when one considers that they make up perhaps as much as 95% of the Earth's crust.

The basic building block of all silicate minerals is the **silica tetrahedron**, which consists of one silicon atom and four oxygen atoms (➤ Figure 3-9). These atoms are arranged so that the four oxygen atoms surround a silicon atom, which occupies the space between the oxygen atoms; thus, a four-faced pyramidal structure is formed. The silicon atom has a positive charge of 4, while each of the four oxygen atoms has a negative charge of 2, resulting in an ion group with a total negative charge of 4 $(SiO_4)^{-4}$.

Because the silica tetrahedron has a negative charge, it does not exist in nature as an isolated ion group; rather it combines with positively charged ions or shares its oxygen atoms with other silica tetrahedra. In the simplest silicate minerals, the silica tetrahedra exist as single units bonded to positively charged ions. In minerals containing isolated tetrahedra, the silicon to oxygen ratio is 1:4, and the negative charge of the silica ion is balanced by positive ions (➤ Figure 3-10a). Olivine [$(Mg,Fe)_2SiO_4$], for example, has

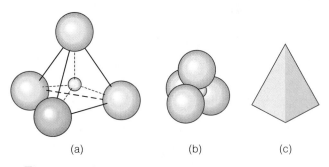

➤ **FIGURE 3-9** The silica tetrahedron. (*a*) Expanded view showing oxygen atoms at the corners of a tetrahedron and a small silicon atom at the center. (*b*) View of the silica tetrahedron as it really exists with the oxygen atoms touching. (*c*) The silica tetrahedron represented diagramatically; the oxygen atoms are at the four points of the tetrahedron.

either two magnesium (Mg^{+2}) ions, two iron (Fe^{+2}) ions, or one of each to offset the −4 charge of the silica ion.

Silica tetrahedra may also be arranged so that they join together to form chains of indefinite length (Figure 3-10b).

			Formula of negatively charged ion group	Silicon to oxygen ratio	Example
(a)	Isolated tetrahedra		$(SiO_4)^{-4}$	1:4	Olivine
(b)	Continuous chains of tetrahedra	Single chain	$(SiO_3)^{-2}$	1:3	Pyroxene group
		Double chain	$(Si_4O_{11})^{-6}$	4:11	Amphibole group
(c)	Continuous sheets		$(Si_4O_{10})^{-4}$	2:5	Micas
(d)	Three-dimensional networks	Too complex to be shown by a simple two-dimensional drawing	$(SiO_2)^0$	1:2	Quartz

➤ FIGURE 3-10 Structures of some of the common silicate minerals shown by various arrangements of silica tetrahedra: (*a*) isolated tetrahedra; (*b*) continuous chains; (*c*) continuous sheets; and (*d*) networks. The arrows adjacent to single-chain, double-chain, and sheet silicates indicate that these structures continue indefinitely in the directions shown.

Single chains, as in the pyroxene minerals, form when each tetrahedron shares two of its oxygens with adjacent tetrahedra; the result is a silicon to oxygen ratio of 1:3. Enstatite, a pyroxene group mineral, reflects this ratio in its chemical formula, $MgSiO_3$. Individual chains, however, possess a net −2 electrical charge, so they are balanced by positive ions, such as Mg^{+2}, that link parallel chains together (Figure 3-10b).

The amphibole group of minerals is characterized by a double-chain structure in which alternate tetrahedra in two parallel rows are cross-linked (Figure 3-10b). The formation of double chains results in a silicon to oxygen ratio of 4:11, so that each double chain possesses a −6 electrical charge. Mg^{+2}, Fe^{+2}, and Al^{+2} are usually involved in linking the double chains together.

In sheet structure silicates, three oxygens of each tetrahedron are shared by adjacent tetrahedra (Figure 3-10c). Such structures result in continuous sheets of silica tetrahedra with silicon to oxygen ratios of 2:5. Continuous sheets also possess a negative electrical charge that is satisfied by positive ions located between the sheets. This particular structure is what accounts for the characteristic sheet structure of the *micas*, such as biotite and muscovite, and the *clay minerals*.

Three-dimensional frameworks of silica tetrahedra form when all four oxygens of the silica tetrahedron are shared by adjacent tetrahedra (Figure 3-10d). Such sharing of oxygen atoms results in a silicon to oxygen ratio of 1:2, which is electrically neutral. Quartz is a common framework silicate.

Two subgroups of silicates are recognized, ferromagnesian and nonferromagnesian silicates. The **ferromagnesian silicates** are those containing iron (Fe), magnesium (Mg), or both. These minerals are commonly dark colored and more dense than nonferromagnesian silicates. Some of the common ferromagnesian silicate minerals are olivine, the pyroxenes, the amphiboles, and biotite (➤ Figure 3-11).

The **nonferromagnesian silicates** lack iron and magnesium, are generally light colored, and are less dense than ferromagnesian silicates (➤ Figure 3-12). The most common minerals in the Earth's crust are nonferromagnesian silicates known as *feldspars*. Feldspar is a general name for two distinct groups, each of which includes several species (Figure 3-12b and c). The *potassium feldspars* are represented by microcline and orthoclase ($KAlSi_3O_8$). The second group of feldspars, the *plagioclase feldspars,* range from calcium-rich ($CaAl_2Si_2O_8$) to sodium-rich ($NaAlSi_3O_8$) varieties.

Quartz (SiO_2) is another common nonferromagnesian silicate. It is a framework silicate that can usually be recognized by its glassy appearance and hardness (Figure 3-12a). Another fairly common nonferromagnesian silicate is muscovite, which is a mica (Figure 3-12d).

(a)   (b)

(c)   (d)

➤ **FIGURE 3-11** Common ferromagnesian silicates: (a) olivine; (b) augite, a pyroxene group mineral; (c) hornblende, an amphibole group mineral; and (d) biotite mica. (Photo courtesy of Sue Monroe.)

(a)   (b)

(c)   (d)

➤ **FIGURE 3-12** Common nonferromagnesian silicates: (a) quartz; (b) the potassium feldspar orthoclase; (c) plagioclase feldspar; and (d) muscovite mica. (Photo courtesy of Sue Monroe.)

Various clay minerals also possess the sheet structure typical of the micas, but their crystals are so small that they can be seen only with extremely high magnification. These clay minerals are important constituents of several types of rocks and are essential components of soils.

Other Mineral Groups

In addition to silicates, several other mineral groups are recognized (Table 3-3). One of these, the *native elements* such as gold and silver, has already been discussed. Another group consists of the **carbonate minerals,** all of which contain the negatively charged carbonate ion $(CO_3)^{-2}$. An example is calcium carbonate, the mineral calcite, which is the main constituent of the sedimentary rock limestone. A number of other carbonate minerals are known, but only dolomite $[CaMg (CO_3)_2]$, the mineral in the rock dolostone, need concern us.

PHYSICAL PROPERTIES OF MINERALS

All minerals possess characteristic physical properties that are determined by their internal structure and chemical composition. Some of these properties are what set gemstones apart from other minerals (see Perspective 3-1). Many physical properties are remarkably constant for a given mineral species, but some, especially color, may vary. Though a professional geologist may use sophisticated techniques in studying and identifying minerals, most common minerals can be identified by using the following physical properties.

Color and Luster

Although the color of many minerals varies because of minute amounts of impurities, some generalizations can be made. Ferromagnesian silicates are typically black, brown, or dark green, although olivine is olive green (Figure 3-11). Nonferromagnesian silicates, on the other hand, can vary considerably in color, but are only rarely dark (Figure 3-12). Minerals that have the appearance of metals are rather consistent in color.

Luster (not to be confused with color) is the appearance of a mineral in reflected light. Two basic types of luster are recognized: *metallic* and *nonmetallic* (➤ Figure 3-13). They are distinguished by observing the quality of light reflected from a mineral and determining if it has the appearance of a metal or a nonmetal. Several types of nonmetallic luster are recognized, including glassy or vitreous, greasy, waxy, brilliant (as in diamond), and dull or earthy.

➤ **FIGURE 3-13** Luster is the appearance of a mineral in reflected light. Galena (left), the ore of lead, has the appearance of a metal and is said to have a metallic luster, whereas orthoclase has a nonmetallic luster.

GEMSTONES

Gemstones of one kind or another have been sought for thousands of years. Archaeological evidence indicates that 75,000 years ago people in Spain and France were carving objects from bone, ivory, horn, and various stones. The ancient Egyptians mined turquoise more than 5,000 years ago, and by 3400 B.C. they were using rock crystal (colorless quartz), amethyst (purple quartz), lapis lazuli (a rock composed of a variety of minerals), and several other stones to make ornaments. Turquoise remains a popular gemstone with the Native Americans of the southwestern United States.

Most gemstones are minerals, more rarely rocks, that are cut and polished for jewelry. Gemstones are desired because of one or several features such as their brilliance, beauty, durability, and scarcity. They become even more desirable when some kind of lore is associated with them. Relating gemstones to one's birth month gives them added appeal to many people.

Many minerals and rocks are attractive but fail to qualify as gemstones. Cut and polished fluorite is beautiful, but it is fairly common and too soft to be durable. Diamond, on the other hand, meets most of the criteria for a gemstone. In fact, only about two dozen gemstones are used widely. Some are recognized as precious, and others are considered to be semiprecious.

The precious gemstones include diamond (▷ Figure 1); ruby and sapphire, which are red and blue varieties of the mineral corundum, respectively; emerald, a bright green variety of the mineral beryl; and precious opal. Some of the semiprecious gemstones are garnet, jade, tourmaline, topaz, aquamarine (light bluish-green beryl), turquoise, and several varieties of quartz such as amethyst, agate, and tiger's eye. In addition, amber, a fossil resin from coniferous trees that often contains insects, is included among the semiprecious gemstones. Recall that insects preserved in amber played a pivotal role in the novel and movie *Jurassic Park*.

Transparent gemstones are most often cut and faceted in various ways to enhance the quality of reflected or refracted light. The brilliance of diamonds is maximized by faceting the back of the gemstone so that as much light as possible is reflected (Figure 1). Most opaque and translucent gemstones are not faceted. Rather they are cut and polished into dome-shaped cabochons to emphasize their most interesting features, or they are simply polished by tumbling.

Throughout history gemstones have been used for personal adornment and as symbols of wealth. Gemstones and other minerals, rocks, and fossils have also served as religious symbols and talismans, or they have been worn or carried for their presumed mystical powers. Many people own small gemstones, but most of the truly magnificent ones are in museums or collections of crown jewels.

▷ **FIGURE 1** Gem-quality diamonds. Their hardness, brilliance, beauty, durability, and scarcity make them the most sought after gemstones.

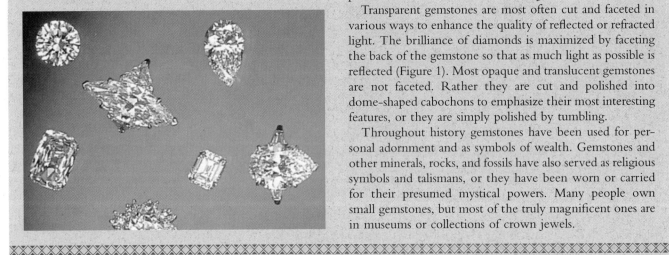

Crystal Form

As previously noted, mineral crystals are rare, so many mineral specimens will not show the perfect crystal form typical of that mineral species. Some minerals do typically occur as crystals. For example, 12-sided crystals of garnet are common, as are 6- and 12-sided crystals of pyrite (Figure 3-6). Minerals that grow in cavities or are precipitated from circulating hot water (hydrothermal solutions) in cracks and crevices in rocks also commonly occur as crystals.

Although crystal form can be very useful for identifying minerals, a number of minerals have the same crystal form. For example, pyrite (FeS_2), galena (PbS), and halite (NaCl) all occur as cubic crystals. However, such minerals can usually be easily identified by other properties such as color, luster, hardness, and density.

Cleavage and Fracture

Cleavage is a property of individual mineral crystals. Not all minerals possess cleavage, but those that do tend to break, or split, along a smooth plane or planes of weakness determined by the strength of the bonds within the mineral crystal.

Cleavage can be characterized in terms of quality (perfect, good, poor), direction, and angles of intersection of cleavage planes. Biotite, a common ferromagnesian silicate, has perfect cleavage in one direction (➤ Figure 3-14a). The fact that biotite preferentially cleaves along a number of closely spaced, parallel planes is related to its structure; it is a sheet silicate with the sheets of silica tetrahedra weakly bonded to one another by iron and magnesium ions (Figure 3-10c).

Feldspars possess two directions of cleavage that intersect at right angles (Figure 3-14b), and the mineral halite has three directions of cleavage, all of which intersect at right angles (Figure 3-14c). Calcite also possesses three directions of cleavage, but none of the intersection angles is a right angle, so cleavage fragments of calcite are rhombohedrons (Figure 3-14d). Minerals with four directions of cleavage include fluorite and diamond (Figure 3-14e). Ironically, diamond, the hardest mineral, can be easily cleaved. A few minerals such as sphalerite, an ore of zinc, have six directions of cleavage (Figure 3-14f).

Cleavage is a very important diagnostic property of minerals, and its recognition is essential in distinguishing between some minerals. The pyroxene mineral augite and the amphibole mineral hornblende look much alike: both are generally dark green to black, have the same hardness, and possess two directions of cleavage. But the cleavage planes of augite intersect at about 90°, whereas the cleavage planes of hornblende intersect at angles of 56° and 124° (➤ Figure 3-15).

In contrast to cleavage, *fracture* is mineral breakage along irregular surfaces. Any mineral can be fractured if enough force is applied, but the fracture surfaces will not be smooth; they are commonly uneven or conchoidal (smoothly curved).

Hardness

Hardness is the resistance of a mineral to abrasion. An Austrian geologist, Friedrich Mohs, devised a relative hardness scale for 10 minerals. He arbitrarily assigned a hardness value of 10 to diamond, the hardest mineral known, and

---

➤ FIGURE 3-14 Several types of mineral cleavage: (*a*) one direction; (*b*) two directions at right angles; (*c*) three directions at right angles; (*d*) three directions, not at right angles; (*e*) four directions; and (*f*) six directions.

(a) Cleavage in one direction — Cleavage plane — Micas—biotite and muscovite

(b) Cleavage in two directions at right angles — Potassium feldspars, plagioclase feldspars

(c) Cleavage in three directions at right angles — Halite, galena

(d) Cleavage in three directions, not at right angles — Calcite, dolomite

(e) Cleavage in four directions — Fluorite, diamond

(f) Cleavage in six directions — Sphalerite

➤ FIGURE 3-15 Cleavage in augite and hornblende. (*a*) Augite crystal and cross section of crystal showing cleavage. (*b*) Hornblende crystal and cross section of crystal showing cleavage.

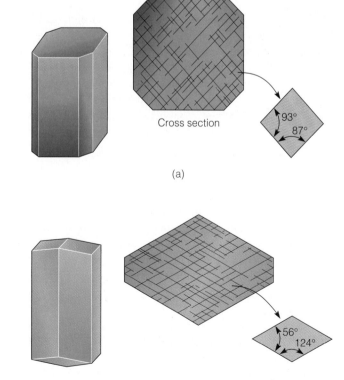

Cross section

93°
87°

(a)

56°
124°

(b)

TABLE 3-4 Mohs Hardness Scale		
Hardness	Mineral	Hardness of Some Common Objects
10	Diamond	
9	Corundum	
8	Topaz	
7	Quartz	
		Steel file (6½)
6	Orthoclase	
		Glass (5½–6)
5	Apatite	
4	Fluorite	
3	Calcite	Copper penny (3) Fingernail (2½)
2	Gypsum	
1	Talc	

lesser values to the other minerals. Relative hardness can be determined easily by using the Mohs hardness scale (◉ Table 3-4). For example, quartz will scratch fluorite but cannot be scratched by fluorite, gypsum can be scratched by a fingernail, and so on. Hardness is controlled mostly by internal structure. Both graphite and diamond are composed of carbon, but the former has a hardness of 1 to 2 whereas the latter has a hardness of 10.

Specific Gravity

The *specific gravity* of a mineral is the ratio of its weight to the weight of an equal volume of water. A mineral with a specific gravity of 3.0 is three times as heavy as water. Like all ratios, specific gravity is not expressed in units such as grams per cubic centimeter—it is a dimensionless number.

Specific gravity varies in minerals depending upon their composition and structure. Among the common silicates, the ferromagnesian silicates have specific gravities ranging from 2.7 to 4.3, whereas the nonferromagnesian silicates vary from 2.6 to 2.9. Obviously, the ranges of values overlap somewhat, but for the most part ferromagnesian silicates have greater specific gravities than nonferromagnesian silicates. In general, the metallic minerals, such as galena (7.58) and hematite (5.26), are heavier than nonmetals. Structure as a control of specific gravity is illustrated by the native element carbon (C): the specific gravity of graphite varies from 2.09 to 2.33; that of diamond is 3.5.

Other Properties

A number of other physical properties characterize some minerals. Talc has a distinctive soapy feel, graphite writes on paper, halite tastes salty, and magnetite is magnetic. Some minerals are plastic and, when bent into a new shape, will retain that shape, whereas others are flexible and, if bent, will return to their original position when the forces that bent them are removed.

The minerals calcite and dolomite can be identified by a simple chemical test in which a drop of dilute hydrochloric acid is applied to the mineral specimen. If the mineral is calcite, it will react vigorously with the acid and release carbon dioxide, which causes the acid to bubble or effervesce. Dolomite, on the other hand, will not react with hydrochloric acid unless it is powdered.

◉ IMPORTANT ROCK-FORMING MINERALS

Rocks are generally defined as solid aggregates of grains of one or more minerals. Two important exceptions to this definition are natural glass such as obsidian (see Chapter 4) and the sedimentary rock coal (see Chapter 7). Although it is true that many minerals occur in various kinds of rocks, only a few varieties are common enough to be designated as **rock-forming minerals**. Most of the others occur in such small amounts that they can be disregarded in the identification and classification of rocks; these are generally called *accessory minerals*.

Most common rocks are composed of silicate minerals. The igneous rock basalt is made up largely of ferromagnesian silicates, such as pyroxene group minerals and olivine (Figure 3-11), and plagioclase feldspar, a nonferromagnesian silicate (Figure 3-12c). Granite, another igneous rock, consists mostly of the nonferromagnesian silicates potassium feldspar and quartz (▷ Figure 3-16). Both of these rocks contain a variety of accessory minerals, most of which are silicates as well. The minerals noted in the preceding examples are also common in metamorphic rocks, and many sedimentary rocks are composed of quartz, various feldspars, and clay minerals. Among the nonsilicate minerals, only the carbonates calcite ($CaCO_3$) and dolomite [$CaMg(CO_3)_2$]—the primary constituents of the sedimentary rocks limestone and dolostone, respectively—are particularly common as rock-forming minerals.

⬡ MINERAL RESOURCES AND RESERVES

Geologists of the U.S. Geological Survey and the U.S. Bureau of Mines define a **resource** as follows:

> A concentration of naturally occurring solid, liquid, or gaseous material in or on the Earth's crust in such form and amount that economic extraction of a commodity from the concentration is currently or potentially feasible.

Accordingly, resources include such substances as metals (*metallic resources*); sand, gravel, crushed stone, and sulfur (*nonmetallic resources*); and uranium, coal, oil, and natural gas (*energy resources*). An important distinction, however, must be made between a resource, which is the total amount of a commodity whether discovered or undiscovered, and a **reserve,** which is that part of the resource base that can be extracted economically.

(b)

(a)

▷ **FIGURE 3-16** The igneous rock granite is composed largely of potassium feldspar and quartz, lesser amounts of plagioclase feldspar, and accessory minerals such as biotite mica. (*a*) Hand specimen of granite. (*b*) Photomicrograph showing the various minerals.

The amount of mineral resources used has steadily increased since Europeans settled North America. In the United States alone, nearly 90 million metric tons of iron and steel were consumed in 1990, and about 70 million metric tons of cement and more than 5 million metric tons of aluminum were used. According to one estimate, the yearly per capita used of mineral resources by North Americans is about 14 metric tons; much of this is bulk items such as crushed stone and sand and gravel. It is no exaggeration to say that our industrialized societies are totally dependent on mineral resources. Unfortunately, they are being used at rates far faster than they form. Thus, mineral resources are nonrenewable, meaning that once the resources from a deposit have been exhausted, new supplies or suitable substitutes must be found.

For some mineral resources, adequate supplies are available for indefinite periods (sand and gravel, for example), whereas for others, supplies are limited or must be imported from other parts of the world (◉ Table 3-5). The United States is almost totally dependent on imports of manganese, an essential element in the manufacture of steel. Even though the United States is a leading producer of gold, it still depends on imports for more than half of its gold needs. More than half of the crude oil used in the United States is imported, much of it from the Middle East where more than 50% of the Earth's proven reserves exist. A poignant reminder of our dependence on the availability of resources was the United States' response to the takeover of Kuwait by Iraq in August 1990.

In terms of mineral and energy resources, Canada is more self-reliant than the United States. It meets most of its domestic needs for mineral resources, although it must import phosphate, chromium, manganese, and bauxite, the ore of aluminum. Canada also produces more crude oil and natural gas than it uses, and it is the world leader in the production and export of uranium.

What constitutes a resource as opposed to a reserve depends on several factors. For example, iron-bearing minerals occur in many rocks, but in quantities or ways that make their recovery uneconomical. As a matter of fact, most minerals that are concentrated in economic quantities are mined in only a few areas; 75% of all the metals mined in the world come from about 150 locations. Geographic location is also an important consideration. A mineral resource in a remote region may not be mined because transportation costs are too high, and what may be considered a resource in the United States or Canada may be a reserve in a less-developed country where labor costs are low and it can be economically mined. The market price of a commodity is, of course, important in evaluating a potential resource. From 1935 to 1968, the U.S. government maintained the price of gold at $35 per troy ounce (= 31.1 g). When this restriction was removed and the price of gold became subject to supply and demand, the price rose (it reached an all-time high of $843 per troy ounce during January 1980). As a result, many marginal deposits became reserves and many abandoned mines were reopened.

Technological developments can also change the status of a resource. The rich iron ores of the Great Lakes region of the United States and Canada had been depleted by World War II. However, the development of a method of separating the iron from previously unusable rocks and shaping it into pellets that are ideal for use in blast furnaces made it feasible to mine poorer grade ores.

Most of the largest and richest mineral deposits have probably already been discovered and, in some cases,

Mineral Resource	Uses	World Reserves	U.S. Reserves	Canadian Reserves	Major Producing Countries
Bauxite	Ore of aluminum	21,559,000	38,000	0	Australia, Guinea, Jamaica
Chromium	Alloys, electroplating	418,900	0	0	South Africa, CIS,* India, Turkey
Copper	Alloys, electric wires	321,000	55,000	12,000	Chile, US, Canada, CIS
Gold	Jewelry, circuitry in computers and communications equipment, dentistry	42	5	1.8	South Africa, US, CIS, Australia, Canada
Iron ore	Iron and steel	64,648,000	3,800,000	4,600,000	CIS, Brazil, Australia, China
Lead	Storage batteries, solder, pipes	70,440	11,000	7,000	CIS, US, Mexico
Manganese	Iron and steel production	812,800	0	0	CIS, South Africa, Gabon, Australia, Brazil
Nickel	Stainless steel	48,660	30	8,130	CIS, Canada, New Caledonia
Silver	Jewelry, photography, dentistry	780	190	26.8	Mexico, US, Peru, CIS, Canada
Tin	Coating on metals, tin cans, alloys, solder	59,300	20	60	China, Brazil, Indonesia, Malaysia
Titanium	Alloys, white pigment in paint, paper, plastics	288,600	8,100	27,000	Australia, Norway, CIS
Zinc	Iron and steel alloys, rubber products, medicines	143,910	20,000	21,000	Canada, Australia, CIS, China, Peru

*Commonwealth of Independent States (includes much of the former Soviet Union).

SOURCES: *World resources 1992–1993: A guide to the global environment* (New York: Oxford University Press); *Minerals yearbook 1990*, vols. 1 and 3 (Metals and Minerals, U.S. Department of the Interior, Bureau of Mines); *Canada Year Book 1994*.

depleted. To ensure continued supplies of essential minerals, geologists are using increasingly sophisticated geophysical and geochemical mineral exploration techniques. The U.S. Geological Survey and the U.S. Bureau of Mines continually assess the status of resources in view of changing economic and political conditions and developments in science and technology. In the following chapters, we will discuss the origin and distribution of various mineral resources and reserves.

CHAPTER SUMMARY

1. All matter is composed of chemical elements, each of which consists of atoms. Individual atoms consist of a nucleus, containing protons and neutrons, and electrons that circle the nucleus in electron shells.
2. Atoms are characterized by their atomic number (the number of protons in the nucleus) and their atomic mass number (the number of protons plus the number of neutrons in the nucleus).
3. Bonding is the process whereby atoms are joined to other atoms. If atoms of different elements are bonded, they form a compound. Ionic and covalent bonds are most common

in minerals, but metallic and van der Waals bonds occur in a few.
4. Most minerals are compounds, but a few, including gold and silver, are composed of a single element and are called native elements.
5. All minerals are crystalline solids, meaning that they possess an orderly internal arrangement of atoms.
6. Some minerals vary in chemical composition because atoms of different elements can substitute for one another provided that the electrical charge is balanced and the atoms are about the same size.

7. Most of the more than 3,500 known minerals are silicates. Ferromagnesian silicates contain iron (Fe) and magnesium (Mg), and nonferromagnesian silicates lack these elements.

8. In addition to silicates, several other mineral groups are recognized, including native elements and carbonates.

9. The physical properties of minerals such as color, hardness, cleavage, and crystal form are controlled by composition and structure.

10. A few minerals are common enough constituents of rocks to be designated rock-forming minerals.

11. Many resources are concentrations of minerals of economic importance.

12. Reserves are that part of the resource base that can be extracted economically.

IMPORTANT TERMS

atom	crystalline solid	isotope	reserve
atomic mass number	electron	mineral	resource
atomic number	electron shell	native element	rock
bonding	element	neutron	rock-forming mineral
carbonate mineral	ferromagnesian silicate	nonferromagnesian silicate	silica
cleavage	ion	nucleus	silica tetrahedron
compound	ionic bond	proton	silicate
covalent bond			

REVIEW QUESTIONS

1. The atomic number of an element is determined by the:
   a. ____ number of electrons in its outermost shell;
   b. ____ number of protons in its nucleus;
   c. ____ diameter of its most common isotope;
   d. ____ number of neutrons plus electrons in its nucleus;
   e. ____ total number of neutrons orbiting the nucleus.

2. The two most abundant elements in the Earth's crust are:
   a. ____ iron and magnesium;
   b. ____ carbon and potassium;
   c. ____ sodium and nitrogen;
   d. ____ silicon and oxygen;
   e. ____ sand and clay.

3. The sharing of electrons by adjacent atoms is a type of bonding called:
   a. ____ van der Waals;
   b. ____ covalent;
   c. ____ silicate;
   d. ____ tetrahedral;
   e. ____ ionic.

4. Many minerals break along closely spaced planes and are said to possess:
   a. ____ specific gravity;
   b. ____ cleavage;
   c. ____ covalent bonds;
   d. ____ fracture;
   e. ____ double refraction.

5. The chemical formula for olivine is $(Mg,Fe)_2SiO_4$, which means that:
   a. ____ magnesium and iron can substitute for one another;
   b. ____ magnesium is more common than iron;
   c. ____ magnesium is heavier than iron;
   d. ____ all olivine contains both magnesium and iron;
   e. ____ more magnesium than iron occurs in the Earth's crust.

6. The basic building block of all silicate minerals is the:
   a. ____ silicon sheet;
   b. ____ oxygen silicon cube;
   c. ____ silica tetrahedron;
   d. ____ silicate double chain;
   e. ____ silica framework.

7. Those chemical elements having eight electrons in their outermost electron shell are the:
   a. ____ noble gases;
   b. ____ native elements;
   c. ____ carbonates;
   d. ____ halides;
   e. ____ isotopes.

8. Minerals are solids possessing an orderly internal arrangement of atoms, meaning that they are:
   a. ____ amorphous substances;
   b. ____ crystalline;
   c. ____ composed of at least three different elements;
   d. ____ composed of a single element;
   e. ____ ionic compounds.

9. Calcite and dolomite are:
   a. ____ oxide minerals of great value;
   b. ____ ferromagnesian silicates possessing a distinctive sheet structure;
   c. ____ common rock-forming carbonate minerals;
   d. ____ minerals used in the manufacture of pencil leads;
   e. ____ important energy resources.

10. How does a crystalline solid differ from a liquid and a gas?

11. An atom of the element magnesium is shown below. If the two electrons in its outer electron shell are lost, what is the electrical charge of the magnesium ion?

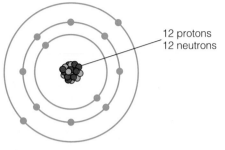

12 protons
12 neutrons

12. What is the atomic mass number of the magnesium atom shown above?
13. Compare ionic and covalent bonding.
14. Define compound and native element.
15. What accounts for the fact that some minerals have a range of chemical compositions?
16. Why are the angles between the same crystal faces on all specimens of a mineral species always the same?
17. What is a silicate mineral? How do the two subgroups of silicate minerals differ from one another?
18. In sheet silicates, individual sheets composed of silica tetrahedra possess a negative electrical charge. How is this negative charge satisfied?

POINTS TO PONDER

1. Why must the United States, a natural resource–rich nation, import a large part of the resources it needs? What are some of the problems created by dependence on imports?

2. Explain how the composition and structure of minerals control such mineral properties as hardness, cleavage, color, and specific gravity.

Chapter 4

IGNEOUS ROCKS AND INTRUSIVE IGNEOUS ACTIVITY

OUTLINE

El Capitan, meaning "The Chief," in Yosemite National Park, California, is composed of intrusive igneous rock. It rises more than 900 m above the valley floor and is the tallest unbroken cliff in the world. (Photo courtesy of Richard L. Chambers.)

About 45 to 50 million years ago, several small masses of molten rock were intruded into the Earth's crust in what is now northeastern Wyoming. These cooled and solidified, forming igneous rock bodies. The best known of these, Devil's Tower, was established as our first national monument by President Theodore Roosevelt in 1906. Devil's Tower is a remarkable landform. It rises nearly 260 m above its base (▷ Figure 4-1) and is visible from 48 km away.

Devil's Tower and other similar nearby bodies are important in the legends of the Cheyenne and Lakota Sioux Indians. These Native Americans call Devil's Tower

Mateo Tepee, which means "Grizzly Bear Lodge." It was also called the "Bad God's Tower," and reportedly, "Devil's Tower" is a translation of this phrase. According to one legend, the tower formed when the Great Spirit caused it to rise up from the ground, carrying with it several children who were trying to escape from a gigantic grizzly bear. Another legend tells of six brothers and a woman who were also being pursued by a grizzly bear. The youngest brother carried a small rock, and when he sang a song, the rock grew to the present size of Devil's Tower. In both legends, the bear's attempts to reach the Indians left deep scratch marks in the tower's rocks.

Geologists have a less dramatic explanation for the tower's origin. The near vertical striations (the bear's scratch marks) are simply the lines formed by the intersections of columnar joints. Columnar joints form in response to cooling and contraction in some igneous bodies and in some lava flows. Many of the columns are six sided, but columns with four, five, and seven sides occur as well. The larger columns measure about 2.5 m across. A pile of rubble at the tower's base is an accumulation of columns that have fallen from the tower.

Geologists agree that Devil's Tower originated as a small body that cooled and solidified from molten rock, and that subsequent erosion exposed it in its present form. The type of igneous body and the extent of its modification by erosion are debatable. Some geologists think that Devil's Tower is the eroded remnant of a more extensive body of intrusive rock, whereas others think it is simply the remnant of the rock that solidified in a pipelike conduit of a volcano and that it has been little modified by erosion.

▷ FIGURE 4-1 Devil's Tower in Wyoming exhibits well-developed columnar jointing. (Photo courtesy of R. V. Dietrich.)

INTRODUCTION

Rocks resulting from volcanic eruptions are widespread, but they probably represent only a small portion of the total rocks formed by the cooling and crystallization of molten rock material called magma. Most magma cools below the Earth's surface and forms bodies of rock known as *plutons.* The same types of magmas are involved in both volcanism and the origin of plutons, although some magmas are more mobile and more commonly reach the surface. Plutons typically underlie areas of extensive volcanism and were the sources of the overlying lavas and fragmental materials ejected from volcanoes during explosive eruptions. Furthermore, like volcanism, most plutonism occurs at or near plate margins. In this chapter we are concerned primarily with the textures, composition, and classification of igneous rocks and

with plutonic or intrusive igneous activity. Volcanism will be discussed in Chapter 5.

MAGMA AND LAVA

Magma is molten rock material below the Earth's surface, and **lava** is magma at the Earth's surface. Magma is less dense than the solid rock from which it was derived, so it tends to move upward toward the surface. Some magma is erupted onto the surface as **lava flows,** and some is forcefully ejected into the atmosphere as particles called **pyroclastic materials** (from the Greek *pyro* = fire and *klastos* = broken).

Igneous rocks (from the Latin *ignis* = fire) form when magma cools and crystallizes or when pyroclastic materials such as volcanic ash become consolidated. Magma extruded onto the Earth's surface as lava and pyroclastic materials forms **volcanic** or **extrusive igneous** rocks, whereas magma that crystallizes within the Earth's crust forms **plutonic** or **intrusive igneous** rocks.

Composition

Recall from Chapter 3 that the most abundant minerals in the Earth's crust are silicates, composed of silicon, oxygen, and the other elements listed in Table 3-2. Accordingly, when crustal rocks melt and form magma, the magma is typically silica rich and also contains considerable aluminum, calcium, sodium, iron, magnesium, and potassium as well as many other elements in lesser quantities. Not all magmas originate by melting of crustal rocks, however; some are derived from upper mantle rocks that are composed largely of ferromagnesian silicates. A magma from this source contains comparatively less silica and more iron and magnesium.

Although silica is the primary constituent of nearly all magmas, silica content varies and serves to distinguish **felsic, intermediate,** and **mafic** magmas (◉ Table 4-1). A felsic magma, for example, contains more than 65% silica and considerable sodium, potassium, and aluminum, but little calcium, iron, and magnesium. In contrast to felsic magmas, mafic magmas are silica poor, and contain proportionately more calcium, iron, and magnesium. As one would expect, intermediate magmas have mineral compositions intermediate between those of mafic and felsic magmas (Table 4-1).

Temperature

No direct measurements of temperatures of magma below the Earth's surface have been made. Erupting lavas generally have temperatures in the range of 1,000° to 1,200°C, although temperatures of 1,350°C have been recorded above Hawaiian lava lakes where volcanic gases reacted with the atmosphere.

Most direct temperature measurements have been taken at volcanoes characterized by little or no explosive activity where geologists can safely approach the lava. Therefore, little is known of the temperatures of felsic lavas, because eruptions of such lavas are rare, and when they do occur, they tend to be explosive. The temperatures of some lava domes, most of which are bulbous masses of felsic magma,

have been measured at a distance by using an instrument called an optical pyrometer. The surfaces of these domes have temperatures up to 900°C, but the exterior of a dome is probably much cooler than its interior.

When Mount St. Helens erupted in 1980, it ejected felsic magma as particulate matter in pyroclastic flows. Two weeks later, these flows still had temperatures between 300° and 420°C.

Viscosity

Magma is also characterized by its **viscosity,** or resistance to flow. The viscosity of some liquids, such as water, is very low; they are highly fluid and flow readily. The viscosity of some other liquids is so high that they flow much more slowly. Motor oil and syrup flow readily when they are hot, but become stiff and flow very slowly when they are cold. Thus, one might expect that temperature controls the viscosity of magma, and such an inference is partly correct. We can generalize and say that hot lava flows more readily than cooler lava.

Magma viscosity is also strongly controlled by silica content. In a felsic lava, numerous networks of silica tetrahedra retard flow, because the strong bonds of the networks must be ruptured for flow to occur. Mafic lavas, on the other hand, contain fewer silica tetrahedra networks and consequently flow more readily. Felsic lavas form thick, slow-moving flows, whereas mafic lavas tend to form thinner flows that move rather rapidly over great distances. One such flow in Iceland in 1783 flowed about 80 km, and some ancient flows in the state of Washington can be traced for more than 500 km.

◉ IGNEOUS ROCKS

All intrusive and many extrusive igneous rocks form when minerals crystallize from magma. The process of crystallization involves the formation and subsequent growth of crystal nuclei. The atoms in a magma are in constant motion, but when cooling begins, some atoms bond to form small groups, or nuclei, whose arrangement of atoms corresponds to the arrangement in mineral crystals. As other atoms in the liquid chemically bond to these nuclei, they do so in an ordered geometric arrangement, and the nuclei grow into crystalline *mineral grains,* the individual particles that comprise a rock. During rapid cooling, the rate of nuclei formation exceeds the rate of growth, and an aggregate of many small grains results (▶ Figure 4-2a). With slow cooling, the rate of growth exceeds the rate of nucleation, so relatively large grains form (Figure 4-2b).

Textures

Several textures of igneous rocks are related to the cooling history of a magma or lava. Rapid cooling, as occurs in lava flows or some near-surface intrusions, results in a fine-

TABLE 4-1 The Most Common Types of Magmas	
Type of Magma	Silica Content (%)
Mafic	45–52
Intermediate	53–65
Felsic	>65

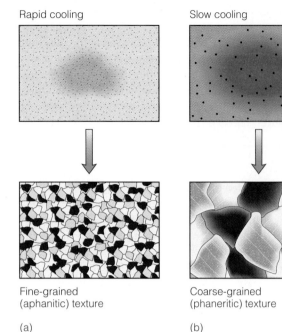

Rapid cooling   Slow cooling

Fine-grained
(aphanitic) texture

(a)

Coarse-grained
(phaneritic) texture

(b)

➤ **FIGURE 4-2** The effect of the cooling rate of a magma on nucleation and growth of crystals. (*a*) Rapid cooling results in many small grains and a fine-grained or aphanitic texture. (*b*) Slow cooling results in a coarse-grained or phaneritic texture.

grained texture termed **aphanitic.** In an aphanitic texture, individual mineral grains are too small to be observed without magnification (➤ Figure 4-3a). In contrast, igneous rocks with a coarse-grained or **phaneritic** texture have mineral grains that are easily visible without magnification (Figure 4-3b). Such large mineral grains indicate slow cooling and generally an intrusive origin; a phaneritic texture can develop in the interiors of some thick lava flows as well.

Rocks with **porphyritic** textures have a somewhat more complex cooling history. Such rocks have a combination of mineral grains of markedly different sizes. The larger grains are *phenocrysts,* and the smaller ones are referred to as *groundmass* (Figure 4-3c). Suppose that a magma begins cooling slowly as an intrusive body, and that some mineral crystal nuclei form and begin to grow. Suppose further that before the magma has completely crystallized, the remaining liquid phase and solid mineral grains within it are extruded onto the Earth's surface where it cools rapidly, forming an aphanitic texture. The resulting igneous rock would have large mineral grains (phenocrysts) suspended in a finely crystalline groundmass, and the rock would be characterized as a *porphyry.*

A lava may cool so rapidly that its constituent atoms do not have time to become arranged in the ordered, three-dimensional frameworks typical of minerals. As a result of such rapid cooling, a *natural glass* such as *obsidian* forms (➤ Figure 4-4a). Even though obsidian is not composed of minerals, it is still considered to be an igneous rock.

(a)

(b)

(c)

➤ **FIGURE 4-3** Textures of igneous rocks. (*a*) Aphanitic or fine-grained texture in which individual minerals are too small to be seen without magnification. (*b*) Phaneritic or coarse-grained texture in which minerals are easily discerned without magnification. (*c*) Porphyritic texture consisting of minerals of markedly different sizes. (Photos courtesy of Sue Monroe.)

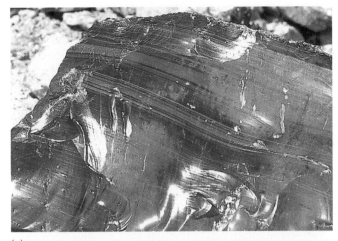

Some magmas contain large amounts of water vapor and other gases. These gases may be trapped in cooling lava where they form numerous small holes or cavities called **vesicles**; rocks possessing numerous vesicles are termed *vesicular*, as in vesicular basalt (Figure 4-4b).

A **pyroclastic** or **fragmental texture** characterizes igneous rocks formed by explosive volcanic activity. Ash may be discharged high into the atmosphere and eventually settle to the surface where it accumulates; if it is turned into solid rock, it is considered to be a pyroclastic igneous rock.

Composition

Magmas are characterized as mafic (45−52% silica), intermediate (53−65% silica), or felsic (>65% silica) (see Table 4-1). The parent magma plays a significant role in determining the mineral composition of igneous rocks, yet the same magma can yield different igneous rocks because its composition can change as a consequence of the sequence in which minerals crystallize, crystal settling, assimilation, and magma mixing.

Bowen's Reaction Series. During the early part of this century, N. L. Bowen hypothesized that mafic, intermediate, and felsic magmas could all derive from a parent mafic magma. He knew that minerals do not all crystallize simultaneously from a cooling magma, but rather crystallize in a predictable sequence. Based on his observations and laboratory experiments, Bowen proposed a mechanism, now called **Bowen's reaction series,** to account for the derivation of intermediate and felsic magmas from a basaltic (mafic) magma (▷ Figure 4-5). Bowen's reaction series consists of two branches: a *discontinuous branch* and a *continuous branch*. Crystallization of minerals occurs along both branches simultaneously, but for convenience we will discuss them separately.

In the discontinuous branch, which contains only ferromagnesian minerals, one mineral changes to another over specific temperature ranges (Figure 4-5). As the temperature decreases, it reaches a range where a given mineral begins to crystallize. Once a mineral forms, it reacts with the remaining liquid magma (the melt) such that it forms the next mineral in the sequence. For example, olivine [(Mg, Fe)$_2$SiO$_3$] is the first ferromagnesian mineral to crystallize. As the magma continues to cool, it reaches the temperature range at which pyroxene is stable; a reaction occurs between the olivine and the remaining melt, and pyroxene forms.

A similar reaction takes place between pyroxene and the melt as further cooling occurs, and the pyroxene structure is rearranged to form amphibole. Further cooling causes a reaction between the amphibole and the melt, and its structure is rearranged so that the sheet structure typical of biotite mica forms. Although the reactions just described tend to convert one mineral to the next in the series, the reactions are not always complete. Olivine might have a rim of pyroxene, indicating an incomplete reaction. If a magma cools rapidly enough, the early-formed minerals do not have time to react with the melt, and all the ferromagnesian minerals in the discontinuous branch can be in one rock. In any case, by the time biotite has crystallized, essentially all magnesium and iron present in the original magma have been used up.

Plagioclase feldspars are the only minerals in the continuous branch of Bowen's reaction series (Figure 4-5). Calcium-rich plagioclase crystallizes first. As cooling of the magma proceeds, calcium-rich plagioclase reacts with the melt, and plagioclase containing proportionately more sodium crystallizes until all of the calcium and sodium are used up. In many cases, cooling is too rapid for a complete transformation from calcium-rich to sodium-rich plagioclase to occur. Plagioclase forming under these conditions is *zoned,* meaning that it has a calcium-rich core surrounded by zones progressively richer in sodium.

Magnesium and iron on the one hand and calcium and sodium on the other are used up as crystallization occurs along the two branches in Bowen's reaction series. Accordingly, any magma left over is enriched in potassium, aluminum, and silicon. These elements combine to form potassium

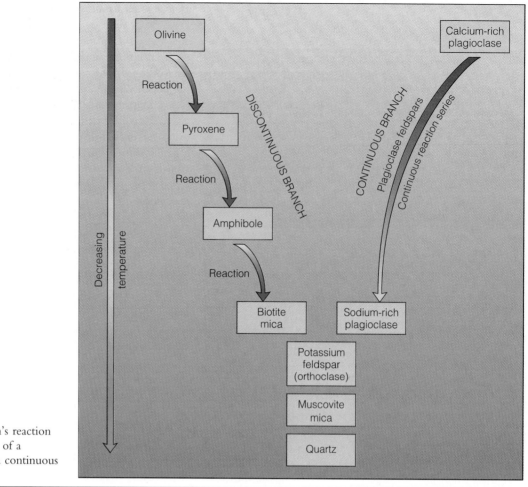

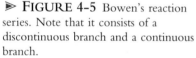

➤ **FIGURE 4-5** Bowen's reaction series. Note that it consists of a discontinuous branch and a continuous branch.

feldspar ($KAlSi_3O_8$), and if the water pressure is high, the sheet silicate muscovite mica will form. Any remaining magma is predominantly silicon and oxygen (silica) and forms the mineral quartz (SiO_2). The crystallization of potassium feldspar and quartz is not a true reaction series because they form independently rather than from a reaction of the orthoclase with the remaining melt.

Crystal Settling. A magma's composition may change by **crystal settling,** which involves the physical separation of minerals by crystallization and gravitational settling (➤ Figure 4-6). Olivine, the first ferromagnesian mineral to form in the discontinuous branch of Bowen's reaction series, has a specific gravity greater than that of the remaining magma and tends to sink downward in the melt. Accordingly, the remaining melt becomes relatively rich in silica, sodium, and potassium, because much of the iron and magnesium were removed when minerals containing these elements crystallized.

Although crystal settling does occur, it does not do so on the scale envisioned by Bowen. In some thick, tabular, intrusive igneous bodies called sills, the first formed minerals in the reaction series are indeed concentrated. The lower parts of these bodies contain more olivine and pyroxene than

the upper parts, which are less mafic. But, even in these bodies, crystal settling has yielded very little felsic magma from an original mafic magma.

If felsic magma could be derived on a large scale from mafic magma as Bowen thought, there should be far more mafic magma than felsic magma. For crystal settling to yield

➤ **FIGURE 4-6** Differentiation by crystal settling. Early-formed ferromagnesian minerals have a specific gravity greater than that of the magma so they settle and accumulate in the lower part of the magma chamber.

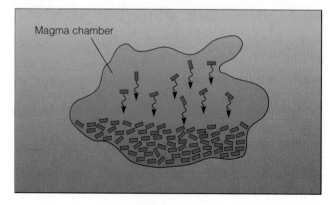

Magma chamber

a particular volume of granite (a felsic igneous rock), about 10 times as much mafic magma would have to be present initially. If this were so, then mafic intrusive igneous rocks should be much more common than felsic ones. Just the opposite is the case, however, so it appears that mechanisms other than crystal settling must account for the large volume of felsic magma. Partial melting of mafic oceanic crust and silica-rich sediments of continental margins during subduction yields magma richer in silica than the source rock. Furthermore, magma rising through the continental crust can absorb some felsic materials by *assimilation* and become more enriched in silica.

Assimilation. The composition of a magma can be changed by **assimilation,** a process whereby a magma reacts with preexisting rock, called *country rock,* with which it comes in contact (➢ Figure 4-7). The walls of a volcanic conduit or magma chamber are, of course, heated by the adjacent magma, which may reach temperatures of 1,300°C. Some of these rocks can be partly or completely melted, provided their melting temperature is less than that of the magma. Because the assimilated rocks seldom have the same composition as the magma, the composition of the magma is changed.

The fact that assimilation occurs can be demonstrated by *inclusions,* incompletely melted pieces of rock that are fairly common within igneous rocks. Many inclusions were simply wedged loose from the country rock as the magma forced its way into preexisting fractures (Figures 4-7 and ➢ 4-8).

No one doubts that assimilation occurs, but its effect on the bulk composition of most magmas must be slight. The reason is that the heat for melting must come from the magma itself, and this would have the effect of cooling the magma. Consequently, only a limited amount of rock can be

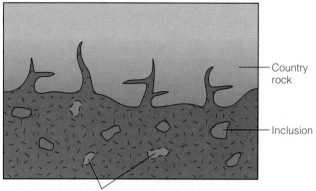

➢ **FIGURE 4-7** As magma moves upward, fragments of country rock are dislodged and settle into the magma. If they have a lower melting temperature than the magma, they may be incorporated into the magma by assimilation. Incompletely assimilated pieces of country rock are inclusions.

assimilated by a magma, and that amount is usually insufficient to bring about a major compositional change.

Neither crystal settling nor assimilation can produce a significant amount of felsic magma from a mafic one. But the two processes operating concurrently can change the composition of a mafic magma much more than either process acting alone. Some geologists think that this is one way that many intermediate magmas form where oceanic lithosphere is subducted beneath continental lithosphere.

Magma Mixing. The fact that a single volcano can erupt lavas of different composition indicates that magmas of differing composition must be present. It seems likely that

➢ **FIGURE 4-8** Dark-colored inclusions in granitic rock in California. (Photo courtesy of David J. Matty.)

some of these magmas would come into contact and mix with one another. If this is the case, we would expect that the composition of the magma resulting from **magma mixing** would be a modified version of the parent magmas. Suppose that a rising mafic magma mixes with a felsic magma of about the same volume (➤ Figure 4-9). The resulting "new" magma would have a more intermediate composition.

Classification of Igneous Rocks

Most igneous rocks are classified on the basis of their textures and composition (➤ Figure 4-10). Notice in Figure 4-10 that all of the rocks, except peridotite, constitute pairs; the members of a pair have the same composition but different textures. Basalt and gabbro, andesite and diorite, and rhyolite and granite are compositional (mineralogical) equivalents, but basalt, andesite, and rhyolite are aphanitic and most commonly extrusive, whereas gabbro, diorite, and granite have phaneritic textures that generally indicate an intrusive origin.

The igneous rocks shown in Figure 4-10 are also differentiated by composition. Reading across the chart from rhyolite to andesite to basalt, for example, the relative proportions of nonferromagnesian and ferromagnesian minerals differ. The differences in composition are gradual, however, so that a compositional continuum exists. In other words, there are rocks whose compositions are intermediate between rhyolite and andesite, and so on.

Ultramafic Rocks. Ultramafic rocks (<45% silica) are composed largely of ferromagnesian silicate minerals. The ultramafic rock *peridotite* contains mostly olivine, lesser amounts of pyroxene, and generally a little plagioclase feldspar (Figure 4-10). Another ultramafic rock (pyroxenite) is composed predominantly of pyroxene. Because these minerals are dark colored, the rocks are generally black or dark green. Peridotite is thought to be the rock type composing the upper mantle, but ultramafic rocks are rare at the Earth's surface. Ultramafic rocks are generally thought to have originated by concentration of the early-formed ferromagnesian minerals that separated from mafic magmas.

Basalt-Gabbro. *Basalt* and *gabbro* (45−52% silica) are the fine-grained and coarse-grained rocks, respectively, that crystallize from mafic magmas (➤ Figure 4-11). Both have the same composition—mostly calcium-rich plagioclase and pyroxene, with smaller amounts of olivine and amphibole (Figure 4-10). Because they contain a large proportion of ferromagnesian minerals, basalt and gabbro are dark colored;

➤ **FIGURE 4-9** Magma mixing. Two rising magmas mix and produce a magma with a composition different from either of the parent magmas.

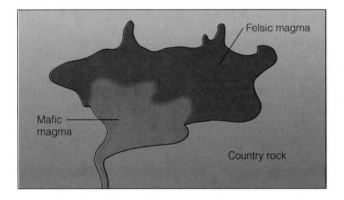

➤ **FIGURE 4-10** Classification of igneous rocks. The diagram illustrates the relative proportions of the chief mineral components of common igneous rocks.

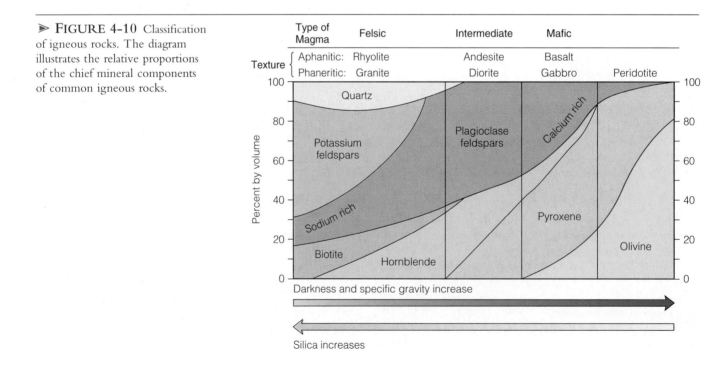

those that are porphyritic typically contain calcium plagioclase or olivine phenocrysts.

Basalt is generally considered to be the most common extrusive igneous rock. Extensive basalt lava flows were erupted in vast areas in Washington, Oregon, Idaho, and northern California (see Chapter 5). Oceanic islands such as Iceland, the Galapagos, the Azores, and the Hawaiian Islands are composed mostly of basalt. Furthermore, the upper part of the oceanic crust is composed almost entirely of basalt.

Gabbro is much less common than basalt, at least in the continental crust or where it can be easily observed. Small intrusive bodies of gabbro do occur in the continental crust, but intermediate to felsic intrusive rocks such as diorite and granite are much more common. The lower part of the oceanic crust is composed of gabbro, however.

Andesite-Diorite. Magmas intermediate in composition (53–65% silica) crystallize to form *andesite* and *diorite,* which are compositionally equivalent fine- and coarse-grained igneous rocks. Andesite and diorite are composed predominantly of plagioclase feldspar, with the typical ferromagnesian component being amphibole or biotite (Figure 4-10).

Andesite is a common extrusive igneous rock formed from lavas erupted in volcanic island arcs at convergent plate margins. The volcanoes of the Andes Mountains of South America and the Cascade Range in western North America are composed in part of andesite. Intrusive bodies composed of diorite are fairly common in the continental crust.

Rhyolite-Granite. *Rhyolite* and *granite* (>65% silica) crystallize from felsic magmas and are therefore silica-rich rocks (➤ Figure 4-12). They consist largely of potassium feldspar, sodium-rich plagioclase, and quartz, with perhaps some biotite and rarely amphibole (Figure 4-10). Because nonferromagnesian minerals predominate, these rocks are generally light colored. Rhyolite is fine-grained, although most often

(a)

(b)

➤ **FIGURE 4-11** Mafic igneous rocks: (*a*) basalt and (*b*) gabbro. (Photos courtesy of Sue Monroe.)

➤ **FIGURE 4-12** Felsic igneous rocks: (*a*) granite and (*b*) rhyolite. (Photos courtesy of Sue Monroe.)

(a)

(b)

it contains phenocrysts of potassium feldspar or quartz, and granite is coarse-grained. Granite porphyry is also fairly common.

Rhyolite lava flows are much less common than andesite and basalt flows. Recall that the greatest control of viscosity in a magma is the silica content. Thus, if a felsic magma rises to the surface, it begins to cool, the pressure on it decreases, and gases are released explosively, usually yielding rhyolitic pyroclastic materials. The rhyolitic lava flows that do occur are thick and highly viscous and move only short distances.

Granitic rocks are by far the most common intrusive igneous rocks, although they are restricted to the continents. Most granitic rocks were intruded at or near convergent plate margins during episodes of mountain building. When these mountainous regions are uplifted and eroded, the vast bodies of granitic rocks forming their cores are exposed. The granitic rocks of the Sierra Nevada of California form a composite body measuring about 640 km long and 110 km wide, and the granitic rocks of the Coast Ranges of British Columbia, Canada, are much more voluminous.

Pegmatite. Pegmatite is a very coarsely crystalline igneous rock. It contains minerals measuring at least 1 cm across, and many crystals are much larger (▷ Figure 4-13). The name pegmatite refers to texture rather than a specific composition, but most pegmatites are composed largely of quartz, potassium feldspar, and sodium-rich plagioclase—a composition similar to granite. Many pegmatites are associated with granite batholiths and appear to represent the minerals that formed from the fluid and vapor phases that remained after most of the granite crystallized.

▷ **FIGURE 4-13** Pegmatite is a textural term for very coarse-grained igneous rock; most pegmatites have a composition close to that of granite. The mineral grains in this specimen measure 2 to 3 cm.

The water-rich vapor phase that exists after most of a magma has crystallized as granite has properties that differ from the magma from which it separated. It has a lower density and viscosity and commonly invades the country rock where it crystallizes. The water-rich vapor phase ordinarily contains a number of elements that rarely enter into the common minerals that form granite.

The formation and growth of mineral crystal nuclei in pegmatites are similar to those processes in magma, but with

▷ **FIGURE 4-14** (*a*) Obsidian and (*b*) pumice. (Photos courtesy of Sue Monroe.)

(a)   (b)

one critical difference: the vapor phase from which pegmatites crystallize inhibits the formation of nuclei. Some nuclei do form, however, and because the appropriate ions in the liquid can move easily and attach themselves to a growing crystal, individual mineral grains have the opportunity to grow to very large sizes, several meters long in some cases.

Other Igneous Rocks. Some igneous rocks, including tuff, volcanic breccia, obsidian, and pumice, are identified solely by their textures. Much of the fragmental material erupted by volcanoes is *ash,* a designation for pyroclastic materials less than 2.0 mm in diameter; much ash consists of broken pieces or shards of volcanic glass. The consolidation of ash forms the pyroclastic rock *tuff.* Some ash flows are so hot that as they come to rest, the ash particles fuse together and form a *welded tuff.* Consolidated deposits of larger pyroclasts, such as cinders, blocks, and bombs, are *volcanic breccia.*

Both *obsidian* and *pumice* are varieties of volcanic glass. Obsidian may be black, dark gray, red, or brown, with the color depending on the presence of tiny particles of iron minerals (➢ Figure 4-14a). Analyses of numerous samples indicate that most obsidian has a high silica content and is compositionally similar to rhyolite.

Pumice is a variety of volcanic glass containing numerous bubble-shaped vesicles that develop when gas escapes through lava and forms a froth (Figure 4-14b). Some pumice forms as crusts on lava flows, and some forms as particles erupted from explosive volcanoes. If pumice falls into water, it can be carried great distances because it is so porous and light that it floats.

⊛ INTRUSIVE IGNEOUS BODIES: PLUTONS

Intrusive igneous bodies called **plutons** form when magma cools and crystallizes within the Earth's crust (➢ Figure 4-15). Geologists face a special challenge in studying the origins of plutons for, unlike extrusive or volcanic activity which can be observed, intrusive igneous activity can be studied only indirectly. Although plutons can be observed after erosion has exposed them at the surface, we cannot duplicate the conditions that existed deep in the crust when they formed, except in small-scale laboratory experiments.

➢ FIGURE 4-15 Block diagram showing the various types of plutons. Notice that some of these plutons cut across the layering in the country rock and are thus discordant, whereas others parallel the layering and are concordant.

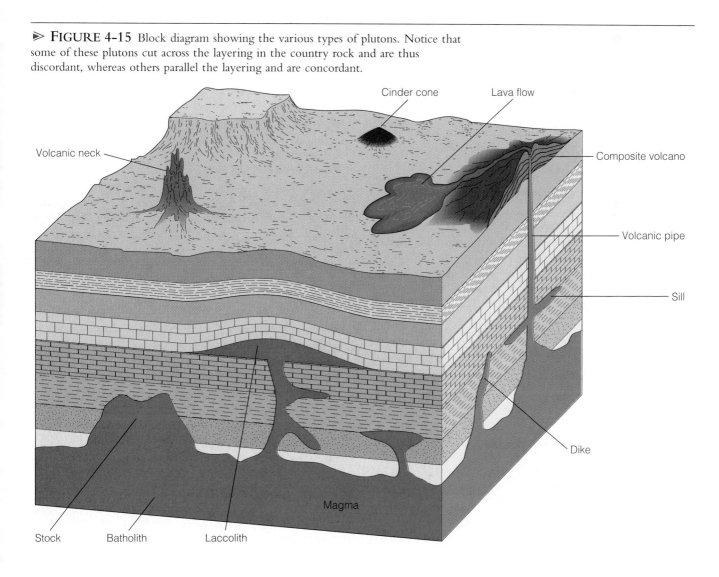

Several types of plutons are recognized, all of which are defined by their geometry (three-dimensional shape) and their relationship to the country rock (Figure 4-15). Geometrically, plutons may be characterized as massive or irregular, tabular, cylindrical, or mushroom shaped. Plutons are also described as concordant or discordant. A **concordant** pluton, such as a sill, has boundaries that are parallel to the layering in the intruded rock or what is commonly called the *country rock*. A **discordant** pluton, such as a dike, has boundaries that cut across the layering of the country rock (Figure 4-15).

Dikes and Sills

Both **dikes** and **sills** are tabular or sheetlike plutons, but dikes are discordant whereas sills are concordant (Figure 4-15). Dikes are common intrusive features (Figure 4-15). Many are small bodies measuring 1 or 2 m across, but they range from a few centimeters to more than 100 m thick. Dikes are emplaced within preexisting zones of weakness where fractures already exist or where the fluid pressure is great enough for them to form their own fractures during emplacement.

Erosion of the Hawaiian volcanoes exposes dikes in rift zones, the large fractures that cut across these volcanoes. The Columbia River basalts in Washington issued from long fissures, and the magma that cooled in the fissures formed dikes. Some of the large historic fissure eruptions are underlain by dikes; for example, dikes underlie both the Laki fissure eruption of 1783 in Iceland and the Eldgja fissure, also in Iceland, where eruptions occurred in A.D. 950 from a fissure 300 km long.

Sills are concordant plutons, many of which are a meter or less thick, although some are much thicker (Figure 4-15). A well-known sill in the United States is the Palisades sill that forms the Palisades along the west side of the Hudson River in New York and New Jersey. It is exposed for 60 km along the river and is up to 300 m thick.

Most sills have been intruded into sedimentary rocks, but eroded volcanoes also reveal that sills are commonly injected into piles of volcanic rocks. In fact, some of the inflation of volcanoes preceding eruptions may be caused by the injection of sills.

In contrast to dikes, which follow zones of weakness, sills are emplaced when the fluid pressure is so great that the intruding magma actually lifts the overlying rocks. Because emplacement requires fluid pressure exceeding the force exerted by the weight of the overlying rocks, sills are typically shallow intrusive bodies.

Laccoliths

Laccoliths are similar to sills in that they are concordant, but instead of being tabular, they have a mushroomlike geometry (Figure 4-15). They tend to have a flat floor and are domed up in their central part. Like sills, laccoliths are rather shallow intrusive bodies that actually lift up the overlying strata. In this case, however, the overlying rock layers are arched upward over the pluton (Figure 4-15). Most lacco-

liths are rather small bodies. The best-known laccoliths in the United States are in the Henry Mountains of southeastern Utah.

Volcanic Pipes and Necks

The conduit connecting the crater of a volcano with an underlying magma chamber is a **volcanic pipe** (Figure 4-15). In other words, it is the structure through which magma rises to the surface. When a volcano ceases to erupt, it is eroded as it is attacked by water, gases, and acids. The volcanic mountain eventually erodes away, but the magma that solidified in the pipe is commonly more resistant to weathering and erosion and is often left as an erosional remnant, a **volcanic neck** (Figure 4-15). A number of volcanic necks are present in the southwestern United States, especially in Arizona and New Mexico (see Perspective 4-1), and others are recognized elsewhere.

Batholiths and Stocks

Batholiths are the largest intrusive bodies. By definition they must have at least 100 km^2 of surface area, and most are much larger than this (Figure 4-15). **Stocks** have the same general features as batholiths but are smaller, although some stocks are simply the exposed parts of much larger intrusions. Batholiths are generally discordant, and most consist of multiple intrusions. In other words, a batholith is a large composite body produced by repeated, voluminous intrusions of magma in the same area. The coastal batholith of Peru, for instance, was emplaced over a period of 60 to 70 million years and consists of perhaps as many as 800 individual plutons.

The igneous rocks composing batholiths are mostly granitic, although diorite may also occur. Most batholiths are emplaced along convergent plate margins. A good example is the Coast Range batholith of British Colombia, Canada. It was emplaced over a period of millions of years. Later uplift and erosion exposed this huge composite pluton at the Earth's surface. Other large batholiths in North America include the Idaho batholith and the Sierra Nevada batholith in California (▷ Figure 4-16).

A number of mineral resources occur in rocks of batholiths and stocks and in the country rocks adjacent to them. For example, silica-rich igneous rocks, such as granite, are the primary source of gold, which forms from mineral-rich solutions moving through cracks and fractures of the igneous body. The copper deposits at Butte, Montana, are in rocks near the margins of the granitic rocks of the Boulder batholith. Near Salt Lake City, Utah, copper is mined from the mineralized rocks adjacent to the Bingham stock, a composite pluton composed of granite and granite porphyry.

⊚ MECHANICS OF BATHOLITH EMPLACEMENT

Geologists realized long ago that the emplacement of batholiths posed a space problem; that is, what happened to

SHIPROCK, NEW MEXICO

According to Navajo legend, a young man named Nayenezgani asked his grandmother where the mythical birdlike creatures known as Tse'na'hale lived. She replied, "They dwell at Tsaebidahi," which means Winged Rock or Rock with Wings. We know Winged Rock as Shiprock, a volcanic neck rising nearly 550 m above the surrounding plain (➤ Figure 1). Radiating outward from this conical volcanic neck are three dikes. Navajo legend holds that Winged Rock represents a giant bird that brought the Navajo people from the north and that the dikes are snakes that have turned to stone.

Shiprock is the most impressive of many volcanic necks exposed in the Four Corners region of the southwestern United States. (Four Corners is a designation for the point where the boundaries of Colorado, Utah, Arizona, and New Mexico converge.) Shiprock is visible from as far as 160 km and was a favorite with rock climbers for many years until the Navajo put a stop to all climbing on the reservation.

The country rock penetrated by this volcanic neck includes ancient metamorphic and igneous rocks and about 1,000 m of overlying sedimentary rocks. The rock unit exposed at the surface is the Mancos Shale, a sedimentary rock unit composed mostly of mud that was deposited in an arm of the sea that existed in North America during the Cretaceous Period. Absolute dating of one of the dikes indicates that the magma that solidified to form the dike was emplaced about 27 million years ago.

Shiprock is one of several volcanic necks in the Navajo volcanic field that formed as a result of explosive volcanic eruptions. During these eruptions, volcanic materials along with large pieces of country rock torn from the vent walls were hurled high into the air and fell randomly around the area. The material composing Shiprock itself is characterized as a tuff-breccia consisting of fragmental volcanic debris along with inclusions of various sedimentary rocks and some granite and metamorphic rocks. Because Shiprock now stands about 550 m above the surrounding plain, at least that much erosion must have occurred to expose it in its present form. We can only speculate on how much higher and larger it was when it was part of an active volcano.

The dikes radiating from Shiprock (Figure 1) formed when magma ascended rather quietly and was emplaced in the country rock. The fractures along which this magma rose, however, may have formed as a result of the explosive emplacement of the tuff-breccia that filled the volcanic vent. The dike on the northeast side of Shiprock extends for more than 2,900 m outward from the vent and averages 2.3 m thick. Because the dike rock, like the material composing the volcanic neck, is more resistant to erosion than the adjacent Mancos Shale, the dikes stand as near-vertical walls above the surrounding plain (Figure 1).

➤ **FIGURE 1** Shiprock, a volcanic neck in northwestern New Mexico, rises nearly 550 m above the surrounding plain. This view shows one of the dikes radiating from Shiprock. (Photo courtesy of Frank Hanna.)

the rock that formerly occupied the space now occupied by a granite batholith? One proposed answer was that no displacement had occurred, but rather that the granite had been formed in place by alteration of the country rock through a process called *granitization*. According to this view, granite did not originate as a magma, but rather from hot, ion-rich solutions that simply altered the country rock and transformed it into granite. Granitization is a solid-state

▷ FIGURE 4-16 View of granitic rocks in part of the Sierra Nevada batholith in Yosemite National Park, California.

phenomenon so it is essentially an extreme type of metamorphism (see Chapter 8).

Most geologists think that only small quantities of granite are formed by granitization and that it cannot account for the huge granite batholiths of the world. These geologists think an igneous origin for granite is clear, but then they must deal with the space problem. One solution is that these large igneous bodies melted their way into the crust. In other words, they simply assimilated the country rock as they moved upward (Figure 4-7). The presence of inclusions, especially near the tops of such intrusive bodies, indicates that assimilation does occur. Nevertheless, as we noted previously, assimilation is a limited process because magma is cooled as country rock is assimilated; calculations indicate that far too little heat is available in a magma to assimilate the huge quantities of country rock necessary to make room for a batholith.

Most geologists now agree that batholiths were emplaced by forceful injection as magma moved upward toward the surface. Recall that granite is derived from viscous felsic magma and therefore rises slowly. It appears that the magma deforms and shoulders aside the country rock, and as it rises further, some of the country rock fills the space beneath the magma. A somewhat analogous situation occurs when large masses of sedimentary rock called rock salt rise through the overlying rocks to form salt domes.

Salt domes are recognized in several areas of the world, including the Gulf Coast of the United States. Layers of rock salt exist at some depth, but salt is less dense than most other types of rock materials. When under pressure, it rises toward the surface even though it remains solid, and as it moves upward, it pushes aside and deforms the country rock (▷ Figure 4-17). Natural examples of rock salt flowage are known, and it can easily be demonstrated experimentally. In the arid Middle East, for example, salt moving upward in the manner described actually flows out at the surface.

Some batholiths do indeed show evidence of having been emplaced forcefully by shouldering aside and deforming the country rock. This mechanism probably occurs in the deeper parts of the crust where temperature and pressure are high and the country rocks are easily deformed in the manner described. At shallower depths, however, the crust is more rigid and tends to deform by fracturing. In this environment, batholiths may be emplaced by **stoping,** a process in which magma detaches and engulfs pieces of country rock (▷ Figure 4-18). According to this concept, magma moves upward along fractures and the planes separating layers of country rock. Eventually, pieces of country rock are detached and settle into the magma. No new room is created during stoping, the magma simply fills the space formerly occupied by country rock (Figure 4-18).

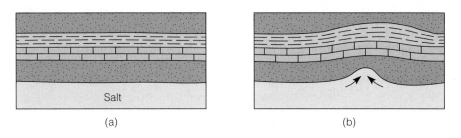

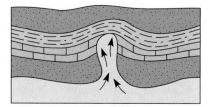

(a)   (b)   (c)

➢ **FIGURE 4-17** Three stages in the origin of a salt dome. Rock salt is a low-density sedimentary rock that (*a*) when deeply buried (*b*) tends to rise toward the surface, (*c*) pushing aside and deforming the country rock and forming a dome. Salt domes are thought to originate in much the same manner as batholiths are intruded into the Earth's crust.

(a)   (b)

➢ **FIGURE 4-18** Emplacement of a batholith by stoping. (*a*) The magma is injected into the country rock along fractures and planes between layers. (*b*) Blocks of country rock are detached and engulfed in the magma by stoping. Some of these blocks may be assimilated.

CHAPTER SUMMARY

1. Magma is molten rock material below the Earth's surface; lava is magma that reaches the surface. The silica content of magma varies and serves to differentiate felsic, intermediate, and mafic magmas.

2. The viscosity of magma depends on its temperature and especially its composition. Silica-rich (felsic) magma is more viscous than silica-poor (mafic) magma.

3. Minerals crystallize from magma and lava when small crystal nuclei form and grow.

4. Volcanic rocks generally have aphanitic textures because of rapid cooling, whereas slow cooling and phaneritic textures characterize plutonic rocks. Igneous rocks with a porphyritic texture have mineral crystals of markedly different sizes. Other igneous rock textures include glassy, vesicular, and pyroclastic.

5. The composition of igneous rocks is determined largely by the composition of the parent magma. It is possible, however, for an individual magma to yield igneous rocks of differing compositions.

6. Under ideal cooling conditions, a mafic magma yields a sequence of different minerals that are stable within specific temperature ranges. This sequence, called Bowen's reaction series, consists of a discontinuous branch and a continuous branch.
   a. The discontinuous branch contains only ferromagnesian minerals, each of which reacts with the melt to form the next mineral in the sequence.
   b. The continuous branch involves changes only in plagioclase feldspar as sodium replaces calcium in the crystal structure.

7. The ferromagnesian minerals that form first in Bowen's reaction series can settle and become concentrated near the base of a magma chamber or intrusive body. Such settling of iron- and magnesium-rich minerals causes a chemical change in the remaining melt.

8. A magma can be changed compositionally when it assimilates country rock, but this process usually has only a

limited effect. Magma mixing may also bring about compositional changes in magmas.

9. Most igneous rocks are classified on the basis of their textures and composition. Two fundamental groups of igneous rocks are recognized: volcanic or extrusive rocks, and plutonic or intrusive rocks.
   a. Common volcanic rocks include tuff, rhyolite, andesite, and basalt.
   b. Common plutonic rocks include granite, diorite, and gabbro.

10. Pegmatites are very coarse-grained igneous rocks, most of which have an overall composition similar to that of granite. Crystallization from a vapor-rich phase left over after the crystallization of granite accounts for the very large mineral crystals in pegmatites.

11. Plutons are igneous bodies that formed in place or were intruded into the Earth's crust. Various types of plutons are classified by their geometry and whether they are concordant or discordant.

12. Common plutons include dikes (tabular geometry, discordant); sills (tabular geometry, concordant); volcanic necks (cylindrical geometry, discordant); laccoliths (mushroom shaped, concordant); and batholiths and stocks (irregular geometry, discordant).

13. By definition batholiths must have at least 100 km² of surface area; stocks are similar to batholiths but smaller. Many batholiths are large composite bodies consisting of many plutons emplaced over a long period of time.

14. Most batholiths appear to have formed in the cores of mountain ranges during episodes of mountain building.

15. Some geologists think that granite batholiths are emplaced when felsic magma moves upward and shoulders aside and deforms the country rock. The upward movement of rock salt and the formation of salt domes provide a somewhat analogous situation.

IMPORTANT TERMS

aphanitic	igneous rock	pluton	stock
assimilation	intermediate magma	plutonic (intrusive igneous)	stoping
batholith	laccolith	rock	vesicle
Bowen's reaction series	lava	porphyritic	viscosity
concordant	lava flow	pyroclastic materials	volcanic neck
crystal settling	mafic magma	pyroclastic (fragmental)	volcanic pipe
dike	magma	texture	volcanic (extrusive igneous)
discordant	magma mixing	sill	rock
felsic magma	phaneritic		

REVIEW QUESTIONS

1. The first minerals to crystallize from a mafic magma are:
   a. ____ quartz and potassium feldspar;
   b. ____ calcium-rich plagioclase and olivine;
   c. ____ biotite and muscovite;
   d. ____ amphibole and pyroxene;
   e. ____ andesite and basalt.

2. Volcanic rocks can usually be distinguished from plutonic rocks by:
   a. ____ color;
   b. ____ composition;
   c. ____ iron-magnesium content;
   d. ____ the size of their mineral grains;
   e. ____ specific gravity.

3. An example of a concordant pluton having a tabular geometry is a:
   a. ____ sill;
   b. ____ batholith;
   c. ____ volcanic neck;
   d. ____ lava flow;
   e. ____ dike.

4. An igneous rock possessing a combination of mineral grains with markedly different sizes is:
   a. ____ a natural glass;
   b. ____ the product of very rapid cooling;
   c. ____ formed by explosive volcanism;
   d. ____ a porphyry;
   e. ____ a tuff.

5. Which of the following minerals is likely to be separated from a mafic magma by crystal settling?
   a. ____ sodium-rich plagioclase;
   b. ____ muscovite;
   c. ____ quartz;
   d. ____ olivine;
   e. ____ potassium feldspar.

6. The process whereby a magma reacts with and incorporates preexisting rock is:
   a. ____ crystal differentiation;
   b. ____ granitization;
   c. ____ plutonism;
   d. ____ magma mixing;
   e. ____ assimilation.

7. Igneous rocks composed largely of

ferromagnesian minerals are characterized as:

a. ____ pyroclastic;
b. ____ ultramafic;
c. ____ intermediate;
d. ____ felsic;
e. ____ mafic.

8. Which of the following pairs of igneous rocks have the same mineral composition?

a. ____ granite-tuff;
b. ____ andesite-rhyolite;
c. ____ pumice-diorite;
d. ____ basalt-gabbro;
e. ____ peridotite-andesite.

9. Batholiths are composed mostly of what type of rock?

a. ____ granitic;
b. ____ gabbro;
c. ____ basalt;
d. ____ andesite;
e. ____ periodotite.

10. What are the two major kinds of igneous rocks? How do they differ?

11. Describe the process whereby mineral crystals form and grow. Why are volcanic rocks generally aphanitic?

12. Compare the continuous and discontinuous branches of Bowen's reaction series.

13. Describe how the composition of a magma can be changed by crystal settling; by assimilation. Cite evidence indicating that both of these processes occur.

14. How do dikes and sills differ? How is each emplaced?

15. Describe the sequence of events in the formation of a volcanic neck.

16. Briefly explain where and how batholiths form.

17. What are pegmatites? Explain why some pegmatites contain very large mineral crystals.

18. Why are felsic lava flows so much more viscous than mafic lava flows?

POINTS TO PONDER

1. In the discontinuous branch of Bowen's reaction series, olivine forms in a specific temperature range, but as the magma continues to cool, it reacts with the remaining melt and changes to pyroxene. Pyroxene in turn changes to amphibole, and amphibole changes to biotite with continued cooling. How is it possible to have any of these minerals other than biotite in an igneous rock?

2. Two rock specimens have the following compositions:

Specimen 1: 15% biotite, 15% sodium-rich plagioclase, 60% potassium feldspar, and 10% quartz.

Specimen 2: 10% olivine, 55% pyroxene, 5% hornblende, and 30% calcium-rich plagioclase.

How would these two rocks differ in color and specific gravity? What was the viscosity of the magmas from which these rocks crystallized?

VOLCANISM

OUTLINE

Parícutin volcano in Mexico soon after it formed. On February 20, 1943, a farmer noticed fumes emanating from a crack in his cornfield, and a few minutes later ash and cinders were erupted. Within a month a cone 300 m high had formed. Lava flows broke through the flanks and base of the volcano and covered two nearby towns. Activity ceased in 1952.

On June 15 1991, Mount Pinatubo in the Philippines erupted violently, discharging 3 to 5 km³ of particulate matter, mostly volcanic ash, into the atmosphere ($\triangleright$ Figure 5-1). It was the world's largest volcanic eruption in more than half a century. During the eruption, much of the volcano's summit was destroyed as it collapsed following the ejection of material from below the mountain. The summit was

replaced by a caldera (a large oval depression) measuring 2 km in diameter.

Mount Pinatubo is but one of about 550 known active volcanoes, at least several of which are erupting at any given time. Before the 1991 eruptions, Mount Pinatubo had been quiet for 600 years, but on April 2, 1991, farmers noticed a small steam explosion from a line of vents on the volcano. Philippine and American geologists quickly installed instruments to monitor earthquake activity and to see if the volcano's slope changed as a result of bulging that commonly precedes an eruption. By late May, molten rock injected into the mountain formed a lava dome, and emissions of sulfur dioxide gas increased markedly, indicating an impending eruption.

Just four days before the major eruption, emissions of ash and gases increased even more, and the lava dome grew rapidly. Fearing an imminent explosion, officials ordered the evacuation of everyone within 24 km of the volcano. When the eruption occurred on June 15, most people in the zone of immediate danger had been evacuated, including thousands of U.S. military personnel and their families at Clark Air Force Base. Nevertheless, an estimated 281 people perished during the eruption, 83 were killed by later volcanic mudflows, and 358 died of illness; in addition, 42,000 homes were destroyed.

During Mount Pinatubo's eruptions, about 20 million tons of sulfur dioxide (SO_2) were ejected into the atmosphere where it reacted with water vapor to form tiny droplets of sulfuric acid. In fact, one weather study indicates that the cool summer of 1992 can be attributed to the presence of these gases in the atmosphere. According to this study, summer temperatures were at a 10-year low in the Northern Hemisphere and a 15-year low in the Southern Hemisphere.

Mount Pinatubo remained active throughout the rest of 1991 and well into 1992. On April 4, 1992, a steam explosion produced an ash cloud 1,200 m high, and in June 1992, mudflows roared down the mountain's slopes again. As recently as February 1994, small steam explosions were occurring, and officials still consider the area within 10 km of the summit a danger zone.

$\triangleright$ FIGURE 5-1 Mount Pinatubo erupting on June 12, 1991. Within 17 minutes of the eruption, this huge cloud of volcanic ash and steam had risen about 18 km.

$\circledcirc$ INTRODUCTION

Erupting volcanoes are the most impressive manifestations of the dynamic processes operating within the Earth. During many eruptions, molten rock rises to the surface and flows as incandescent streams or is ejected into the atmosphere in fiery displays that are particularly impressive at night ($\triangleright$ Figure 5-2). In some parts of the world, volcanic eruptions are commonplace events. The residents of the Philippines, Iceland, Hawaii, and Japan are fully cognizant of volcanoes and their effects ($\circledcirc$ Table 5-1).

Ironically, eruptions of volcanoes are constructive processes when considered in the context of Earth history. The Hawaiian Islands and Iceland, for example, owe their existence to volcanism. Oceanic crust is continually produced by

▶ **FIGURE 5-2** Lava fountains such as this one in Hawaii are particularly impressive at night.

volcanism at spreading ridges, and volcanic eruptions during the early history of the Earth released gases that are thought to have formed the atmosphere and surface waters.

✦ VOLCANISM

Volcanism refers to the processes whereby magma and its associated gases rise through the Earth's crust and are extruded onto the surface or into the atmosphere. Currently, more than 550 volcanoes are *active*—that is, they have erupted during historic time. Well-known examples of active volcanoes include Mauna Loa and Kilauea on the island of Hawaii, Mount Etna on Sicily, Fujiyama in Japan, and Mount St. Helens in Washington. Only two other bodies in the solar system are thought to possess active volcanoes: Io, a moon of Jupiter, and perhaps Triton, one of Neptune's moons (see Chapter 2 Prologue).

In addition to active volcanoes, numerous *dormant volcanoes* exist that have not erupted recently but may do so again. Mount Vesuvius in Italy had not erupted in human memory until A.D. 79 when it erupted and destroyed the cities of Herculaneum and Pompeii. Some volcanoes have not erupted during recorded history and show no evidence of doing so again; thousands of these *extinct* or *inactive* volcanoes are known.

Volcanic Gases

Samples of gases taken from present-day volcanoes indicate that 50 to 80% of all volcanic gases are water vapor. Lesser amounts of carbon dioxide, nitrogen, sulfur gases, especially sulfur dioxide and hydrogen sulfide, and very small amounts of carbon monoxide, hydrogen, and chlorine are also commonly emitted. In areas of recent volcanism, such as Lassen Volcanic National Park in California, gases continue to be emitted. One cannot help but notice the rotten-egg odor of hydrogen sulfide gas in such areas.

When magma rises toward the surface, the pressure is reduced and the contained gases begin to expand. In felsic magmas, however, which are highly viscous, expansion is inhibited and gas pressure increases. Eventually, the pressure may become great enough to cause an explosion and produce pyroclastic materials such as ash. In contrast, low-viscosity mafic magmas allow gases to expand and escape easily. Accordingly, mafic magmas generally erupt rather quietly.

The amount of gases contained in magmas varies, but is rarely more than a few percent by weight. Even though volcanic gases constitute a small proportion of a magma, they can be dangerous and, in some cases, have had far-reaching climatic effects. Most volcanic gases quickly dissipate in the atmosphere and pose little danger to humans, but on several occasions these gases have caused numerous fatalities. In 1783, toxic gases, probably sulfur dioxide, erupted from Laki fissure in Iceland had devastating effects. About 75% of the nation's livestock died, and the haze resulting from the gas caused lower temperatures and crop failures; about 24% of Iceland's population died as a result of the ensuing Blue Haze Famine. The country suffered its coldest winter in 225 years in 1783–1784, with temperatures 4.8°C below the long-term average.

The effects of the 1783 Laki fissure eruption were felt far beyond Iceland. The eruption produced what Benjamin Franklin called a "dry fog" that was responsible for dimming the intensity of sunlight in Europe. The severe winter of 1783–1784 in Europe and eastern North America is attributed to the presence of this "dry fog" in the upper atmosphere.

The particularly cold spring and summer of 1816 are attributed to the 1815 eruption of Tambora in Indonesia, the largest and most deadly eruption during historic time. The eruption of Mayon volcano in the Philippines during the previous year may have contributed to the cool spring and summer of 1816 as well. Another large historic eruption that had widespread climatic effects was the eruption of Krakatau in 1883.

TABLE 5-1 Some Notable Volcanic Eruptions

Date	Volcano	Deaths
Aug. 24, 79	Mt. Vesuvius, Italy	3,360 killed in Pompeii and Herculaneum.
1586	Kelut, Java	Mudflows kill 10,000.
Dec. 16, 1631	Mt. Vesuvius, Italy	3,500 killed.
Aug. 4, 1672	Merapi, Java	3,000 killed by mudflows and pyroclastic flows.
Dec. 10, 1711	Awu, Indonesia	3,000 killed by pyroclastic flows.
Sept. 22, 1760	Makian, Indonesia	Eruption kills 2,000; island evacuated for seven years.
June 8, 1783	Lakagigar, Iceland	Largest historic lava flows: 12 km^3; 9,350 die.
July 26, 1783	Asama, Japan	Pyroclastic flows and floods kill 1,200+.
May 21, 1792	Unzen, Japan	14,500 die in debris avalanche and tsunami.
Apr. 10, 1815	Tambora, Indonesia	92,000 killed; another 80,000 reported to have died from famine and disease.
Oct. 8, 1822	Galunggung, Java	4,011 die in pyroclastic flows and mudflows.
Mar. 2, 1856	Awu, Indonesia	Pyroclastic flows kill 2,806.
Aug. 27, 1883	Krakatau, Indonesia	36,417 die; most killed by tsunami.
June 7, 1892	Awu, Indonesia	1,532 die in pyroclastic flows.
May 8, 1902	Mt. Pelée, Martinique	St. Pierre destroyed by pyroclastic flow; 28,000 killed.
Oct. 24, 1902	Santa María, Guatemala	5,000 killed.
June 6, 1912	Novarupta, Alaska	Largest 20th-century eruption: about 33 km^3 of pyroclastic materials erupted; no fatalities.
May 19, 1919	Kelut, Java	Mudflows kill 5,110, devastate 104 villages.
Jan. 21, 1951	Lamington, New Guinea	2,942 killed by pyroclastic flows.
Mar. 17, 1963	Agung, Indonesia	1,148 killed.
Aug. 12, 1976	Soufrière, Guadeloupe	74,000 residents evacuated.
May 18, 1980	Mt. St. Helens, Washington	63 killed; 600 km^2 of forest devastated.
Mar. 28, 1982	El Chichón, Mexico	Pyroclastic flows kill 1,877.
Nov. 13, 1985	Nevado del Ruiz, Colombia	Mudflows kill 23,000.
Aug. 21, 1986	Oku volcanic field, Cameroon	1,746 asphyxiated by cloud of CO_2 released from Lake Nyos.
June 1991	Unzen, Japan	43 killed; at least 8,500 fled.
June 1991	Mt. Pinatubo, Philippines	~281 killed during initial eruption; 83 killed by later mudflows; 358 died of illness; 84,000 evacuated.
Feb. 2, 1993	Mt. Mayon, Philippines	At least 70 killed; 60,000 evacuated.

SOURCE: American Geological Institute Data Sheets, except for last three entries.

More recently, in 1986, in the African nation of Cameroon 1,746 people died when a cloud of carbon dioxide engulfed them. The gas accumulated in the waters of Lake Nyos, which occupies a volcanic crater. No agreement exists on what caused the gas to suddenly burst forth from the lake, but once it did, it flowed downhill along the surface because it was denser than air. In fact, the density and velocity of the gas cloud were great enough to flatten vegetation, including trees, a few kilometers from the lake. Unfortunately, thousands of animals and many people, some as far as 23 km from the lake, were asphyxiated.

Lava Flows and Pyroclastic Materials

Lava flows are frequently portrayed in movies and on television as fiery streams of incandescent rock material posing a great danger to humans. Actually, lava flows are the least dangerous manifestation of volcanism, although they may destroy buildings and cover agricultural land. Most lava flows do not move particularly fast, and because they are fluid, they follow existing low areas. So once a flow erupts from a volcano, determining the path it will take is fairly easy, and anyone in areas likely to be affected can be evacuated.

Two types of lava flows, both of which were named for Hawaiian flows, are generally recognized. A **pahoehoe** (pronounced pah-hoy-hoy) flow has a ropy surface almost like taffy (⊳ Figure 5-3a). The surface of an **aa** (pronounced ah-ah) flow is characterized by rough, jagged angular blocks and fragments (Figure 5-3b). Pahoehoe flows are less viscous than aa flows; indeed, the latter are viscous enough to break up into blocks and move forward as a wall of rubble.

Columnar joints are common in many lava flows, especially mafic flows, but they also occur in other kinds of flows and in some intrusive igneous rocks (⊳ Figure 5-4). A lava

(a)

(b)

➤ **FIGURE 5-3** (*a*) Pahoehoe flow in the east rift zone of Kilauea volcano in 1972. (*b*) An aa flow in the east rift zone of Kilauea volcano, Hawaii in 1983. The flow front is about 2.5 m high.

(a)

(b)

➤ **FIGURE 5-4** (*a*) Columnar joints in a lava flow at Devil's Postpile National Monument, California. (*b*) Surface view of the same columnar joints showing their polygonal pattern. The straight lines and polish resulted from glacial ice moving over this surface.

flow contracts as it cools and produces forces that cause fractures called *joints* to open up. On the surface of a flow, these joints commonly form polygonal (often six-sided) cracks. These cracks also extend downward into the flow, forming parallel columns with their long axes perpendicular to the principal cooling surface. Excellent examples of columnar joints can be seen at Devil's Postpile National Monument in California (Figure 5-4), Devil's Tower National Monument in Wyoming (see Chapter 4 Prologue), the Giant's Causeway in Ireland, and many other areas.

Much of the igneous rock in the upper part of the oceanic crust is of a distinctive type; it consists of bulbous masses of basalt resembling pillows, hence the name **pillow lava**. It was long recognized that pillow lava forms when lava is rapidly chilled beneath water, but its formation was not observed until 1971. Divers near Hawaii saw pillows form when a blob of lava broke through the crust of an underwater lava flow and cooled almost instantly, forming a glassy exterior. The remaining fluid inside then broke through the crust of the pillow, resulting in an accumulation of interconnected pillows (➤ Figure 5-5).

Much pyroclastic material is erupted as **ash**, a designation for pyroclastic particles measuring less than 2.0 mm (➤ Figure 5-6). Ash may be erupted in two ways: an ash fall

▷ **FIGURE 5-5** These bulbous masses of pillow lava form when magma is erupted under water.

▷ **FIGURE 5-6** Pyroclastic materials: volcanic ash being erupted from Mount Ngauruhoe, New Zealand, during Jaunary 1974.

or an ash flow. During an ash fall, ash is ejected into the atmosphere and settles to the surface over a wide area. In 1947, ash erupted from Mount Hekla in Iceland fell 3,800 km away on Helsinki, Finland. Ash is also erupted in ash flows, which are coherent clouds of ash and gas that commonly flow along or close to the land surface. Such flows can move at more than 100 km per hour, and some of them cover vast areas.

Pyroclastic materials larger than ash are also erupted by explosive volcanoes. Particles measuring from 2 to 64 mm are known as *lapilli,* and any particle larger than 64 mm is called a *bomb* or *block* depending on its shape. Bombs have twisted, streamlined shapes that indicate they were erupted as globs of fluid that cooled and solidified during their flight through the air (▷ Figure 5-7). Blocks are angular pieces of rock ripped from a volcanic conduit or pieces of a solidified crust of a magma. Because of their large size, volcanic bomb and block accumulations are not nearly as widespread as ash deposits; instead, they are confined to the immediate area of eruption.

Volcanoes

Conical mountains formed around a vent where lava and pyroclastic materials are erupted are **volcanoes**. Volcanoes, which are named for *Vulcan,* the Roman deity of fire, come in many shapes and sizes, but geologists recognize several major categories, each of which has a distinctive eruptive style. One must realize, however, that each volcano has a unique overall history of eruptions and development.

▷ **FIGURE 5-7** Pyroclastic materials: volcanic bombs collected in Hawaii. The bombs' streamlined shape indicates they were erupted as globs of magma that cooled and solidified as they descended.

Most volcanoes have a circular depression or **crater** at their summit. Craters form as a result of the extrusion of gases and lava from a volcano and are connected via a conduit to a magma chamber below the surface. It is not unusual, though, for magma to erupt from vents on the flanks of large volcanoes where smaller, parasitic cones develop. Mount Etna on Sicily has some 200 smaller vents on its flanks.

Some volcanoes are characterized by a **caldera** rather than a crater. Craters are generally less than 1 km in diameter, whereas calderas exceed this dimension and have steep sides. One of the best-known calderas in the United States is the misnamed Crater Lake in Oregon—Crater Lake is actually a caldera (➤ Figure 5-8). It formed about 6,600 years ago after voluminous eruptions partially drained the magma chamber. This drainage left the summit of the mountain, Mount Mazama, unsupported, and it collapsed into the magma chamber, forming a caldera more than 1,200 m deep and measuring 9.7 by 6.5 km. Many calderas probably formed when a summit collapsed during particularly large, explosive eruptions as in the case of Crater Lake, but a few apparently formed when the top of the original volcano was blasted away.

Shield Volcanoes. **Shield volcanoes** resemble the outer surface of a shield lying on the ground with the convex side up (➤ Figure 5-9). They have low, rounded profiles with gentle slopes ranging from about 2 to 10 degrees. Their low slopes reflect the fact that they are composed mostly of mafic flows that had low viscosity, so the flows spread out and formed thin layers. Shield volcanoes have a summit crater or caldera and a number of smaller cones on their flanks through which lava is erupted (Figure 5-9). A vent opened on the flank of Kilauea and grew to more than 250 m high between June 1983 and September 1986.

Eruptions from shield volcanoes, sometimes called *Hawaiian-type volcanoes*, are quiet compared to those of volcanoes such as Mount St. Helens; lavas most commonly rise to the surface with little explosive activity, so they usually pose little danger to humans. Lava fountains, some up to 400 m high, contribute some pyroclastic materials to shield volcanoes (Figure 5-2), but otherwise these volcanoes are composed largely of basalt lava flows; flows comprise more than 99% of the Hawaiian volcanoes above sea level.

Shield volcanoes are most common in oceanic areas, such as the Hawaiian Islands and Iceland, but some are also present on the continents—for example, in East Africa. The island of Hawaii consists of five huge shield volcanoes, two of which, Kilauea and Mauna Loa, are active much of the time. These Hawaiian volcanoes are the largest in the world. Mauna Loa is nearly 100 km across at the base and stands more than 9.5 km above the surrounding sea floor. Its volume is estimated at about 50,000 km³. By contrast, the largest volcano in the continental United States, Mount Shasta in northern California, has a volume of only about 205 km³.

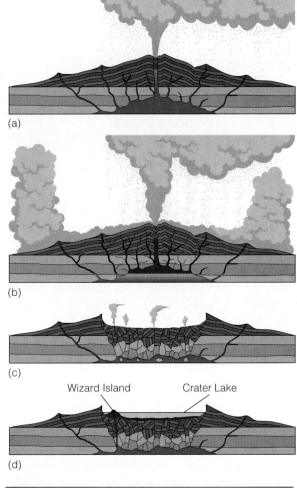

(e)

➤ **FIGURE 5-8** The sequence of events leading to the origin of Crater Lake, Oregon. (*a–b*) Ash clouds and ash flows partly drain the magma chamber beneath Mount Mazama. (*c*) The summit collapses, forming the caldera. (*d*) Post-caldera eruptions partly cover the caldera floor, and the small volcano known as Wizard Island forms. (*e*) View from the rim of Crater Lake showing Wizard Island.

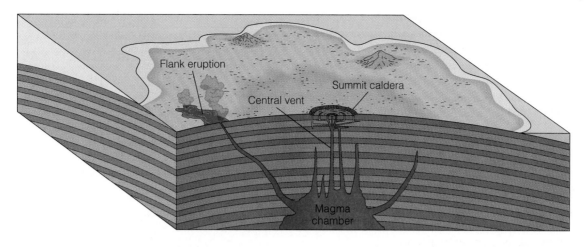

Flank eruption

Summit caldera

Central vent

Magma chamber

▷ **FIGURE 5-9** A shield volcano. Each layer shown consists of numerous, thin basalt lava flows.

Cinder Cones. Volcanic peaks composed of pyroclastic materials that resemble cinders are known as **cinder cones** (▷ Figure 5-10). They form when pyroclastic materials are ejected into the atmosphere and fall back to the surface to accumulate around the vent, thus forming small, steep-sided cones. The slope angle may be as much as 33 degrees, depending on the angle that can be maintained by the irregularly shaped pyroclastic materials. Cinder cones are rarely more than 400 m high, and many have a large, bowl-shaped crater.

Many cinder cones form on the flanks or within the calderas of larger volcanic mountains and appear to represent the final stages of activity, particularly in areas formerly characterized by basalt lava flows. Wizard Island in Crater Lake, Oregon, is a small cinder cone that formed after the summit of Mount Mazama collapsed to form the caldera (Figure 5-8). Cinder cones are common in the southern Rocky Mountain states, particularly New Mexico and Arizona, and many others occur in northern California, Oregon, and Washington.

In 1973, on the Icelandic island of Heimaey, the town of Vestmannaeyjar was threatened by a new cinder cone. The

▷ **FIGURE 5-10** (*a*) Cinder cones are composed of layers of angular pyroclastic materials. (*b*) The town of Vestmannaeyjar in Iceland was threatened by lava flows from Eldfell, a cinder cone that formed in 1973. Within two days of the initial eruption on January 23, the new volcano had grown to about 100 m high. Another cinder cone called Helgafell is also visible.

Central vent filled with rock fragments

Crater

Layers of pyroclastic materials

(a)

(b)

ERUPTIONS OF CASCADE RANGE VOLCANOES

During the summer of 1914, Mount Lassen in northern California began erupting without warning and culminated with the "Great Hot Blast," a huge steam explosion on May 22, 1915. Mount Lassen is one of 15 large volcanoes in the Cascade Range of northern California, Oregon, Washington, and southern British Columbia, Canada. After Mount Lassen's eruptions, the Cascade volcanoes remained quiet for 63 years. Then, on March 16, 1980, following an inactive period of 123 years, Mount St. Helens in southern Washington (▷ Figure 1) showed signs of renewed activity, and on May 18 it erupted violently, causing the worst volcanic disaster in U.S. history.

The awakening of Mount St. Helens came as no surprise to geologists of the U.S. Geological Survey (USGS) who warned in 1978 that it was an especially dangerous volcano. Although no one could predict precisely when Mount St. Helens would erupt, the USGS report included maps showing areas where damage from an eruption could be expected. Forewarned with such data, local officials were better prepared to formulate policies when the eruption did occur.

On March 27, 1980, Mount St. Helens began erupting steam and ash and continued to do so during the rest of March and most of April. By late March, a visible bulge had developed on its north face as molten rock was injected into the mountain, and the bulge continued to expand at about 1.5 m per day. On May 18, an earthquake shook the area,

the unstable bulge collapsed, and the pent-up volcanic gases below expanded rapidly, creating a tremendous northward-directed lateral blast that blew out the north side of the mountain (▷ Figure 2). The lateral blast accelerated from 350 to 1,080 km/hr, obliterating virtually everything in its path. Some 600 km² of forest were completely destroyed; trees were snapped off at their bases and strewn about the countryside, and trees as far as 30 km from the bulge were seared by the intense heat. Tens of thousands of animals were killed; roads, bridges, and buildings were destroyed; and 63 people perished.

Shortly after the lateral blast, volcanic ash and steam erupted forming a cloud above the volcano 19 km high. The ash cloud drifted east-northeast, and the resulting ash fall at Yakima, Washington, 130 km to the east, caused almost total darkness at midday. Detectable amounts of ash were deposited over a huge area. Flows of hot gases and volcanic ash raced down the north flank of the mountain, causing steam explosions when they encountered bodies of water or moist ground. Steam explosions continued for weeks, and at least one occurred a year later.

Snow and glacial ice on the upper slopes of Mount St. Helens melted and mixed with ash and other surface debris to form thick, pasty volcanic mudflows. The largest and most destructive mudflow surged down the valley of the North Fork of the Toutle River. Ash and mudflows displaced water in lakes and streams and flooded downstream areas. Ash and other particles carried by the flood

---

initial eruption began on January 23, and within two days a cinder cone, later named Eldfell, rose to about 100 m above the surrounding area (Figure 5-10). Pyroclastic materials from the volcano buried parts of the town, and by February a massive aa lava flow was advancing toward the town. The flow's leading edge ranged from 10 to 20 m thick, and its central part was as much as 100 m thick. By spraying the leading edge of the flow with sea water, which caused it to cool and solidify, the residents of Vestmannaeyjar successfully diverted the flow before it did much damage to the town.

Composite Volcanoes. **Composite volcanoes**, also called *stratovolcanoes*, are composed of both pyroclastic layers and lava flows (▷ Figure 5-11). Typically, both materials have an intermediate composition, and the flows cool to form andesite. Recall that lava of intermediate composition is more viscous than mafic lava. In addition to lava flows and pyroclastic layers, a significant proportion of a composite volcano is made up of **lahars** (volcanic mudflows). Some lahars form when rain falls on layers of loose pyroclastic

materials and creates a muddy slurry that moves downslope. On November 13, 1985, mudflows resulting from a rather minor eruption of Nevado del Ruiz in Colombia killed about 23,000 people. In the Philippines, 83 of the 722 victims of the June 1991 eruptions of Mount Pinatubo were killed by lahars (Table 5-1).

Composite volcanoes are steep sided near their summits, perhaps as much as 30 degrees, but the slope decreases toward the base where it is generally less than 5 degrees. Mayon volcano in the Philippines is one of the most perfectly symmetrical composite volcanoes on Earth. Its concave slopes rise ever steeper to the summit with its central vent through which lava and pyroclastic materials are periodically erupted (Figure 5-11b). Mayon erupted for the twelfth time this century in February 1993.

Composite volcanoes are the typical large volcanoes of the continents and island arcs. Familiar examples include Fujiyama in Japan and Mount Vesuvius in Italy as well as Mount St. Helens and many of the other volcanic peaks in the Cascade Range of western North America (See Perspective 5-1).

➤ **FIGURE 1** View of Mount St. Helens from the southwest in 1978.

➤ **FIGURE 2** The eruption of Mount St. Helens on May 18, 1980. The lateral blast occurred when the bulge on the north face of the mountain collapsed and reduced the pressure on the molten rock within the mountain.

waters were deposited in stream channels; many kilometers from Mount St. Helens, the navigation channel of the Columbia River was reduced from 12 m to less than 4 m as a result of such deposition.

Although the damage resulting from the eruption of Mount St. Helens was significant and the deaths were tragic, it was not a particularly large or deadly eruption compared with some historic eruptions. For example, the 1902 eruption of Mount Pelée on the island of Martinique killed 28,000 people, and the 1815 eruption of Tambora in Indonesia resulted in an estimated 172,000 deaths (Table 5-1). The Tambora eruption produced at least 80 times more ash than the 0.9 km^3 that spewed forth from Mount St. Helens.

➤ **FIGURE 5-11** (*a*) Composite volcanoes are the typical, large volcanic mountains on continents. They are composed of lava flows, pyroclastic layers, and volcanic mudflows. (*b*) Mayon volcano in the Philippines is one of the most nearly symmetrical composite volcanoes in the world.

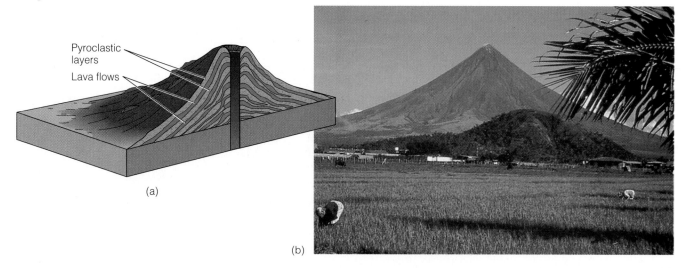

Pyroclastic layers

Lava flows

(a)

(b)

Lava Domes. If the upward pressure in a volcanic conduit is great enough, the most viscous magmas move upward and form bulbous, steep-sided **lava domes** (➤ Figure 5-12). Lava domes are generally composed of felsic lavas although some are of intermediate composition. Because such magma is so viscous, it moves upward very slowly; the lava dome that formed in Santa María volcano in Guatemala in 1922 took two years to grow to 500 m high and 1,200 m across. Lava domes contribute significantly to many composite volcanoes. Beginning in 1980, a number of lava domes were emplaced in the crater of Mount St. Helens; most of these were destroyed during subsequent eruptions. Since 1983, Mount St. Helens has been characterized by sporadic dome growth.

In June 1991, a dome in Japan's Unzen volcano collapsed, causing a flow of debris and hot ash that killed 43 people in a nearby town. Lava domes are also often responsible for extremely explosive eruptions. In 1902, viscous magma accumulated beneath the summit of Mount Pelée on the island of Martinique. Eventually, the pressure within the mountain increased to the point that it could no longer be contained, and the side of the mountain blew out in a tremendous explosion. When this occurred, a mobile, dense cloud of pyroclastic materials and gases called a **nuée ardente** (French for "glowing cloud") was ejected and raced downhill at about 100 km/hr, engulfing the city of St. Pierre (➤ Figure 5-13). This nuée ardente had internal temperatures of 700°C and incinerated everything in its path. Of the 28,000 residents of St. Pierre, only two survived, a prisoner in a cell below the ground surface and a man on the surface who was terribly burned by the nuée ardente.

Monitoring Volcanoes and Forecasting Eruptions

According to the U.S. Geological Survey, nearly 500 million people live near the volcanoes on the margins of the Earth's tectonic plates. Many of these volcanoes have erupted explosively during historic time and have the potential to do so again. As a matter of fact, volcanic eruptions are not as unusual as one might think; 376 separate outbursts occurred between 1975 and 1985. Fortunately, none of these compared to the 1815 eruption of Tambora; nevertheless, fatalities occurred in several instances, the worst being in 1985 in Colombia where about 23,000 perished in mudflows generated by an eruption (Table 5-1). Only a few of these potentially dangerous volcanoes are monitored, including some in Italy, Japan, New Zealand, Russia, and the Cascade Range.

Many of the methods for monitoring active volcanoes were developed at the Hawaiian Volcano Observatory.

➤ **FIGURE 5-12** A cross section showing the internal structure of a lava dome. Lava domes form when a viscous mass of magma, generally of felsic composition, is forced up through a volcanic conduit.

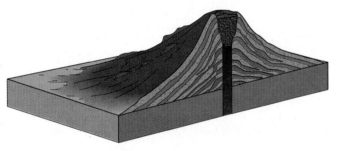

➤ **FIGURE 5-13** St. Pierre, Martinique after it was destroyed by a nuée ardente erupted from Mount Pelée in 1902. Only two of the city's 28,000 inhabitants survived.

These methods involve recording and analyzing various changes in both the physical and chemical attributes of volcanoes. Tiltmeters are used to detect changes in the slopes of a volcano when it inflates as magma is injected into it, while a geodimeter uses a laser beam to measure horizontal distances, which also change when a volcano inflates (➢ Figure 5-14). Geologists also monitor gas emissions and changes in the local magnetic and electrical fields of volcanoes.

Of critical importance in volcano monitoring and eruption forecasting are a sudden increase in earthquake activity and the detection of *harmonic tremor*. Harmonic tremor is continuous ground motion as opposed to the sudden jolts produced by earthquakes. It precedes all eruptions of Hawaiian volcanoes and also preceded the eruption of Mount St. Helens. Such activity indicates that magma is moving below the surface.

The analysis of data gathered during monitoring is not by itself sufficient to forecast eruptions; the past history of a particular volcano must also be known. To determine the eruptive history of a volcano, the record of previous eruptions as preserved in rocks must be studied and analyzed. Indeed, prior to 1980, Mount St. Helens was considered one of the most likely Cascade volcanoes to erupt because detailed studies indicated that it has had a record of explosive activity for the past 4,500 years.

For the better monitored volcanoes, such as those in Hawaii, it is now possible to make accurate short-term

➢ **FIGURE 5-14** Volcano monitoring. These diagrams show three stages in a typical eruption of a Hawaiian volcano: (*a*) The volcano begins to inflate; (*b*) inflation reaches its peak; (*c*) the volcano erupts and then deflates, returning to its normal shape.

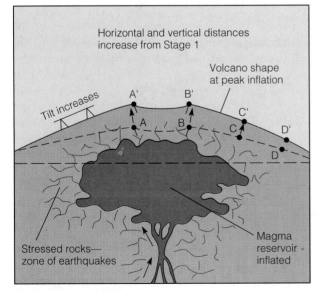

(a) Stage 1

(b) Stage 2

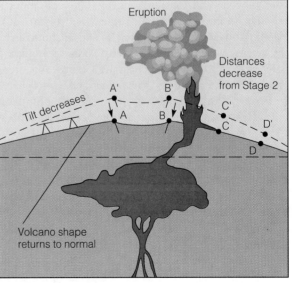

(c) Stage 3

▷ **FIGURE 5-15** The Columbia River basalts.

forecasts of eruptions. In 1960 the warning signs of an eruption of Kilauea were recognized soon enough to evacuate the residents of a small village that was subsequently buried by lava flows. Unfortunately, current forecasting is limited to just a few months in the future.

For some volcanoes little or no information is available for making predictions. On January 14, 1993, for example, Colombia's Galeras volcano erupted without warning, killing 6 of 10 volcanologists on a field trip and three Colombian tourists. Ironically, the volcanologists were attending a conference on improving methods for predicting volcanic eruptions.

Fissure Eruptions

During the Miocene and Pliocene epochs (between about 17 million and 5 million years ago), some 164,000 km^2 of eastern Washington and parts of Oregon and Idaho were covered by overlapping basalt lava flows. These Columbia River basalts, as they are called, are now well exposed in the walls of the canyons eroded by the Snake and Columbia rivers (▷ Figure 5-15). These lavas, which were erupted from long fissures, were so fluid that volcanic cones failed to develop. Such **fissure eruptions** yield flows that spread out over large areas and form **basalt plateaus** (Figure 5-15). The Columbia River basalt flows have an aggregate thickness of about 1,000 m, and some individual flows cover huge areas—the Roza flow, which is 30 m thick, advanced along a front about 100 km wide and covered 40,000 km^2.

Fissure eruptions and basalt plateaus are not common, although several large areas with these features are known. Currently, this type of activity is occurring only in Iceland. A number of volcanic mountains are present in Iceland, but the bulk of the island is composed of basalt flows erupted from

fissures. Two large fissure eruptions, one in A.D. 930 and the other in 1783, account for about half of the magma erupted in Iceland during historic time. The 1783 eruption occurred along the Laki fissure, which is more than 30 km long; lava flowed several tens of kilometers from the fissure, covering more than 560 km^2, and in one place filled a valley to a depth of about 200 m.

Pyroclastic Sheet Deposits

More than 100 years ago, geologists were aware of vast areas covered by felsic volcanic rocks a few meters to hundreds of meters thick. It seemed improbable that these could have formed as vast lava flows, but it also seemed equally unlikely that they were ash fall deposits. Based on observations of historic pyroclastic flows, such as the nuée ardente erupted by Mount Pelée in 1902, it now seems probable that these ancient rocks originated as pyroclastic flows, hence the name **pyroclastic sheet deposits.** They cover far greater areas than any observed during historic time and apparently erupted from long fissures rather than from a central vent. The pyroclastic materials of many of these flows were so hot they fused together to form *welded tuff.*

It now appears that major pyroclastic flows issue from fissures formed during the origin of calderas. The Yellowstone Tuff, for instance, was erupted during the formation of a large caldera in the area of present-day Yellowstone National Park in Wyoming. Similarly, the Bishop Tuff of eastern California appears to have been erupted shortly before the formation of the Long Valley caldera. Interestingly, earthquake activity in the Long Valley caldera and nearby areas beginning in 1978 may indicate that magma is moving upward beneath part of the caldera. Thus, the possibility of future eruptions in that area cannot be discounted.

✪ DISTRIBUTION OF VOLCANOES

Rather than being distributed randomly around the Earth, volcanoes occur in well-defined zones or belts. More than 60% of all active volcanoes are in the **circum-Pacific belt** that nearly encircles the margins of the Pacific Ocean basin (▷ Figure 5-16). This belt includes the volcanoes along the west coast of South America, those in Central America, Mexico, and the Cascade Range, and the Alaskan volcanoes in the Aleutian Island arc. The belt continues on the western side of the Pacific Ocean basin where it extends through Japan, the Philippines, Indonesia, and New Zealand. Mount Pinatubo and Mayon volcano, two Philippine volcanoes that have erupted since June 1991, are in this belt. The circum-Pacific belt also includes the southernmost active volcano, Mount Erebus in Antarctica, and a large caldera at Deception Island that erupted during 1970.

About 20% of all active volcanoes are in the **Mediterranean belt** (Figure 5-16). Included in this belt are the famous

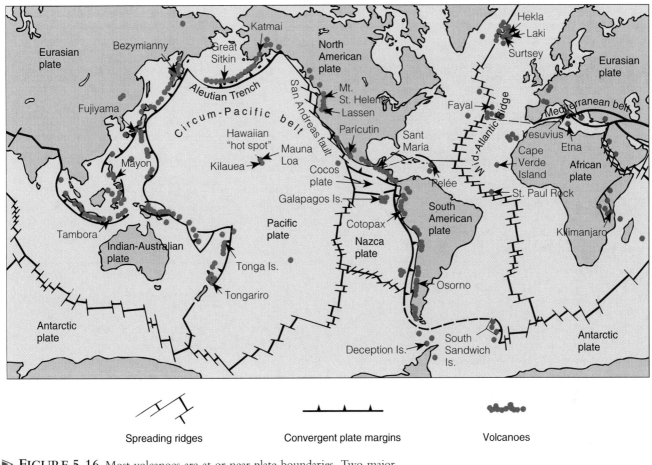

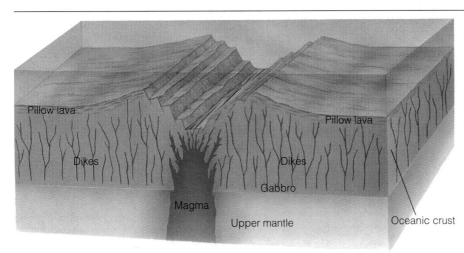

Spreading ridges   **Convergent plate margins**   **Volcanoes**

▷ **FIGURE 5-16** Most volcanoes are at or near plate boundaries. Two major volcano belts are recognized: the circum-Pacific belt contains about 60% of all active volcanoes, about 20% are in the Mediterranean belt, and most of the rest are located along mid-oceanic ridges.

Italian volcanoes such as Mount Etna, Stromboli, and Mount Vesuvius.

Most of the large volcanoes in the circum-Pacific and Mediterranean belts are composite volcanoes, but a number of them have had lava domes emplaced in their craters or calderas. The fact that most of the volcanoes in these two belts are composite volcanoes is significant. Recall that such volcanoes are composed of lava flows and pyroclastic layers of intermediate and felsic composition whereas those within the ocean basins are composed primarily of mafic lavas.

Most of the rest of the active volcanoes are at or near the mid-oceanic ridges (Figure 5-16). The longest of these

▷ **FIGURE 5-17** Intrusive and extrusive igneous activity at a spreading ridge. The oceanic crust is composed largely of vertical dikes of basaltic composition and gabbro that appears to have crystallized in the upper part of a magma chamber. The upper part of the oceanic crust consists of submarine lavas, especially pillow lavas.

ridges is the Mid-Atlantic Ridge, which is near the middle of the Atlantic Ocean basin and curves around the southern tip of Africa where it continues as the Indian Ridge. Branches of the Indian Ridge extend into the Red Sea and East Africa. Mount Kilimanjaro in Africa is on this latter branch (Figure 5-16). Most of the volcanism along the mid-oceanic ridges is submarine, and much of it goes undetected; but in a few places, such as Iceland, it occurs above sea level.

Volcanism is occurring in a few other areas at present, most notably on and near the island of Hawaii (Figure 5-16). Only two volcanoes are currently active on the island, Mauna Loa and Kilauea, although a submarine volcano named Loihi exists about 32 km to the south; Loihi rises more than 3,000 m above the sea floor, but its summit is still about 940 m below sea level.

 PLATE TECTONICS AND IGNEOUS ACTIVITY

At this point, two questions might be asked regarding volcanoes: (1) What accounts for the alignment of volcanoes in belts? (2) Why do magmas erupted within ocean basins and magmas erupted at or near continental margins have different compositions? In addition, plutons emplaced within the ocean basins are invariably mafic, mostly gabbro, whereas the vast batholiths emplaced at continental margins are composed of felsic and intermediate rocks such as granite and diorite. Recall from Chapter 1 that the outer part of the Earth is divided into large plates, which are sections of the lithosphere. Most igneous activity occurs at spreading ridges where plates diverge or along subduction zones where plates converge.

Igneous Activity at Spreading Ridges

Spreading ridges are areas where new oceanic lithosphere is produced by igneous activity as plates diverge from one another. Mafic magma originates beneath these spreading ridges; some of the magma is erupted at the surface as basalt lava flows and/or pyroclastic materials, but much is simply emplaced at depth as vertical dikes and gabbro plutons (▷ Figure 5-17). In fact, the oceanic crust is composed largely of such mafic rocks.

The fact that volcanism occurs at spreading ridges is undisputed, but how magma originates beneath the ridges is not fully understood. One explanation is related to the manner in which the Earth's temperature increases with depth. We know from deep mines and deep drill holes that a temperature increase, called *the geothermal gradient,* occurs and that, on average, it is about 25°C/km. Accordingly, rocks at depth are hot, but remain solid because their melting temperature rises with increasing pressure.

Beneath spreading ridges, the temperature locally exceeds the melting temperature, at least in part, because pressure decreases. That is, rifting probably causes a decrease in pressure on the hot rocks at depth, thus initiating melting (▷ Figure 5-18a). Furthermore, the presence of water can

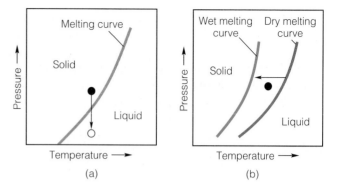

▷ **FIGURE 5-18** (a) Melting temperature rises with increasing pressure, so a decrease in pressure on already hot rocks can initiate melting. (b) The melting curve shifts to the left when water is present because it provides an additional agent to break chemical bonds.

also decrease the melting temperature beneath spreading ridges because water aids thermal energy in breaking the chemical bonds in minerals (Figure 5-18b).

Another explanation for spreading-ridge igneous activity is that localized, cylindrical plumes of hot mantle material, called **mantle plumes,** rise beneath ridges and spread outward in all directions. Perhaps localized concentrations of radioactive minerals within the crust and upper mantle decay and generate the heat responsible for the melting associated with these hot mantle plumes.

The lavas erupted at spreading ridges are invariably mafic and cool to form basalt. But the upper mantle, from which these lavas are derived, is composed of ultramafic rock, probably peridotite, which consists largely of ferromagnesian silicates and lesser amounts of nonferromagnesian silicates. To explain how mafic magma (45–52% silica) originates from ultramafic rock (≤45% silica), geologists propose that the magma is formed from source rock that only partially melts. This phenomenon of *partial melting* occurs because various minerals have different melting temperatures. Recall the sequence of minerals in Bowen's reaction series (Figure 4-5). The order in which these minerals melt is the opposite of their order of crystallization. Accordingly, quartz, potassium feldspar, and sodium-rich plagioclase melt before most of the ferromagnesian silicates and the calcic varieties of plagioclase. So when ultramafic rock begins to melt, the minerals richest in silica melt first followed by those containing less silica. Accordingly, if melting is not complete, a mafic magma containing proportionately more silica than the source rock results. Once this mafic magma is formed, some of it rises to the surface where it is erupted, cools, and crystallizes to form basalt.

Igneous Activity at Subduction Zones

Subduction occurs where an oceanic plate and a continental plate converge, or where two oceanic plates converge. In either case, a belt of composite volcanoes and plutons occurs near the leading edge of the overriding plate (▷ Figure

5-19). As the subducted plate descends toward the astheno-sphere, it eventually reaches a depth where the temperature is high enough for partial melting to occur, and magma is generated. Additionally, the wet oceanic crust descends to a depth at which dewatering occurs, and as the water rises into the overlying mantle, it enhances melting and magma forms (Figure 5-19b).

Partial melting is one phenomenon accounting for the fact that magmas generated at subduction zones are interme-diate and felsic in composition. Recall that partial melting of ultramafic rock of the upper mantle yields mafic magma. Likewise, partial melting of oceanic crust, which has a mafic composition, may yield magma richer in silica than the source rock. Additionally, some of the silica-rich sediments and sedimentary rocks of continental margins are probably carried downward with the subducted plate and contribute their silica to the magma. Also, mafic magma rising through the lower continental crust may be contaminated with felsic materials, which change its composition.

Intermediate and felsic magmas are typically produced at convergent plate margins where subduction occurs. The intermediate magma that is erupted is more viscous than mafic magma and tends to form composite volcanoes. Much felsic magma is intruded into the continental crust where it forms various plutons, especially batholiths, but some is erupted as pyroclastic materials or emplaced as lava domes, thus accounting for the explosive eruptions that characterize convergent plate margins.

Intraplate Volcanism

Mauna Loa and Kilauea on the island of Hawaii and Loihi just to the south are within the interior of a rigid plate far from any spreading ridge or subduction zone (Figure 5-17). It is postulated that a mantle plume creates a local "hot spot" beneath Hawaii. The magma is mafic and relatively fluid, so it builds up shield volcanoes.

Even though these Hawaiian volcanoes are unrelated to spreading ridges or subduction zones, the evolution of the Hawaiian Islands is related to plate tectonics. Notice in Figure 2-21 that the ages of the rocks composing the islands in the Hawaiian chain increase toward the northwest; Kauai formed 3.8 to 5.6 million years ago, whereas Hawaii began forming less than one million years ago, and Loihi began forming even more recently. Continuous motion of the Pacific plate over the "hot spot," now beneath Hawaii, has created the various islands in succession.

Mantle plumes and "hot spots" have also been proposed to explain volcanism in a few other areas. A mantle plume may be beneath Yellowstone National Park in Wyoming. Some source of heat at depth is responsible for the present-day hot springs and geysers such as Old Faithful, but many geologists think that the source of heat is a body of intruded magma that has not yet completely cooled rather than a mantle plume.

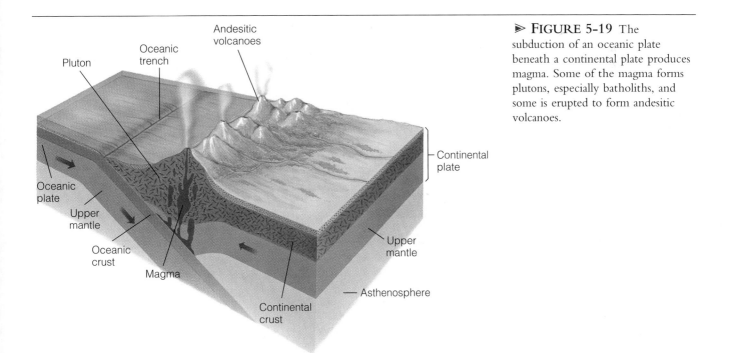

➤ **FIGURE 5-19** The subduction of an oceanic plate beneath a continental plate produces magma. Some of the magma forms plutons, especially batholiths, and some is erupted to form andesitic volcanoes.

1. Volcanism is the process whereby magma and its associated gases erupt at the surface. Some magma erupts as lava flows, and some is ejected explosively as pyroclastic materials.

2. Only a few percent by weight of a magma consists of gases, most of which is water vapor. Sulfur gases emitted during large eruptions can have far-reaching climatic effects.

3. The surface of an aa lava flow consists of rough, angular blocks, whereas a pahoehoe flow has a smoothly wrinkled surface.

4. Columnar joints form in some lava flows when they cool. Pillow lavas form under water and consist of interconnected bulbous masses.

5. Volcanoes are conical mountains built up around a vent where lava flows and/or pyroclastic materials are erupted.

6. Shield volcanoes have low, rounded profiles and are composed mostly of mafic flows that have cooled and formed basalt. Cinder cones form where pyroclastic materials that resemble cinders are erupted and accumulate as small, steep-sided cones. Composite volcanoes are composed of lava flows of intermediate composition, layers of pyroclastic materials, and volcanic mudflows.

7. Viscous masses of lava, generally of felsic composition, are forced up through the conduits of some volcanoes and form bulbous, steep-sided lava domes. Volcanoes with lava domes are dangerous because they erupt explosively and frequently eject nuée ardentes.

8. The summits of volcanoes are characterized by a circular or oval crater or a much larger caldera. Many calderas form by summit collapse when an underlying magma chamber is partly drained.

9. Fluid mafic lava erupted from long fissures (fissure eruptions) spreads over large areas to form basalt plateaus.

10. Pyroclastic flows erupted from fissures formed during the origin of calderas cover vast areas. Such eruptions of pyroclastic materials form sheetlike deposits.

11. Most active volcanoes are distributed in linear belts. The circum-Pacific belt and Mediterranean belt contain more than 80% of all active volcanoes.

12. Volcanism and plutonism occurs at speading ridges where plates diverge and at convergent plate margins where subduction occurs. Partial melting of a subducted plate generates intermediate and felsic magmas.

13. Magma derived by partial melting of the upper mantle beneath spreading ridges accounts for the mafic plutons and lavas of ocean basins. Melting in these areas may be caused by reduction in pressure and/or hot mantle plumes.

14. The two active volcanoes on the island of Hawaii and one just to the south are thought to lie above a hot mantle plume. The Hawaiian Islands developed as a series of volcanoes formed on the Pacific plate as it moved over the mantle plume.

IMPORTANT TERMS

aa	columnar joint	lava dome	pillow lava
ash	composite volcano	mantle plume	pyroclastic sheet deposit
basalt plateau	(stratovolcano)	Mediterranean belt	shield volcano
caldera	crater	nuée ardente	volcanism
cinder cone	fissure eruption	pahoehoe	volcano
circum-Pacific belt	lahar		

REVIEW QUESTIONS

1. Which of the following is most dangerous to humans?
   a. _____ nuée ardente;
   b. _____ lava flows;
   c. _____ volcanic bombs;
   d. _____ pahoehoe;
   e. _____ pillow lava.
2. Most calderas form by:
   a. _____ summit collapse;
   b. _____ explosions;
   c. _____ fissure eruptions;
   d. _____ forceful injection;
   e. _____ erosion of lava domes.
3. Basalt plateaus form as a result of:
   a. _____ repeated eruptions of cinder cones;
   b. _____ widespread ash falls;
   c. _____ accumulation of thick layers of pyroclastic materials;
   d. _____ the origin of lahars on composite volcanoes;
   e. _____ eruptions of fluid lava from long fissures.

4. The most commonly emitted volcanic gas is:
   a. ____ carbon dioxide;
   b. ____ hydrogen sulfide;
   c. ____ nitrogen;
   d. ____ chlorine;
   e. ____ water vapor.
5. Much of the upper part of the oceanic crust is composed of interconnected bulbous masses of igneous rock called:
   a. ____ pillow lava;
   b. ____ lapilli;
   c. ____ pyroclastic material;
   d. ____ parasitic cones;
   e. ____ blocks.
6. Shield volcanoes have low slopes because they are composed of:
   a. ____ mostly pyroclastic layers;
   b. ____ lahars and viscous lava flows;
   c. ____ fluid mafic lava flows;
   d. ____ felsic magma;
   e. ____ pillow lavas.

7. The volcanic conduit of a lava dome is most commonly plugged by:
   a. ____ mafic magma;
   b. ____ columnar joints;
   c. ____ viscous, felsic magma;
   d. ____ volcanic mudflows;
   e. ____ spatter cones.
8. Most active volcanoes are in:
   a. ____ the Mediterranean belt;
   b. ____ the Hawaiian Islands;
   c. ____ Iceland;
   d. ____ the circum-Pacific belt;
   e. ____ the oceanic ridge belt.
9. The magma generated beneath spreading ridges is mostly:
   a. ____ mafic;
   b. ____ felsic;
   c. ____ intermediate;
   d. ____ all of these;
   e. ____ answers (a) and (b) only.
10. Explain how pyroclastic materials and volcanic gases can affect climate.

11. What accounts for the fact that volcanic ash can cover vast areas, whereas pyroclastic materials such as cinders are not very widely distributed?
12. Explain how most calderas form.
13. What kinds of warning signs enable geologists to forecast eruptions?
14. Why do shield volcanoes have such low slopes?
15. How do pahoehoe and aa lava flows differ?
16. Draw a cross section of a composite volcano. Indicate its constituent materials, and show how and where a flank eruption might occur.
17. Why do composite volcanoes occur in belts near convergent plate margins? Are such volcanoes present at all convergent plate margins?
18. Why are lava domes so dangerous?

POINTS TO PONDER

1. During this century, two Cascade Range volcanoes have erupted. What kinds of evidence would indicate that some of the other volcanoes in this range might erupt in the future?

2. What geologic events would have to occur in order for a chain of volcanoes to form along the east coasts of Canada and the United States?

Chapter 6

WEATHERING, EROSION, AND SOIL

OUTLINE

PROLOGUE
INTRODUCTION
MECHANICAL WEATHERING
Frost Action
Pressure Release
Thermal Expansion and Contraction
Salt Crystal Growth
Activities of Organisms
CHEMICAL WEATHERING
Solution
Oxidation

● *Perspective 6-1:* Acid Rain
Hydrolysis
FACTORS CONTROLLING THE
    RATE OF CHEMICAL
    WEATHERING
Particle Size
Climate
Parent Material
SOIL
THE SOIL PROFILE

FACTORS CONTROLLING SOIL
    FORMATION
Climate
Parent Material
Organic Activity
Relief and Slope
Time
SOIL DEGRADATION
WEATHERING AND MINERAL
    RESOURCES
CHAPTER SUMMARY

Weathering and erosion of sedimentary rocks are responsible for the scenery in Bryce Canyon National Park, Utah.

Since 1950, about half of the original 8.8 million km^2 of tropical rainforest has disappeared as a result of logging and clearing for agriculture. Currently, more than 52 million acres are being cleared annually, an area equivalent to the state of Utah. Scientists find these figures disturbing because clearing tropical rainforests or deforestation has far-reaching implications.

In many developing nations in the tropics, population growth has far surpassed the number of jobs created in the depressed economies. And with all fertile soils already devoted to agriculture, poor farmers are increasingly moving into areas of poor soils, namely, the rainforests. There they clear the land for crops and use the wood for lumber and fuel.

In addition, lumbering in rainforests has become increasingly profitable. As a result, roads have been built into formerly inaccessible areas, heavy machinery has been brought in, and trees are being cut using modern lumbering practices.

Deforestation has several adverse effects. One is the removal of the vegetation itself. If left undisturbed, cleared areas may eventually return to their original condition, but these areas are commonly used for crops or grazing, thus

▷ FIGURE 6-1 Soil erosion on a bare surface in Madagascar that was once covered by lush forest.

effectively eliminating the possibility of regrowth. Furthermore, trees in rainforests are particularly effective at absorbing soil nutrients through their roots before the nutrients can be leached out of the soil by the abundant rainfall. In fact, most of the nutrients in rainforests are in the trees and are slowly released when they die and decompose.

Soil erosion rates generally increase when rainforest is cleared because the surface is left unprotected (➤ Figure 6-1). Without the trees the soil also becomes more compacted and less absorbent, inhibiting the infiltration of water into the soil. As a result, surface runoff increases, and the erosion of gullies becomes more common. Flooding is also more frequent because the trees with their huge water-holding capacity are no longer present.

Other negative effects of deforestation include a decrease in biodiversity and possibly climatic changes. The rainforests are the home of millions of species of plants and animals; indeed, more species are found in the tropics than anywhere else on Earth. Many of these species will become extinct because of habitat destruction.

The relationship between deforestation and climate is somewhat speculative, but some have suggested that deforestation exacerbates the greenhouse effect and could contribute to global warming. More carbon dioxide (CO_2), an important greenhouse gas, is present in the vegetation of the tropical rainforests than in all other vegetation on Earth. Without vegetation to hold this huge reservoir of carbon dioxide, more will remain in the atmosphere and perhaps contribute to global warming.

(a)   (b)

➤ FIGURE 6-2 Differential weathering has yielded these odd shapes and surfaces. (a) Camel Rock near Santa Fe, New Mexico. (b) This intricate, uneven weathering surface at Pebble Beach, California, is an example of honeycomb weathering. (Photos courtesy of Sue Monroe.)

INTRODUCTION

Weathering is the physical breakdown (disintegration) and chemical alteration (decomposition) of rocks and minerals at or near the Earth's surface. It is the process whereby rocks and minerals are physically and chemically altered so that they are more nearly in equilibrium with a new set of environmental conditions. Many rocks form within the Earth's crust where little or no water or oxygen is present and where temperatures, pressures, or both are high. At or near the surface, however, the rocks are exposed to low temperatures and pressures and are attacked by atmospheric gases, water, acids, and organisms.

Geologists are interested in the phenomenon of weathering because it is an essential part of the rock cycle (see Figure 1-11). The **parent material**, or rock being weathered, is broken down into smaller pieces, and some of its constituent minerals are dissolved or altered and removed from the weathering site. The removal of the weathered materials is known as **erosion**. Running water, wind, or glaciers commonly **transport** the weathered materials elsewhere, where they are deposited as sediment, which may become sedimentary rock. Whether they are eroded or not, weathered rock materials can be further modified to form a soil. Thus, weathering provides the raw materials for both sedimentary rocks and soils. Weathering is also important in the origin of some mineral resources such as aluminum ores, and it is responsible for the enrichment of other deposits of economic importance.

Weathering is such a pervasive phenomenon that many people take it for granted or completely overlook it. Nevertheless, it occurs continuously although its rate and impact vary from area to area or even within the same area. Rocks do not weather at the same rate, even in a single rock layer, because of slight differences in composition and structure.

Weathering is more intense on fractures than on adjacent areas of unfractured rock, for example. As a consequence of these variations, **differential weathering** occurs, which means that rocks weather at different rates yielding uneven surfaces and peculiar shapes (➤ Figure 6-2).

Two types of weathering are recognized, *mechanical* and *chemical*. Both types occur simultaneously at the weathering site, during erosion and transport, and even in the environments where weathered materials are deposited.

MECHANICAL WEATHERING

Mechanical weathering occurs when physical forces break rock materials into smaller pieces that retain the chemical composition of the parent material. Granite, for example, may be mechanically weathered to yield smaller pieces of granite, or disintegration may liberate individual mineral grains from it (➤ Figure 6-3). The physical processes responsible for mechanical weathering include *frost action*, *pressure release*, *thermal expansion and contraction*, *salt crystal growth*, and the *activities of organisms*.

Frost Action

Frost action involves the repeated freezing and thawing of water in cracks and crevices in rocks. When water seeps into a crack and freezes, it expands by about 9% and exerts great force on the walls of the crack, thereby widening and extending it by **frost wedging**. As a result of repeated freezing and thawing, pieces of rock are eventually detached from the parent material (➤ Figure 6-4). Frost wedging is particularly effective if the crack is convoluted. If the crack is a simple wedge-shaped opening, much of the force of expansion is released upward toward the surface. The debris

➤ **FIGURE 6-3** Mechanically weathered granite. The sandy material consists of small pieces of granite (rock fragments) and minerals such as quartz and feldspars liberated from the parent material.

➤ **FIGURE 6-4** Frost wedging occurs when water seeps into cracks and expands as it freezes. Repeated freezing and thawing pry loose angular pieces of rock.

the mass. Frost heaving is particularly evident where water freezes beneath roadways and sidewalks.

Pressure Release

Pressure release is a mechanical weathering process that is especially evident in rocks that formed as deeply buried intrusive bodies such as batholiths, but it occurs in other types of rocks as well. When a batholith forms, the magma crystallizes under tremendous pressure (the weight of the overlying rock) and is stable under these pressure conditions. When the batholith is uplifted and the overlying rock is stripped away by erosion, the pressure is reduced. But the rock contains energy that is released by expansion and the formation of **sheet joints**, large fractures that more or less parallel the rock surface (➤ Figure 6-6). Slabs of rock bounded by sheet joints may slip, slide, or spall (break) off of the host rock—a process called **exfoliation**—and accumulate as talus. The large rounded domes of rock resulting from this process are **exfoliation domes**; examples are found in Yosemite National Park in California and Stone Mountain in Georgia (➤ Figure 6-7).

Thermal Expansion and Contraction

During **thermal expansion and contraction**, the volume of solids, such as rocks, changes in response to heating and cooling. In a desert, where the temperature may vary as much as 30°C in one day, rocks expand when heated and contract as they cool. Expansion and contraction, however, do not occur uniformly throughout rocks. For one thing, a rock is a poor conductor of heat, so its outside heats up more than its inside. Consequently, the surface expands more than the interior, creating stresses that may cause fracturing. Furthermore, dark minerals absorb heat faster than light-colored

produced by frost wedging in mountains commonly accumulates as large cones of **talus** lying at the bases of slopes (➤ Figure 6-5).

Frost action is most effective in areas where temperatures commonly fluctuate above and below freezing, as in the high mountains of the western United States and Canada. In the tropics and in areas where water is permanently frozen, frost action is of little or no importance.

In the phenomenon known as **frost heaving**, a mass of sediment or soil undergoes freezing, expansion, and actual lifting, followed by thawing, contraction, and lowering of

➤ **FIGURE 6-5** Talus in the Canadian Rocky Mountains.

≽ **FIGURE 6-6** Sheet joints in granite in the Sierra Nevada of California.

≽ **FIGURE 6-7** Exfoliation domes in Yosemite National Park, California.

minerals, so differential expansion occurs even between the mineral grains of some rocks.

Experiments in which rocks are heated and cooled repeatedly to simulate years of such activity indicate that thermal expansion and contraction is not an important agent of mechanical weathering. But thermal expansion and contraction may be a significant mechanical weathering process on the Moon where extreme temperature changes occur quickly.

Daily temperature variation is the most common cause of alternate expansion and contraction, but these changes occur over periods of hours. In contrast, fire can cause very rapid expansion. During a forest fire, rocks may heat very rapidly, especially near the surface, because they conduct heat so poorly. The heated surface layer expands more rapidly than the interior, and thin sheets paralleling the rock surface become detached.

Salt Crystal Growth

Under some circumstances, salt crystals forming from solution can cause disaggregation of rocks. Growing crystals exert enough force to widen cracks and crevices or dislodge particles in porous, granular rocks such as sandstones. Even in crystalline rocks such as granite, **salt crystal growth** can pry loose individual mineral grains. To the extent that salt crystal growth produces forces that expand openings in rocks, it is similar to frost wedging. Most salt crystal growth occurs in hot arid areas, although it probably affects rocks in some coastal regions as well.

Activities of Organisms

Animals, plants, and bacteria all participate in the mechanical and chemical alteration of rocks. Burrowing animals, such as worms, reptiles, rodents, and many others, constantly mix soil and sediment particles and bring material from depth to the surface where further weathering may occur. Even materials ingested by worms are further reduced in size, and animal burrows allow gases and water to have easier access to greater depths. The roots of plants, especially large bushes and trees, wedge themselves into cracks in rocks and further widen them (▷ Figure 6-8).

CHEMICAL WEATHERING

Chemical weathering is the process whereby rock materials are decomposed by chemical alteration of the parent material. A number of clay minerals, for example, form as the chemically altered products of other minerals. Some minerals are completely decomposed during chemical weathering, but others, which are more resistant, are simply liberated from the parent material. Such weathering is accomplished by the action of atmospheric gases, especially oxygen, and water and acids. Organisms also play an important role in chemical weathering. Rocks that have lichens (composite organisms consisting of fungi and algae) growing on their surfaces undergo more extensive chemical alteration than lichen-free rocks. Plants remove ions from soil water and reduce the chemical stability of soil minerals, and their roots release organic acids.

Solution

During **solution** the ions of a substance become dissociated from one another in a liquid, and the solid substance dissolves. Water is a remarkable solvent because its molecules have an asymmetric shape; they consist of one oxygen atom with two hydrogen atoms arranged so that the angle between the two hydrogens is about 104 degrees (▷ Figure 6-9). Because of this asymmetry, the oxygen end of the molecule retains a slight negative electrical charge, whereas the hydrogen end retains a slight positive charge. When a soluble substance such as the mineral halite (NaCl) comes in contact with a water molecule, the positively charged sodium ions are attracted to the negative end of the water molecule, and the negatively charged chloride ions are attracted to the positively charged end of the water molecule

▷ **FIGURE 6-8** The contribution of organisms to mechanical weathering. Tree roots enlarge cracks in rocks.

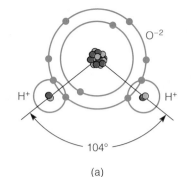

(a)

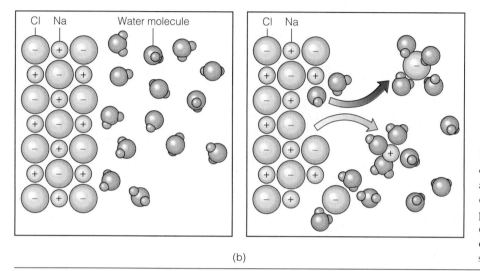

(b)

FIGURE 6-9 (a) The structure of a water molecule. The asymmetric arrangement of the hydrogen atoms causes the molecule to have a slight positive electrical charge at its hydrogen end and a slight negative charge at its oxygen end. (b) The dissolution of sodium chloride (NaCl) in water.

(Figure 6-9). Thus, ions are liberated from the crystal structure, and the solid dissolves.

Most minerals are not very soluble in pure water because the attractive forces of water molecules are not sufficient to overcome the forces between particles in minerals. The mineral calcite ($CaCO_3$), the major constituent of the sedimentary rock limestone and the metamorphic rock marble, is practically insoluble in pure water, but rapidly dissolves if a small amount of acid is present. An easy way to make water acidic is by dissociating the ions of carbonic acid as follows:

$$H_2O + CO_2 \rightleftharpoons H_2CO_3 \rightleftharpoons H^+ + HCO_3^-$$
water   carbon   carbonic   hydrogen   bicarbonate
dioxide   acid   ion   ion

According to this chemical equation, water and carbon dioxide combine to form *carbonic acid*, a small amount of which dissociates to yield hydrogen and bicarbonate ions. The concentration of hydrogen ions determines the acidity of a solution; the more hydrogen ions present, the stronger the acid.

There are several sources of carbon dioxide that may combine with water and react to form acid solutions. The atmosphere is mostly nitrogen and oxygen, but about 0.03% is carbon dioxide, causing rain to be slightly acidic. Human activities have added materials to the atmosphere that contribute to the problem of acid rain (see Perspective 6-1). Carbon dioxide is also produced in soil by the decay of organic matter and the respiration of organisms, so groundwater in humid areas is slightly acidic. Arid regions have sparse vegetation, so groundwater has a limited source of carbon dioxide and tends to be alkaline (that is, it has a low concentration of hydrogen ions).

Whatever the source of carbon dioxide, once an acidic solution is present, calcite rapidly dissolves according to the following reaction:

$$CaCO_3 + H_2O + CO_2 \rightleftharpoons Ca^{++} + 2HCO_3^-$$
calcite   water   carbon   calcium   bicarbonate
dioxide   ion   ion

In many places, the dissolution of the calcite in limestone and marble has had dramatic effects, ranging from small cavities to large caverns such as Mammoth Cave in Kentucky and Carlsbad Caverns in New Mexico.

Oxidation

Oxidation refers to reactions with oxygen to form oxides or, if water is present, hydroxides. For example, iron rusts when it combines with oxygen to form the iron oxide hematite:

ACID RAIN

Atmospheric pollution is one of the consequences of industrialization. Several of the most industrialized nations, such as the United States, Canada, and Russia, have actually reduced their emissions into the atmosphere, but many developing nations continue to increase theirs. Some of the consequences of atmospheric pollution include smog, possible disruption of the ozone layer, global warming, and *acid rain*.

As we noted previously, water and carbon dioxide in the atmosphere react to form carbonic acid that dissociates and yields hydrogen ions and bicarbonate ions. The net effect of this reaction is that all rainfall is slightly acidic. Thus, acid rain is the direct consequence of the self-cleansing nature of the atmosphere; that is, many suspended particles of gases in the atmosphere are soluble in water and are removed from the atmosphere during precipitation events.

Several natural processes, including volcanism and the activities of soil bacteria, introduce gases into the atmosphere that cause acid rain. Human activities, however, produce added atmospheric stress. For example, the burning of fossil fuels (oil, natural gas, and coal) has added carbon dioxide to the atmosphere. Nitrogen oxide (NO) from internal combustion engines and nitrogen dioxide (NO_2), which is formed in the atmosphere from NO, react to form nitric acid (HNO_3). Although carbon dioxide and nitrogen gases contribute to acid rain, the greatest culprit is sulfur dioxide (SO_2), which is primarily released by burning coal that contains sulfur. Once in the atmosphere, sulfur dioxide reacts with oxygen to form sulfuric acid (H_2SO_4), the main component of acid rain.

The phenomenon of acid rain was first recognized in England by Robert Angus Smith in 1872, about a century after the beginning of the Industrial Revolution. It was not until 1961, however, that acid rain become a public environmental concern. At that time, it was realized that acid rain is corrosive and irritating, kills vegetation, and has a detrimental effect on surface waters. Since then, the effects of acid rain have been recognized in Europe, especially in Eastern Europe where so much coal is burned, the eastern United States, and southeastern Canada. During the last 10 years, the developed countries have made efforts to reduce the impact of acid rain; in the United States, the Clean Air Act of 1990 outlined specific steps to reduce the emissions of pollutants that cause acid rain.

The areas most affected by acid rain invariably lie downwind from coal-burning power plants or other industries that emit sulfur gases. Chemical plants and smelters (plants where metal ores are refined) discharge large quantities of sulfur gases and other substances such as heavy metals. The effect of acid rain in these areas may be modified by the existing geology. If an area is underlain by limestone or alkaline soils, for example, the acid rain tends to be neutralized by the limestone or soil. Areas underlain by granite, on the other hand, are acidic to begin with and have little or no effect on the rain.

The effects of acid rain vary. Small lakes become more acid as they lose the ability to neutralize the acid rainfall. As the lakes increase in acidity, various types of organisms disappear, and, in some cases, all lifeforms eventually die. Acid rain also causes increased weathering of limestone and marble (recall that both are soluble in weak acids) and, to a lesser degree, sandstone. Such effects are particularly visible on buildings, monuments, and tombstones; a notable example is Gettysburg National Military Park in Pennsylvania, which lies in an area that receives some of the most acidic rain in the country.

Although the effects on vegetation in the immediate areas of industries emitting sulfur gas are apparent, some people have questioned whether acid rain has much effect on forests and crops distant from these sources. Nevertheless, many forests in the eastern United States show signs of stress that cannot be attributed to other causes. In Germany's Black Forest, the needles of firs, spruce, and pines are turning yellow and falling off.

Currently, about 20 million tons of sulfur dioxide are released yearly into the atmosphere in the United States, mostly from coal-burning power plants. Power plants built before 1975 have no emission controls and must be addressed if emissions are to be reduced to an acceptable level. The most effective way to reduce emissions from these older plants is with flue-gas desulfurization (FGD), a process that removes up to 90% of sulfur dioxide from exhaust gases. There are drawbacks to FGD, however. One is that some plants are simply too old to be profitably upgraded; the 85-year-old Phelps Dodge copper smelter in Douglas, Arizona, closed in 1987 for this reason. Other problems with FGD include disposal of sulfur wastes, the lack of control on nitrogen gas emissions, and reduced efficiency of the power plant, which must burn several percent more coal.

Other ways to control emissions include the conservation of electricity; the less electricity used, the lower the emissions of pollutants. Natural gas contains practically no sulfur, but converting to this alternate energy source would require the installation of expensive new furnaces in existing plants.

Acid rain is a global problem that knows no national boundaries. Wind currents may blow pollutants from the source in one country to another where the effects are felt. Developed nations have the economic resources to reduce emissions, but many underdeveloped nations cannot afford to do so. Furthermore, many nations have access to only high-sulfur coal and cannot afford to install FGD devices. Nevertheless, acid rain can be controlled only by the cooperation of all nations contributing to the problem.

$$4Fe + 3O_2 \rightarrow 2Fe_2O_3$$

iron   oxygen   iron oxide
(hematite)

Of course, atmospheric oxygen is abundantly available for oxidation reactions, but oxidation is generally a slow process unless water is present. Most oxidation is carried out by oxygen dissolved in water.

Oxidation is very important in the alteration of ferromagnesian minerals such as olivine, pyroxenes, amphiboles, and biotite. Iron in these minerals combines with oxygen to form the reddish iron oxide hematite (Fe_2O_3) or the yellowish or brown hydroxide limonite [$FeO(OH) \cdot nH_2O$]. The yellow, brown, and red colors of many soils and sedimentary rocks are caused by the presence of small amounts of hematite or limonite.

An oxidation reaction of particular concern in some areas is the oxidation of iron- and sulfur-bearing minerals such as pyrite (FeS_2). Pyrite is commonly associated with coal, so in mine tailings* pyrite oxidizes to form sulfuric acid (H_2SO_4) and iron oxide. Acid soils and waters in coal-mining areas are produced in this manner and present a serious environmental hazard (▷ Figure 6-10).

Hydrolysis

Hydrolysis is the chemical reaction between the hydrogen (H^+) ions and hydroxyl (OH^-) ions of water and a mineral's ions. In hydrolysis, hydrogen ions actually replace positive ions in minerals. Such replacement changes the composition of minerals by liberating soluble salts and iron that may then be oxidized.

As an illustration of hydrolysis, consider the chemical alteration of feldspars. All feldspars are framework silicates, but when altered, they yield compounds in solution and clay minerals, such as kaolinite, which are sheet silicates.

The chemical weathering of potassium feldspar by hydrolysis occurs as follows:

$$2KAlSi_3O_8 + 2H^+ + 2HCO_3^- + H_2O \rightarrow$$

orthoclase   hydrogen   bicarbonate   water
ion   ion

$$Al_2Si_2O_5(OH)_4 + 2K^+ + 2HCO_3^- + 4SiO_2$$

clay (kaolinite)   potassium   bicarbonate   silica
ion   ion

In this reaction hydrogen ions attack the ions in the orthoclase structure, and some liberated ions are incorporated in a developing clay mineral, while others simply go into solution. On the right side of the equation is excess silica that would not fit into the crystal structure of the clay mineral.

---

*Tailings are the rock debris of mining; they are considered too poor for further processing and are left as heaps on the surface.

▷ **FIGURE 6-10** The oxidation of pyrite in mine tailings forms acid water as in this small stream. More than 11,000 km of U.S. streams, mostly in the Appalachian region, are contaminated by abandoned coal mines that leak sulfuric acid.

FACTORS CONTROLLING THE RATE OF CHEMICAL WEATHERING

Chemical weathering processes operate on the surfaces of particles; that is, chemically weathered rocks or minerals are altered from the outside inward. Several factors including particle size, climate, and parent material control the rate of chemical weathering.

Particle Size

Because chemical weathering affects particle surfaces, the greater the surface area, the more effective the weathering. It is important to realize that small particles have larger surface areas compared to their volume than do large particles. Notice in ▷ Figure 6-11 that a block measuring 1 m on a side has a total surface area of 6 m^2, but when the block is broken into particles measuring 0.5 m on a side, the total surface area increases to 12 m^2. And if these particles are all reduced to 0.25 m on a side, the total surface area increases to 24 m^2. Note that while the surface area in this example increases, the total volume remains the same at 1 m^3.

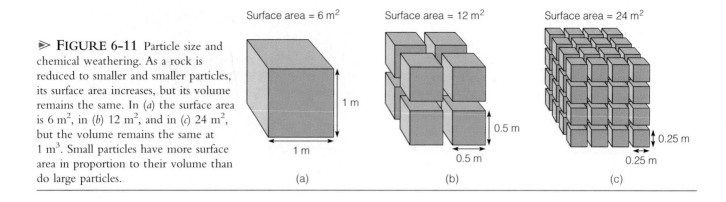

► **FIGURE 6-11** Particle size and chemical weathering. As a rock is reduced to smaller and smaller particles, its surface area increases, but its volume remains the same. In (a) the surface area is 6 m², in (b) 12 m², and in (c) 24 m², but the volume remains the same at 1 m³. Small particles have more surface area in proportion to their volume than do large particles.

Surface area = 6 m² (a) 1 m 1 m

Surface area = 12 m² (b) 0.5 m 0.5 m

Surface area = 24 m² (c) 0.25 m 0.25 m

We can make two important statements regarding the block in Figure 6-11. First, as it is split into a number of smaller blocks, its total surface area increases. Second, the smaller any single block is, the more surface area it has compared to its volume. We can conclude that mechanical weathering, which reduces the size of particles, contributes to chemical weathering by exposing more surface area.

Climate

Most chemical processes occur more rapidly at high temperatures and in the presence of liquids. Accordingly, it is not surprising that chemical weathering is more effective in the tropics than in arid and arctic regions because temperatures and rainfall are high and evaporation rates are low. In

► **FIGURE 6-12** Spheroidal weathering. (a) The rectangular blocks outlined by joints are attacked by chemical weathering processes, (b) but the corners and edges are weathered most rapidly. (c) When a block has been weathered so that it is spherical, its entire surface is weathered evenly, and no further change in shape occurs. (d) Spheroidal weathering of granite in Australia.

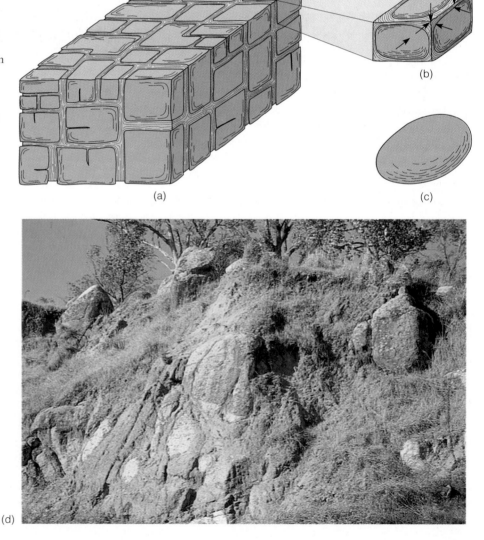

(a)

(b)

(c)

(d)

addition, vegetation and animal life are much more abundant in the tropics than in arid or cold regions. Consequently, the effects of weathering extend to depths of several tens of meters in the tropics, but commonly extend only centimeters to a few meters deep in arid and arctic regions. One should realize, though, that chemical weathering goes on everywhere, except perhaps where Earth materials are permanently frozen.

Parent Material

Some rocks are chemically more stable than others and are not altered as rapidly by chemical processes. The metamorphic rock quartzite, composed of quartz, is an extremely stable substance that alters very slowly compared to most other rock types. In contrast, rocks such as granite, which contain large amounts of feldspar minerals, decompose rapidly because feldspars are chemically unstable. Ferromagnesian minerals are also chemically unstable and, when chemically weathered, yield clays, iron oxides, and ions in solution. In fact, the stability of common minerals is just the opposite of their order of crystallization in Bowen's reaction series (see Figure 4-5): the minerals that form last in this series are chemically stable, whereas those that form early are easily altered by chemical processes because they are most out of equilibrium with their conditions of formation.

One manifestation of chemical weathering is **spheroidal weathering** (➤ Figure 6-12). In spheroidal weathering, a stone, even one that is rectangular to begin with, weathers to form a spheroidal shape because that is the most stable shape it can assume. On a rectangular stone, the corners are attacked by weathering processes from three sides, and the edges are attacked from two sides, but the flat surfaces are weathered more or less uniformly (Figure 6-12a). Consequently, the corners and edges are altered more rapidly, the material sloughs off them, and a more spherical shape develops. Once a spherical shape is present, all surfaces are weathered at the same rate.

Spheroidal weathering is often observed in granitic rock bodies cut by joints. Fluids follow the joint surfaces and reduce rectangular joint-bounded blocks to a spherical shape (Figure 6-12d).

◉ SOIL

In most places the land surface is covered by a layer of unconsolidated rock and mineral fragments called **regolith**. Regolith may consist of volcanic ash, sediment deposited by wind, streams, or glaciers, or weathered rock material formed in place as a residue. Some regolith that consists of weathered material, water, air, and organic matter and can support plant growth is recognized as **soil**.

A good, fertile soil for gardening or farming is about 45% weathered rock material including sand, silt, and clay, but an essential constituent of such soils is **humus**. Humus, which gives many soils their dark color, is derived by bacterial decay of organic matter. It contains more carbon and less nitrogen than the original material and is resistant to further bacterial decay. Although a fertile soil may contain only a

(a)

(b)

➤ **FIGURE 6-13** (*a*) Residual soil developed on bedrock near Denver, Colorado. (*b*) Transported soil developed on a windblown dust deposit.

small amount of humus, it is an essential source of plant nutrients and enhances moisture retention.

Some weathered materials in soils are simply sand- and silt-sized mineral grains, especially quartz, but other weathered materials may be present as well. These solid particles are important because they hold soil particles apart, allowing oxygen and water to circulate more freely. Clay minerals are also important constituents of soils and aid in the retention of water as well as supplying nutrients to plants. Soils with excess clay minerals, however, drain poorly and are sticky when wet and hard when dry.

Soils are commonly characterized as *residual* or *transported*. If a body of rock weathers and the weathering residue accumulates over it, the soil so formed is residual, meaning that it formed in place (➤ Figure 6-13a). In contrast, transported soil develops on weathered material that has been eroded and transported from the weathering site and deposited elsewhere, such as on a stream's floodplain. Many fertile transported soils of the Mississippi River valley and the Pacific Northwest developed on deposits of windblown dust called *loess* (Figure 6-13b).

THE SOIL PROFILE

Soil-forming processes begin at the surface and work downward, so the upper layer of soil is more altered from the parent material than the layers below. Observed in vertical cross section, a soil consists of distinct layers or **soil horizons** that differ from one another in texture, structure, composition, and color (▷ Figure 6-14). Starting from the top, the horizons typical of soils are designated O, A, B, and C, but the boundaries between horizons are transitional rather than sharp.

The O horizon, which is generally only a few centimeters thick, consists of organic matter. The remains of plant materials are clearly recognizable in the upper part of the O horizon, but its lower part consists of humus.

Horizon A, called *top soil,* contains more organic matter than the layers below. It is also characterized by intense biological activity because plant roots, bacteria, fungi, and animals such as worms are abundant. Threadlike soil bacteria give freshly plowed soil its earthy aroma. In soils developed over a long period of time, the A horizon consists mostly of clays and chemically stable minerals such as quartz. Water percolating down through horizon A dissolves the soluble minerals that were originally present and carries them away or downward to lower levels in the soil by a process called **leaching** (Figure 6-14).

Horizon B, or *subsoil,* contains fewer organisms and less organic matter than horizon A. Horizon B is known as the **zone of accumulation** because soluble minerals leached from horizon A accumulate as irregular masses. If horizon A is stripped away by erosion leaving horizon B exposed, plants do not grow as well, and if horizon B is clayey, it is harder when dry and stickier when wet than other soil horizons.

Horizon C, the lowest soil layer, consists of partially altered to unaltered parent material (Figure 6-14). In horizons A and B, the composition and texture of the parent material have been so thoroughly altered that the parent material is no longer recognizable. In contrast, rock fragments and mineral grains of the parent material retain their identity in horizon C. Horizon C contains little organic matter.

FACTORS CONTROLLING SOIL FORMATION
Climate

It has long been acknowledged that climate is the single most important factor influencing soil type and depth. Intense chemical weathering in the tropics yields deep soils from which most of the soluble minerals have been removed by leaching. In arctic and desert climates, on the other hand, soils tend to be thin, contain significant quantities of soluble minerals, and are composed mostly of materials derived by mechanical weathering.

A very general classification recognizes three major soil types characteristic of different climatic settings. Soils that develop in humid regions such as the eastern United States and much of Canada are **pedalfers**, a name derived from the Greek word *pedon* meaning soil and the chemical symbols for aluminum (Al) and iron (Fe). Because these soils form where abundant moisture is present, most of the soluble minerals have been leached from horizon A. Although it may be gray, horizon A is generally dark colored because of abundant organic matter, and aluminum-rich clays and iron oxides tend to accumulate in horizon B.

Pedocals are soils characteristic of arid and semiarid regions and are found in much of the western United States, especially the Southwest. Pedocal derives its name in part from the first three letters of calcite. These soils contain less organic matter than pedalfers, so horizon A is generally lighter colored and contains more unstable minerals because of less intense chemical weathering. As soil water evaporates, calcium carbonate leached from above commonly precipitates in horizon B where it forms irregular masses of *caliche.* Precipitation of sodium salts in some desert areas where soil water evaporation is intense yields *alkali soils* that cannot support plants.

Laterite is a soil formed in the tropics where chemical weathering is intense and leaching of soluble minerals is complete. Such soils are red, commonly extend to depths of several tens of meters, and are composed largely of aluminum hydroxides, iron oxides, and clay minerals; even quartz, a chemically stable mineral, is generally leached out (▷ Figure 6-15).

Although laterites support lush vegetation, they are not very fertile. The native vegetation is sustained by nutrients derived mostly from the surface layer of organic matter. When these soils are cleared of their native vegetation, the existing surface accumulation of organic matter is rapidly oxidized, and there is little to replace it. Consequently, when societies practicing slash-and-burn agriculture clear these soils, they can raise crops for only a few years at best. Then the soil is depleted of plant nutrients, the clay-rich laterite bakes brick hard in the tropical sun, and the farmers move on to another area where the process is repeated.

▷ **FIGURE 6-14** The soil horizons in a fully developed or mature soil.

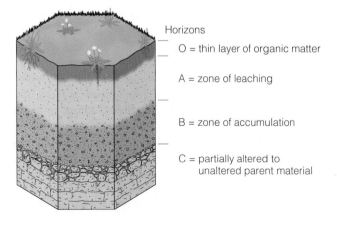

Horizons

O = thin layer of organic matter

A = zone of leaching

B = zone of accumulation

C = partially altered to unaltered parent material

➤ FIGURE 6-15 Laterite, shown here in Madagascar, is a deep, red soil that forms in response to intense chemical weathering in the tropics.

One aspect of laterites is of great economic importance. If the parent material is rich in aluminum, aluminum hydroxides may accumulate in horizon B as *bauxite*, the ore of aluminum. Because such intense chemical weathering currently does not occur in North America, the United States and Canada are dependent on foreign sources for aluminum ores (see Table 3-5). Some aluminum ores do exist in Arkansas, Alabama, and Georgia, which had a tropical climate about 50 million years ago, but currently it is cheaper to import aluminum ore than to mine these deposits.

Parent Material

The same rock type can yield different soils in different climatic regimes, and in the same climatic regime the same soils can develop on different rock types. It is apparent that climate is more important than parent material in determining the type of soil that develops. Nevertheless, rock type does exert some control. For example, the metamorphic rock quartzite will have a thin soil over it because it is chemically stable, whereas an adjacent body of granite will have a much deeper soil.

Soil that develops on basalt will be rich in iron oxides because basalt contains abundant ferromagnesian minerals, but rocks lacking such minerals will not yield an iron oxide–rich soil no matter how thoroughly they are weathered. Also, weathering of a pure quartz sandstone will yield no clay, whereas weathering of clay will yield no sand.

Organic Activity

Soils depend on organisms for their fertility, and in return they provide a suitable habitat for many organisms. Earthworms—as many as one million per acre—ants, sowbugs, termites, centipedes, millipedes, and nematodes, along with various types of fungi, algae, and single-celled animals, make their homes in the soil. All of these contribute to the formation of soils and provide humus when they die and are decomposed by bacterial action.

Much humus in soils is provided by grasses or leaf litter that microorganisms decompose to obtain food. In so doing, they break down organic compounds within plants and release nutrients back into the soil. Additionally, organic acids produced by decaying soil organisms are important in further weathering of parent materials and soil particles.

Burrowing animals constantly churn and mix soils, and their burrows provide avenues for gases and water. Soil organisms, especially some types of bacteria, are extremely important in changing atmospheric nitrogen into a form of soil nitrogen suitable for use by plants.

Relief and Slope

Relief is the difference in elevation between high and low points in a region. Because climate changes with elevation, relief affects soil-forming processes largely through elevation. Slope affects soils in two ways. One is simply *slope angle*: the steeper the slope, the less opportunity for soil development because weathered material is eroded faster than soil-forming processes can work. The other slope control is the direction a slope faces. In the Northern Hemisphere, north-facing slopes receive less sunlight than south-facing slopes. If a north-facing slope is steep, it may receive no sunlight at all. Consequently, north-facing slopes have soils with cooler internal temperatures, may support different vegetation, and, if in a cold climate, remain frozen longer.

Time

The properties of a soil are determined by the factors of climate and organisms altering parent material through time; the longer these processes have operated, the more fully developed the soil will be. If a soil is weathered for extended periods of time, however, its fertility decreases as plant nutrients are leached out, unless new materials are delivered. For example, agricultural lands adjacent to major streams such as the Nile River in Egypt have their soils replenished during yearly floods. In areas of active tectonism, uplift and erosion provide fresh materials that are transported to adjacent areas where they contribute to soils.

How much time is needed to develop a centimeter of soil or a fully developed soil a meter or so deep? No definitive answer can be given because weathering proceeds at vastly different rates depending on climate and parent material, but an overall average might be about 2.5 cm per century. However, a lava flow a few centuries old in Hawaii may have a well-developed soil on it, whereas a flow the same age in Iceland will have considerably less soil. Given the same climatic conditions, soil will develop faster on unconsolidated sediment than on solid bedrock, a general term for the rock underlying soil or sediment.

Under optimum conditions of soil formation, the soil-forming process occurs at a rapid rate in the context of geologic time. From the human perspective, though, soil

▷ FIGURE 6-16 During the 1930s, the drought-stricken southern Great Plains in the western United States were hard hit by wind erosion. Huge dust storms were common. This one was photographed at Hugoton, Kansas, on April 14, 1935, also known as Black Sunday.

formation is a slow process; consequently, soil is regarded as a nonrenewable resource.

⊛ SOIL DEGRADATION

Any decrease in soil productivity or loss of soil to erosion is referred to as **soil degradation.** Between 1945 and 1990, the soils of 17% of the world's vegetated land were degraded to some extent as a result of human activities. In North America, 5.3% of the soil has been degraded, and the figures are much higher for all other continents.

Three types of soil degradation are recognized: erosion, chemical deterioration, and physical deterioration. Most soil erosion occurs by the action of wind and water. When the natural vegetation is removed and a soil is pulverized by plowing, the fine particles are easily blown away (▷ Figure 6-16). Falling rain also disrupts soil particles and carries soil with it when it runs off at the surface. This is particularly devastating on steep slopes where the vegetation has been removed by overgrazing, deforestation, or construction. Two types of erosion by water are recognized: sheet erosion and rill erosion.

Sheet erosion is more or less evenly distributed over the surface and removes thin layers of soil. **Rill erosion** occurs when running water scours small channels. If these channels can be eliminated by plowing, they are *rills*, but if they are too deep (about 30 cm) to be plowed over, they are *gullies* (▷ Figure 6-17). Where gullying becomes extensive, croplands can no longer be tilled and must be abandoned.

If soil losses to erosion are minimal, soil-forming processes can keep pace, and the soil remains productive. Should the

▷ FIGURE 6-17 Rill erosion in a field during a rainstorm. This rill was later plowed over.

loss rate exceed the formation rate, however, the most productive upper layer of soil, horizon A, is removed exposing horizon B. The Soil Conservation Service of the U.S. Department of Agriculture estimates that 25% of the cropland in the United States is eroding faster than soil-forming processes can replace it. Such losses are problems, of course, but there are additional consequences. The eroded soil is transported elsewhere, perhaps onto neighboring cropland, onto roads, or into channels. Sediment accumulates in canals and irrigation ditches, and agricultural fertilizers and insecticides are carried into streams and lakes.

A soil undergoes chemical deterioration when its nutrients are depleted and its productivity decreases. Loss of soil nutrients is most notable in countries where soils are overused in an attempt to maintain agricultural productivity. Other causes include insufficient use of chemical fertilizers and clearing soils of their natural vegetation. Chemical deterioration of soils occurs on all continents, but is most serious in South America where it accounts for 29% of all soil degradation.

Other types of chemical deterioration are pollution and *salinization,* which occurs when the concentration of salts increases in a soil, making it unfit for agriculture. Pollution can be caused by improper disposal of domestic and industrial wastes, oil and chemical spills, and the concentration of insecticides and pesticides in soils. Soil pollution is a particularly serious problem in Europe.

Physical deterioration of soils results when soil particles are compacted under the weight of heavy machinery and livestock, especially cattle. When soils have been compacted, they are more costly to plow, and plants have a more difficult time emerging from the soil. Furthermore, water does not readily infiltrate, so more runoff occurs; this in turn accelerates the rate of water erosion.

In North America, the rich prairie soils of the midwestern United States and the Great Plains of the United States and Canada are suffering significant soil degradation. Nevertheless, this degradation, which is characterized as moderate, is less serious than in many other parts of the world where it is considered severe or extreme. Other areas of concern are the central valleys of California, an area in Washington State, and some parts of Mississippi and Missouri where water erosion rates are high.

Problems experienced during the past have stimulated the development of methods to minimize soil erosion on agricultural lands. Crop rotation, contour plowing, and the construction of terraces have all proved helpful (➤ Figure 6-18). So has no-till planting in which the residue from the harvested crop is left on the ground to protect the surface from the ravages of wind and water.

◉ WEATHERING AND MINERAL RESOURCES

In a preceding section, we discussed intense chemical weathering in the tropics and the origin of *bauxite,* the chief ore of aluminum. Such an accumulation of valuable minerals formed by the selective removal of soluble substances is a

➤ **FIGURE 6-18** Contour plowing, which involves plowing parallel to the contours of the land, can be an effective soil conservation practice. The furrows and ridges are perpendicular to the direction that water would otherwise flow downhill and thus inhibit erosion.

residual concentration. It represents an insoluble residue of chemical weathering. In addition to bauxite, a number of other residual concentrations are economically important; for example, ore deposits of iron, manganese, clays, nickel, phosphate, tin, diamonds, and gold.

Some limestones contain small amounts of iron carbonate minerals. When the limestone is dissolved during chemical weathering, a residual concentration of insoluble iron oxides accumulates. Some of the sedimentary iron deposits (see Chapter 7) of the Lake Superior region were enriched by chemical weathering when the soluble constituents that were originally present were carried away. Residual concentrations of insoluble manganese oxides form in a similar fashion from manganese-rich source rocks.

Most commercial clay deposits were formed by hydrothermal alteration of granitic rocks or by sedimentary processes. However, some have formed in place as residual concentrations. A number of kaolinite deposits in the southern United States were formed by the chemical weathering of feldspars in pegmatites and of clay-bearing limestones and dolostones. Kaolinite is a type of clay mineral used in the manufacture of paper and ceramics.

A gossan is a yellow to reddish deposit composed largely of iron hydroxides that formed by the alteration of iron- and sulfur-bearing minerals such as pyrite (FeS_2). The dissolution of such minerals forms sulfuric acid, which causes other metallic minerals to dissolve, and these tend to be carried downward toward the water table.

Gossans have been used occasionally as sources of iron, but they are far more important as indicators of underlying ore deposits. One of the oldest known underground mines exploited such ores about 3,400 years ago in what is now southern Israel.

CHAPTER SUMMARY

1. Mechanical and chemical weathering are processes whereby parent material is disintegrated and decomposed so that it is more nearly in equilibrium with new physical and chemical conditions. The products of weathering include solid particles, soluble compounds, and ions in solution.

2. The residue of weathering can be further modified to form soil, or it can be deposited as sediment, which might become sedimentary rock.

3. Mechanical weathering includes such processes as frost action, pressure release, thermal expansion and contraction, salt crystal growth, and the activities of organisms. Particles liberated by mechanical weathering retain the chemical composition of the parent material.

4. Solution, oxidation, and hydrolysis are chemical weathering processes; they result in a chemical change of the weathered products. Clay minerals, various ions in solution, and soluble compounds are formed during chemical weathering.

5. Chemical weathering proceeds most rapidly in hot, wet environments, but it occurs in all areas, except perhaps where water is permanently frozen.

6. Mechanical weathering aids chemical weathering by breaking parent material into smaller pieces, thereby exposing more surface area.

7. Mechanical and chemical weathering produce regolith, some of which is soil if it consists of solids, air, water, and humus and supports plant growth.

8. Soils are characterized by horizons that are designated, in descending order, as O, A, B, and C; soil horizons differ from one another in texture, structure, composition, and color.

9. The factors controlling soil formation include climate, parent material, organic activity, relief and slope, and time.

10. Soils called pedalfers develop in humid regions such as the eastern United States and much of Canada. Arid and semiarid regions' soils are pedocals, many of which contain irregular masses of caliche in horizon B.

11. Laterite is a soil resulting from intense chemical weathering as in the tropics. Such soils are deep and red and are sources of aluminum ores if derived from aluminum-rich parent material.

12. Soil degradation is a problem in some areas. Human practices such as construction, agriculture, and deforestation can accelerate soil degradation.

13. Intense chemical weathering is responsible for the origin of residual concentrations, many of which contain valuable minerals such as iron, lead, copper, and clay.

IMPORTANT TERMS

chemical weathering
differential weathering
erosion
exfoliation
exfoliation dome
frost action
frost heaving
frost wedging
humus

hydrolysis
laterite
leaching
mechanical weathering
oxidation
parent material
pedalfer
pedocal
pressure release

regolith
rill erosion
salt crystal growth
sheet erosion
sheet joint
soil
soil degradation
soil horizon

solution
spheroidal weathering
talus
thermal expansion and
    contraction
transport
weathering
zone of accumulation

REVIEW QUESTIONS

1. The type of soil typical of arid and semiarid regions is:
   a. ____ laterite;
   b. ____ pedocal;
   c. ____ gossan;
   d. ____ bauxite;
   e. ____ pedalfer.

2. Limestone, which is composed of the mineral calcite ($CaCO_3$), is nearly insoluble in pure water but dissolves rapidly if _____ is present.
   a. ____ carbonic acid;
   b. ____ silicon dioxide;
   c. ____ calcium sulfate;

d. ____ residual manganese;

e. ____ clay.

3. Talus is an accumulation of:

   a. ____ calcium carbonate in horizon B of pedocals;

   b. ____ angular rock fragments at the base of a slope;

   c. ____ valuable minerals formed by selective removal of soluble substances;

   d. ____ debris produced mostly by the activities of organisms;

   e. ____ soil produced by intense weathering in the tropics.

4. When the ions in a substance become dissociated, the substance has been:

   a. ____ weathered mechanically;

   b. ____ altered to clay:

   c. ____ dissolved;

   d. ____ oxidized;

   e. ____ converted to soil.

5. The process whereby hydrogen and hydroxyl ions of water replace ions in minerals is:

   a. ____ residual concentration;

   b. ____ oxidation;

   c. ____ laterization;

   d. ____ hydrolysis;

   e. ____ carbonization.

6. The soil and unconsolidated rock material covering the Earth's surface in most places is:

   a. ____ regolith;

   b. ____ laterite;

   c. ____ humus;

   d. ____ parent material;

   e. ____ talus.

7. Horizon B of a soil is also known as the:

   a. ____ top soil;

   b. ____ humus layer;

   c. ____ alkali zone;

   d. ____ zone of accumulation;

   e. ____ organic-rich layer.

8. The chief ore of aluminum is:

   a. ____ caliche;

   b. ____ pedalfer;

   c. ____ subsoil;

   d. ____ gossan;

   e. ____ bauxite.

9. The removal of thin layers of soil by water over a more or less continuous surface is:

   a. ____ gullying;

   b. ____ sheet erosion;

   c. ____ weathering;

   d. ____ leaching;

   e. ____ exfoliation.

10. How does mechanical weathering differ from and contribute to chemical weathering?

11. Describe the process whereby soluble minerals such as halite (NaCl) are dissolved.

12. Why are most minerals not very soluble in pure water?

13. What is an acid solution, and why are acid solutions important in chemical weathering?

14. What role do hydrogen ions play in the hydrolysis process?

15. Explain why particle size is an important factor in chemical weathering.

16. Draw a soil profile and list the characteristics of each soil horizon.

17. What is the significance of climate and parent material in the development of soil?

18. What kinds of weathering would you expect to occur on the Moon? On Venus? Explain your answer.

POINTS TO PONDER

1. Consider the following: A soil is 1.5 m thick, new soil forms at the rate of 2.5 cm per century, and the erosion rate is 4 mm per year. How much soil will be left after 100 years?

2. How do human practices contribute to soil degradation? What can be done to minimize the impact of such practices on soils?

Chapter 7

SEDIMENT AND SEDIMENTARY ROCKS

OUTLINE

Sedimentary rocks exposed in John Day Fossil Beds National Monument, Oregon. These rocks, composed mostly of mud, erode easily and form smooth, rounded hills.

About 50 million years ago, two large lakes existed in what are now parts of Wyoming, Utah, and Colorado. The sedimentary rocks that formed in these lakes, called the Green River Formation, contain the fossilized remains of millions of fish, plants, and insects and are a potential source of large quantities of oil, combustible gases, and other substances.

Thousands of fossilized fish skeletons are found on single surfaces within the Green River Formation, indicating that mass mortality must have occurred repeatedly (▷ Figure 7-1). The cause of these events is not known with certainty, but some geologists have suggested that blooms of blue-green algae produced toxic substances that killed the fish. Others propose that rapidly changing water temperatures or excessive salinity at times of increased evaporation was responsible. Whatever the cause, the fish died by the thousands and settled to the lake bottom where their decomposition was inhibited because the water contained little or no oxygen. One area of the formation in Wyoming where fossil plants are particularly abundant has been designated as Fossil Butte National Monument.

The Green River Formation is also well known for its huge deposits of oil shale (▷ Figure 7-2). Oil shale consists of small clay particles and an organic substance known as *kerogen*. When the appropriate extraction processes are used, liquid oil and combustible gases can be produced from the kerogen of oil shale. To be designated as a true oil shale, the rock must yield a minimum of 10 gallons of oil per ton of rock.

The use of oil shale as a source of fuel is not new, nor is oil shale restricted to the Green River Formation. During the Middle Ages, people in Europe used oil shale as solid fuel for domestic purposes, and during the 1850s, small oil shale industries existed in the eastern United States; the latter were discontinued when drilling and pumping of oil began in 1859. Oil shales occur on all continents, but the Green River Formation contains the most extensive deposits and has the potential to yield huge quantities of oil.

Oil can be produced from oil shale by a process in which the rock is heated to nearly 500°C in the absence of oxygen, and hydrocarbons are driven off as gases and recovered by condensation. During this process, 25 to 75% of the organic matter of oil shale can be converted to oil and combustible gases. The Green River Formation oil shales yield from 10 to 140 gallons of oil per ton of rock processed, and the total amount of oil recoverable with present processes is estimated at 80 billion barrels. Currently no oil is produced from oil shale in the United States, because conventional drilling and pumping is less expensive. Nevertheless, the Green River oil shale constitutes one of the largest untapped sources of oil in the world. If more effective processes are developed, it could eventually yield even more than the currently estimated 80 billion barrels.

One should realize, though, that at the current and expected consumption rates of oil in the United States, oil production from oil shale will not solve all of our energy needs. Furthermore, the large-scale mining that would be necessary would have considerable environmental impact. What would be done with the billions of tons of processed rock? Can such large-scale mining be conducted with minimal disruption of wildlife habitats and groundwater systems? Where will the huge volumes of water necessary for processing come from—especially in an area where water is already in short supply?

These and other questions are currently being considered by scientists and industry. Perhaps at some future time, the Green River Formation will provide some of our energy needs.

▷ FIGURE 7-1 Fossil fish from the Green River Formation of Wyoming. (Photo courtesy of Sue Monroe.)

▷ FIGURE 7-2 Layers of oil shale of the Green River Formation are exposed along these hillsides.

INTRODUCTION

Mechanical and chemical weathering disintegrate and decompose rocks, yielding the raw materials for both soils and **sedimentary rocks.** All **sediment** is derived from preexisting rocks and may be *detrital,* consisting of rock fragments and mineral grains liberated during mechanical weathering, or it may be *chemical,* consisting of minerals formed from the materials that were dissolved during chemical weathering. Once sediment has been derived from parent material, it is commonly eroded and transported to another location where it is deposited as an aggregate of loose solids, such as sand on a beach.

Most sedimentary rocks formed from sediment that was transformed into solid rock, but a few sedimentary rocks skipped the unconsolidated sediment stage. Coral reefs, for example, form as solids when the reef organisms extract dissolved mineral matter from seawater for their skeletons. But if a reef is broken apart during a storm, the solid pieces of reef material deposited on the sea floor are sediment.

One important criterion for classifying sedimentary particles is their size (● Table 7-1). *Gravel* refers to any sedimentary particle larger than 2 mm, whereas *sand* is any particle, regardless of composition, that measures 1/16 to 2 mm. Gravel- and sand-sized particles are large enough to be observed with the unaided eye or with low-power magnification, but silt- and clay-sized particles are too small to be observed except with very high magnification. Gravel generally consists of rock fragments, whereas sand, silt, and clay particles are mostly individual mineral grains. We should note, though, that *clay* has two meanings: in textural terms, clay refers to sedimentary grains less than 1/256 mm in size, and in compositional terms, clay refers to certain types of sheet silicate minerals (see Figure 3-10). However, most clay-sized particles in sedimentary rocks are, in fact, clay minerals.

SEDIMENT TRANSPORT AND DEPOSITION

Sediment can be transported by any geologic agent possessing enough energy to move particles of a given size. Glaciers are very effective agents of transport and can move particles of any size. Wind, on the other hand, can transport only sand-sized and smaller sediment. Waves and marine currents also transport sediment, but by far the most effective way to erode sediment from the weathering site and transport it elsewhere is by streams.

During sediment transport, *abrasion* reduces the size of particles, and the sharp corners and edges are worn smooth as gravel and sand particles collide with one another and the particles become **rounded** (➤ Figure 7-3a). Transport also results in **sorting,** which refers to the size distribution in an

➤ **FIGURE 7-3** Rounding and sorting of sedimentary particles. (*a*) A deposit consisting of well-sorted and well-rounded gravel. (*b*) Poorly sorted, angular gravel. (Photos courtesy of R. V. Dietrich.)

(a)

(b)

TABLE 7-1 Classification of Sedimentary Particles	
Size	Sediment Name
>2 mm	Gravel
¹⁄₁₆–2 mm	Sand
¹⁄₂₅₆–¹⁄₁₆ mm	Silt ⎫
<¹⁄₂₅₆ mm	Clay ⎭ Mud★

★Mixtures of silt and clay are generally referred to as mud.

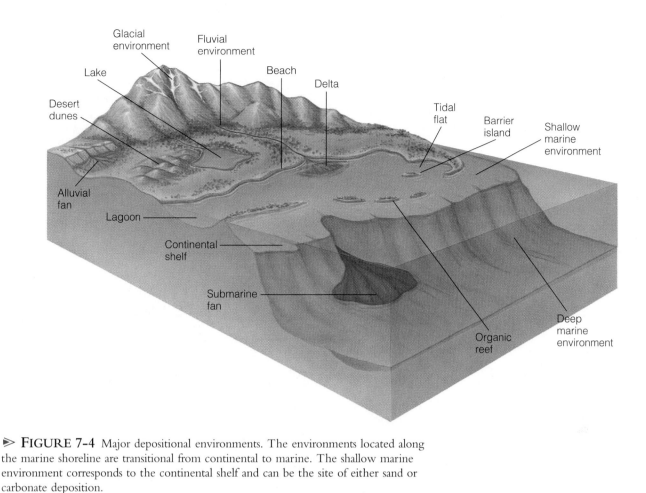

⫸ FIGURE 7-4 Major depositional environments. The environments located along the marine shoreline are transitional from continental to marine. The shallow marine environment corresponds to the continental shelf and can be the site of either sand or carbonate deposition.

aggregate of sediment. If all the particles are about the same size, the sediment is characterized as well sorted, but if a wide range of grain sizes occurs, it is poorly sorted (Figure 7-3b). Both rounding and sorting are important properties used to determine the origin of sedimentary rocks; they are discussed more fully in a later section.

Sediment may be transported a considerable distance from its source area, but eventually it is deposited. Some of the sand and mud being deposited at the mouth of the Mississippi River at the present time came from such distant places as Ohio, Minnesota, and Wyoming. Any geographic area in which sediment is deposited is a **depositional environment**. Although no completely satisfactory classification of depositional environments exists, geologists generally recognize three major depositional settings: continental, transitional, and marine, each with several specific depositional environments (⫸ Figure 7-4).

LITHIFICATION: SEDIMENT TO SEDIMENTARY ROCK

At present, calcium carbonate mud is accumulating in the shallow waters of Florida Bay, and sand is being deposited in river channels, on beaches, and in sand dunes. Such deposits of sediment might be compacted and/or cemented and thereby converted into sedimentary rock; the process of transforming sediment into sedimentary rock is **lithification**.

When sediment is deposited, it consists of solid particles and *pore spaces*, which are the voids between particles. The amount of pore space varies depending on the depositional process, the size of the sediment grains, and sorting. When sediment is buried, **compaction**, resulting from the pressure exerted by the weight of overlying sediments, reduces the amount of pore space, and thus the volume of the deposit (⫸ Figure 7-5). When deposits of mud, which can have as much as 80% water-filled pore space, are buried and compacted, water is squeezed out, and the volume can be reduced by up to 40%. Sand has up to 50% pore space, although it is generally somewhat less, and it, too, can be compacted so that the sand grains fit more tightly together.

Compaction alone is generally sufficient for lithification of mud, but for sand and gravel deposits **cementation** is necessary to convert the sediment into sedimentary rock (Figure 7-5). Recall from Chapter 6 that calcium carbonate ($CaCO_3$) readily dissolves in water containing a small amount of carbonic acid, and that chemical weathering of feldspars and other silicate minerals yields silica (SiO_2) in solution. These compounds may be precipitated in the pore spaces of sediments, where they act as a cement that effectively binds the sediment together (Figure 7-5).

Calcium carbonate and silica are the most common cements in sedimentary rocks, but iron oxides and hydroxides, such as hematite (Fe_2O_3) and limonite [$FeO(OH)$], respectively, also form a chemical cement in some rocks.

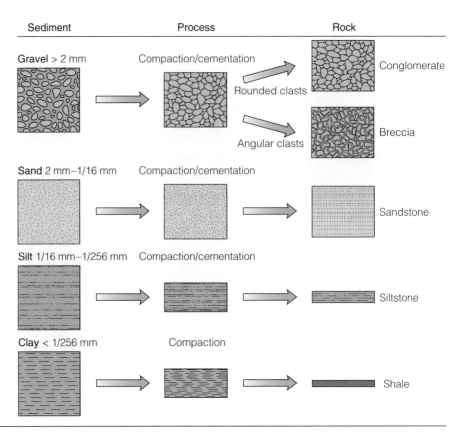

► FIGURE 7-6 These sedimentary rocks at Jemez Pueblo, New Mexico are red because they contain iron oxide cement. (Photo courtesy of Sue Monroe.)

Much of the iron oxide cement is derived from the oxidation of iron in ferromagnesian minerals present in the original deposit, although some is carried in by circulating groundwater. The yellow, brown, and red sedimentary rocks exposed in the southwestern United States are colored by small amounts of iron oxide or hydroxide cement (⊳ Figure 7-6).

SEDIMENTARY ROCKS

Even though about 95% of the Earth's crust is composed of igneous and metamorphic rocks, sedimentary rocks are the most common at or near the surface. About 75% of the surface exposures on continents consist of sediments or sedimentary rocks, and they cover most of the sea floor. Sedimentary rocks are generally classified as *detrital* or *chemical* (◉ Table 7-2).

Detrital Sedimentary Rocks

Detrital sedimentary rocks consist of *detritus*, the solid particles of preexisting rocks. Such rocks have a *clastic texture*,

meaning that they are composed of fragments or particles also known as *clasts*. Several varieties of detrital sedimentary rocks are recognized, each of which is characterized by the size of its constituent particles (Table 7-2).

Conglomerate and Sedimentary Breccia. Both *conglomerate* and *sedimentary breccia* consist of gravel-sized particles (Table 7-2; ⊳ Figure 7-7a). Usually, the particles measure a few millimeters to a few centimeters, but boulders several meters in diameter are sometimes present. The only difference between conglomerate and sedimentary breccia is the shape of the gravel particles; conglomerate consists of rounded gravel, whereas sedimentary breccia is composed of angular gravel called *rubble*.

Conglomerate is a fairly common rock type, but sedimentary breccia is rather rare. The reason is that gravel-sized particles become rounded very quickly during transport. So, if a sedimentary breccia is encountered, one can conclude that the rubble that composes it was not transported very far. High-energy transport agents such as rapidly flowing streams and waves are needed to transport gravel, so gravel tends to

TABLE 7-2 Classification of Sedimentary Rocks

Detrital Sedimentary Rocks

Sediment Name and Size	Description	Rock Name	
Gravel (>2 mm)	Rounded gravel particles	Conglomerate	
	Angular gravel particles	Sedimentary breccia	
Sand (1/16–2mm)	Mostly quartz sand	Quartz sandstone	
	Quartz with >25% feldspar	Arkose	
Mud (<1/16 mm)	Mostly silt	Siltstone	
	Silt and clay	Mudstone*	] Mudrocks
	Mostly clay	Claystone*	

Chemical Sedimentary Rocks

Texture	Composition	Rock Name	
Crystalline	Calcite ($CaCO_3$)	Limestone	] Carbonates
Crystalline	Dolomite [$CaMg(CO_3)_2$]	Dolostone	
Crystalline	Gypsum ($CaSO_4 \cdot 2H_2O$)	Rock gypsum	] Evaporites
Crystalline	Halite (NaCl)	Rock salt	

Biochemical Sedimentary Rocks

Texture	Composition	Rock Name
Clastic	Calcium carbonate ($CaCO_3$) shells	Limestone (various types such as chalk and coquina)
Usually crystalline	Altered microscopic shells of silicon dioxide (SiO_2)	Chert
—	Mostly carbon from altered plant remains	Coal

*Mudrocks possessing the property of fissility, meaning they break along closely spaced planes, are commonly called *shale.*

(a)

(b)

➤ **FIGURE 7-7** Detrital sedimentary rocks: (*a*) conglomerate; (*b*) sandstone. (Photos courtesy of Sue Monroe.)

be deposited in high-energy environments such as stream channels and beaches.

Sandstone. The term *sand* is simply a size designation, so *sandstone* may be composed of grains of any type of mineral or rock fragment. Most sandstones consist primarily of the mineral quartz (Figure 7-7b) with small amounts of a number of other minerals. Geologists recognize several types of sandstones, each characterized by its composition. *Quartz sandstone*, composed mostly of quartz, is the most common, but *arkose*, which contains more than 25% feldspars, is also fairly common (Table 7-2).

Mudrocks. The *mudrocks* include all detrital sedimentary rocks composed of silt- and clay-sized particles. Among the mudrocks we can differentiate between *siltstone, mudstone,* and *claystone.* Siltstone, as the name implies, is composed of silt-sized particles; mudstone contains a mixture of silt- and clay-sized particles; and claystone is composed mostly of clay (Table 7-2). Some mudstones and claystones are designated

as *shale* if they are fissile, which means they break along closely spaced parallel planes.

Mudrocks comprise about 40% of all detrital sedimentary rocks, making them the most common of these rocks. Turbulence in water keeps silt and clay suspended and must therefore be at a minimum if they are to settle. Consequently, deposition occurs in low-energy depositional environments where currents are weak such as in the quiet offshore waters of lakes and in lagoons.

Chemical and Biochemical Sedimentary Rocks

Chemical sedimentary rocks originate from the materials taken into solution during chemical weathering (Table 7-2). These dissolved materials are transported to lakes and the oceans where they become concentrated. They can be extracted from lake or ocean water to form minerals either by inorganic chemical processes or by the chemical activities of organisms. Some rocks formed by lithification of these minerals have a *crystalline texture,* meaning they consist of a mosaic of interlocking crystals, whereas others have a clastic texture. Rocks formed by the activities of organisms are referred to as **biochemical sedimentary rocks**. In any case, aggregates of minerals accumulate that become lithified by compaction and cementation just as in detrital sedimentary rocks.

Limestone-Dolostone. Calcite (the main component of limestone) and dolomite (the mineral comprising the rock dolostone) are both carbonate minerals; calcite is a calcium carbonate ($CaCO_3$), whereas dolomite [$CaMg(CO_3)_2$] is a calcium magnesium carbonate. Thus, limestone and dolostone are **carbonate rocks**. Recall from Chapter 6 that calcite readily dissolves in water containing a small amount of acid, but the chemical reaction leading to dissolution is reversible, so solid calcite can be precipitated from solution. Accordingly, some limestone, although probably not very much, results from inorganic chemical reactions.

Because organisms play such a significant role in their origin, most limestones are conveniently classified as *biochemical sedimentary rocks* (Table 7-2). For example, the limestone known as *coquina* consists entirely of broken shells cemented by calcium carbonate, and *chalk* is a soft variety of biochemical limestone composed largely of microscopic shells of organisms (➤ Figure 7-8).

The near-absence of recent dolostone and evidence from chemistry and studies of rocks indicate that most dolostone was originally limestone that has been changed to dolostone. Many geologists think most dolostones originated when magnesium replaced some of the calcium in calcite.

Evaporites. Evaporites include such rocks as *rock salt* and *rock gypsum,* which form by inorganic chemical precipitation of minerals from solution (Table 7-2; ➤ Figure 7-9). In Chapter 6 we noted that some minerals are dissolved during chemical weathering, but a solution can hold only a certain volume of dissolved mineral matter. If the volume of a solution is reduced by evaporation, the amount of dissolved

(a)   (b)

➤ FIGURE 7-8 Two types of limestones. (*a*) Coquina is composed of the broken shells of organisms. (*b*) Fossilliferous limestone. (Photos courtesy of Sue Monroe.)

(a)   (c)

➤ FIGURE 7-9 (*a*) Core of rock salt from a well in Michigan. (*b*) Rock gypsum.
(*c*) Chert. (Photos courtesy of Sue Monroe.)

mineral matter increases in proportion to the volume of the solution and eventually reaches the saturation limit, the point at which precipitation must occur.

Rock salt, composed of the mineral halite (NaCl), is simply sodium chloride that was precipitated from seawater or, more rarely, lake water (Figure 7-9a). Rock gypsum, the most common evaporite rock, is composed of the mineral gypsum ($CaSO_4 \cdot H_2O$), which also precipitates from evaporating solutions (Figure 7-9b). A number of other evaporite rocks and minerals are known, but most of these are rare.

Chert. *Chert* is a hard rock composed of microscopic crystals of quartz (SiO_2) (Table 7-2; Figure 7-9c). It is found in several varieties including *flint*, which is black because of inclusions of organic matter, and *jasper*, which is red or brown because of iron oxide inclusions. Because chert lacks cleavage and can be shaped to form sharp cutting edges, many cultures have used it for the manufacture of tools, spear points, and arrowheads.

Chert occurs as irregular masses or *nodules* in other rocks, especially limestones, and as distinct layers of rock called *bedded chert*. Most nodules in limestones are clearly secondary in origin; that is, they have replaced part of the host rock, apparently by being precipitated from solution.

Bedded chert can be precipitated directly from seawater, but because so little silica is dissolved in seawater, it seems unlikely that most bedded cherts are inorganic chemical precipitates. It appears that many bedded cherts are biochemical, resulting from accumulations of shells of silica-secreting, single-celled organisms such as radiolarians and diatoms.

Coal. *Coal* is a biochemical sedimentary rock composed of the compressed, altered remains of organisms, especially land plants (Table 7-2; ▷ Figure 7-10). It forms in swamps and bogs where the water is deficient in oxygen or where organic matter accumulates faster than it decomposes. The bacteria that decompose vegetation in swamps can exist without oxygen, but their wastes must be oxidized, and because no oxygen is present, the wastes accumulate and kill the bacteria. Thus, bacterial decay ceases and plant materials are not completely destroyed. These partly altered plant remains accumulate as layers of organic muck. When buried, this organic muck becomes *peat*, which looks rather like coarse pipe tobacco. Where peat is abundant, as in Ireland and Scotland, it is burned as a fuel. Peat that is buried more deeply and compressed, especially if it is heated too, is altered to a type of dark brown coal called *lignite*, in which plant remains are still clearly visible. During the change from organic muck to coal, volatile elements of the vegetation such as oxygen, hydrogen, and nitrogen are partly vaporized and driven off, enriching the residue in carbon; lignite contains about 70% carbon as opposed to about 50% in peat.

Bituminous coal, which contains about 80% carbon, is a higher grade coal than lignite. It is dense and black and has been so thoroughly altered that plant remains can only rarely

▷ **FIGURE 7-10** Coal is a biochemical sedimentary rock composed of the altered remains of land plants. (Photo courtesy of Sue Monroe.)

be seen. The highest grade coal is *anthracite*, which is a metamorphic type of coal (see Chapter 8). It contains up to 98% carbon and, when burned, yields more heat per unit volume than other types of coal.

ENVIRONMENTAL ANALYSIS

When geologists investigate sedimentary rocks in the field, they are observing the products of events that occurred during the past. The only record of these events is preserved in the rocks, so geologists must evaluate those aspects of sedimentary rocks that allow inferences to be made about the original processes and the environment of deposition. Sedimentary textures such as sorting and rounding can give clues to the depositional process. Windblown dune sands, for example, tend to be well sorted and well rounded. Other aspects of sedimentary rocks that are important in environmental analysis include *sedimentary structures* and *fossils*.

Sedimentary Structures

When sediments are deposited, they contain a variety of features known as **sedimentary structures** that formed as a result of physical and biological processes operating in the depositional environment. Among the most common sedimentary structures are distinct layers known as **strata** or **beds** (▷ Figure 7-11). Beds vary in thickness from less than a millimeter up to many meters. Individual beds are separated from one another by **bedding planes** and are distinguished from one another by differences in composition, grain size, color, or a combination of features (Figure 7-11). Almost all sedimentary rocks show some kind of bedding; a few, such as limestones that formed as coral reefs, lack this feature, however.

In **graded bedding,** grain size decreases upward within a single bed (▷ Figure 7-12). Most graded bedding appears to have formed from turbidity current deposition, although

▷ **FIGURE 7-11** Like these sandstones in Montana, most sedimentary rocks show some kind of layering or bedding.

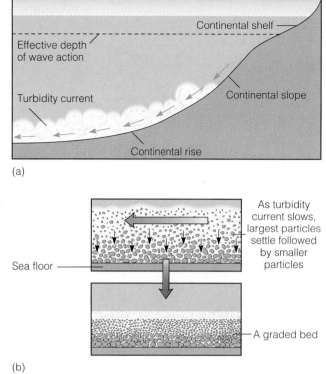

(a)

(b)

▷ **FIGURE 7-12** (*a*) Turbidity currents flow downslope along the sea floor (or lake bottom) because of their density. (*b*) Graded bedding formed by deposition from a turbidity current.

some forms in stream channels during the waning stages of floods. *Turbidity currents* are underwater flows of sediment-water mixtures that are denser than sediment-free water. Such flows move downslope along the bottom of the sea or a lake until they reach the relatively level sea or lake floor. There, they rapidly slow down and begin depositing transported sediment, the coarsest first followed by progressively smaller particles (Figure 7-12).

Many sedimentary rocks are characterized by **cross-bedding**; cross-beds are arranged at an angle to the surface upon which they are deposited (▷ Figure 7-13). Cross-bedding is common in sedimentary rocks that originated in stream channels and shallow marine environments and as desert dunes. Invariably, cross-beds result from transport by wind or water currents and deposition on the downcurrent sides of dunelike structures. Cross-beds are inclined downward, or dip, in the direction of flow. Because their orientation depends on the direction of flow, cross-beds are good indicators of ancient current directions or **paleocurrents** (Figure 7-13).

In sand deposits one can commonly observe small-scale, ridgelike **ripple marks** on bedding planes. Two common types of ripple marks are recognized. One type is asymmetrical in cross section with a gentle upstream slope and a steep downstream slope. Known as *current ripple marks* (▷ Figure 7-14a), they are formed by currents that move in one direction as in a stream channel. Like cross-bedding, current ripple marks are good paleocurrent indicators. In contrast, the to-and-fro motion of waves produces ripples that tend to be symmetrical in cross section. These are known as *wave-formed ripple marks* (Figure 7-14b); they form mostly in the shallow nearshore waters of oceans and lakes.

Mud cracks are found in clay-rich sediment that has dried out (▷ Figure 7-15). When such sediment dries, it shrinks and forms intersecting fractures (mud cracks). These features in ancient sedimentary rocks indicate that the sediment was deposited where periodic drying was possible as on a river floodplain, near a lake shore, or where muddy deposits are exposed on marine shorelines at low tide.

Fossils

Fossils are the remains or traces of ancient organisms (▷ Figure 7-16). These remains are mostly the hard skeletal parts of organisms such as shells, bones, and teeth, but under exceptional conditions, even the soft-part anatomy may be preserved. For example, several frozen woolly mammoths have been discovered in Alaska and Siberia with hair, flesh, and internal organs preserved. The remains of organisms are known as *body fossils* to distinguish them from *trace fossils* such as tracks, trails, and burrows (Figure 7-16), which are indications of ancient organic activity.

For any potential fossil to be preserved, it must escape the ravages of such destructive processes as running water, waves, scavengers, exposure to the atmosphere, and bacterial decay. Obviously, the soft parts of organisms are devoured or decomposed most rapidly, but even the hard skeletal parts will be destroyed unless they are buried and protected in

➤ **FIGURE 7-13** Cross-bedding forms when the beds are inclined with respect to the surface upon which they accumulate. Cross-beds indicate ancient current directions by their dip, to the right in this case.

(a)

(b)

➤ **FIGURE 7-14** (*a*) Current ripple marks on the bed of a stream in California. (*b*) Wave-formed ripples on ancient rocks in Montana.

▷ FIGURE 7-15 Mud cracks form in clay-rich sediments when they dry and shrink.

mud, sand, or volcanic ash. Even if buried, skeletal elements may be dissolved by groundwater or destroyed by alteration of the host rock. Nevertheless, fossils are quite common. The remains of microscopic plants and animals are the most common, but these require specialized methods of recovery, preparation, and study and are not sought out by casual fossil collectors. Shells of marine animals are also very common and easily collected in many areas, and even the bones and teeth of dinosaurs are much more common than most people realize.

If it were not for fossils, we would have no knowledge of such extinct animals as trilobites and dinosaurs. Thus, fossils constitute our only record of ancient life. They are not simply curiosities, however, but have several practical uses. In many geologic studies, it is necessary to correlate or determine age equivalence of sedimentary rocks in different areas. Such correlations are most commonly demonstrated with fossils; we will discuss correlation more fully in Chapter 17. Fossils are also useful in determining environments of deposition.

Environment of Deposition

The sedimentary rocks in the geologic record acquired their various properties, in part, as a result of the physical, chemical, and biological processes that operated in the original depositional environment. One of geologists' major tasks is to determine the specific depositional environment of sedimentary rocks. Based on their knowledge of how sedimentary structures originate and present-day processes, such as sediment transport and deposition by streams, geologists can make inferences regarding the depositional environments of ancient sedimentary rocks.

While conducting field studies, geologists commonly make some preliminary interpretations. For example, some sedimentary particles in limestones most commonly form in shallow marine environments where currents are vigorous. Large-scale cross-bedding is typical of, but not restricted to, desert dunes. Fossils of land plants and animals can be washed

(a)

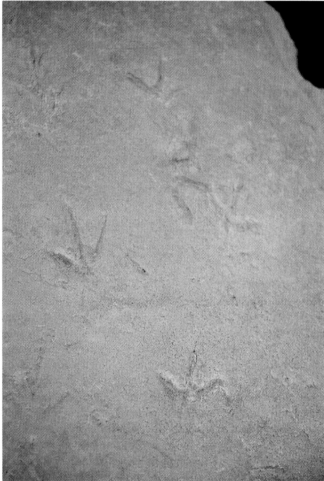

(b)

▷ FIGURE 7-16 (a) Body fossils consist of the actual remains of organisms. Fossil horse teeth are preserved in this rock. (b) Trace fossils are an indication of ancient organic activity. These bird tracks are preserved in mudrock of the Green River Formation of Wyoming. (Photos courtesy of Sue Monroe.)

into transitional environments, but most of them are preserved in deposits of continental environments. Fossil shells of such marine-dwelling animals as corals obviously indicate marine depositional environments.

▷ **FIGURE 7-17** Outcrop of the Navajo Sandstone in Zion National Park, Utah. The characteristics of this rock unit, including the large cross-beds, indicate it was deposited as desert dunes.

Evaluation of rock properties such as sorting of sedimentary particles by size is also useful in environmental interpretation. Sorting results from processes that selectively transport and deposit particles. Windblown dunes are composed of well-sorted sand, because wind cannot transport gravel and it blows silt and clay beyond the areas of sand accumulation. Glaciers and mudflows, on the other hand, are unselective because their energy allows them to transport many different-sized particles, and their deposits tend to be poorly sorted.

Much environmental interpretation is done in the laboratory where the data and rock samples collected during field work can be more fully analyzed. The analyses include microscopic and chemical examination of rock samples, identification of fossils, and graphic representations showing the three-dimensional shapes of rock units and their relationships to other rock units. In addition, the features of sedimentary rocks are compared with those of sediments from present-day depositional environments; the contention is that features in ancient rocks, such as ripple marks, formed during the past in response to the same processes responsible for them now. Finally, when all data have been analyzed, an environmental interpretation is made.

The following example illustrates how environmental interpretations are made. The Navajo Sandstone of the southwestern United States covers a vast area, perhaps as much as 500,000 km^2. It has an irregular three-dimensional shape, reaches a thickness of about 300 m in the area of Zion National Park, Utah, and consists mostly of well-sorted sand grains measuring about 0.2 to 0.5 mm in diameter. Some of the sandstone beds also possess tracks of dinosaurs and other land-dwelling animals, ruling out the possibility of a marine origin for the rock unit. These features and the fact that the Navajo Sandstone has cross-beds up to 30 m high (▷ Figure 7-17) and ripple marks, both of which appear to have formed in sand dunes, lead to the conclusion that the sandstone represents an ancient desert dune deposit. The cross-beds are generally inclined downward, or dip, to the southwest, indicating that the wind blew mostly from the northeast.

SEDIMENTS, SEDIMENTARY ROCKS, AND NATURAL RESOURCES

Sediments and sedimentary rocks or the materials they contain have a variety of uses. Sand and gravel are essential to the construction industry, pure clay deposits are used for ceramics, and limestone is used in the manufacture of cement and in blast furnaces where iron ore is refined to make steel. Rock salt (NaCl) is the source of common table salt, sylvite, a potassium chloride (KCl), is used in the manufacture of fertilizers, dyes, and soaps, and gypsum ($CaSO_4 \cdot 2H_2O$) is used to make wallboard. Sand composed mostly of quartz, or what is called silica sand, is used for a variety of purposes including the manufacture of glass, refractory bricks for blast furnaces, and molds for casting iron, aluminum, and copper alloys.

The tiny island nation of Nauru, with one of the highest per capita incomes in the world, has an economy based almost entirely on mining and exporting phosphate-bearing sedimentary rock that is used in fertilizers. More than half of Florida's mineral value in 1989 came from mining phosphorus from phosphate rock. In addition to fertilizers, phosphorus from phosphate is used in metallurgy, preserving foods, ceramics, and matches.

Dolostones in Missouri are the host rocks for ores of lead and zinc. Diatomite is a sedimentary rock composed of the microscopic silica (SiO_2) skeletons of single-celled plants. This lightweight, porous rock is used in gas purification and to filter a number of fluids such as molasses, fruit juices, water, and sewage. The United States is the world leader in diatomite production, mostly from mines in California, Oregon, and Washington.

Historically, most of the coal mined in the United States has been bituminous coal from the coal fields of the Appalachian coal basin. These coal deposits formed in coastal swamps during the Pennsylvanian Period between 286 and 320 million years ago. Huge lignite and subbituminous coal deposits also exist in the western United States, and these are becoming increasingly important resources.

Anthracite coal is an especially desirable resource because it burns hot with a smokeless flame. Unfortunately, it is the least common type of coal, so most coal used for heating buildings and for generating electrical energy is bituminous (Figure 7-10). Bituminous coal is also used to make *coke*, a hard, gray substance consisting of the fused ash of bituminous coal; coke is prepared by heating the coal and driving off the volatile matter. Coke is used to fire blast furnaces during the production of steel. Synthetic oil and gas and a number of other products are also made from bituminous coal and lignite.

Petroleum and Natural Gas

Both petroleum and natural gas are *hydrocarbons*, meaning that they are composed of hydrogen and carbon. Hydrocarbons form from the remains of microscopic organisms that exist in the seas and in some large lakes. When these organisms die, their remains settle to the sea or lake floor where little oxygen is available to decompose them. They are then buried under layers of sediment. As the depth of burial increases, they are heated and transformed into petroleum and natural gas. The rock in which the hydrocarbons formed is generally called the *source rock*.

For petroleum and natural gas to occur in economic quantities, they must migrate from the source rock into some kind of rock in which they can be trapped. The rock in which petroleum and natural gas accumulate is known as *reservoir rock* (➤ Figure 7-18). Effective reservoir rocks contain considerable pore space where appreciable quantities of hydrocarbons can accumulate. Furthermore, the reservoir rocks must possess high *permeability*, or the capacity to transmit fluids; otherwise hydrocarbons cannot be extracted in reasonable quantities. In addition, some kind of impermeable *cap rock* must be present over the reservoir rock to prevent the hydrocarbons from migrating upward (Figure 7-18).

Many hydrocarbon reservoirs consist of nearshore marine sandstones in proximity with fine-grained, organic-rich source rocks. Such oil and gas traps are called *stratigraphic traps* because they owe their existence to variations in the strata (Figure 7-18a). Ancient coral reefs are also good stratigraphic traps. Indeed, some of the oil in the Persian Gulf region is trapped in ancient reefs (see Perspective 7-1). *Structural traps* result when rocks are deformed by folding, fracturing, or both (Figure 7-18b).

In the Gulf Coast region, hydrocarbons are commonly found in structures adjacent to salt domes. A vast layer of

➤ **FIGURE 7-18** Oil and natural gas traps. The arrows in both diagrams indicate the migration of hydrocarbons. (*a*) Two examples of stratigraphic traps. (*b*) Two examples of structural traps, one formed by folding, the other by faulting.

(a)

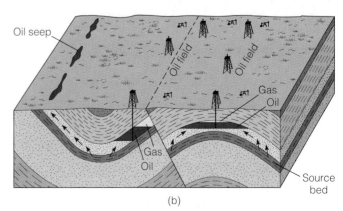

(b)

PERSIAN GULF PETROLEUM

Even though petroleum was discovered as early as 1908 in Iran, the Persian Gulf region did not become a significant petroleum-producing area until the economic recovery after World War II. Following the war, Western Europe and Japan in particular became dependent on Gulf oil and still rely heavily on this region for most of their supply. The United States is also dependent on imports from the Gulf, but receives significant quantities of petroleum from other sources such as Mexico and Venezuela. Currently, fully 40% of all petroleum imports in the world come from the Gulf countries.

Although large concentrations of petroleum occur in many areas of the world, more than 50% of all proven reserves are in the Gulf region (▷ Figure 1)! Furthermore, some of the oil fields are gigantic; at least 20 are expected to yield more than five billion barrels of oil each. Several factors account for the prolific quantities of oil in the Gulf region. By the beginning of the Mesozoic Era, all of the continents had joined together to form the supercontinent Pangaea. However, they were arranged such that present-day Africa and Eurasia were separated by the Tethys Sea (▷ Figure 2). What is now the Gulf region was a broad, stable, marine shelf extending eastward from Africa. Geologists refer to such a shelf as a passive continental margin as opposed to an active margin characterized by plate convergence, volcanism, earthquake activity, and strong deformation. The U.S. Gulf Coast, which is also a passive continental margin, is another area of significant petroleum reserves.

During the Mesozoic Era, and particularly the Cretaceous Period when most of the petroleum formed, this continental margin lay near the equator where countless microorganisms lived in the surface waters. The remains of these organisms accumulated with the bottom sediments and were buried, beginning the complex process of oil generation and formation of source beds.

▷ **FIGURE 1** The top 10 countries in proven oil reserves in 1992. Numbers indicate billions of barrels of oil.

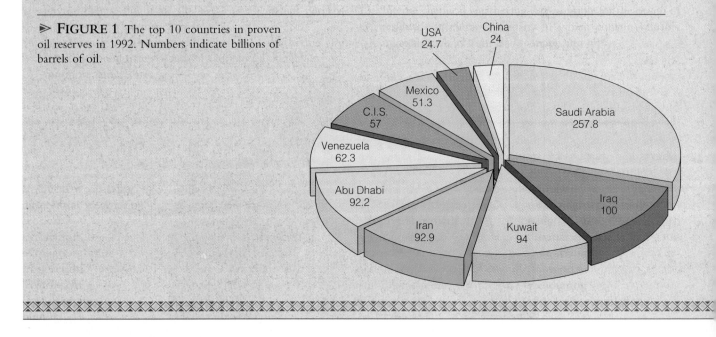

rock salt was precipitated in this region during the Jurassic Period as the ancestral Gulf of Mexico formed when North America separated from North Africa. Rock salt is a low-density sedimentary rock, and when deeply buried beneath denser sediments such as sand and mud, it rises toward the surface in pillars known as *salt domes* (see Figure 4-17). As the rock salt rises, it penetrates and deforms the overlying rock layers, forming structures along its margins that may trap petroleum and gas.

Other sources of petroleum that will probably become increasingly important in the future include *oil shales* and *tar sands*. The United States has about two-thirds of all known oil shales, although large deposits also occur in South America, and all continents have some oil shale. The richest deposits in the United States are in the Green River Formation of Colorado, Utah, and Wyoming (see the Prologue).

Tar sand is a type of sandstone with viscous, asphaltlike hydrocarbons filling the pore spaces. This substance is the sticky residue of once-liquid petroleum from which the volatile constituents have been lost. Liquid petroleum can be recovered from tar sand, but to do so, large quantities of rock must be mined and processed. Since the United States has

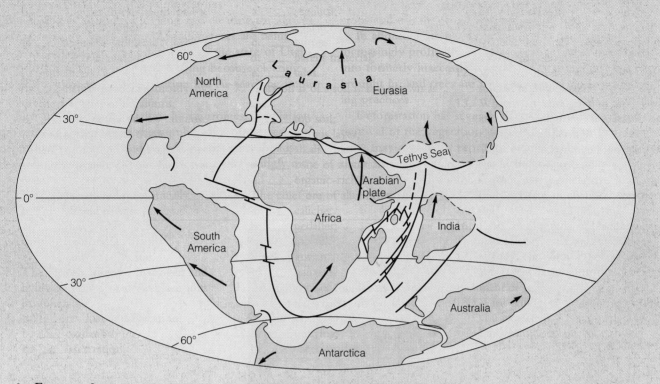

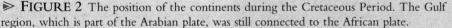

▷ **FIGURE 2** The position of the continents during the Cretaceous Period. The Gulf region, which is part of the Arabian plate, was still connected to the African plate.

Broad passive continental margins such as the one that existed in the Gulf region during the Mesozoic Era are particularly susceptible to fluctuations in sea level, so they are periodically exposed and then covered by shallow seas. Several such events occurred during which some of the reservoir rocks formed as extensive, thick sandstones. Other important reservoirs consist of algal reef limestones and reefs composed of the shells of clams. In any case, these reservoir rocks are geographically extensive because the shelf upon which they were deposited was 2,000 to 3,000 km wide and at least twice as long. Overlying the reservoir rocks are cap rocks that include widespread shale and evaporite units.

Many nations including the United States are heavily dependent on imports of Gulf oil, a dependence that will increase in the future. Most geologists think that all of the truly gigantic oil fields have already been found, but concede that some significant discoveries are yet to be made. One must view these potential discoveries in the proper perspective, however. The discovery of an oil field comparable to that of the North Slope of Alaska (about 10 billion barrels), for example, constitutes only about a two-year supply for the United States at the current consumption rate.

few tar sand deposits, it cannot look to this source as a significant future energy resource. The Athabaska tar sands in Alberta, Canada, however, are one of the largest deposits of this type. These deposits are currently being mined, and it is estimated that they contain several hundred billion barrels of recoverable petroleum.

Uranium

Most of the uranium used in nuclear reactors in North America comes from the complex potassium-, uranium-, vanadium-bearing mineral *carnotite* found in some sedimen-

tary rocks. Some uranium is also derived from *uraninite* (UO_2), a uranium oxide that occurs in granitic rocks and hydrothermal veins. Uraninite is easily oxidized and dissolved in groundwater, transported elsewhere, and chemically reduced and precipitated in the presence of organic matter.

The richest uranium ores in the United States are widespread in the Colorado Plateau area of Colorado and adjoining parts of Wyoming, Utah, Arizona, and New Mexico. These ores, consisting of fairly pure masses and encrustations of carnotite, are associated with plant remains in sandstones that formed in ancient stream channels. Although most of

▷ **FIGURE 7-19** Outcrop of banded iron formation in northern Michigan.

these ores are associated with fragmentary plant remains, some petrified trees also contain large quantities of uranium.

Large reserves of low-grade uranium ore also occur in the Chattanooga Shale. The uranium is finely disseminated in this black, organic-rich mudrock that underlies large parts of several states including Illinois, Indiana, Ohio, Kentucky, and Tennessee.

Banded Iron Formation

Banded iron formation is a chemical sedimentary rock of great economic importance. It consists of alternating thin layers of chert and iron minerals, mostly the iron oxides hematite and magnetite (▷ Figure 7-19). Banded iron formations are present on all the continents and account for most of the iron ore mined in the world today. Vast banded iron formations are present in the Lake Superior region of the United States and Canada and in the Labrador trough of eastern Canada.

The origin of banded iron formations is not fully understood, and none are currently forming. Fully 92% of all banded iron formations were deposited in shallow seas during the Proterozoic Eon, between 2.5 and 2.0 billion years ago. Iron is a highly reactive element that in the presence of oxygen combines to form rustlike oxides that are not readily soluble in water. During early Earth history, however, little oxygen was present in the atmosphere, and thus little was dissolved in seawater. Soluble reduced iron (Fe^{+2}) and silica, however, were present in seawater.

Geological evidence indicates that abundant photosynthesizing organisms were present about 2.5 billion years ago. These organisms, such as bacteria, release oxygen as a byproduct of respiration; thus, they released oxygen into seawater and caused large-scale precipitation of iron oxides and silica as banded iron formations.

CHAPTER SUMMARY

1. Sediment consists of mechanically weathered solid particles and minerals extracted from solution by inorganic chemical processes and the activities of organisms.
2. Sedimentary particles are designated in order of decreasing size as gravel, sand, silt, and clay.

3. Sedimentary particles are rounded and sorted during transport although the degree of rounding and sorting depends on particle size, transport distance, and depositional process.
4. Any area where sediment is deposited is a depositional

environment. Major depositional settings are continental, transitional, and marine, each of which includes several specific depositional environments.

5. Compaction and cementation are the processes of sediment lithification in which sediment is converted into sedimentary rock. Silica and calcium carbonate are the most common chemical cements, but iron oxide and iron hydroxide cements are important in some rocks.

6. Sedimentary rocks are generally classified as detrital or chemical:
   a. Detrital sedimentary rocks consist of solid particles derived from preexisting rocks.
   b. Chemical sedimentary rocks are derived from ions in solution by inorganic chemical processes or the biochemical activities of organisms. A subcategory called biochemical sedimentary rocks is recognized.

7. Carbonate rocks contain minerals with the carbonate ion $(CO_3)^{-2}$ as in limestone and dolostone. Dolostone probably forms when magnesium partly replaces the calcium in limestone.

8. Evaporites include rock salt and rock gypsum, both of which form by inorganic precipitation of minerals from evaporating water.

9. Coal is a type of biochemical sedimentary rock composed of the altered remains of land plants.

10. Sedimentary structures such as bedding, cross-bedding, and ripple marks commonly form in sediments when or shortly after they are deposited. Such features preserved in sedimentary rocks help geologists determine ancient current directions and depositional environments.

11. Sediments and sedimentary rocks are the host materials for most fossils. Fossils provide the only record of prehistoric life and are useful for correlation and environmental interpretations.

12. Depositional environments of ancient sedimentary rocks are determined by studying sedimentary textures and structures, examining fossils, and making comparisons with present-day depositional processes.

13. Many sediments and sedimentary rocks including sand, gravel, evaporites, coal, and banded iron formations are important natural resources. Most oil and natural gas are found in sedimentary rocks.

IMPORTANT TERMS

bed
biochemical sedimentary
   rock
carbonate rock
cementation
chemical sedimentary rock

compaction
cross-bedding
depositional environment
detrital sedimentary rock
evaporite
fossil

graded bedding
lithification
mud crack
paleocurrent
ripple mark
rounding

sediment
sedimentary rock
sedimentary structure
sorting
strata

REVIEW QUESTIONS

1. If an aggregate of sediment consists of particles that are all about the same size, it is said to be:
   a. _____ well sorted;
   b. _____ poorly rounded;
   c. _____ completely abraded;
   d. _____ sandstone;
   e. _____ lithified.

2. The process whereby dissolved mineral matter precipitates in the pore spaces of sediment and binds it together is:
   a. _____ compaction;
   b. _____ rounding;
   c. _____ bedding;
   d. _____ weathering;
   e. _____ cementation.

3. The most abundant detrital sedimentary rocks are:
   a. _____ limestones;
   b. _____ sandstones;
   c. _____ evaporites;
   d. _____ mudrocks;
   e. _____ arkoses.

4. Most limestones have a large component of calcite that was originally extracted from seawater by:
   a. _____ inorganic chemical reactions;
   b. _____ organisms;
   c. _____ evaporation;
   d. _____ chemical weathering;
   e. _____ lithification.

5. Dolostone is formed by the addition of _____ to limestone.
   a. _____ calcium;
   b. _____ carbonate;
   c. _____ magnesium;
   d. _____ iron;
   e. _____ sodium.

6. The most common evaporite rock is:
   a. _____ rock gypsum;
   b. _____ chert;
   c. _____ bituminous coal;
   d. _____ rock salt;
   e. _____ siltstone.

7. Which of the following can be used to determine paleocurrent direction?
   a. _____ mud cracks;

b. ____ graded bedding;

c. ____ cross-bedding;

d. ____ turbidity currents;

e. ____ grain size.

8. Turbidity current deposition is responsible for most:

a. ____ bedding planes;

b. ____ graded bedding:

c. ____ wave-formed ripple marks;

d. ____ mud cracks;

e. ____ limestone deposition.

9. Traps for petroleum and natural gas resulting from variations in the properties of sedimentary rocks are _____ traps.

a. ____ reservoir;

b. ____ stratigraphic;

c. ____ cap rock;

d. ____ structural;

e. ____ salt dome.

10. Explain why the sediment in wind-blown sand dunes is better sorted than that in glacial deposits.

11. What are the common chemical cements in sedimentary rocks, and how do they form?

12. Why is quartz the predominant mineral in most sandstones? When a sandstone contains at least 25% feldspar, what is it called?

13. What are the common evaporites, and how do they originate?

14. Briefly describe the origin of coal.

15. Name three sedimentary structures and explain how they form.

16. How can fossils be used to interpret ancient depositional environments?

17. What kinds of data do geologists use to determine depositional environment?

18. What is oil shale, and how can liquid oil be extracted from it?

◆◆-◆◆-◆◆-◆◆-◆◆-◆◆-◆◆-◆◆-◆◆-◆◆-◆◆-◆◆-◆◆-◆◆ ◆◆-◆◆-◆◆-◆◆-◆◆-◆◆-◆◆-◆◆-◆◆-◆◆-◆◆-◆◆-◆◆-◆◆

POINTS TO PONDER

1. As a field geologist, you encounter a rock unit consisting of well-sorted, well-rounded sandstone. In addition, the unit has large cross-beds and contains reptile footprints. What can you infer about the depositional environment?

2. The Earth's crust is estimated to contain 51% feldspars, 24% ferromagnesian silicates (biotite, pyroxenes, amphiboles), 12% quartz, and 13% other minerals. Considering these abundances, how can you explain the fact that quartz is by far the most common mineral in sandstones?

Chapter 8

METAMORPHISM AND METAMORPHIC ROCKS

OUTLINE

Marble quarry in northcentral Vermont. (Photo courtesy of R. V. Dietrich.)

Marble is a metamorphic rock formed from limestone or dolostone. Its homogeneity, softness, and various textures have made marble a favorite rock of sculptors throughout history. As the value of authentic marble sculptures has increased through the years, the number of forgeries has also increased. With the price of some marble sculptures in the millions of dollars, private collectors and museums need some means of assuring the authenticity of the work they are buying. Aside from the monetary considerations, it is important that forgeries not become part of the historical and artistic legacy of human endeavor.

Experts have traditionally relied on the artistic style and weathering characteristics to determine whether a marble sculpture is authentic or a forgery. Because marble is not very resistant to weathering, forgers have resorted to a variety of methods to produce the weathered appearance of an authentic ancient work. Now, however, using new techniques, geologists can distinguish a naturally weathered marble surface from one that has been artificially altered.

Although marbles result when the agents of metamorphism (heat, pressure, and fluid activity) are applied to carbonate rocks, the type of marble formed depends, in part, on the original composition of the parent carbonate rock as well as the type and intensity of metamorphism. Therefore, one way to authenticate a marble sculpture is to determine the origin of the marble itself. During the Preclassical, Greek, and Roman periods, the islands of Naxos, Thasos, and Paros in the Aegean Sea as well as the Greek mainland, Turkey, and Italy, were all sites of major marble quarries.

To identify the source of the marble in various sculptures, geologists employ a wide variety of analytical techniques. These include hand specimen and thin-section analysis of the marble, trace element analysis by X-ray fluorescence, stable isotopic ratio analysis for carbon and oxygen, and other more esoteric techniques.

The J. Paul Getty Museum in Malibu, California, has employed some of these techniques to help authenticate an ancient Greek kouros (a sculptured figure of a Greek youth) thought to have been carved around 530 B.C. (▷ Figure 8-1). The kouros was offered to the Getty Museum in 1984 for a reported price of $9 million. Some of its stylistic features, however, caused some experts to question its authenticity. Consequently, the museum had a variety of geochemical and mineralogical tests performed in an effort to determine the authenticity of the kouros.

Isotopic analysis of the weathered surface and fresh interior of the kouros confirmed that the marble probably came from the Cape Vathy quarries on the island of Thasos in the Aegean Sea. But these results did not prove the age of the kouros—it might still have been a forgery carved from marble taken from an archaeological site on the island.

The kouros was carved from dolomitic marble and its surface is covered with a complex thin crust (0.01 to 0.05 mm thick) consisting mostly of whewellite, a calcium oxalate monohydrate mineral. In order to be sure that the crust is the result of long-term weathering, and not a modern forgery, dolomitic marble samples were subjected to a variety of forgery techniques to try and replicate the surface of the kouros. Samples were soaked or boiled in

▷ **FIGURE 8-1** This Greek kouros, which stands 206 cm tall, has been the object of an intensive authentication study by the Getty Museum. Using several geological tests, experts have determined that the kouros was carved from dolomitic marble that probably came from the Cape Vathy quarries on Thasos.

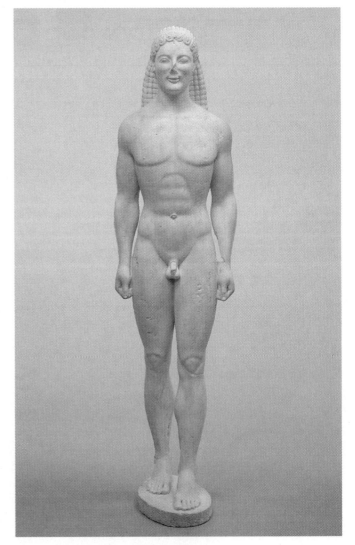

various mixtures for periods of time ranging from hours to months, and their surfaces treated and retreated to try and match the appearance of the weathered surface of the kouros. Such tests yielded only a few examples that appeared similar to the surface of kouros. Even those samples, however, were different when examined under high magnification or subjected to geochemical analysis. In fact, all of the samples clearly showed that they were the result of recent alteration and not long-term weathering processes.

While the scientific tests have been unable to unequivocably prove authenticity, they have shown that the weathered surface layer of the kouros bears more similarities to naturally occurring weathered surfaces than to known artificially produced surfaces. Furthermore, there is no evidence indicating that the surface alteration of the kouros is of modern origin.

In spite of intensive study by scientists, archaeologists, and art historians, opinion is still divided as to the authenticity of the Getty kouros. Most scientists accept that the kouros was carved sometime around 530 B.C., but most art historians are doubtful. Pointing to inconsistencies in its style of sculpture for that period, they argue that it is a modern forgery.

Regardless of the ultimate conclusion on the Getty kouros, geological testing to authenticate marble sculptures is now an important part of many museums' curatorial functions. In addition, a large body of data about the characteristics and origin of marble is being amassed as more sculptures and quarries are analyzed.

INTRODUCTION

Metamorphic rocks (from the Greek *meta* meaning change and *morpho* meaning shape) are the third major group of rocks. They result from the transformation of other rocks by metamorphic processes that usually occur beneath the Earth's surface (see Figure 1-11). During metamorphism, rocks are subjected to sufficient heat, pressure, and fluid activity to change their mineral composition and/or texture, thus forming new rocks. These transformations take place in the solid state, and the type of metamorphic rock formed depends on the original composition and texture of the parent rock, the agents of metamorphism, and the amount of time the parent rock was subjected to the effects of metamorphism.

A large portion of the Earth's continental crust is composed of metamorphic and igneous rocks. Together, they form the crystalline basement rocks that underlie the sedimentary rocks of a continent's surface. This basement rock is widely exposed in regions of the continents known as *shields,* which have been very stable during the past 600 million years (▷ Figure 8-2). Metamorphic rocks also constitute a sizable portion of the crystalline core of large mountain ranges. Some of the oldest known rocks, dated at 3.96 billion years from the Canadian Shield, are metamorphic, indicating they formed from even older rocks.

Why is it important to study metamorphic rocks? For one thing, they provide information about geological processes operating within the Earth and about the way these processes have varied through time. Furthermore, metamorphic rocks such as marble and slate are used as building materials, and certain metamorphic minerals are economically important. For example, talc is used in cosmetics, in the manufacture of paint, and as a lubricant, while asbestos is used for insulation and fireproofing (see Perspective 8-1).

THE AGENTS OF METAMORPHISM

The three agents of metamorphism are heat, pressure, and fluid activity. During metamorphism, the original rock undergoes change so as to achieve equilibrium with its new environment. The changes may result in the formation of new minerals and/or a change in the texture of the rock by the reorientation of the original minerals. In some instances the change is minor, and features of the parent rock can still be recognized. In other cases the rock changes so much that the identity of the parent rock can be determined only with great difficulty, if at all.

In addition to heat, pressure, and fluid activity, time is also important to the metamorphic process. Chemical reactions proceed at different rates and thus require different amounts of time to complete. Reactions involving silicate compounds are particularly slow, and because most metamorphic rocks are composed of silicate minerals, it is thought that metamorphism is a slow geologic process.

Heat

Heat is an important agent of metamorphism because it increases the rate of chemical reactions that may produce minerals different from those in the original rock. The heat may come from intrusive magmas or result from deep burial in the Earth's crust such as occurs during subduction along a convergent plate boundary.

When rocks are intruded by bodies of magma, they are subjected to intense heat that affects the surrounding rock; the most intense heating usually occurs adjacent to the magma body and gradually decreases with distance from the intrusion. The zone of metamorphosed rocks that forms in the country rock adjacent to an intrusive igneous body is usually rather distinct and easy to recognize.

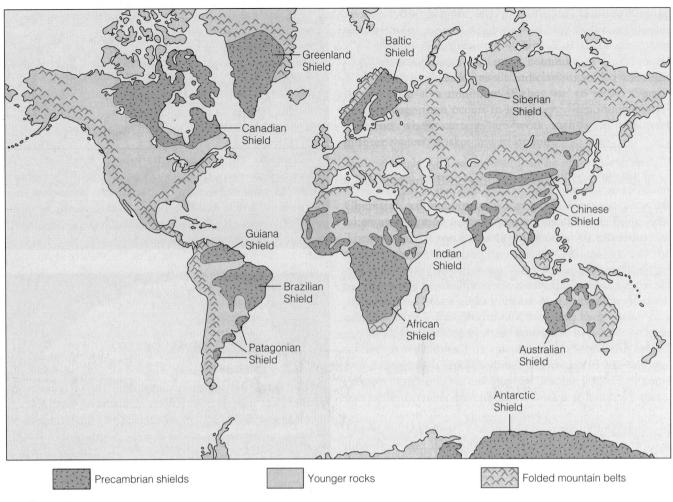

▶ **FIGURE 8-2** Shields of the world. Shields are the exposed portion of the crystalline basement rocks that underlie each continent; these areas have been very stable during the past 600 million years.

| Precambrian shields | Younger rocks | Folded mountain belts |

Pressure

When rocks are buried, they are subjected to increasingly greater **lithostatic pressure;** this pressure, which results from the weight of the overlying rocks, is applied equally in all directions (▶ Figure 8-3). As rocks are subjected to increasing lithostatic pressure with depth, the mineral grains within a rock may become more closely packed. Under such conditions, the minerals may *recrystallize;* that is, they may form smaller and denser minerals.

In addition to the lithostatic pressure resulting from burial, rocks may also experience **differential pressures** (▶ Figure 8-4). In this case, the pressures are not equal on all sides, and the rock is consequently distorted. Differential pressures typically occur during deformation associated with mountain building and can produce distinctive metamorphic textures and features.

Fluid Activity

In almost every region where metamorphism occurs, water and carbon dioxide (CO_2) are present in varying amounts along mineral grain boundaries or in the pore spaces of

▶ **FIGURE 8-3** Lithostatic pressure is applied equally in all directions in the Earth's crust due to the weight of the overlying rocks. Thus, pressure increases with depth.

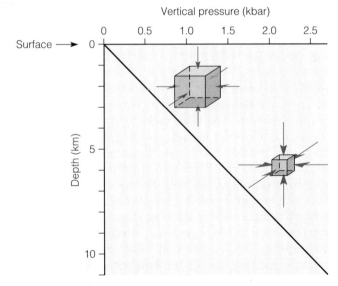

1 kilobar (kbar) = 1,000 bars
Atmospheric pressure at sea level = 1 bar

ASBESTOS

Asbestos (from the Latin, meaning unquenchable) is a general term applied to any silicate mineral that easily separates into flexible fibers (▷ Figure 1).

The combination of such features as non-combustibility and flexibility makes asbestos an important industrial material of considerable value. In fact, asbestos has more than 3,000 known uses, including brake linings, fireproof fabrics, and heat insulators.

Asbestos can be divided into two broad groups, serpentine and amphibole asbestos. *Chrysotile*, which is a hydrous magnesium silicate with the chemical formula $Mg_3Si_2O_5(OH)_4$, is the fibrous form of serpentine asbestos; it is the most valuable type and constitutes the bulk of all commercial asbestos. Chrysotile's strong, silky fibers are easily spun and can withstand temperatures up to 2,750°C.

The vast majority of chrysotile asbestos occurs in serpentine, a type of rock formed by the alteration of ultramafic igneous rocks such as peridotite under low- and medium-grade metamorphic conditions. Serpentine is thought to form from the alteration of olivine by hot, chemically active, residual fluids emanating from cooling magma. The chrysotile asbestos forms veinlets of fiber within the serpentine and may comprise up to 20% of the rock.

At least five varieties of amphibole asbestos are known, but *crocidolite,* a sodium-iron amphibole with the chemical formula $Na_2(Fe^{+3})_2(Fe^{+2})_3Si_8O_{22}(OH)_2$, is the most common. Crocidolite, which is also known as blue asbestos, is a long, coarse, spinning fiber that is stronger but more brittle than chrysotile and also less resistant to heat. The other varieties of amphibole asbestos have little commercial value and are used chiefly for insulation.

Crocidolite is found in such metamorphic rocks as slates and schists. It is thought that crocidolite forms by the solid-state alteration of other minerals within the high-temperature and high-pressure environment that results from deep burial.

In spite of its widespread use, the federal Environmental Protection Agency (EPA) has instituted a gradual ban on all new asbestos products. The ban was imposed because some forms of asbestos can cause lung cancer and scarring of the lungs if its fibers are inhaled. Because the EPA apparently paid little attention to the issue of risks versus benefits when it enacted this rule, the U.S. Fifth Circuit Court of Appeals overturned the EPA ban on asbestos in 1991.

The threat of lung cancer has also resulted in legislation mandating the removal of asbestos already in place in all public buildings, including all public and private schools. Recently, however, important questions have been raised concerning the threat posed by asbestos and the additional potential hazards that may arise from its improper removal.

The current policy (1993) of the EPA mandates that all forms of asbestos be treated as identical hazards. Yet studies indicate that only the amphibole forms constitute a known health hazard. Chrysotile, whose fibers tend to be curly, does not become lodged in the lungs. Furthermore, its fibers are generally soluble and disappear in tissue. In contrast, crocidolite has long, straight, thin fibers that penetrate the lungs and stay there. These fibers irritate the lung tissue and, over a long period of time, can lead to lung cancer. Thus, crocidolite, and not chrysotile, is overwhelmingly responsible for asbestos-related lung cancer. Because about 95% of the asbestos in place in the United States is chrysotile, many people are questioning whether the dangers from asbestos have been somewhat exaggerated.

The problem of asbestos contamination is a good example of how geology affects our lives and why a basic knowledge of science is important. As we have indicated, only the amphibole forms of asbestos have been shown to be a health hazard, and then only when its fibers are inhaled. Yet legislation has been passed requiring all asbestos to be removed from public buildings regardless of whether it is one of the hazardous amphibole varieties or benign chrysotile.

▷ **FIGURE 1** Hand specimen of chrysotile from Thetford, Quebec, Canada. Chrysotile is the fibrous form of serpentine asbestos.

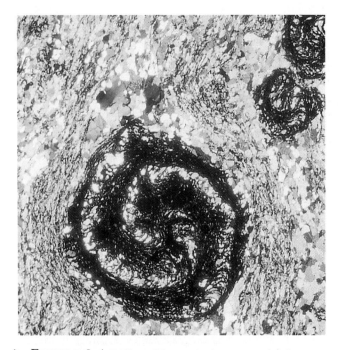

▷ **FIGURE 8-4** Differential pressure is pressure that is unequally applied to an object. Rotated garnets are a good example of the effects of differential pressure applied to a rock during metamorphism. These rotated garnets come from a calcareous schist of the Waits River Formation, north of Springfield, Vermont. (Photo courtesy of John L. Rosenfeld, University of California, Los Angeles.)

rocks. These fluids which may contain ions in solution enhance metamorphism by increasing the rate of chemical reactions. Under dry conditions, most minerals react very slowly, but when even small amounts of fluid are introduced, reaction rates increase, mainly because ions can move readily through the fluid and thus enhance chemical reactions and the formation of new minerals.

The following reaction provides a good example of how new minerals can be formed by **fluid activity**. Here, seawater moving through hot basaltic rock of the oceanic crust transforms olivine into the metamorphic mineral serpentine:

$$2Mg_2SiO_4 + 2H_2O \rightarrow Mg_3Si_2O_5(OH)_4 + MgO$$

| olivine | water | serpentine | carried away in solution |

The chemically active fluids that are part of the metamorphic process come primarily from three sources. The first is water trapped in the pore spaces of sedimentary rocks as they form. A second is the volatile fluid within magma. The third source is the dehydration of water-bearing minerals such as gypsum ($CaSO_4 \cdot 2H_2O$) and some clays.

⊚ TYPES OF METAMORPHISM

Three major types of metamorphism are recognized: *contact metamorphism* in which magmatic heat and fluids act to produce change; *dynamic metamorphism,* which is principally the result of high differential pressures associated with intense deformation; and *regional metamorphism,* which occurs within a large area and is caused primarily by mountain-building forces. Even though we will discuss each type of metamorphism separately, the boundary between them is not always distinct and depends largely on which of the three metamorphic agents was dominant.

Contact Metamorphism

Contact metamorphism takes place when a body of magma alters the surrounding country rock. At shallow depths an intruding magma raises the temperature of the surrounding rock, causing thermal alteration. Furthermore, the release of hot fluids into the country rock by the cooling intrusion can also aid in the formation of new minerals.

Important factors in contact metamorphism are the initial temperature and size of the intrusion as well as the fluid content of the magma and/or country rock. The initial temperature of an intrusion is controlled, in part, by its composition: mafic magmas are hotter than felsic magmas (see Chapter 4) and hence have a greater thermal effect on the rocks directly surrounding them. The size of the intrusion is also important. In the case of small intrusions, such as dikes and sills, usually only those rocks in immediate contact with the intrusion are affected. Because large intrusions, such as batholiths, take a long time to cool, the increased temperature in the surrounding rock may last long enough for a larger area to be affected.

Fluids also play an important role in contact metamorphism. Many magmas are wet and contain hot, chemically active fluids that may emanate into the surrounding rock. These fluids can react with the rock and aid in the formation of new minerals. In addition, the country rock may contain pore fluids that, when heated by the magma, also increase reaction rates.

Temperatures can reach nearly 900°C adjacent to an intrusion, but they gradually decrease with distance. The effects of such heat and the resulting chemical reactions usually occur in concentric zones known as **aureoles** (▷ Figure 8-5). The boundary between an intrusion and its aureole may be either sharp or transitional (▷ Figure 8-6).

Metamorphic aureoles vary in width depending on the size, temperature, and composition of the intrusion as well as the composition of the surrounding country rock. Typically, large intrusive bodies have several metamorphic zones, each characterized by distinctive mineral assemblages indicating the decrease in temperature with distance from the intrusion (Figure 8-5). The zone closest to the intrusion, and hence subject to the highest temperatures, may contain high-temperature metamorphic minerals (that is, minerals in equilibrium with the higher temperature environment) such as sillimanite. The outer zones may be characterized by lower-temperature metamorphic minerals such as chlorite, talc, and epidote.

The formation of new minerals by contact metamorphism

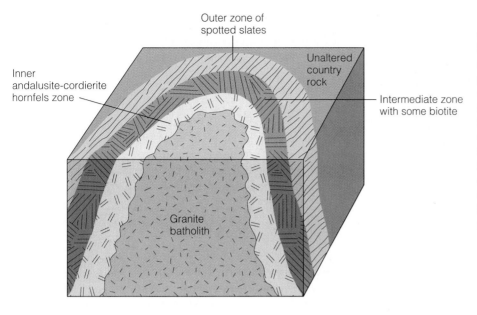

Outer zone of
spotted slates

Unaltered
country
rock

Inner
andalusite-cordierite
hornfels zone

Intermediate zone
with some biotite

Granite
batholith

➤ FIGURE 8-5 A metamorphic aureole typically surrounds many igneous intrusions. The metamorphic aureole associated with this idealized granite batholith contains three zones of mineral assemblages reflecting the decreases in temperature with distance from the intrusion. An andalusite-cordierite hornfels forms the inner zone adjacent to the batholith. This is followed by an intermediate zone of extensive recrystallization in which some biotite develops; farthest from the intrusion is the outer zone, which is characterized by spotted slates.

➤ FIGURE 8-6 A sharp, clearly defined boundary occurs between the intruding light-colored igneous rock on the left and the dark-colored metamorphosed country rock on the right. The intrusion is part of the Peninsular Ranges Batholith, east of San Diego, California. (Photo courtesy of David J. Matty.)

depends not only on proximity to the intrusion, but also on the composition of the country rock. Shales, mudstones, impure limestones, and impure dolostones are particularly susceptible to the formation of new minerals by contact metamorphism, whereas pure sandstones or pure limestones typically are not.

Two types of contact metamorphic rocks are generally recognized: those resulting from baking of country rock and those altered by hot solutions. Many of the rocks resulting from contact metamorphism have the texture of porcelain; that is, they are hard and fine grained. This is particularly true for rocks with a high clay content, such as shale. Such texture results because the clay minerals in the rock are baked, just as a clay pot is baked when fired in a kiln.

During the final stages of cooling when an intruding magma begins to crystallize, large amounts of hot, watery solutions are often released. These solutions may react with the country rock and produce new metamorphic minerals. This process, which usually occurs near the Earth's surface, is called *hydrothermal alteration,* and may result in valuable mineral deposits. Geologists think that many of the world's ore deposits result from the migration of metallic ions in hydrothermal solutions. Examples include copper, gold, iron ores, tin, and zinc in various localities including Australia, Canada, China, Cyprus, Finland, Russia, and the western United States.

Dynamic Metamorphism

Most **dynamic metamorphism** is associated with fault (fractures along which movement has occurred) zones where rocks are subjected to high differential pressures. The metamorphic rocks resulting from pure dynamic metamorphism are called *mylonites,* and they are typically restricted to narrow zones adjacent to faults. Mylonites are hard, dense, fine-grained rocks, many of which are characterized by thin laminations. Tectonic settings where mylonites occur include the Moine Thrust Zone in northwest Scotland and portions of the San Andreas fault in California.

Regional Metamorphism

Most metamorphic rocks result from **regional metamorphism,** which occurs over a large area and is usually caused by tremendous temperatures, pressures, and deformation within the deeper portions of the Earth's crust. Regional metamorphism is most obvious along convergent plate margins where rocks are intensely deformed and recrystallized during convergence and subduction. Within these metamorphic rocks, there is usually a gradation of metamorphic intensity from areas that were subjected to the most intense pressures and/or highest temperatures to areas of lower pressures and temperatures. Such a gradation in metamor-

TABLE 8-1 Metamorphic Zones and Their Mineral Assemblages for Different Country Rock Types

Metamorphic Grade	Metamorphic Zone for Clay-Rich Rocks	Mineral Assemblage Produced for Different Country Rocks		
		Mudrocks	*Limestones*	*Mafic Igneous Rocks*
Increasing				
Low	Chlorite	Chlorite,★ quartz, muscovite, plagioclase	Chlorite,★ calcite or dolomite, plagioclase	Chlorite,★ plagioclase
	Biotite	Biotite,★ quartz, plagioclase		
Medium	Garnet	Garnet,★ mica, quartz, plagioclase	Garnet,★ epidote, hornblende, calcite	Garnet,★ chlorite, epidote, plagioclase
	Staurolite	Staurolite,★ mica, garnet, quartz, plagioclase	Garnet, hornblende,★ plagioclase	
High	Kyanite	Kyanite,★ mica, garnet, quartz, plagioclase		
metamorphism				Hornblende,★ plagioclase
	Sillimanite	Sillimanite,★ garnet, mica, quartz, plagioclase	Garnet, augite,★ plagioclase	

★Index mineral.

phism can be recognized by the metamorphic minerals that are present.

Regional metamorphism is not just confined to convergent margins. It also occurs in areas where plates diverge, though usually at much shallower depths because of the high geothermal gradient associated with these areas.

From field studies and laboratory experiments, certain minerals are known to form only within specific temperature and pressure ranges. Such minerals are known as **index minerals** because their presence allows geologists to recognize low-, intermediate-, and high-grade metamorphic zones (◉ Table 8-1). A typical progression of index minerals forming primarily in rocks that were originally clay rich involves the sequential formation of the following minerals: chlorite → biotite → amphibole → staurolite → sillimanite.

Different rock compositions, though, develop different index minerals. When sandy dolomites are metamorphosed

they produce an entirely different set of index minerals. Therefore, a specific set of index minerals commonly forms in particular rock types as metamorphism progresses.

CLASSIFICATION OF METAMORPHIC ROCKS

For purposes of classification, metamorphic rocks are commonly divided into two groups: those exhibiting a *foliated texture* and those with a *nonfoliated texture* (◉ Table 8-2).

Foliated Metamorphic Rocks

Rocks subjected to heat and differential pressure during metamorphism typically have minerals arranged in a parallel fashion that gives them a **foliated texture** (▷ Figure 8-7). The size and shape of the mineral grains determine whether

TABLE 8-2 Classification of Common Metamorphic Rocks

Texture	Metamorphic Rock	Typical Minerals	Metamorphic Grade	Characteristics of Rocks	Parent Rock
Foliated	Slate	Clays, micas, chlorite	Low	Fine-grained, splits easily into flat pieces	Mudrocks, claystones, volcanic ash
	Phyllite	Fine-grained quartz, micas, chlorite	Low to medium	Fine-grained, glossy or lustrous sheen	Mudrocks
	Schist	Micas, chlorite, quartz, talc, hornblende, garnet, staurolite, graphite	Low to high	Distinct foliation, minerals visible	Mudrocks, carbonates, mafic igneous rocks
	Gneiss	Quartz, feldspars, hornblende, micas	High	Segregated light and dark bands visible	Mudrocks, sandstones, felsic igneous rocks
	Amphibolite	Hornblende, plagioclase	Medium to high	Dark-colored, weakly foliated	Mafic igneous rocks
	Migmatite	Quartz, feldspars, hornblende, micas	High	Streaks or lenses of granite intermixed with gneiss	Felsic igneous rocks mixed with sedimentary rocks
Nonfoliated	Marble	Calcite, dolomite	Low to high	Interlocking grains of calcite or dolomite, reacts with HCl	Limestone or dolostone
	Quartzite	Quartz	Medium to high	Interlocking quartz grains, hard, dense	Quartz sandstone
	Greenstone	Chlorite, epidote, hornblende	Low to high	Fine-grained, green color	Mafic igneous rocks
	Hornfels	Micas, garnets, andalusite, cordierite, quartz	Low to medium	Fine-grained, equidimensional grains, hard, dense	Mudrocks
	Anthracite	Carbon	High	Black, lustrous, subconcoidal fracture	Coal

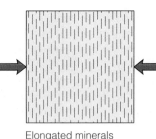

Random arrangement of elongated minerals before pressure is applied to two sides

Elongated minerals arranged in a parallel fashion as a result of pressure applied to two sides

(a)

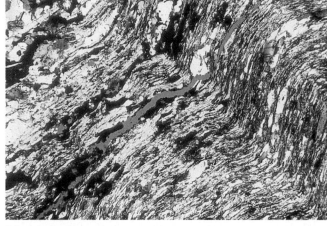

(b)

➤ **FIGURE 8-7** (*a*) When rocks are subjected to differential pressure, the mineral grains are typically arranged in a parallel fashion, producing a foliated texture. (*b*) Photomicrograph of a metamorphic rock with a foliated texture showing the parallel arrangement of mineral grains.

(a)

(b)

➤ **FIGURE 8-8** (*a*) Hand specimen of slate. (*b*) This panel of Arvonia Slate from Albemarne Slate Quarry, Virginia, shows bedding (upper right to lower left) at an angle to the slaty cleavage. (Photo (*a*) courtesy of Sue Monroe; photo (*b*) courtesy of R. V. Dietrich.)

the foliation is fine or coarse. If the foliation is such that the individual grains cannot be recognized without magnification, the rock is said to be slate (➤ Figure 8-8). A coarse foliation results when granular minerals such as quartz and feldspar are segregated into roughly parallel and streaky zones that differ in composition and color as in a gneiss (➤ Figure 8-10). Foliated metamorphic rocks can be arranged in order of increasingly coarse grain size and perfection of foliation.

Slate is a very fine-grained metamorphic rock that commonly exhibits *slaty cleavage* (Figure 8-8b). Slate is the result of low-grade regional metamorphism of shale or, more rarely, volcanic ash. Because it can easily be split along cleavage planes into flat pieces, slate is an excellent rock for roofing and floor tiles, billiard and pool table tops, and blackboards. The different colors of most slates are caused by minute amounts of graphite (black), iron oxide (red and purple), and/or chlorite (green).

Phyllite is similar in composition to slate, but is coarser grained. The minerals, however, are still too small to be identified without magnification. Phyllite can be distin-

guished from slate by its glossy or lustrous sheen. It represents an intermediate grain size between slate and schist.

Schist is most commonly produced by regional metamorphism. The type of schist formed depends on the intensity of metamorphism and the character of the parent rock (➤ Figure 8-9). Metamorphism of many rock types can yield schist, but most schist appears to have formed from clay-rich sedimentary rocks (Table 8-2).

➤ **FIGURE 8-9** Garnet-mica schist. (Photo courtesy of Sue Monroe.)

➤ **FIGURE 8-10** Gneiss is characterized by segregated bands of light and dark minerals. This folded gneiss is exposed near Wawa, Ontario, Canada.

All schists contain more than 50% platy and elongated minerals, all of which are large enough to be clearly visible. Their mineral composition imparts a *schistosity* or *schistose foliation* to the rock that usually produces a wavy type of parting when split. Schistosity is common in low- to high-grade metamorphic environments, and each type of schist is known by its most conspicuous mineral or minerals, such as mica schist, chlorite schist, or talc schist.

Gneiss is a metamorphic rock that is streaked or has segregated bands of light and dark minerals. Gneisses are composed mostly of granular minerals such as quartz and/or feldspar with lesser percentages of platy or elongated minerals such as micas or amphiboles (Figure 8-10). Quartz and feldspar are the principal light-colored minerals, while biotite and hornblende are the typical dark-colored minerals. Most gneiss breaks in an irregular manner, much like coarsely crystalline nonfoliated rocks.

Most gneiss probably results from recrystallization of clay-rich sedimentary rocks during regional metamorphism (Table 8-2). Gneiss also can form from igneous rocks such as granite or older metamorphic rocks.

Another fairly common foliated metamorphic rock is *amphibolite*. It is a dark-colored rock and composed mainly of hornblende and plagioclase. The alignment of the hornblende crystals produces a slightly foliated texture. Many amphibolites result from medium- to high-grade metamorphism of such ferromagnesian mineral-rich igneous rocks as basalt.

In some areas of regional metamorphism, exposures of "mixed rocks" having both igneous and high-grade metamorphic characteristics are present. In these rocks, called *migmatites,* streaks or lenses of granite are usually intermixed with high-grade ferromagnesian-rich metamorphic rocks, imparting a wavy appearance to the rock (➤ Figure 8-11).

Most migmatites are thought to be the product of extremely high-grade metamorphism, and several models for

their origin have been proposed. Part of the problem in determining the origin of migmatites is explaining how the granitic component formed. According to one model, the granitic magma formed in place by the partial melting of rock during intense metamorphism. Such an origin is possible providing that the host rocks contained quartz and feldspars and that water was present. Another possibility is that the granitic components formed by the redistribution of minerals by recrystallization in the solid state, that is, by pure metamorphism.

Nonfoliated Metamorphic Rocks

In some metamorphic rocks, the mineral grains do not show a discernible preferred orientation. Instead, these rocks consist of a mosaic of roughly equidimensional minerals and are characterized as having a **nonfoliated texture** (➤ Figure 8-12). Most nonfoliated metamorphic rocks result from contact or regional metamorphism of rocks in which no platy or elongate minerals are present. Frequently, the only indication that a granular rock has been metamorphosed is the large grain size resulting from recrystallization. Nonfoliated metamorphic rocks are generally of two types: those composed mainly of only one mineral, for example, marble or quartzite; and those in which the different mineral grains are too small to be seen without magnification, such as greenstone and hornfels.

Marble is a well-known metamorphic rock composed predominantly of calcite or dolomite; its grain size ranges from fine to coarsely granular (➤ Figure 8-13). Marble results from either contact or regional metamorphism of limestones or dolostones (Table 8-2). Pure marble is snowy white or bluish, but varieties of all colors exist because of the presence of mineral impurities in the parent sedimentary rock. The softness of marble, its uniform texture, and its

> **FIGURE 8-11** Migmatites consist of high-grade metamorphic rock intermixed with streaks or lenses of granite. This Precambrian(?) migmatite is exposed at Thirty Thousand Islands of Georgian Bay, Lake Huron, Ontario, Canada. (Photo by Ed Bartram, courtesy of R. V. Dietrich.)

> **FIGURE 8-12** Nonfoliated textures are characterized by a mosaic of roughly equidimensional minerals as in this photomicrograph of marble.

various colors have made it the favorite rock of builders and sculptors throughout history (see the Prologue).

Quartzite is a hard, compact rock formed from quartz sandstone under medium-to-high-grade metamorphic conditions during contact or regional metamorphism (➢ Figure 8-14). Because recrystallization is so complete, metamorphic quartzite is of uniform strength and therefore usually breaks across the component quartz grains rather than around them when it is struck. Pure quartzite is white, but iron and other impurities commonly impart a reddish or other color to it. Quartzite is commonly used as foundation material for road and railway beds.

The name *greenstone* is applied to any compact, dark-green, altered, mafic igneous rock that formed under low-to-high-grade metamorphic conditions. The green color results from the presence of chlorite, epidote, and hornblende.

Hornfels is a fine-grained, nonfoliated metamorphic rock resulting from contact metamorphism; it is composed of various equidimensional mineral grains. The composition of hornfels is directly dependent upon the composition of the parent rock, and many compositional varieties are known. The majority of hornfels, however, are apparently derived from contact metamorphism of clay-rich sedimentary rocks or impure dolostones.

Anthracite is a black, lustrous, hard coal that contains a high percentage of fixed carbon and a low percentage of other elements. It usually forms from the metamorphism of various types of coals by heat and pressure and is thus considered by many geologists to be a metamorphic rock.

Metamorphism →

▷ **FIGURE 8-13** Marble results from the metamorphism of the sedimentary rocks limestone and dolostone. (Photos courtesy of Sue Monroe.)

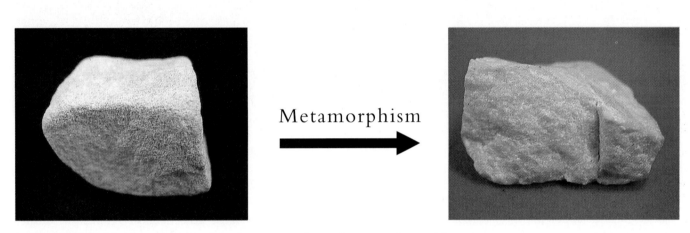

Metamorphism →

▷ **FIGURE 8-14** Quartzite results from the metamorphism of quartz sandstone. (Photos courtesy of Sue Monroe.)

⌘ METAMORPHIC ZONES

The first systematic study of metamorphic zones was conducted during the late 1800s by George Barrow and other British geologists working in the Dalradian schists of the southwestern Scottish Highlands. Here, clay-rich sedimentary rocks have been subjected to regional metamorphism, and the resulting metamorphic rocks can be divided into different zones based on the presence of distinctive silicate mineral assemblages. These mineral assemblages, each recognized by the presence of one or more index minerals, indicate different degrees of metamorphism. The index minerals Barrow and his associates chose to represent increasing metamorphic intensity were chlorite, biotite, garnet, staurolite, kyanite, and sillimanite (Table 8-1). Note that these are the metamorphic minerals produced from clay-rich sedimentary rocks. Other mineral assemblages and index minerals are produced from rocks with different original compositions (Table 8-1).

The successive appearance of metamorphic index miner-

als indicates gradually increasing or decreasing intensity of metamorphism. Going from lower- toward higher-grade zones, the first appearance of a particular index mineral indicates the location of the minimum temperature and pressure conditions needed for the formation of that mineral. When the locations of the first appearances of that index mineral are connected on a map, the result is a line of equal metamorphic intensity or an *isograd*. The region between isograds is known as a **metamorphic zone**. By noting the occurrence of metamorphic index minerals, geologists can construct a map showing the metamorphic zones of an entire area (▷ Figure 8-15).

Numerous studies of different metamorphic rocks have demonstrated that while the texture and composition of any rock may be altered by metamorphism, the overall chemical composition may be little changed. Thus, the different mineral assemblages found in increasingly higher grade metamorphic rocks derived from the same parent rock result from changes in temperature and pressure (Table 8-1).

➢ **FIGURE 8-15** Metamorphic zones in northern Michigan. The zones in this region are based on the presence of distinctive silicate mineral assemblages resulting from the metamorphism of sedimentary rocks during an interval of mountain building and minor granitic intrusion during the Proterozoic Eon, about 1.5 billion years ago.

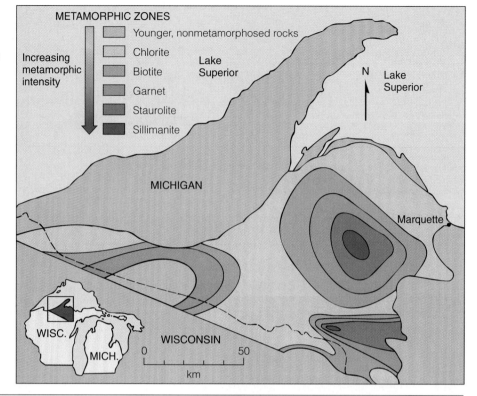

METAMORPHISM AND PLATE TECTONICS

Although metamorphism is associated with all three types of plate boundaries (see Figure 1-10), it is most common along convergent plate margins. Metamorphic rocks form at convergent plate boundaries because temperature and pressure increase as a result of plate collisions.

➢ Figure 8-16 illustrates the various temperature-pressure regimes that are produced along an oceanic-continental convergent plate boundary. When an oceanic plate collides with a continental plate, tremendous pressure is generated as the oceanic plate is subducted. Because rock is a poor heat conductor, the cold descending oceanic plate heats

➢ **FIGURE 8-16** Various temperature-pressure conditions are produced along an oceanic-continental convergent plate boundary.

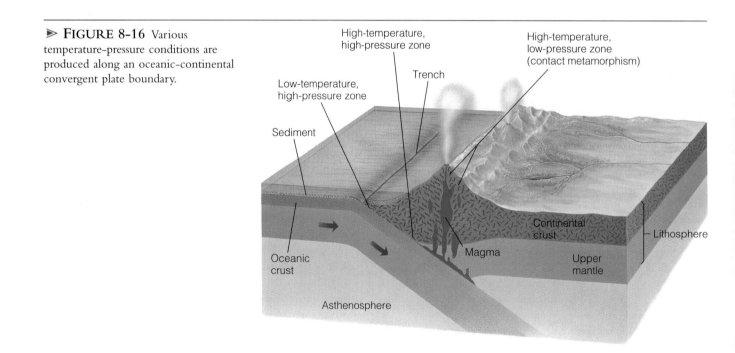

TABLE 8-3 The Main Ore Deposits Resulting from Contact Metamorphism

Ore Deposit	Major Mineral	Formula	Use
Copper	Bornite Chalcopyrite	Cu_5FeS_4 $CuFeS_2$	Important sources of copper, which is used in various aspects of manufacturing, transportation, communications, and construction.
Iron	Hematite Magnetite	Fe_2O_3 Fe_3O_4	Major sources of iron for manufacture of steel, which is used in nearly every form of construction, manufacturing, transportation, and communications.
Lead	Galena	PbS	Chief source of lead, which is used in batteries, pipes, solder, and elsewhere where resistance to corrosion is required.
Tin	Cassiterite	SnO_2	Principal source of tin, which is used for tin plating, solder, alloys, and chemicals.
Tungsten	Scheelite Wolframite	$CaWO_4$ $(Fe, Mn)WO_4$	Chief sources of tungsten, which is used in hardening metals and manufacturing carbides.
Zinc	Sphalerite	$(Zn, Fe)S$	Major source of zinc, which is used in batteries and in galvanizing iron and making brass.

very slowly, and metamorphism is caused mostly by increasing pressure with depth. As subduction continues, both temperature and pressure increase with depth and can result in high-grade metamorphic rocks. Eventually, the descending plate begins to melt and generates a magma that moves upward. This rising magma may alter the surrounding rock by contact metamorphism, producing migmatites in the deeper portions of the crust and hornfels at shallower depths. Such an environment is characterized by high temperatures and low to medium pressures.

While metamorphism is most common along convergent plate margins, many divergent plate boundaries are characterized by contact metamorphism. Rising magma at mid-oceanic ridges heats the adjacent rocks, producing contact metamorphic minerals and textures. In addition to contact metamorphism, fluids emanating from the rising magma—and from the reaction of the magma and sea water—very commonly produce hydrothermal solutions that may precipitate minerals of economic value.

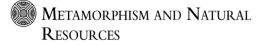

METAMORPHISM AND NATURAL RESOURCES

Many metamorphic rocks and minerals are valuable natural resources. While these resources include various types of ore deposits, the two most familiar and widely used metamorphic rocks, as such, are marble and slate, which, as previously discussed, have been used for centuries in a variety of ways.

Many ore deposits result from contact metamorphism during which hot, ion-rich fluids migrate from igneous intrusions into the surrounding rock, thereby producing rich ore deposits. The most common sulfide ore minerals associated with contact metamorphism are bornite, chalcopyrite, galena, pyrite, and sphalerite, while two common oxide ore minerals are hematite and magnetite. Tin and tungsten are also important ores associated with contact metamorphism (● Table 8-3).

Other economically important metamorphic minerals include talc for talcum powder; graphite for pencils and dry lubricants; garnets and corundum, which are used as abrasives or gemstones, depending on their quality; and andalusite, kyanite, and sillimanite, all of which are used in the manufacture of high-temperature porcelains and temperature-resistant minerals for products such as sparkplugs and the linings of furnaces.

CHAPTER SUMMARY

1. Metamorphic rocks result from the transformation of other rocks, usually beneath the Earth's surface, as a consequence of one or a combination of three agents: heat, pressure, and fluid activity.

2. Heat for metamorphism comes from intrusive magmas or deep burial. Pressure is either lithostatic or differential. Fluids trapped in sedimentary rocks or emanating from intruding magmas can enhance chemical changes and the formation of new minerals.

3. The three major types of metamorphism are contact, dynamic, and regional.

4. Metamorphic rocks are classified primarily according to their texture. In a foliated texture, platy minerals have a preferred orientation. A nonfoliated texture does not exhibit any discernible preferred orientation of the mineral grains.

5. Foliated metamorphic rocks can be arranged in order of grain size and/or perfection of their foliation. Slate is very fine grained, followed by phyllite and schist; gneiss displays segregated bands of minerals. Amphibolite is another fairly common foliated metamorphic rock.

6. Marble, quartzite, greenstone, and hornfels are common nonfoliated metamorphic rocks.

7. Metamorphic rocks can be arranged into metamorphic zones based on the conditions of metamorphism.

8. Metamorphism can occur along all three kinds of plate boundaries, but most commonly occurs at convergent plate margins.

9. Metamorphic rocks formed near the Earth's surface along an oceanic-continental plate boundary result from low-temperature, high-pressure conditions. As a subducted oceanic plate descends, it is subjected to increasingly higher temperatures and pressures that result in higher-grade metamorphism.

10. Many metamorphic rocks and minerals, such as marble, slate, graphite, talc, and asbestos, are valuable natural resources.

IMPORTANT TERMS

aureole
contact metamorphism
differential pressure
dynamic metamorphism

fluid activity
foliated texture
heat

index mineral
lithostatic pressure
metamorphic rock

metamorphic zone
nonfoliated texture
regional metamorphism

REVIEW QUESTIONS

1. The metamorphic rock formed from limestone is:
   a. ____ quartzite;
   b. ____ hornfels;
   c. ____ marble;
   d. ____ slate;
   e. ____ greenstone.

2. From which of the following rock groups can metamorphic rocks form?
   a. ____ plutonic;
   b. ____ sedimentary;
   c. ____ metamorphic;
   d. ____ volcanic;
   e. ____ all of these.

3. Which of the following is not an agent of metamorphism?
   a. ____ foliation;
   b. ____ heat;
   c. ____ pressure;
   d. ____ fluid activity;
   e. ____ none of these.

4. Pressure exerted equally in all directions on an object is:
   a. ____ differential;

   b. ____ directional;
   c. ____ lithostatic;
   d. ____ shear;
   e. ____ none of these.

5. In which type of metamorphism are magmatic heat and fluids the primary agents of change?
   a. ____ contact;
   b. ____ dynamic;
   c. ____ regional;
   d. ____ local;
   e. ____ thermodynamic.

6. Concentric zones surrounding an igneous intrusion are:
   a. ____ metamorphic layers;
   b. ____ thermodynamic rings;
   c. ____ aureoles;
   d. ____ hydrothermal regions;
   e. ____ none of these.

7. Which type of metamorphism produces the majority of metamorphic rocks?
   a. ____ contact;
   b. ____ dynamic;
   c. ____ regional;

   d. ____ lithostatic;
   e. ____ lithospheric.

8. Which of the following metamorphic rocks displays a foliated texture?
   a. ____ marble;
   b. ____ quartzite;
   c. ____ greenstone;
   d. ____ schist;
   e. ____ hornfels.

9. Mixed rocks containing the characteristics of both igneous and high-grade metamorphic rocks are:
   a. ____ mylonites;
   b. ____ migmatites;
   c. ____ amphibolites;
   d. ____ hornfels;
   e. ____ greenstones.

10. Metamorphic zones:
   a. ____ are characterized by distinctive mineral assemblages;
   b. ____ are separated from each other by isograds;
   c. ____ reflect a metamorphic grade;

d. _____ all of these;

e. _____ none of these.

11. Along what type of plate boundary is metamorphism most common?

a. _____ convergent;

b. _____ divergent;

c. _____ transform;

d. _____ mantle plume;

e. _____ static.

12. Metamorphic rocks form a significant proportion of:

a. _____ shields;

b. _____ the cores of mountain ranges;

c. _____ oceanic crust;

d. _____ answers (a) and (b);

e. _____ answers (b) and (c).

13. Where does contact metamorphism occur, and what type of changes does it produce?

14. What are aureoles? How can they be used to determine the effects of metamorphism?

15. What is regional metamorphism, and under what conditions does it occur?

16. Describe the two types of metamorphic texture, and explain how they may be produced.

17. Name the three common nonfoliated rocks, and describe their characteristics.

18. What is the difference between a metamorphic zone and a metamorphic facies?

◆◆

POINTS TO PONDER

1. What specific features of foliated metamorphic rocks make them unsuitable as the foundation rock for a dam? Are there any metamorphic rocks that would make a good foundation? Why?

2. If you were in charge of the EPA, how would you formulate a policy that balances the risks versus the benefits of removing asbestos from public buildings? What role would geologists play in this policy?

EARTHQUAKES AND THE EARTH'S INTERIOR

OUTLINE

This house in Pacific Palisades, California was destroyed by the Northridge earthquake of January 17, 1994.

In the early morning hours of January 17, 1994, southern California was rocked by a devastating earthquake measuring 6.7 on the Richter Magnitude Scale. The earthquake was centered on the city of Northridge just north of Los Angeles. When the initial shock was over, about 55 people had been killed and thousands were injured. Nine freeways had been severely damaged, thousands of homes and buildings were either destroyed or damaged (▷ Figure 9-1), and 250 ruptured gas lines ignited numerous fires. Thousands of aftershocks followed the main earthquake; many of them added to the damage, which has been estimated at $15 to $30 billion.

Geologists know that southern California is riddled with active faults capable of producing strong earthquakes, but it was a previously unknown fault 15 km beneath Northridge that caused the tragic January 1994 earthquake. Many people are aware that movements on the San Andreas fault and its subsidiary faults have been responsible for many earthquakes in southern California. But only recently have geologists begun to realize that a network of interconnected faults that do not break the surface may be causing much of the earthquake activity in the Los Angeles area. Earthquakes on these hidden faults, just as on the obvious ones, are related to movements between the North American and Pacific plates.

How did the Los Angeles area fare during and after this most recent earthquake? Older, unreinforced masonry buildings and more modern wood-frame apartments built over ground-floor garages generally sustained the most damage. Structures built to the stricter building standards in force during the last five years typically escaped unscathed or with only minor damage.

The state transportation department (Caltrans) instituted a program of reinforcing bridges and freeway overpasses soon after the 1971 Sylmar earthquake and began a second phase of reinforcing structures after the 1989 Loma Prieta earthquake. Most of the reinforced structures suffered little or no damage during the Northridge earthquake, but several awaiting reinforcing collapsed, including part of the Santa Monica Freeway, the busiest highway in the world.

Regulations designed to protect utility lines have not yet been implemented, in part because of cost. As a consequence, numerous power and gas lines were ruptured, and three water aqueducts were severed, cutting off water to at least 40,000 people and power to an estimated 3.1 million residents.

Emergency measures had been well planned and rescue operations went well. Shelters were established, people fed and clothed, and disaster relief offices opened in the area shortly after the earthquake. Based on their experiences in other earthquakes, rescue agencies had invested in better rescue equipment including high-pressure air bags that can lift up to 72 tons, fiber-optics search cameras, and specially trained dogs that can sniff out buried victims. This experience and the up-to-date equipment helped rescue workers locate and extricate victims from the earthquake wreckage.

The Northridge earthquake was tragic, but it was not the "Big One" that Californians have been waiting for. And even though rescue and relief agencies operated efficiently, the earthquake reminds us that much still remains to be done in terms of earthquake prediction and preparedness.

INTRODUCTION

As one of nature's most frightening and destructive phenomena, earthquakes have always aroused a sense of fear. Even when an earthquake begins, there is no way to tell how strong the shaking will be or how long it will last. It is estimated that more than 13 million people have died as a result of earthquakes during the past 4,000 years, and approximately 1 million of these deaths occurred during the last century (◉ Table 9-1).

An **earthquake** is defined as the vibration of the Earth caused by the sudden release of energy beneath the Earth's surface, usually as a result of displacement of rocks along fractures known as faults. Following an earthquake, adjustments along a fault commonly generate a series of earthquakes referred to as **aftershocks.** Most of these are smaller than the main shock, but they can cause considerable damage to already weakened structures. Indeed, much of the

destruction from the 1755 earthquake in Lisbon, Portugal, was caused by aftershocks. After a small earthquake, aftershock activity usually ceases within a few days, but it may persist for months following a large earthquake.

Early humans and cultures explained earthquakes in a more imaginative and colorful way, often attributing them to the movements of some organism on which the Earth rested. In Japan, the organism was a giant catfish; in Mongolia, a giant frog; in China, an ox; in India, a giant mole; in parts of South America, a whale; and to the Algonquin Indians of North America, an immense tortoise.

The Greek philosopher Aristotle offered what he considered to be a natural explanation for earthquakes. He believed that atmospheric winds were drawn into the Earth's interior where they caused fires and swept around the various subterranean cavities trying to escape. It was this movement of underground air that caused earthquakes and occasional

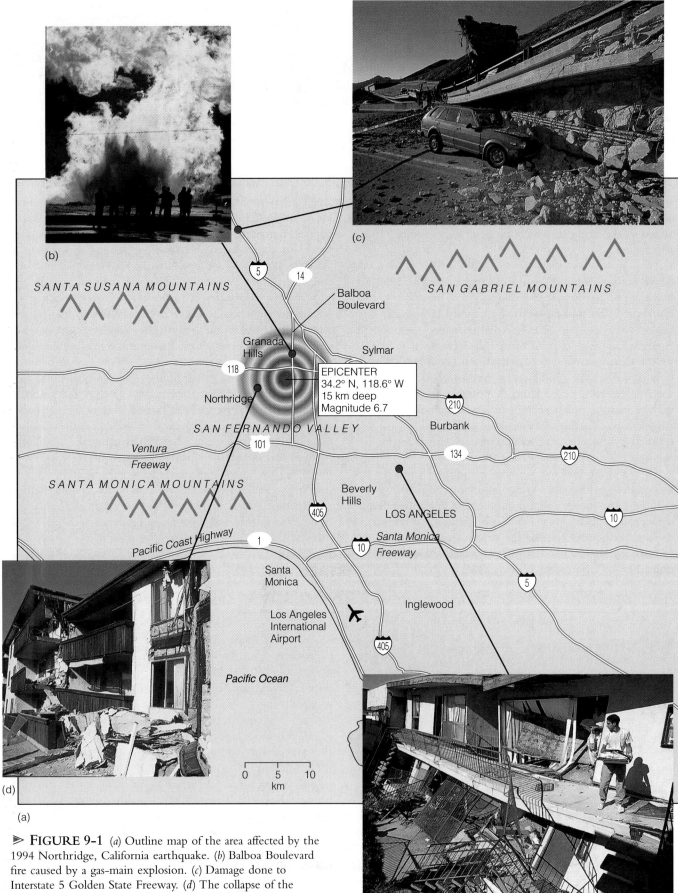

(b)

(c)

SANTA SUSANA MOUNTAINS

SAN GABRIEL MOUNTAINS

Balboa
Boulevard

Granada
Hills

Sylmar

118

EPICENTER
34.2° N, 118.6° W
15 km deep
Magnitude 6.7

Northridge

210

SAN FERNANDO VALLEY

Burbank

Ventura
Freeway

101

134

210

SANTA MONICA MOUNTAINS

Beverly
Hills

405

LOS ANGELES

Santa Monica
Freeway

10

Pacific Coast Highway

1

Santa
Monica

10

Inglewood

5

Los Angeles
International
Airport

405

Pacific Ocean

0   5   10
km

(d)

(a)

(e)

➤ **FIGURE 9-1** (*a*) Outline map of the area affected by the
1994 Northridge, California earthquake. (*b*) Balboa Boulevard
fire caused by a gas-main explosion. (*c*) Damage done to
Interstate 5 Golden State Freeway. (*d*) The collapse of the
Northridge Meadows apartments killed 16 people. (*e*) Severe
damage to a Sherman Oaks apartment building.

TABLE 9-1 Some Significant Earthquakes

Year	Location	Magnitude (Estimated before 1935)	Deaths (Estimated)
1556	China (Shanxi Province	8.0	1,000,000
1755	Portugal (Lisbon)	8.6	70,000
1811–12	USA (New Madrid, Missouri)	7.5	20
1886	USA (Charleston, South Carolina)	7.0	60
1906	USA (San Francisco, California)	8.3	700
1923	Japan (Tokyo)	8.3	143,000
1964	USA (Alaska)	8.6	131
1971	USA (San Fernando, California)	6.6	65
1976	China (Tangshan)	8.0	242,000
1985	Mexico (Mexico City)	8.1	9,500
1988	Armenia	7.0	25,000
1989	USA (Loma Prieta, California)	7.1	63
1990	Iran	7.3	40,000
1992	Turkey	6.8	570
1992	Egypt (Cairo)	5.9	550
1993	India	6.4	30,000
1994	USA (Northridge, California)	6.7	55

volcanic eruptions. Today, geologists know that the majority of earthquakes result from faulting associated with plate movements.

ELASTIC REBOUND THEORY

Based on studies conducted after the 1906 San Francisco earthquake, H. F. Reid of Johns Hopkins University formulated the **elastic rebound theory** to explain how earthquakes occur. Reid studied three sets of measurements taken across the portion of the San Andreas fault that had broken during the 1906 earthquake. The measurements revealed that points on opposite sides of the fault had moved 3.2 m during the 50-year period prior to breakage in 1906, with the west side moving northward (▷ Figure 9-2).

According to Reid, rocks on one side of the fault had moved relative to rocks on the other side, and this movement caused the gradual bending of any straight line that crossed the San Andreas fault such as a fence or road (Figure 9-2). Eventually, the strength of the rocks was exceeded, the rocks on opposite sides of the fault rebounded or "snapped back" to their former undeformed shape, and the energy stored was released as earthquake waves radiating outward from the break (Figure 9-2). Additional field and laboratory studies conducted by Reid and others have confirmed that elastic rebound is the mechanism by which earthquakes are generated.

SEISMOLOGY

Seismology, the study of earthquakes, began emerging as a true science around 1880 with the development of seismo-

graphs that effectively recorded earthquake waves. A *seismograph* is an instrument that detects, records, and measures the various vibrations produced by an earthquake (▷ Figure 9-3). The record made by a seismograph is a *seismogram*.

When an earthquake occurs, energy in the form of *seismic waves* radiates outward in all directions from the point of release (▷ Figure 9-4). Most earthquakes result when rocks in the Earth's crust rupture along a fault because of the buildup of excessive pressure, which is usually caused by plate movement. Once a rupture begins, it moves along the fault at a velocity of several km/sec for as long as conditions for failure exist. The length of the fault along which rupture occurs can range from a few meters to several hundred kilometers. The longer the rupture, the more time it takes for all of the stored energy in the rocks to be released, and therefore the longer the ground will shake.

The location within the crust where rupture initiates, and thus where the energy is released, is referred to as the **focus** or *hypocenter*. The point on the Earth's surface vertically above the focus is the **epicenter**, which is the location that is usually given in news reports on earthquakes (Figure 9-4).

Seismologists recognize three categories of earthquakes based on the depth of their foci. *Shallow-focus* earthquakes have a focal depth of less than 70 km. Earthquakes with foci between 70 and 300 km are referred to as *intermediate focus*, and those with foci greater than 300 km are called *deep focus*. Earthquakes are not evenly distributed among these three categories. Approximately 90% of all earthquake foci occur at a depth of less than 100 km. Shallow-focus earthquakes are, with few exceptions, the most destructive.

There is an interesting relationship between earthquake foci and plate margins. Earthquakes generated along divergent

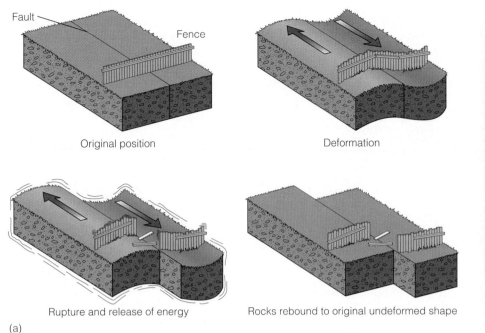

(a)

(b)

> **FIGURE 9-2** (*a*) According to the elastic rebound theory, when rocks are deformed, they store energy and bend. When the inherent strength of the rocks is exceeded, they rupture, releasing the energy in the form of earthquake waves that radiate outward in all directions. Upon rupture, the rocks rebound to their former undeformed shape. (*b*) During the 1906 San Francisco earthquake, this fence in Marin County was displaced 2.5 m.

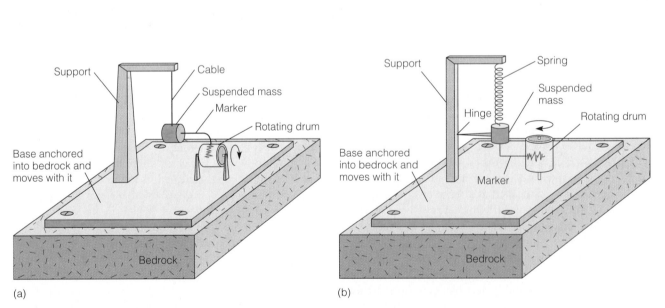

(a)

(b)

> **FIGURE 9-3** (*a*) A horizontal-motion seismograph. Because of its inertia, the heavy mass that contains the marker will remain stationary while the rest of the structure moves along with the ground during an earthquake. As long as the length of the arm is not parallel to the direction of ground movement, the marker will record the earthquake waves on the rotating drum. (*b*) A vertical-motion seismograph. This seismograph operates on the same principle as a horizontal-motion instrument and records vertical ground movement.

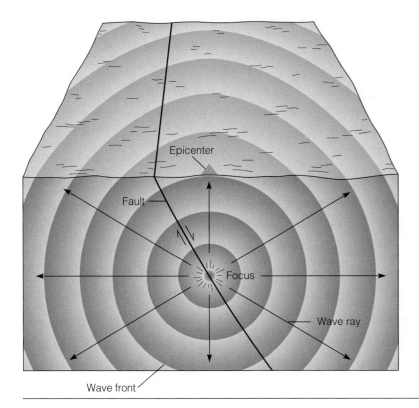

➤ **FIGURE 9-4** The focus of an earthquake is the location where rupture begins and energy is released. The place on the Earth's surface vertically above the focus is the epicenter. Seismic wave fronts move outward in all directions from their source, the focus of an earthquake. Wave rays are lines drawn perpendicular to wave fronts.

or transform plate boundaries are always shallow focus, while almost all intermediate- and deep-focus earthquakes occur within the circum-Pacific belt along convergent margins (➤ Figure 9-5).

The Frequency and Distribution of Earthquakes

Most earthquakes (almost 95%) occur in seismic belts that correspond to plate boundaries where stresses develop as plates converge, diverge, and slide past each other. Earthquake activity distant from plate margins is minimal, but can be devastating when it occurs. The relationship between plate margins and the distribution of earthquakes is readily apparent when the locations of earthquake epicenters are superimposed on a map showing the boundaries of the Earth's plates (Figure 9-5).

The majority of all earthquakes (approximately 80%) occur in the *circum-Pacific belt*, a zone of seismic activity nearly encircling the Pacific Ocean basin. Most of these earthquakes result from convergence along plate margins. The second major seismic belt is the *Mediterranean-Asiatic belt* where approximately 15% of all earthquakes occur. This belt extends westerly from Indonesia through the Himalayas, across Iran and Turkey, and westerly through the Mediterranean region of Europe. The devastating earthquake that struck Armenia in 1988 killing 25,000 people and the 1990 earthquake in Iran that killed 40,000 are recent examples of the destructive earthquakes that strike this region (Table 9-1).

The remaining 5% of earthquakes occur mostly in the interiors of plates and along oceanic spreading ridge systems. Most of these earthquakes are not very strong although several major intraplate earthquakes are worthy of mention. For example, the 1811 and 1812 earthquakes near New Madrid, Missouri, killed approximately 20 people and nearly destroyed the town of New Madrid. So strong were these earthquakes that they were felt from the Rocky Mountains to the Atlantic Ocean and from the Canadian border to the Gulf of Mexico. Another major intraplate earthquake struck Charleston, South Carolina, on August 31, 1886, killing 60 people and causing $23 million in property damage (➤ Figure 9-6).

The cause of intraplate earthquakes is not well understood, but geologists think they arise from localized stresses caused by the compression that most plates experience along their margins. The release of these stresses and hence the resulting intraplate earthquakes are due to local factors. Interestingly, many intraplate earthquakes are associated with very ancient and presumed inactive faults that are reactivated at various intervals.

More than 150,000 earthquakes strong enough to be felt by someone are recorded every year by the worldwide network of seismograph stations. In addition, seismologists estimate that about 900,000 earthquakes occur annually that are recorded by seismographs, but are too small to be individually cataloged. These small earthquakes result from the energy released as continual adjustments occur between the Earth's various plates.

Seismic Waves

The shaking and destruction resulting from earthquakes are caused by two different types of seismic waves: *body waves*, which travel through the Earth and are somewhat like sound

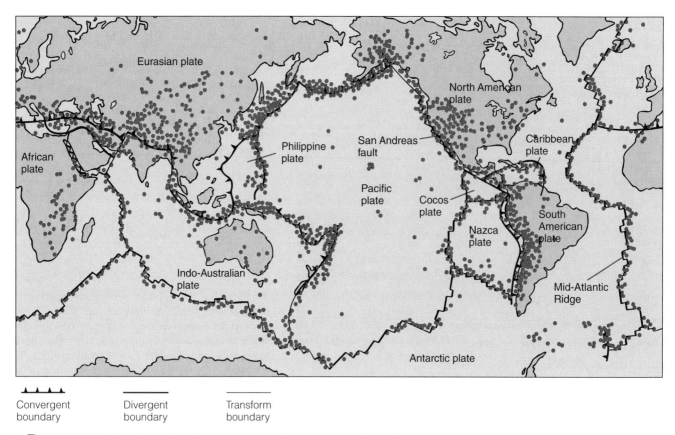

Convergent boundary   Divergent boundary   Transform boundary

➤ **FIGURE 9-5** The relationship between earthquake epicenters and plate boundaries. Approximately 80% of earthquakes occur within the circum-Pacific belt, 15% within the Mediterranean–Asiatic belt, and the remaining 5% within the interiors of plates or along oceanic spreading ridge systems. Each dot represents a single earthquake epicenter.

➤ **FIGURE 9-6** Damage done to Charleston, South Carolina, by the earthquake of August 31, 1886. This earthquake is the largest reported in the eastern United States.

waves; and *surface waves*, which travel only along the ground surface and are analogous to ocean waves.

An earthquake generates two types of body waves: P-waves and S-waves (➤ Figure 9-7). **P-waves** or *primary waves* are the fastest seismic waves and can travel through solids, liquids, and gases. P-waves are compressional, or push-pull, waves and are similar to sound waves in that they move material forward and backward along a line in the same direction that the waves themselves are moving (Figure 9-7b). Thus, the material P-waves travel through is expanded and compressed as the wave moves through it and returns to its original size and shape after the wave passes by.

S-waves or *secondary waves* are somewhat slower than P-waves and can only travel through solids. S-waves are *shear waves* because they move the material perpendicular to the direction of travel, thereby producing shear stresses in the material they move through (Figure 9-7c). Because liquids (as well as gases) are not rigid, they have no shear strength and S-waves cannot be transmitted through them.

The velocities of P- and S-waves are determined by the density and elasticity of the materials through which they travel. For example, seismic waves travel more slowly through rocks of greater density, but more rapidly through rocks with greater elasticity. *Elasticity* is a property of solids, such as rocks, and means that once they have been deformed by an applied force, they return to their original shape when the force is no longer present. Because P-wave velocity is greater than S-wave velocity in all materials, P-waves always arrive at seismic stations first.

Surface waves travel along the surface of the ground, or just below it, and are slower than body waves. Unlike the sharp jolting and shaking that body waves cause, surface waves generally produce a rolling or swaying motion, much like the experience of being on a boat.

Locating an Earthquake

The various seismic waves travel at different speeds and thus arrive at a seismograph at different times. As ➤ Figure 9-8 illustrates, the first waves to arrive are the P-waves, which travel at nearly twice the velocity of the S-waves that follow. Both the P- and S-waves travel directly from the focus to the seismograph through the interior of the Earth. The last

➤ **FIGURE 9-7** Seismic waves. (*a*) Undisturbed material. (*b*) Primary waves (P-waves) compress and expand material in the same direction as the wave movement. (*c*) Secondary waves (S-waves) move material perpendicular to the direction of wave movement.

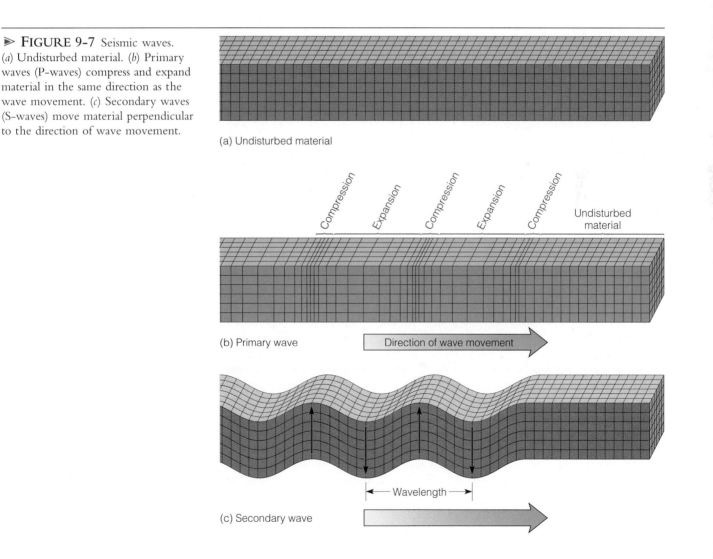

(a) Undisturbed material

Compression Expansion Compression Expansion Compression Undisturbed material

(b) Primary wave   Direction of wave movement

|← Wavelength →|

(c) Secondary wave

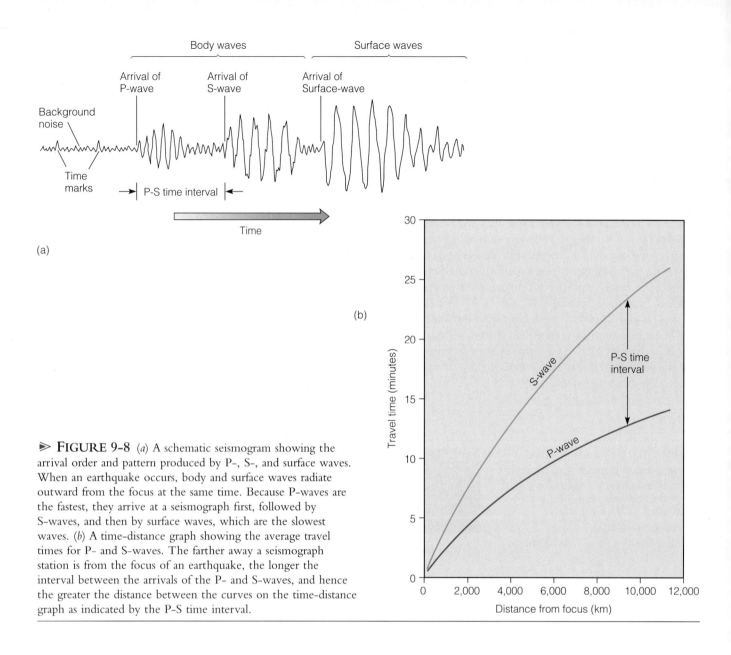

Body waves

Surface waves

Arrival of
P-wave

Arrival of
S-wave

Arrival of
Surface-wave

Background
noise

Time
marks

→| P-S time interval |←

Time

(a)

(b)

Travel time (minutes)

S-wave

P-S time
interval

P-wave

Distance from focus (km)

➤ **FIGURE 9-8** (*a*) A schematic seismogram showing the arrival order and pattern produced by P-, S-, and surface waves. When an earthquake occurs, body and surface waves radiate outward from the focus at the same time. Because P-waves are the fastest, they arrive at a seismograph first, followed by S-waves, and then by surface waves, which are the slowest waves. (*b*) A time-distance graph showing the average travel times for P- and S-waves. The farther away a seismograph station is from the focus of an earthquake, the longer the interval between the arrivals of the P- and S-waves, and hence the greater the distance between the curves on the time-distance graph as indicated by the P-S time interval.

waves to arrive are the surface waves, which are the slowest and also travel the longest route along the Earth's surface.

By accumulating a tremendous amount of data over the years, seismologists have determined the average travel times of P- and S-waves for any specific distance. These P- and S-wave travel times are published as *time-distance graphs* and illustrate that the difference between the arrival times of the P- and S-waves is a function of the distance of the seismograph from the focus (Figure 9-8b).

As ➤ Figure 9-9 demonstrates, the epicenter of any earthquake can be determined by using a time-distance graph and knowing the arrival times of the P- and S-waves at any three seismograph locations. Subtracting the arrival time of the first P-wave from the arrival time of the first S-wave gives the time interval between the arrivals of the two waves for each seismograph location. Each time interval is then plotted on the time-distance graph, and a line is drawn straight down to the distance axis of the graph,

indicating how far away each station is from the focus of the earthquake. Then a circle whose radius equals the distance shown on the time-distance graph from each of the three seismograph locations is drawn on a map (Figure 9-9). The intersection of the three circles is the location of the earthquake's epicenter. A minimum of three locations is needed because two locations will provide two possible epicenters and one location will provide an infinite number of possible epicenters.

Measuring Earthquake Intensity and Magnitude

Geologists measure the strength of an earthquake in two different ways. The first, *intensity*, is a qualitative assessment of the kinds of damage done by an earthquake. The second, *magnitude*, is a quantitative measurement of the amount of energy released by an earthquake. Each method provides geologists with important data about earthquakes and their

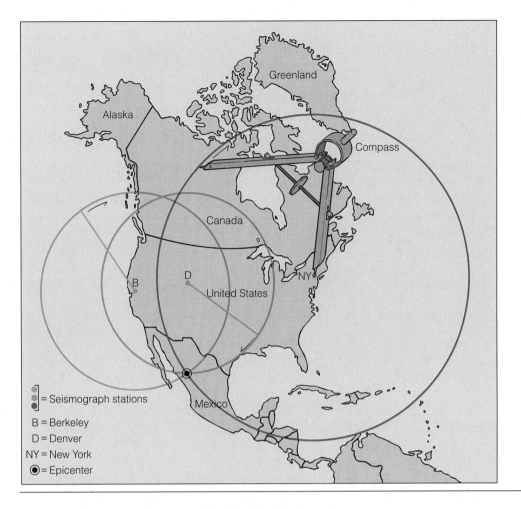

▷ FIGURE 9-9 Three seismograph stations are needed to locate the epicenter of an earthquake. The P-S time interval is plotted on a time-distance graph for each seismograph station to determine the distance that station is from the epicenter. A circle with that radius is drawn from each station, and the intersection of the three circles is the epicenter of the earthquake.

effects. This information can then be used to prepare for future earthquakes.

Intensity is a subjective measure of the kind of damage done by an earthquake as well as people's reaction to it. Since the mid-nineteenth century, geologists have used intensity as a rough approximation of the size and strength of an earthquake. The most common intensity scale used in the United States is the **Modified Mercalli Intensity Scale**, which has values ranging from I to XII (◉ Table 9-2).

While it is generally true that a large earthquake will produce greater intensity values than a small earthquake, many other factors besides the amount of energy released by an earthquake affect its intensity. These include the distance from the epicenter, the focal depth of the earthquake, the population density and local geology of the area, the type of building construction employed, and the duration of shaking.

If earthquakes are to be compared quantitatively, we must use a scale that measures the amount of energy released and is independent of intensity. Such a scale was developed in 1935 by Charles F. Richter, a seismologist at the California Institute of Technology. The **Richter Magnitude Scale** measures earthquake **magnitude**, which is the total amount of energy released by an earthquake at its source. It is an open-ended scale with values beginning at 1. The largest magnitude recorded has been 8.6, and though values greater

than 9 are theoretically possible, they are highly improbable because rocks are not able to store the energy necessary to generate earthquakes of this magnitude.

To avoid large numbers, Richter used a conventional base-10 logarithmic scale to convert the amplitude of the largest recorded seismic wave to a numerical magnitude value. Therefore, each integer increase in magnitude represents a 10-fold increase in wave amplitude. For example, the amplitude of the largest seismic wave for an earthquake of magnitude 6 is 10 times that produced by an earthquake of magnitude 5, 100 times as large as a magnitude 4 earthquake, and 1,000 times that of an earthquake of magnitude 3 (10 × 10 × 10 = 1,000).

While each increase in magnitude represents a 10-fold increase in wave amplitude, each magnitude increase corresponds to a roughly 30-fold increase in the amount of energy released. Thus, the 1964 Alaska earthquake with a magnitude of 8.6 released about 900 times the energy of the 1971 San Fernando Valley earthquake of magnitude 6.6.

We have already mentioned that more than 900,000 earthquakes are recorded around the world each year. These figures can be placed in better perspective by reference to ◉ Table 9-3, which shows that the vast majority of earthquakes have a Richter magnitude of less than 2.5, and that great earthquakes (those with a magnitude greater than 8.0) occur, on average, only once every five years.

TABLE 9-2 Modified Mercalli Intensity Scale

I	Not felt except by a very few under especially favorable circumstances.
II	Felt only by a few people at rest, especially on upper floors of buildings.
III	Felt quite noticeably indoors, especially on upper floors of buildings, but many people do not recognize it as an earthquake. Standing automobiles may rock slightly.
IV	During the day felt indoors by many, outdoors by few. At night some awakened. Sensation like heavy truck striking building, standing automobiles rocked noticeably.
V	Felt by nearly everyone, many awakened. Some dishes, windows, etc. broken, a few instances of cracked plaster. Disturbance of trees, poles, and other tall objects sometimes noticed.
VI	Felt by all, many frightened and run outdoors. Some heavy furniture moved, a few instances of fallen plaster or damaged chimneys. Damage slight.
VII	Everybody runs outdoors. Damage negligible in buildings of good design and construction; slight to moderate in well-built ordinary structures; considerable in poorly built or badly designed structures; some chimneys broken. Noticed by people driving automobiles.
VIII	Damage slight in specially designed structures; considerable in normally constructed buildings with possible partial collapse; great in poorly built structures. Fall of chimneys, monuments, walls. Heavy furniture overturned. Sand and mud ejected in small amounts.
IX	Damage considerable in specially designed structures. Buildings shifted off foundations. Ground noticeably cracked. Underground pipes broken.
X	Some well-built wooden structures destroyed; most masonry and frame structures with foundations destroyed; ground badly cracked. Rails bent. Landslides considerable from river banks and steep slopes. Water splashed over river banks.
XI	Few if any (masonry) structures remain standing. Bridges destroyed. Broad fissures in ground. Underground pipelines completely out of service.
XII	Damage total. Waves seen on ground surfaces. Objects thrown upward into the air.

SOURCE: United States Geological Survey.

TABLE 9-3 Average Number of Earthquakes of Various Magnitudes per Year Worldwide

Magnitude	Effects	Average Number per Year
<2.5	Typically not felt, but recorded.	900,000
2.5–6.0	Usually felt. Minor to moderate damage to structures.	31,000
6.1–6.9	Potentially destructive, especially in populated areas.	100
7.0–7.9	Major earthquakes. Serious damage results.	20
>8.0	Great earthquakes. Usually result in total destruction.	1 every 5 years

SOURCE: Data from *Earthquake Information Bulletin* and Gutenberg and Richter (1949).

THE DESTRUCTIVE EFFECTS OF EARTHQUAKES

The destructive effects of earthquakes are many and varied and include such phenomena as ground shaking, fire, seismic sea waves, and landslides, as well as disruption of vital services, panic, and psychological shock. The amount of property damage, loss of life, and injury depends on the time of day an earthquake occurs, its magnitude, the distance from the epicenter, the geology of the area, the type of construction of the various structures, the population density, and the duration of shaking. Generally speaking, earthquakes occurring during working and school hours in densely populated urban areas are the most destructive.

Ground shaking usually causes more damage and results in more loss of life and injuries than any other earthquake hazard. Structures built on solid bedrock generally suffer less damage than those built on poorly consolidated material such as water-saturated sediments or artificial fill. Structures on poorly consolidated or water-saturated material are subjected to ground shaking of longer duration and greater S-wave amplitude than those on bedrock. In addition, fill and water-saturated sediments tend to liquefy, or behave as a fluid, a process known as *liquefaction*. When shaken, the individual grains lose cohesion and the ground flows. Dramatic examples of damage resulting from liquefaction include Niigata, Japan, where large apartment buildings were tipped to their sides after the water-saturated soil of the hillside collapsed in 1964.

During the Loma Prieta earthquake that caused the third game of the 1989 World Series to be postponed, those districts in the San Francisco–Oakland Bay area built on artificial fill or reclaimed bay mud suffered the most damage. In the Marina district of San Francisco, numerous buildings were destroyed, and a fire, fed by broken gas lines, lit up the night sky. The failure of the columns supporting a portion of the two-tiered Interstate 880 freeway in Oakland sent the upper tier crashing down onto the lower one, killing 42 unfortunate motorists (▷ Figure 9-10). The shaking lasted less than 15 seconds but resulted in 63 deaths, 3,800 injuries, and $6 billion in property damage and left at least 12,000 people homeless.

In addition to the magnitude of an earthquake and the regional geology, the material used and the type of construction also affect the amount of damage done. Adobe and mud-walled structures are the weakest of all and almost always collapse during an earthquake. Unreinforced brick structures and poorly built concrete structures are also particularly susceptible to collapse (▷ Figure 9-11). The 6.4 magnitude earthquake that struck India in 1993 killed about 30,000 people whereas the 6.7 magnitude Northridge earthquake resulted in only 55 deaths. Both earthquakes occurred in densely populated regions, but in India the brick and stone buildings could not withstand ground shaking; most collapsed entombing their occupants.

In many earthquakes, particularly in urban areas, fire is a major hazard (▷ Figure 9-12). Almost 90% of the damage done in the 1906 San Francisco earthquake was caused by fire. The shaking severed many of the electrical and gas lines, which touched off flames and started numerous fires all over the city. Because water mains were ruptured by the earthquake, there was no effective way to fight the fires. Hence, they raged out of control for three days, destroying much of the city.

Seismic sea waves or **tsunami** are destructive sea waves that are usually produced by earthquakes but can also be caused by submarine landslides or volcanic eruptions (▷ Figure 9-13). Tsunami are popularly called tidal waves, although they have nothing to do with tides. Instead, most tsunami result from the sudden movement of the sea floor, which sets up waves within the water that travel outward, much like the ripples that form when a stone is thrown into a pond.

Tsunami travel at speeds of several hundred km/hr and are commonly not felt in the open ocean because their wave height is usually less than 1 m and the distance between wave crests is typically several hundred kilometers. When tsunami approach shorelines, however, the waves slow down and water piles up to heights of up to 65 m.

Following a 1946 tsunami that killed 159 people and caused $25 million in property damage in Hawaii, the U.S. Coast and Geodetic Survey established a Tsunami Early Warning System in Honolulu, Hawaii, in an attempt to minimize tsunami devastation. This system combines seismographs and instruments that can detect earthquake-generated waves. Whenever a strong earthquake occurs anywhere within the Pacific basin, its location is determined, and instruments are checked to see if a tsunami has been generated. If it has, a warning is sent out to evacuate people from low-lying areas that may be affected.

Earthquake-triggered landslides are particularly dangerous in mountainous regions and have been responsible for tremendous amounts of damage and many deaths. For example, the 1970 Peru earthquake caused an avalanche that completely destroyed the town of Yungay, resulting in 25,000 deaths.

⊕ EARTHQUAKE PREDICTION

Can earthquakes be predicted? A successful prediction must include a time frame for the occurrence of the earthquake, its location, and its strength. In spite of the tremendous amount of information geologists have gathered about the cause of earthquakes, successful predictions are still quite rare. Nevertheless, if reliable predictions can be made, they can greatly reduce the number of deaths and injuries.

From an analysis of historic records and the distribution of known faults, *seismic risk maps* can be constructed that indicate the likelihood and potential severity of future earthquakes based on the intensity of past earthquakes (▷ Figure 9-14). Although such maps cannot predict when the next major earthquake will occur, they are useful in helping people plan for future earthquakes.

One long-range prediction technique used in seismically active areas involves plotting the location of major earthquakes and their aftershocks to detect areas that have had major earthquakes in the past but are currently inactive. Such regions are locked and not releasing energy, making these *seismic gaps* prime locations for future earthquakes. Several seismic gaps along the San Andreas fault have the potential for future major earthquakes.

▷ FIGURE 9-10 Interstate 880 in Oakland, California, was damaged during the 1989 Loma Prieta earthquake. The columns supporting the upper deck of the two-tiered highway failed, and the upper deck fell onto the lower deck, killing 42 motorists. Only 1 of the 51 double-deck spans did not collapse.

▷ **FIGURE 9-11** Many of the approximately 242,000 people who died in the 1976 earthquake in Tangshan, China were killed by collapsing structures. Many of the buildings were constructed of unreinforced brick, which has no flexibility, and quickly fell down during the earthquake.

Changes in elevation and tilting of the land surface have frequently preceded earthquakes and may be warnings of impending quakes. Extremely slight changes in the angle of the ground surface can be measured by tiltmeters. Tiltmeters have been placed on both sides of the San Andreas fault to measure tilting of the ground surface that is thought to result from increasing pressure in the rocks. Data from measurements in central California indicate significant tilting occurred immediately preceding small earthquakes. Furthermore, extensive tiltmeter work performed in Japan prior to the 1964 Niigata earthquake clearly showed a relationship between increased tilting and the main shock. While more research is needed, such changes appear to be useful in making short-term earthquake predictions.

Other earthquake precursors include fluctuations in the water level of wells, changes in the Earth's magnetic field, and the electrical resistance of the ground. These fluctuations are thought to result from changes in the amount of pore space in rocks due to increasing pressure. A change in animal behavior prior to earthquakes also is frequently mentioned. It may be that animals are sensing small and subtle changes in the Earth prior to a quake that humans simply do not sense.

Currently, only four nations—the United States, Japan, Russia, and China—have government-sponsored earthquake prediction programs. These programs include laboratory and field studies of the behavior of rocks before, during, and after major earthquakes as well as monitoring activity along major active faults. Most earthquake prediction work in the United States is done by the United States Geological Survey (USGS) and involves a variety of research into all aspects of earthquake-related phenomena.

The Chinese have perhaps one of the most ambitious earthquake prediction programs anywhere in the world, which is understandable considering their long history of destructive earthquakes. The Chinese program on earthquake prediction was initiated soon after two large earthquakes occurred at Xingtai (300 km southwest of Beijing) in 1966. The Chinese program includes extensive study and monitoring of all possible earthquake precursors. In addition,

> **FIGURE 9-12** San Francisco Marina district fire caused by broken gas lines during the 1989 Loma Prieta earthquake.

> **FIGURE 9-13** As a tsunami crashes into the street behind them, residents of Hilo, Hawaii run for their lives. This tsunami was generated by an earthquake in the Aleutian Islands and resulted in considerable property damage to Hilo and the deaths of 154 people.

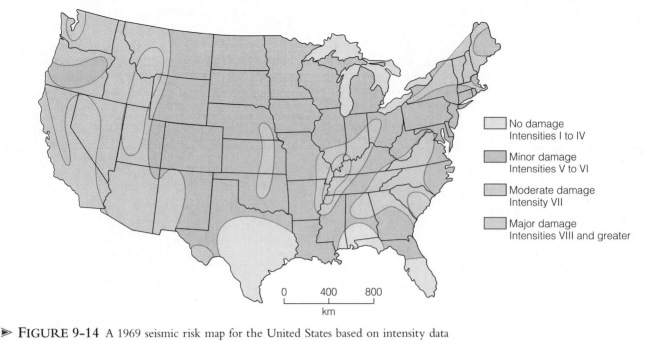

➤ **FIGURE 9-14** A 1969 seismic risk map for the United States based on intensity data collected by the U.S. Coast and Geodetic Survey.

the Chinese also emphasize changes in phenomena that can be observed by seeing and hearing without the use of sophisticated instruments. They have had remarkable success in predicting earthquakes, particularly in the short term, such as the 1975 Haicheng earthquake. They failed, however, to predict the devastating 1976 Tangshan earthquake that killed about 242,000 people.

Great strides are being made toward dependable, accurate earthquake predictions, and studies are underway to assess public reactions to long-, medium-, and short-term earthquake warnings. Unless short-term warnings are actually followed by an earthquake, most people will probably ignore the warnings as they frequently do now for hurricanes, tornadoes, and tsunami.

⊕ EARTHQUAKE CONTROL

If earthquake prediction is still in the future, can anything be done to control earthquakes? Because of the tremendous forces involved, humans are certainly not going to be able to prevent earthquakes. However, there may be ways to dissipate the destructive energy of major earthquakes by releasing it in small amounts that will not cause extensive damage.

During the early to mid-1960s, Denver experienced numerous small earthquakes. This was surprising because Denver had not been prone to earthquakes in the past. In 1962, David M. Evans, a geologist, suggested that the earthquakes in Denver were directly related to the injection of contaminated waste water into a disposal well 3,674 m deep at the Rocky Mountain Arsenal, northeast of Denver. The U.S. Army initially denied that there was any connec-

tion, but a USGS study concluded that the pumping of waste fluids into fractured rocks beneath the disposal well decreased the friction on opposite sides of the fractures and, in effect, lubricated them so that movement occurred, causing the earthquakes that Denver experienced.

Experiments conducted in 1969 at an abandoned oil field near Rangely, Colorado, confirmed the arsenal hypothesis. Water was pumped in and out of abandoned oil wells, the pore-water pressure in these wells was measured, and seismographs were installed in the area to measure any seismic activity. Monitoring showed that small earthquakes were occurring in the area when fluid was injected and that earthquake activity declined when the fluids were pumped out. What the geologists were doing was starting and stopping earthquakes at will, and the relationship between pore-water pressures and earthquakes was established.

Based upon these results, some geologists have proposed that fluids be pumped into the locked segments of active faults to cause small- to moderate-sized earthquakes. They think that this would relieve the pressure on the fault and prevent a major earthquake from occurring. While this plan is intriguing, it also has many potential problems. For instance, there is no guarantee that only a small earthquake might result. Instead a major earthquake might occur, causing tremendous property damage and loss of life. Who would be responsible? Certainly, a great deal more research is needed before such an experiment is performed, even in an area of low population density.

It appears that until such time as earthquakes can be accurately predicted or controlled, the best means of defense is careful planning and preparation.

The Earth's interior has always been an inaccessible, mysterious realm. During most of historic time, it was perceived as an underground world of vast caverns, heat, and sulfur gases, populated by demons. By the 1860s, scientists knew what the average density of the Earth was and that pressure and temperature increase with depth. And even though the Earth's interior is hidden from direct observation, scientists have a reasonably good idea of its internal structure and composition.

Scientists have known for more than 200 years that the Earth's interior is not homogeneous. Sir Isaac Newton (1642–1727) noted in a study of the planets that the Earth's average density is 5.0 to 6.0 g/cm³ (water has a density of 1 g/cm³). In 1797, Henry Cavendish calculated a density value very close to the 5.5 g/cm³ now accepted. The Earth's average density is considerably greater than that of surface rocks, most of which range from 2.5 to 3.0 g/cm³. So, in order for the average density to be 5.5 g/cm³, much of the interior must consist of materials with a density greater than the Earth's average density.

The Earth is generally depicted as consisting of concentric layers that differ in composition and density and are separated from adjacent layers by rather distinct boundaries (see Figure 1-7). Recall that the outermost layer, or the **crust,** is the very thin skin of the Earth. Below the crust and extending about halfway to the Earth's center is the **mantle,** which comprises more than 80% of the Earth's volume. The central part of the Earth consists of a **core,** which is divided into a solid inner core and a liquid outer part.

The behavior and travel times of P- and S-waves within the Earth provide geologists with much information about its internal structure. Seismic waves travel outward as wave fronts from their source areas, although it is most convenient to depict them as *wave rays,* which are lines showing the direction of movement of small parts of wave fronts (Figure 9-4). Any disturbance, such as a passing train or construction equipment, can cause seismic waves, but only those generated by large earthquakes, explosive volcanism, asteroid impacts, and nuclear explosions can travel completely through the Earth.

As we noted earlier, the velocities of P- and S-waves are determined by the density and elasticity of the materials through which they travel. Both the density and elasticity of rocks increase with depth, but elasticity increases faster than density, resulting in a general increase in the velocity of seismic waves. P-waves travel faster than S-waves through all materials, but unlike P-waves, S-waves cannot be transmitted through a liquid because liquids have no shear strength (rigidity)—they simply flow in response to a shear stress.

As a seismic wave travels from one material into another of different density and elasticity, its velocity and direction of travel change. That is, the wave is bent, a phenomenon known as **refraction** (➤ Figure 9-15). Because seismic waves pass through materials of differing density and elastic-

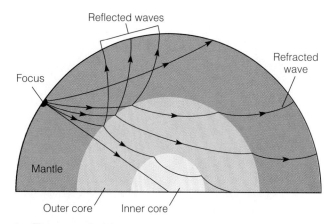

➤ **FIGURE 9-15** Refraction and reflection of P-waves. When seismic waves pass through a boundary separating Earth materials of different density or elasticity, they are refracted, and some of their energy is reflected back to the surface.

ity, they are continually refracted so that their paths are curved; the only exception is that wave rays are not refracted if their direction of travel is perpendicular to a boundary (Figure 9-15). In that case they travel in a straight line.

In addition to refraction, seismic rays are also **reflected,** much as light is reflected from a mirror. Some of the energy of seismic rays that encounter a boundary separating materials of different density or elasticity within the Earth is *reflected* back to the Earth's surface (Figure 9-15). If we know the wave velocity and the time required for it to travel from its source to the boundary and back to the surface, we can calculate the depth of the reflecting boundary. Such information is useful in determining not only the depths of the various layers within the Earth, but also the depths of sedimentary rocks that may contain petroleum.

The Core

In 1906, R. D. Oldham of the Geological Survey of India postulated the existence of a core that transmits seismic waves at a slower rate than shallower Earth materials. We now know that P-wave velocity decreases markedly at a depth of 2,900 km, indicating a major discontinuity now recognized as the core-mantle boundary (➤ Figure 9-16).

The sudden decrease in P-wave velocity at the core-mantle boundary causes P-waves entering the core to be refracted in such a way that very little P-wave energy reaches the Earth's surface in the area between 103° and 143° from an earthquake focus (➤ Figure 9-17a). This area in which little P-wave energy is recorded by seismometers is a **P-wave shadow zone.**

In 1926, the British physicist Harold Jeffreys realized that S-waves were not simply slowed by the core, but were completely blocked by it. So in addition to a P-wave shadow zone, a much larger and more complete **S-wave shadow zone** exists (Figure 9-17b). At locations greater than 103°

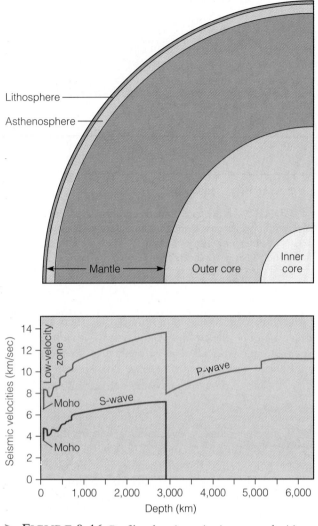

▶ **FIGURE 9-16** Profiles showing seismic wave velocities versus depth. Several discontinuities are shown across which seismic wave velocities change rapidly.

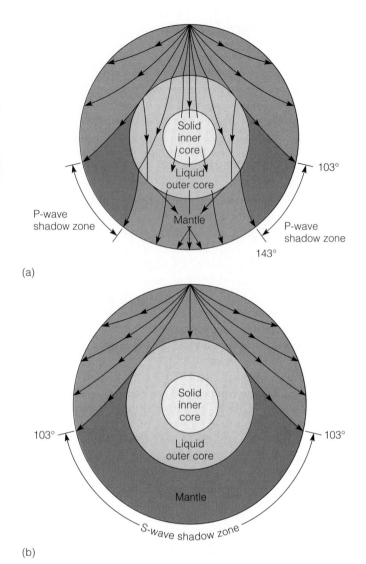

▶ **FIGURE 9-17** (*a*) P-waves are refracted so that no direct P-wave energy reaches the Earth's surface in the P-wave shadow zone. (*b*) The presence of an S-wave shadow zone indicates that S-waves are being blocked within the Earth.

from an earthquake focus, no S-waves are recorded, indicating that S-waves cannot be transmitted through the core. S-waves will not pass through a liquid, so it seems that the outer core must be liquid or behave as a liquid.

We can estimate the core's density and composition by using seismic evidence and laboratory experiments. Furthermore, meteorites, which are thought to represent remnants of the material from which the solar system formed, can be used to make estimates of density and composition. For example, meteorites composed of iron and nickel alloys may represent the differentiated interiors of large asteroids and approximate the density and composition of the Earth's core. The density of the outer core varies from 9.9 to 13.0 g/cm^3. At the Earth's center, the pressure is equivalent to about 3.5 million times normal atmospheric pressure.

The core cannot be composed of the minerals most common at the Earth's surface because even under the tremendous pressures at great depth, they would still not be dense enough to yield an average density of 5.5 g/cm^3 for

the Earth. Both the outer and inner core are thought to be composed largely of iron, but pure iron is too dense to be the sole constituent of the outer core. Thus, it must be "diluted" with elements of lesser density. Laboratory experiments and comparisons with iron meteorites indicate that about 12% of the outer core may consist of sulfur, and perhaps some silicon and small amounts of nickel and potassium.

In contrast, pure iron is not dense enough to account for the estimated density of the inner core. Most geologists think that perhaps 10 to 20% of the inner core also consists of nickel. These metals form an iron–nickel alloy that under the pressure at that depth is thought to be sufficiently dense to account for the density of the inner core. When the core formed during early Earth history, it was probably completely molten and has since cooled to the point that its interior has crystallized.

The Mantle

Another significant discovery about the Earth's interior was made in 1909 when the Yugoslavian seismologist Andrija Mohorovičić detected a discontinuity at a depth of about 30 km. While studying arrival times of seismic waves from Balkan earthquakes, Mohorovičić noticed that two distinct sets of P- and S-waves were recorded at seismic stations a few hundred kilometers from an earthquake's epicenter. He reasoned that one set of waves traveled directly from the epicenter to the seismic station, whereas the other waves had penetrated a deeper layer where they were refracted.

From his observations Mohorovičić concluded that a sharp boundary separating rocks with different properties exists at a depth of about 30 km. He postulated that P-waves below this boundary travel at 8 km/sec, whereas those above the boundary travel at 6.75 km/sec. When an earthquake occurs, some waves travel directly from the focus to the seismic station, while others travel through the deeper layer and some of their energy is refracted back to the surface. Waves traveling through the deeper layer travel farther to a seismic station, but they do so more rapidly than those in the shallower layer. The boundary identified by Mohorovičić separates the crust from the mantle and is now called the **Mohorovičić discontinuity,** or simply the **Moho.** It is present everywhere except beneath spreading ridges, but its depth varies: beneath the continents it averages 35 km, but ranges from 20 to 90 km; beneath the sea floor it is 5 to 10 km deep (Figure 9-16).

Although seismic wave velocity in the mantle generally increases with depth, several discontinuities also exist. Between depths of 100 and 250 km, both P- and S-wave velocities decrease markedly. This layer between 100 and 250 km deep is the **low-velocity zone;** it corresponds closely to the *asthenosphere,* a layer in which the rocks are close to their melting point and thus are less elastic; this decrease in elasticity accounts for the observed decrease in seismic wave velocity. The asthenosphere is an important zone because it may be where some magmas are generated. Furthermore, it lacks strength and flows plastically and is thought to be the layer over which the outer, rigid *lithosphere* moves.

Although the mantle's density, which varies from 3.3 to 5.7 g/cm^3, can be inferred rather accurately from seismic waves, its composition is less certain. The igneous rock *peridotite,* which contains mostly ferromagnesian minerals, is considered the most likely component. Laboratory experiments indicate that it possesses physical properties that would account for the mantle's density and observed rates of seismic wave transmissions. Peridotite also forms the lower parts of igneous rock sequences thought to be fragments of the oceanic crust and upper mantle emplaced on land. In addition, peridotite occurs as inclusions in volcanic rock bodies such as *kimberlite pipes* that are known to have come from great depths. These inclusions are thought to be pieces of the mantle.

THE EARTH'S INTERNAL HEAT

During the nineteenth century, scientists realized that the Earth's temperature in deep mines increases with depth. More recently, the same trend has been observed in deep drill holes, but even in these we can measure temperatures directly down to a depth of only a few kilometers. The temperature increase with depth, or **geothermal gradient,** near the surface is about 25°C/km, although it varies from area to area. In areas of active or recently active volcanism, the geothermal gradient is greater than in adjacent nonvolcanic areas, and temperature rises faster beneath spreading ridges than elsewhere beneath the sea floor.

Unfortunately, the geothermal gradient is not useful for estimating temperatures deep in the Earth. If we were simply to extrapolate from the surface downward, the temperature at 100 km would be so high that in spite of the great pressure, all known rocks would melt. Yet except for pockets of magma, it appears that the mantle is solid rather than liquid because it transmits S-waves. Accordingly, the geothermal gradient must decrease markedly.

Current estimates of the temperature at the base of the crust are 800° to 1,200°C. The latter figure seems to be an upper limit: if it were any higher, melting would be expected. Furthermore, fragments of mantle rock in kimberlite pipes, thought to have come from depths of about 100 to 300 km, appear to have reached equilibrium at these depths and at a temperature of about 1,200°C. At the core-mantle boundary, the temperature is probably between 3,500° and 5,000°C; the wide spread of values indicates the uncertainties of such estimates (see Perspective 9-1). If these figures are reasonably accurate, the geothermal gradient in the mantle is only about 1°C/km.

Considering that the core is so remote and so many uncertainties exist regarding its composition, only very general estimates of its temperature can be made. The maximum temperature at the center of the core is thought to be about 6,500°C, very close to the estimated temperature for the surface of the Sun!

SEISMIC TOMOGRAPHY

The model of the Earth's interior consisting of an iron-rich core and a rocky mantle is probably accurate but is also rather imprecise. Recently, geophysicists have developed a new technique called *seismic tomography* that allows them to develop three-dimensional models of the Earth's interior. In seismic tomography numerous crossing seismic waves are analyzed in much the same way radiologists analyze CAT (computerized axial tomography) scans. In CAT scans, X-rays penetrate the body, and a two-dimensional image of the inside of a patient is formed. Repeated CAT scans, each from a slightly different angle, are computer analyzed and stacked to produce a three-dimensional picture.

In a similar fashion geophysicists use seismic waves to probe the interior of the Earth. From its time of arrival and distance traveled, the velocity of a seismic ray is computed at a seismic station. Only average velocity is determined rather than variations in velocity. In seismic tomography numerous wave rays are analyzed so that "slow" and "fast" areas of wave travel can be detected (➤ Figure 1). Recall that seismic wave velocity is controlled partly by elasticity; cold rocks have greater elasticity and therefore transmit seismic waves faster than hot rocks.

Using this technique, geophysicists have detected areas within the mantle at a depth of about 150 km where seismic velocities are slower than expected. These anomalously hot regions lie beneath volcanic areas and beneath mid-oceanic ridges, where convection cells of rising hot mantle rock are thought to exist. In contrast, beneath the older interior parts of continents, where tectonic activity ceased hundreds of millions or billions of years ago, anomalously cold spots are recognized. In effect, tomographic maps and three-dimensional diagrams show heat variations within the Earth.

Seismic tomography has also yielded additional and sometimes surprising information about the core. For example, the core-mantle boundary is not a smooth surface, but has broad depressions and rises extending several kilometers into the mantle. Of course, the base of the mantle possesses the same features in reverse; geophysicists have termed these features "anticontinents" and "antimountains." It appears that the surface of the core is continually deformed by sinking and rising masses of mantle material.

As a result of seismic tomography, a much clearer picture of the Earth's interior is emerging. It has already given us a better understanding of complex convection within the mantle, including upwelling convection currents thought to be responsible for the movement of the Earth's lithospheric plates.

➤ **FIGURE 1** Numerous earthquake waves are analyzed to detect areas within the Earth that transmit seismic waves faster or slower than adjacent areas. Areas of fast wave travel correspond to "cold" regions (blue), whereas "hot" regions (red) transmit seismic waves more slowly.

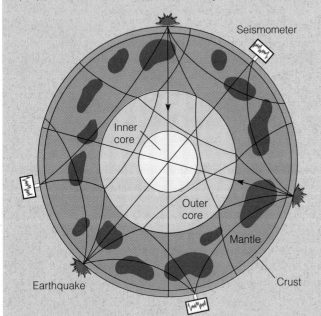

1. Earthquakes are vibrations of the Earth caused by the sudden release of energy, usually along a fault.

2. The elastic rebound theory states that pressure builds in rocks on opposite sides of a fault until the inherent strength of the rocks is exceeded and rupture occurs. When the rocks rupture, stored energy is released as they snap back to their original position.

3. Seismology is the study of earthquakes. Earthquakes are recorded on seismographs, and the record of an earthquake is a seismogram.

4. The focus of an earthquake is the point where energy is released. Vertically above the focus on the Earth's surface is the epicenter.

5. Most earthquakes occur within seismic belts. Approximately 80% of all earthquakes occur in the circum-Pacific belt, 15% within the Mediterranean-Asiatic belt, and the remaining 5% mostly in the interior of the plates or along oceanic spreading ridge systems.

6. The two types of body waves are P-waves and S-waves. Both travel through the Earth, although S-waves do not travel through liquids. P-waves are the fastest waves and are compressional, while S-waves are shear. Surface waves travel along or just below the Earth's surface.

7. The distance to the epicenter of an earthquake can be determined by the use of a time-distance graph of the P- and S-waves. Three seismographs are needed to locate the epicenter of an earthquake.

8. Intensity is a measure of the kind of damage done by an earthquake and is expressed by values from I to XII in the Modified Mercalli Intensity Scale.

9. Magnitude measures the amount of energy released by an earthquake and is expressed in the Richter Magnitude Scale. Each increase in the magnitude number represents about a 30-fold increase in energy released.

10. Ground shaking is the most destructive of all earthquake hazards. The amount of damage done by an earthquake depends upon the geology of the area, the type of building construction, the magnitude of the earthquake, and the duration of shaking. Tsunami are seismic sea waves that are usually produced by earthquakes. They can do a tremendous amount of damage to coastlines.

11. The Earth is concentrically layered into an iron-rich core with a solid inner core and a liquid outer part, a rocky mantle, and an oceanic crust and continental crust.

12. Much of the information about the Earth's interior has been derived from studies of P- and S-waves that travel through the Earth. Laboratory experiments, comparisons with meteorites, and studies of inclusions in volcanic rocks provide additional information.

13. Density and elasticity of Earth materials determine the velocity of seismic waves. Seismic waves are refracted when their direction of travel changes. Wave reflection occurs at boundaries across which the properties of rocks change.

14. The behavior of P- and S-waves within the Earth and the presence of P- and S-wave shadow zones allow geologists to estimate the density and composition of the Earth's interior and to estimate the size and depth of the core and mantle.

15. The Earth's inner core is thought to be composed of iron and nickel, whereas the outer core is probably composed mostly of iron with 10 to 20% sulfur and other substances in lesser quantities. Peridotite is the most likely component of the mantle.

16. The geothermal gradient of 25°C/km cannot continue to great depths, otherwise most of the Earth would be molten. The geothermal gradient for the mantle and core is probably about 1°C/km. The temperature at the Earth's center is estimated to be 6,500°C.

IMPORTANT TERMS

aftershock
core
crust
earthquake
elastic rebound theory
epicenter
focus

geothermal gradient
intensity
low-velocity zone
magnitude
mantle
Modified Mercalli Intensity
    Scale

Mohorovičić discontinuity
    (Moho)
P-wave
P-wave shadow zone
reflection
refraction

Richter Magnitude Scale
seismology
surface wave
S-wave
S-wave shadow zone
tsunami

1. According to the elastic rebound theory:
   a. ____ earthquakes originate deep within the Earth;
   b. ____ earthquakes originate in the asthenosphere where rocks are plastic;
   c. ____ earthquakes occur where the strength of the rock is exceeded;
   d. ____ rocks are elastic and do not rebound to their former position;
   e. ____ none of these.
2. The majority of all earthquakes occur in the:
   a. ____ Mediterranean-Asiatic belt;
   b. ____ interior of plates;
   c. ____ circum-Atlantic belt;
   d. ____ circum-Pacific belt;
   e. ____ along spreading ridges.
3. An epicenter is:
   a. ____ the location where rupture begins;
   b. ____ the point on the Earth's surface vertically above the focus;
   c. ____ the same as the hypocenter;
   d. ____ the location where energy is released;
   e. ____ none of these.
4. A qualitative assessment of the kinds of damage done by an earthquake is expressed by:
   a. ____ seismicity;
   b. ____ dilatancy;
   c. ____ magnitude;
   d. ____ intensity;
   e. ____ none of these.
5. How much more energy is released by a magnitude 5 earthquake than by one of magnitude 2?
   a. ____ 2.5 times;
   b. ____ 3 times;
   c. ____ 30 times;
   d. ____ 1,000 times;
   e. ____ 27,000 times.
6. A tsunami is a:
   a. ____ measure of the energy released by an earthquake;
   b. ____ seismic sea wave;
   c. ____ precursor to an earthquake;
   d. ____ locked portion of a fault;
   e. ____ seismic gap.
7. The average density of the Earth is _____ g/cm^3.
   a. ____ 12.0;
   b. ____ 5.5;
   c. ____ 6.75;
   d. ____ 1.0;
   e. ____ 2.5.
8. When seismic waves travel through materials having different properties, their direction of travel changes. This phenomenon is wave:
   a. ____ elasticity;
   b. ____ energy dissipation;
   c. ____ refraction;
   d. ____ deflection;
   e. ____ reflection.
9. A major seismic discontinuity at a depth of 2,900 km is the:
   a. ____ core-mantle boundary;
   b. ____ oceanic crust–continental crust boundary;
   c. ____ Moho;
   d. ____ inner core–outer core boundary;
   e. ____ lithosphere-asthenosphere boundary.
10. How does the elastic rebound theory explain how energy is released during an earthquake?
11. How is the epicenter of an earthquake determined?
12. What is the relationship between plate boundaries and earthquakes?
13. What is the relationship between plate boundaries and focal depth?
14. Explain the difference between intensity and magnitude and between the Modified Mercalli Intensity Scale and the Richter Magnitude Scale.
15. Explain how tsunami are produced and why they are so destructive.
16. Explain how seismic waves are refracted and reflected.
17. What is the significance of the S-wave shadow zone?

POINTS TO PONDER

1. If the Earth were completely solid and had the same composition and density throughout, how would P- and S-waves behave as they traveled through the Earth?
2. What factors account for higher-than-average heat flow values at spreading ridges? How is heat flow related to the age of crustal rocks?

Chapter 10

DEFORMATION AND MOUNTAIN BUILDING

OUTLINE

Owens Valley and the east face of the Sierra Nevada, California. The Sierra Nevada has been uplifted along a huge fracture so that it now stands more than 3,000 m above the valleys to the east. (Photo © Peter Kreson.)

North America has many scenic mountain ranges, but few are as impressive as the Teton Range of northwestern Wyoming. The Native Americans of the region called these mountains Teewinot, meaning many pinnacles, which is an appropriate name because the range consists of numerous jagged peaks. The loftiest of these, the Grand Teton, rises 4,190 m above sea level. Higher and larger mountain ranges exist in North America, but none rise as abruptly as the Tetons, which ascend nearly vertically more than 2,100 m above the floor of Jackson Hole, the valley to their east. In 1929, when Grand Teton National Park was established, it was largely limited to the peaks within the Teton Range, but in 1950 it was enlarged to include Jackson Hole and now covers 1,260 km².

Mountains began forming in the Teton region about 90 million years ago, but these early mountains were quite different from the present ones. They formed when the Earth's crust was contorted and folded, and they were oriented northwest-southeast. The present-day, north–south–trending Teton Range began forming only about 10 million years ago when a large part of the crust was uplifted along a large fracture called the Teton fault, which parallels the east side of the range.

The Teton Range is made up of a variety of rocks, but most of those exposed are Precambrian-aged metamorphic and plutonic rocks that formed at great depth. These rocks were overlain by a thick sequence of younger sedimentary rocks that are now present only on the west flank of the range. Movement on the Teton fault resulted in uplift of the Teton Range relative to Jackson Hole, the block to the east; the total displacement on this fault is about 6,100 m. As the Teton block was uplifted, the overlying sedimentary rocks were eroded, exposing the underlying metamorphic and plutonic rocks. Displacement of recent sedimentary deposits along the east flank of the Teton Range shows that uplift is continuing today.

Deformation of the Earth's crust was responsible for the origin of the Teton Range, but its spectacular, rugged scenery developed rather recently and in response to a completely different geologic process—glaciation. Currently, the range supports about a dozen small glaciers, but periodically during the last 200,000 years it had many more and much larger glaciers. Glaciers are particularly effective agents of erosion; they deeply scoured the valleys and intricately sculpted the uplifted Teton block, producing excellent examples of glacial landforms. The Grand Teton, which is a glacial feature known as a horn peak, is one of the most prominent of these.

INTRODUCTION

Deformed rocks are a manifestation of the dynamic nature of the Earth. Many ancient rocks are fractured or highly contorted, clearly indicating that forces within the Earth caused deformation during the past. Such deformation is not restricted to the past, however; seismic activity and continuing deformation at convergent, divergent, and transform plate boundaries indicate that deforming forces remain active.

Mountains can form in a variety of ways, some of which involve little or no deformation, but in most mountains the rocks have been complexly deformed by compressive forces at convergent plate boundaries. The Alps of Europe, the Appalachians of North America, the Himalayas of Asia, and many others owe their existence, and in some cases continuing evolution, to plate convergence. In short, deformation and mountain building are closely related phenomena.

A large part of this chapter is devoted to a review of the various types of geologic structures, their descriptive terminology, and the forces responsible for them. The study of deformed rocks has several applications. For instance, geologic structures, such as folds and fractures resulting from deformation, provide a record of the kinds and intensities of forces that operated during the past. By interpreting these structures, geologists can make inferences about Earth history that enable us to satisfy our curiosity about the past or search more efficiently for various natural resources. Understanding the nature of geologic structures helps geologists find and recover resources such as petroleum and natural gas (see Figure 7-18b). Local geologic structures must also be considered when selecting sites for dams, large bridges, and nuclear power plants, especially if the sites are in areas of active deformation.

DEFORMATION

Fractured and contorted rocks such as those in ➤ Figure 10-1 are said to be **deformed**; that is, their original shape or volume or both have been altered by applied forces. If the intensity of the force is greater than the internal strength of the rock, it will be deformed by folding or fracturing.

Three types of deforming forces are recognized: compression, tension, and shear. **Compression** results when rocks are squeezed or compressed by external forces directed toward one another. Rock layers subjected to compression are commonly shortened in the direction of stress by folding or faulting (➤ Figure 10-2a). **Tension** results from forces acting in opposite directions along the same line and tends to

> **FIGURE 10-1** Rock layers deformed by folding and fracturing.

mation, as when they are folded, or they behave as brittle solids and **fracture** (Figure 10-1).

The type of deformation that occurs depends on the kind of force applied, the amount of pressure, the temperature, the rock type, and the length of time the rock is subjected to the stress. A small stress applied over a long period of time will cause plastic deformation. By contrast, a large stress applied rapidly to the same object, as when it is struck by a hammer, will probably result in fracture. Rock type is important because not all rocks respond to stress in the same way. Some rocks exhibit a great amount of plastic deformation, whereas brittle rocks show little or none and most commonly deform by fracturing.

Many rocks show the effects of plastic deformation that must have occurred deep within the Earth's crust where the temperature and pressure are high. At or near the surface, most rocks behave as brittle solids, whereas under conditions of high temperature and high pressure, they commonly deform plastically.

Strike and Dip

According to the principle of original horizontality, sediments are deposited in nearly horizontal layers. Thus, sedimentary rock layers that are steeply inclined must have been tilted following deposition and lithification. Some igneous rocks, especially ash falls and many lava flows, also form nearly horizontal layers. To describe the orientation of deformed rock layers, geologists use the concept of *strike and dip*.

Strike is the direction of a line formed by the intersection of a horizontal plane with an inclined plane, such as a rock layer. In ➤ Figure 10-3, the surface of any of the tilted rock layers constitutes an inclined plane, and the direction of the line formed by the intersection of a horizontal plane with

lengthen rocks or pull them apart (Figure 10-2b). In **shear**, forces act parallel to one another but in opposite directions, resulting in deformation by displacement of adjacent layers along closely spaced planes (Figure 10-2c).

Deformation is characterized as **elastic** if a deformed object returns to its original shape when the forces are relaxed. Squeezing a tennis ball causes it to deform, but once the ball is released, it returns to its original shape. Most rocks show only a very limited amount of elastic behavior. If forces of sufficient intensity are applied to rocks, they are deformed beyond their elastic limit and cannot recover their original shape. Under these conditions, rocks exhibit **plastic defor-**

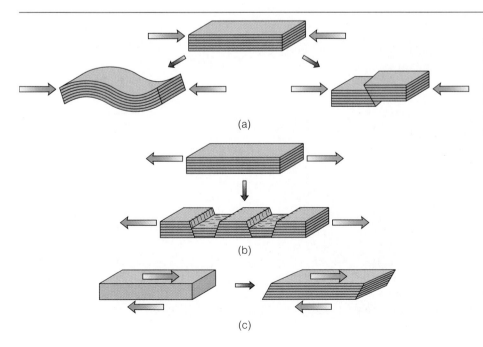

(a)

(b)

(c)

> **FIGURE 10-2** Stress and possible types of resulting deformation. (*a*) Compression causes shortening of rock layers by folding or faulting. (*b*) Tension lengthens rock layers and causes faulting. (*c*) Shear causes deformation by displacement along closely spaced planes.

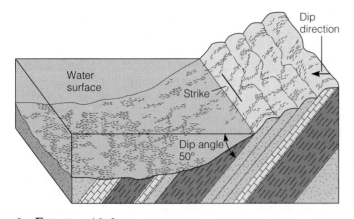

> **FIGURE 10-3** Strike and dip. The strike is formed by the intersection of a horizontal plane (the water surface) with the surface of an inclined plane (the surface of the rock layer). The dip is the maximum angular deviation of the inclined plane from horizontal.

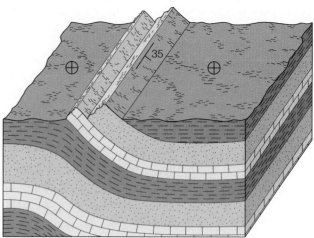

> **FIGURE 10-4** A monocline. Notice the strike and dip symbols. The other symbols indicate horizontal layers.

any of these inclined planes is the strike. The strike line's orientation is determined by using a compass to measure its angle with respect to north. **Dip** is a measure of the maximum angular deviation of an inclined plane from horizontal, so it must be measured perpendicular to the strike direction (Figure 10-3).

Geologic maps indicate strike and dip by using a long line oriented in the strike direction and a short line perpendicular to the strike line and pointing in the dip direction (Figure 10-3). The number adjacent to the strike and dip symbol indicates the dip angle.

Folds

If you place your hands on a tablecloth and move them toward one another, the tablecloth is deformed by compression into a series of up- and down-arched folds. Similarly, rock layers within the Earth's crust commonly respond to compression by folding. Unlike folds in the tablecloth, however, folding in rock layers is permanent; that is, the rocks have been deformed plastically, so that once folded, they stay folded. Most folding probably occurs deep within the crust because rocks at or near the surface are brittle and generally deform by fracturing rather than by folding.

> **FIGURE 10-5** Anticline and syncline in the Calico Mountains of southeastern California.

Monoclines, Anticlines, and Synclines. A **monocline** is a simple bend or flexure in otherwise horizontal or uniformily dipping rock layers (▷ Figure 10-4). An **anticline** is an up-arched fold, while a **syncline** is a down-arched fold (▷ Figure 10-5). A line along the center of a fold is its *axis*, and the rocks on each side of the axis are *limbs*. Because folds most commonly occur as a series of anticlines alternating with synclines, a limb is generally shared by an anticline and an adjacent syncline (▷ Figure 10-6a).

Even where the exposed view has been deeply eroded, anticlines and synclines can easily be distinguished from each other by strike and dip and by the relative ages of the folded rocks. As Figure 10-6b shows, in an eroded anticline, each limb dips outward or away from the center of the fold, where the oldest rocks are. In eroded synclines, on the other hand, each limb dips inward toward the fold's axis, and the youngest rocks coincide with the center of the fold.

In some folds each limb dips at the same angle, in which case the fold is characterized as *symmetrical* (Figure 10-6).

Not uncommonly, however, the limbs dip at different angles, and the fold is *asymmetric* (▷ Figure 10-7a). In an *overturned fold*, both limbs dip in the same direction (Figure 10-7b). Overturned folds are particularly common in mountain ranges that formed by compression at convergent plate boundaries.

Plunging Folds. Folds may be further characterized as *nonplunging* or *plunging*. In the former, the fold *axis*, a line along the folded beds, is horizontal (Figure 10-6a). But it is much more common for the axis to be inclined so that it appears to plunge beneath the surrounding rocks; folds possessing an inclined axis are **plunging folds** (▷ Figure 10-8). To differentiate plunging anticlines from plunging synclines, geologists use exactly the same criteria used for nonplunging folds: that is, all rocks dip away from the fold axis in plunging anticlines, whereas in plunging synclines all rocks dip inward toward the axis. The oldest exposed rocks are in the center of an eroded plunging anticline, whereas the youngest

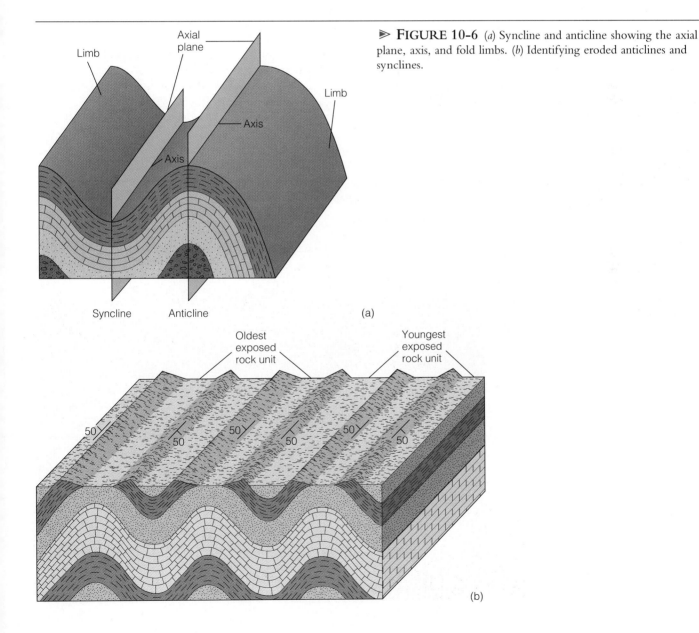

▷ **FIGURE 10-6** (*a*) Syncline and anticline showing the axial plane, axis, and fold limbs. (*b*) Identifying eroded anticlines and synclines.

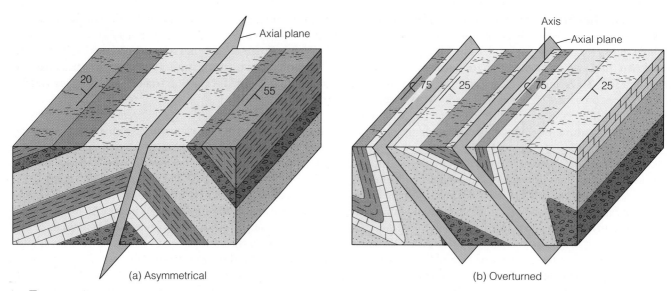

(a) Asymmetrical

(b) Overturned

➤ **FIGURE 10-7** (*a*) An asymmetrical fold. The axial plane is not vertical, and the fold limbs dip at different angles. (*b*) Overturned folds. Both fold limbs dip in the same direction, but one limb is inverted.

➤ **FIGURE 10-8** Plunging folds. (*a*) A block diagram showing surface and cross-sectional views of plunging folds. The long arrow at the center of each fold shows the direction of plunge. (*b*) Surface view of the eroded, plunging Sheep Mountain anticline in Wyoming. This fold plunges toward the observer.

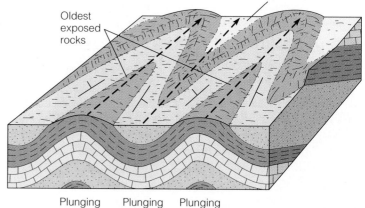

(a)

(b)

exposed rocks are in the center of an eroded plunging syncline (Figure 10-8b).

Domes and Basins. **Domes** and **basins** are the oval to circular equivalents of anticlines and synclines, which tend to be elongated structures (➤ Figure 10-9). Essentially the same criteria used in recognizing anticlines and synclines are used to identify domes and basins. In an eroded dome, the oldest exposed rocks are in the center, whereas in a basin the opposite is true. All rocks in a dome dip away from a central point, and in a basin they all dip inward toward a central point (Figure 10-9).

Some domes and basins are small structures that are easily recognized by their surface exposure patterns, but many are so large that they can be visualized only on geologic maps or aerial photographs. Many of these large-scale structures formed in the continental interior, not by compression, but as a result of vertical uplift of the Earth's crust with little additional folding and faulting.

The Black Hills of South Dakota where a central core of ancient rocks is surrounded by progressively younger rocks is a good example of an eroded dome. One of the best-known large basins is the Michigan basin. Most of it is buried beneath younger rocks, but strike and dip of exposed rocks near the basin margin and thousands of drill holes for oil and natural gas clearly show that the rock layers beneath the surface are deformed into a large basin.

Joints

Joints, which are fractures along which no movement has occurred, are the commonest structures in rocks (➤ Figure 10-10). This lack of movement is what distinguishes joints from *faults*, which do show movement along fracture surfaces. Coal miners originally used the term "joint" long ago for cracks in rocks that appeared to be surfaces where adjacent blocks were "joined" together.

Joints form when brittle rocks are deformed by fracturing. They vary in size from minute fractures to structures of regional extent and are often arranged in parallel or nearly parallel sets. Most joints and joint sets are related to other geologic structures such as faults and large folds. Weathering and erosion of jointed rocks in Utah has produced the spectacular scenery of Arches National Park (see Perspective 10-1 on pages 176–177).

We have already discussed two other types of joints in earlier chapters: columnar joints and sheet joints. Columnar joints form in some lava flows and in some plutons as cooling magma contracts and fractures develop (see Figure 5-4). Sheet joints form in response to unloading during mechanical weathering (see Figure 6-6).

Faults

A **fault** is a fracture along which blocks on opposites sides of the fracture move parallel to the fracture surface, which is a *fault plane* (➤ Figure 10-11a). Notice in Figure 10-11a that the rocks adjacent to the fault plane are designated **hanging wall block** and **footwall block**. The hanging wall block is the block of rock overlying the fault, whereas the footwall block lies beneath the fault plane. These two blocks can be recognized on any fault except a vertical one.

To recognize the various types of fault movement, one must be able to identify hanging wall and footwall blocks and understand the concept of relative movement. Geologists refer to relative movement because one usually cannot tell which block actually moved or if both blocks moved. In Figure 10-11a, for example, it cannot be determined whether the hanging wall block moved up, the footwall block moved down, or both blocks moved. Nevertheless,

---

➤ **FIGURE 10-9** In a dome (*a*), the oldest exposed rocks are in the center and all rocks dip outward from a central point, whereas in a basin (*b*), the youngest exposed rocks are in the center and all rocks dip inward toward a central point.

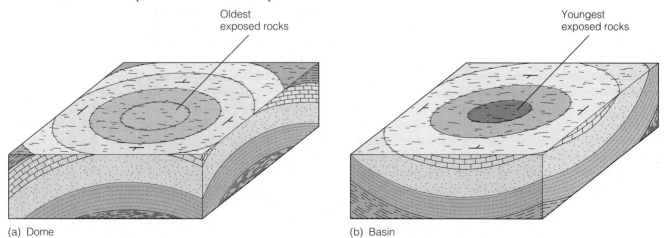

(a) Dome

(b) Basin

> ▷ FIGURE 10-10 (*a*) Rectangular joint pattern in Wales. (Photo courtesy of C. G. Tillman.) (*b*) Intersecting joints forming an "X" pattern at Marquette, Michigan.

(b)

(a)

the hanging wall block appears to have moved downward relative to the footwall block.

Like rock layers, fault planes can also be characterized by their strike and dip. Two basic types of faults are recognized according to whether the blocks on opposite sides of the fault plane have moved parallel to the direction of dip or along the direction of strike.

Dip-Slip Faults. In **dip-slip faults**, all movement is parallel to the dip of the fault plane (Figure 10-11a, b, and c); that is, one block moves up or down relative to the block on the opposite side of the fault plane. Depending on the relative movement of the hanging wall and footwall blocks, two types of dip-slip faults are recognized: normal and reverse.

In **normal faults**, the hanging wall block appears to have moved downward relative to the footwall block (Figure 10-11a and ▷ Figure 10-12). This type of faulting is caused by tension such as occurs when the Earth's crust is stretched and thinned by rifting. The mountain ranges in the Basin and Range Province, a large area in the western United States and northern Mexico, are bounded on one or both sides by large normal faults. A normal fault is also present along the east side of the Sierra Nevada in California where uplift of the block west of the fault has elevated the mountains more than 3,000 m above the lowlands to the east. The Teton Range in Wyoming is also bounded by a normal fault.

The second type of dip-slip fault is a **reverse fault**

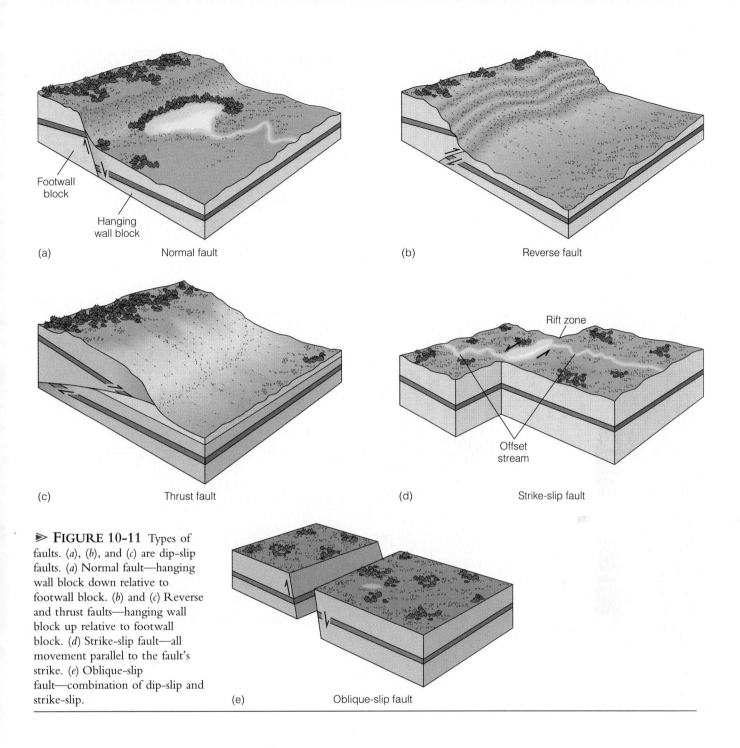

(a) Normal fault

Footwall block

Hanging wall block

(b) Reverse fault

(c) Thrust fault

Rift zone

Offset stream

(d) Strike-slip fault

➤ **FIGURE 10-11** Types of faults. (*a*), (*b*), and (*c*) are dip-slip faults. (*a*) Normal fault—hanging wall block down relative to footwall block. (*b*) and (*c*) Reverse and thrust faults—hanging wall block up relative to footwall block. (*d*) Strike-slip fault—all movement parallel to the fault's strike. (*e*) Oblique-slip fault—combination of dip-slip and strike-slip.

(e) Oblique-slip fault

(Figure 10-11b). A reverse fault with a dip of less than 45° is a *thrust fault* (Figure 10-11c). Reverse and thrust faults are easily distinguished from normal faults because the hanging wall block moves up relative to the footwall block.

Reverse and thrust faults are caused by compression, and both varieties are common in mountain ranges that formed at convergent plate boundaries. A well-known thrust fault is the Lewis overthrust of northern Montana, where a large slab of ancient Precambrian-aged rocks moved at least 75 km eastward on the fault and now rests upon much younger rock of Cretaceous age.

Strike-Slip Faults. Shearing forces are responsible for **strike-slip faulting**, a type of faulting involving horizontal movement in which blocks on opposite sides of a fault plane slide sideways past one another (Figure 10-11d). In other words, all movement is in the direction of the fault plane's strike. One of the best-known strike-slip faults is the San Andreas fault of California. Recall from Chapter 2 that the San Andreas fault is also called a transform fault in plate tectonics terminology.

Strike-slip faults can be characterized as right-lateral or left-lateral, depending on the apparent direction of offset. In

FOLDING, JOINTS, AND ARCHES

Arches National Park in eastern Utah is noted for its panoramic vistas, which include such landforms as Delicate Arch, Double Arch, Landscape Arch, and many others (➤ Figure 1). Unfortunately, the term *arch* is used for a variety of geologic features of different origin, but here we will restrict the term to mean an opening through a wall of rock that is formed by weathering and erosion.

The arches of Arches National Park continue to form as a result of weathering and erosion of the folded and jointed Entrada Sandstone, the rock underlying much of the park. Accordingly, geologic structures play a significant role in the origin of arches. Where the Entrada Sandstone was folded into anticlines, it was stretched so that parallel, vertical joints formed. Weathering and erosion occur most vigorously along joints because these processes can attack the exposed rock from both the top and the sides, whereas only the top is attacked in unjointed rocks.

➤ **FIGURE 1** Delicate Arch in Arches National Park, Utah, formed by weathering and erosion of jointed sedimentary rocks. It is 9.7 m wide and 14 m high.

➤ **FIGURE 10-12** Two small normal faults cutting through layers of volcanic ash in Oregon.

Erosion along joints causes them to enlarge, thereby forming long slender fins of rock between adjacent joints (➤ Figure 2). Some parts of these fins are more susceptible to weathering and erosion than others, and as the sides are attacked, a recess may form. If it does, eventually pieces of the unsupported rock above the recess will fall away, forming an arch as the original recess is enlarged (Figure 2). Thus, arches are remnants of fins formed by weathering and erosion along joints.

Historical observations show that arches continue to form today. For example, in 1940, Skyline Arch was enlarged when a large block fell from its underside. The park also contains many examples of arches that collapsed during prehistoric time. When arches collapse, they leave isolated pinnacles and spires. Arches National Park is well worth visiting; the pinnacles, spires, and arches are impressive features indeed.

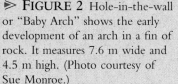

➤ **FIGURE 2** Hole-in-the-wall or "Baby Arch" shows the early development of an arch in a fin of rock. It measures 7.6 m wide and 4.5 m high. (Photo courtesy of Sue Monroe.)

Figure 10-11d, for example, an observer looking at the block on the opposite side of the fault determines whether it appears to have moved to the right or to the left. In this example, movement appears to have been to the left, so the fault is characterized as a *left-lateral strike-slip fault*. Had this been a *right-lateral strike-slip fault*, the block across the fault from the observer would appear to have moved to the right. The San Andreas fault is a right-lateral strike-slip fault.

Oblique-Slip Faults. It is possible for movement on a fault to show components of both dip-slip and strike-slip. Strike-slip movement may be accompanied by a dip-slip component giving rise to a combined movement that includes left-lateral and reverse, or right-lateral and normal (Figure 10-11e). Faults having components of both dip-slip and strike-slip movement are **oblique-slip faults**.

MOUNTAINS

The term *mountain* refers to any area of land that stands significantly higher than the surrounding country. Some mountains are single, isolated peaks, but much more commonly they are parts of a linear association of peaks and/or ridges called *mountain ranges* that are related in age and origin. A *mountain system* is a mountainous region consisting of several or many mountain ranges such as the Rocky Mountains and Appalachians. Mountain systems are complex linear zones of intense deformation and crustal thickening characterized by many of the geologic structures previously discussed.

Mountain systems are indeed impressive features and represent the effects of dynamic processes operating within the Earth. They are large-scale manifestations of tremendous

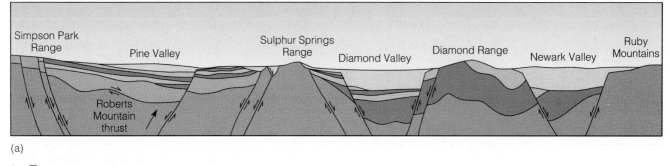

(a)

➤ **FIGURE 10-13** Block-faulting and the origin of horsts and grabens. Cross section of part of the Basin and Range Province in Nevada. The ranges (horsts) and valleys (grabens) are bounded by normal faults.

forces that have produced folded, faulted, and thickened parts of the crust. Furthermore, in some mountain systems, such as the Andes of South America and the Himalayas of Asia, the mountain-building processes remain active today.

Types of Mountains

Mountainous topography can develop in a variety of ways, some of which involve little or no deformation of the Earth's crust. For example, a single volcanic mountain can develop over a hot spot, but more commonly a series of volcanoes develops as a plate moves over the hot spot, as in the case of the Hawaiian Islands.

Block-faulting is yet another way mountains are formed. Block-faulting involves movement on normal faults so that one or more blocks are elevated relative to adjacent areas. A classic example is the large-scale block-faulting currently occurring in the Basin and Range Province of the western United States, a large area centered on Nevada but extending into several adjacent states and northern Mexico. Here, the Earth's crust is being stretched in an east-west direction, and tensional forces produce north-south–oriented, range-bounding faults. Differential movement on these faults has yielded uplifted blocks called *horsts* and down-dropped blocks called *grabens* (➤ Figure 10-13). Horsts and grabens are bounded on both sides by parallel normal faults. Erosion of the horsts has yielded the mountainous topography now present, and the grabens have filled with sediments eroded from the horsts (Figure 10-13).

The processes discussed above can certainly yield mountains. However, the truly large mountain systems of the continents, such as the Alps of Europe and the Appalachians in North America, were produced by compression along convergent plate margins.

◉ MOUNTAIN BUILDING: OROGENESIS

During an episode of mountain building, termed an **orogeny**, intense deformation occurs, generally accompanied by metamorphism and the emplacement of plutons, especially batholiths. *Orogenesis*, the process whereby mountains form, is still not fully understood, but is related to plate movements. In fact, the advent of plate tectonic theory has completely changed the way geologists view the origin of mountains.

Any theory accounting for orogenesis must explain the various characteristics of mountain systems such as their geometry and location; they tend to be long and narrow and to form at or near plate margins. Mountain systems also show intense deformation, especially compression-induced folds and reverse and thrust faults. Furthermore, the deeper, interior parts of *cores* of mountain systems are characterized by granitic plutons and regional metamorphism. The presence of deformed shallow and deep marine sedimentary rocks that have been elevated far above sea level is another feature.

Most orogenesis occurs in response to compressive forces at convergent plate boundaries. Recall from Chapter 2 that three varieties of convergent plate boundaries are recognized: oceanic-oceanic, oceanic-continental, and continental-continental.

Orogenesis at Oceanic-Oceanic Plate Boundaries

Orogenies occurring where oceanic lithosphere is subducted beneath oceanic lithosphere are characterized by the formation of a volcanic island arc and by deformation, igneous activity, and metamorphism. The subducted plate forms the outer wall of an oceanic trench, and the inner wall of the trench consists of a subduction complex or *accretionary wedge* composed of wedge-shaped slices of highly folded and faulted marine sedimentary rocks and oceanic lithosphere scraped from the descending plate (➤ Figure 10-14). This subduction complex is elevated as a result of uplift along faults as subduction continues. In addition, plate convergence results in low-temperature, high-pressure metamorphism.

Deformation also occurs in the island arc system where it is caused largely by the emplacement of plutons of intermediate and felsic composition, and many rocks show evidence of high-temperature, low-pressure metamorphism. As a result, the overriding oceanic plate is thickened as it is intruded by plutons and becomes more continental. The overall effect of island arc orogenesis is the origin of two more-or-less parallel orogenic belts consisting of a deformed volcanic island arc underlain by batholiths and a seaward belt

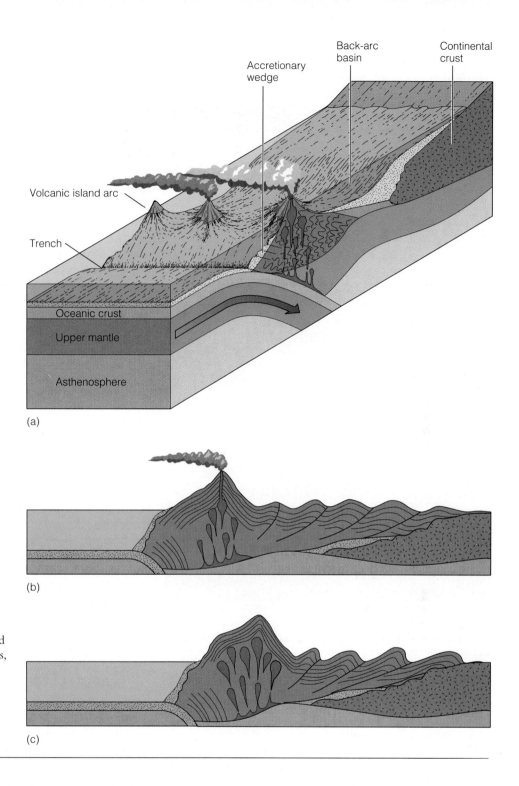

(a)

(b)

(c)

▷ **FIGURE 10-14**
(*a*) Subduction of an oceanic plate beneath an island arc. (*b*) Continued subduction, emplacement of plutons, and beginning of deformation of back-arc basin sediments. (*c*) Thrusting of back-arc basin sediments onto the adjacent continent and suturing of the island arc to the continent.

Labels in figure: Accretionary wedge, Back-arc basin, Continental crust, Volcanic island arc, Trench, Oceanic crust, Upper mantle, Asthenosphere

of deformed trench rocks (Figure 10-14). The Japanese Islands are a good example of this type of deformation.

In the back-arc basin, volcanic rocks derived from the island arc and sediments eroded from the island arc and the adjacent continent are also deformed as the plates continue to converge. The sediments are intensely folded and displaced toward the continent along low-angle thrust faults. Eventually, the entire island arc complex is fused to the edge of the continent, and the back-arc basin sediments are thrust onto the continent and form a thick stack of thrust sheets (Figure 10-14).

Orogenesis at Oceanic-Continental Plate Boundaries

Several mountain systems such as the Alps of Europe formed at oceanic-continental plate boundaries where oceanic lithosphere is subducted. The Andes of western South America are perhaps the best example of a continuing orogeny of this type (▷ Figure 10-15). Among the ranges of the Andes are the highest mountain peaks in the Americas; 49 of these peaks are more than 6,000 m high. The Andes also include many active volcanoes, and the western part of South

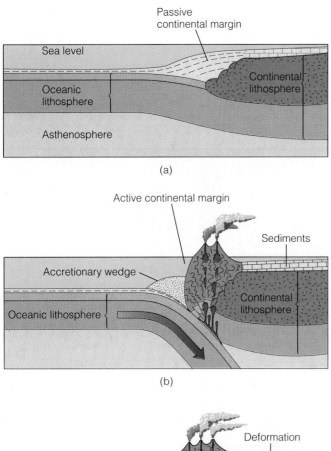

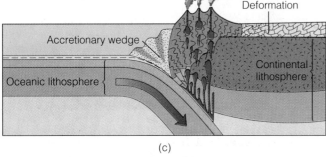

▶ **FIGURE 10-15** Generalized diagrams showing three stages in the development of the Andes of South America. (*a*) Prior to 200 million years ago, the west coast of South America was a passive continental margin. (*b*) Orogenesis began when the west coast of South America became an active continental margin. (*c*) Continued deformation, volcanism, and plutonism.

America is an extremely active segment of the circum-Pacific earthquake belt. Furthermore, one of the Earth's great oceanic trench systems, the Peru-Chile Trench, lies just off the west coast.

Prior to 200 million years ago, the western margin of South America was a passive continental margin, where sediments accumulated on the continental shelf, slope, and rise much as they do now along the east coast of North America (Figure 10-15a). When Pangaea began fragmenting in response to rifting along what is now the Mid-Atlantic Ridge, the South American plate moved westward, and an eastward-moving oceanic plate began subducting beneath the continent (Figure 10-15b). What had been a passive continental margin was now an active one.

As subduction proceeded, rocks of the continental margin and trench were folded and faulted and are now part of an accretionary wedge along the west coast of South America (Figure 10-15c). Subduction also resulted in partial melting of the descending plate, producing an andesitic volcanic arc of composite volcanoes near the edge of the continent. More viscous felsic magmas, mostly of granitic composition, were emplaced as large plutons beneath the volcanic arc (Figure 10-15c). The coastal batholith of Peru, for example, consists of perhaps 800 individual plutons that were emplaced over several tens of millions of years.

As a result of the events just described, the Andes Mountains consist of a central core of granitic rocks capped by andesitic volcanoes. To the west of this central core along the coast are the deformed rocks of the accretionary wedge. And to the east of the central core are sedimentary rocks that have been intensely folded and thrust eastward onto the continent (Figure 10-15c). Present-day subduction, volcanism, and seismicity along South America's western margin indicate that the Andes Mountains are still actively forming.

Orogenesis at Continental-Continental Plate Boundaries

The best example of orogenesis at a continental-continental plate boundary is provided by the Himalayas of Asia. The Himalayas began forming when India collided with Asia about 40 to 50 million years ago. Prior to that time, India was far south of Asia and separated from it by an ocean basin (▶ Figure 10-16a). As the Indian plate moved northward, a subduction zone formed along the southern margin of Asia where oceanic lithosphere was consumed. Partial melting generated magma, which rose to form a volcanic arc, and large granite plutons were emplaced into what is now Tibet (Figure 10-16a). At this stage, the activity along Asia's southern margin was similar to what is now occurring along the west coast of South America (10-15c).

The ocean separating India from Asia continued to close, and India eventually collided with Asia (Figure 10-16b). As a result, two continental plates became welded, or sutured, together. Thus, the Himalayas are now located within a continent rather than along a continental margin. The exact time of India's collision with Asia is uncertain, but between 40 and 50 million years ago, India's rate of northward drift decreased abruptly—from 15 to 20 cm per year to about 5 cm per year. Because continental lithosphere is not dense enough to be subducted, this decrease in rate seems to mark the time of collision and India's resistance to subduction. Consequently, India's leading margin was thrust beneath Asia, causing crustal thickening, thrusting, and uplift. Sedimentary rocks that had been deposited in the sea south of Asia were thrust northward, and two major thrust faults carried rocks of Asian origin onto the Indian plate (Figure 10-16c and d). Rocks deposited in the shallow seas along

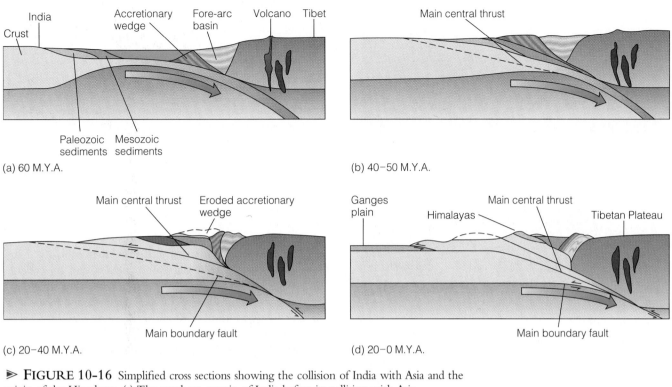

India
Crust
Accretionary wedge
Fore-arc basin
Volcano
Tibet
Paleozoic sediments
Mesozoic sediments

(a) 60 M.Y.A.

Main central thrust

(b) 40–50 M.Y.A.

Main central thrust
Eroded accretionary wedge
Main boundary fault

(c) 20–40 M.Y.A.

Ganges plain
Himalayas
Main central thrust
Tibetan Plateau
Main boundary fault

(d) 20–0 M.Y.A.

▷ FIGURE 10-16 Simplified cross sections showing the collision of India with Asia and the origin of the Himalayas. (*a*) The northern margin of India before its collision with Asia. Subduction of oceanic lithosphere beneath southern Tibet as India approached Asia. (*b*) About 40 to 50 million years ago, India collided with Asia, but because India was too light to be subducted, it was underthrust beneath Asia. (*c*) Continued convergence accompanied by thrusting of rocks of Asian origin onto the Indian subcontinent. (*d*) Since about 10 million years ago, India has moved beneath Asia along the main boundary fault. Shallow marine sedimentary rocks that were deposited along India's northern margin now form the higher parts of the Himalayas. Sediment eroded from the Himalayas has been deposited on the Ganges Plain.

India's northern margin now form the higher parts of the Himalayas. Since its collision with Asia, India has been underthrust about 2,000 km beneath Asia. Currently, India is moving north at a rate of about 5 cm per year.

MICROPLATE TECTONICS AND MOUNTAIN BUILDING

Orogenies at convergent plate boundaries result in material being added to continental margins by a process known as *accretion.* Much of the material accreted to continents during these events is simply eroded older continental crust, but a significant amount of new material is added to continents as well—igneous rocks that formed as a result of subduction and partial melting and the suturing of an island arc to a continent, for example. Although subduction is the predominant influence on tectonic history in many regions of orogenesis, other processes are also involved in mountain building and continental accretion, especially the accretion of microplates (▷ Figure 10-17).

During the late 1970s and 1980, geologists discovered that parts of many mountain systems are composed of small accreted lithospheric blocks that are clearly of foreign origin. These **microplates** differ completely from the rocks of the surrounding mountain system. In fact, many microplates are so different from adjacent rocks that most geologists think they formed elsewhere and were carried great distances as parts of other plates until they collided with other microplates or continents.

Geologic evidence indicates that more than 25% of the entire Pacific coast from Alaska to Baja California consists of accreted microplates, some of which were seamounts, volcanic island arcs, oceanic ridges, or, in some cases, simply displaced parts of a continent. It is estimated that more than 100 different-sized microplates have been added to the western margin of North America during the last 200 million years.

Most microplates so far identified are in mountains of the North American Pacific coast region, but a number of others are suspected to be present in different areas as well. They are more difficult to recognize in older mountain systems, such as the Appalachians, because of greater deformation and erosion. Nevertheless, about a dozen microplates have been identified in the Appalachians, although their boundaries are hard to discern.

The basic plate tectonic reconstructions of orogenies and continental accretion remain unchanged, but the details of these reconstructions are decidedly different in view of

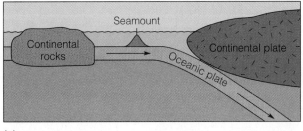

(a)

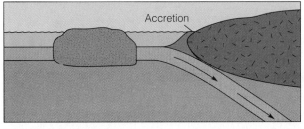

(b)

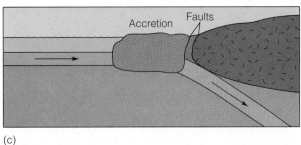

(c)

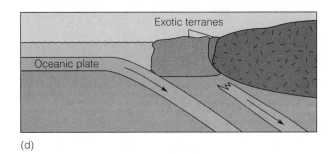

(d)

microplate tectonics. Furthermore, many of these accreted microplates are new additions to a continent rather than reworked older continental material.

THE PRINCIPLE OF ISOSTASY

Gravity studies have revealed that mountains have a low-density "root" projecting deep into the mantle. If it were not for this low-density root, a gravity survey across a mountainous area would show a huge positive gravity anomaly because an excess of mass would be present as shown in ➤ Figure 10-18a. The fact that no such anomaly exists indicates that some of the dense mantle at depth must be displaced by lighter crustal rocks (Figure 10-18b).

According to the **principle of isostasy,** the Earth's crust is in floating equilibrium with the denser mantle below. This phenomenon is easy to understand by analogy to an iceberg. Ice is slightly less dense than water, and thus it floats. According to Archimedes' principle of buoyancy, an iceberg will sink in water until it displaces a volume of water that equals its total weight. When the iceberg has sunk to an equilibrium position, only about 10% of its volume is above

➤ **FIGURE 10-17** Accretion of microplates at a convergent plate boundary. In this example, a seamount and a small block of continental rocks are scraped off a subducting plate and added to the continental margin. Much of the western part of the United States and Canada is composed of accreted microplates.

➤ **FIGURE 10-18** (a) Gravity measurements along the line shown would indicate a positive gravity anomaly over the excess mass of the mountains if the mountains were simply thicker crust resting on denser material below. (b) An actual gravity survey across a mountain region shows no departure from the expected and thus no gravity anomaly. Such data indicate that the mass of the mountains above the surface must be compensated for at depth by low-density material displacing denser material.

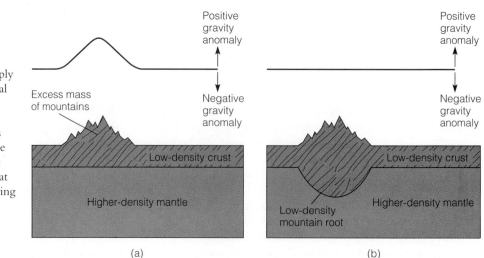

(a)

(b)

water level. If some of the ice above water level melts, the iceberg will rise in order to maintain the same proportion of ice above and below water.

The Earth's crust is similar to an iceberg in that it sinks into the mantle to its equilibrium level. Where the crust is thickest, as beneath mountain ranges, it sinks further down into the mantle but also rises higher above the equilibrium surface (Figure 10-18b). Continental crust being thicker and less dense than oceanic crust stands higher than the ocean basins. Should the crust be loaded, as where widespread glaciers accumulate, it responds by sinking further into the mantle to maintain equilibrium. In Greenland and Antarctica, for example, the surface of the crust has been depressed below sea level by the weight of glacial ice. The crust also responds isostatically to widespread erosion and sediment deposition (▷ Figure 10-19).

Unloading of the Earth's crust causes it to respond by rising upward until equilibrium is again attained. This phenomenon, known as **isostatic rebound,** occurs in areas that are deeply eroded and in areas that were formerly glaciated. Scandinavia, which was covered by a vast ice sheet until about 10,000 years ago, is still rebounding isostatically at a rate of up to 1 m per century. Isostatic rebound has also occurred in eastern Canada, where the land has risen as much as 100 m during the last 6,000 years.

The principle of isostasy implies that the mantle behaves as a liquid, but in preceding discussions, we said that the mantle must be solid because it transmits S-waves, which will not move through a liquid. How can this apparent paradox be resolved? When considered in terms of the short time necessary for S-waves to pass through it, the mantle is indeed solid. When subjected to forces over long periods of time, however, it will yield by flowage, so at these time scales it can be considered a viscous liquid. A familiar substance that has the properties of a solid or a liquid depending on how rapidly deforming forces are applied is silly putty. It will flow under its own weight if given sufficient time, but shatters as a brittle solid if struck a sharp blow.

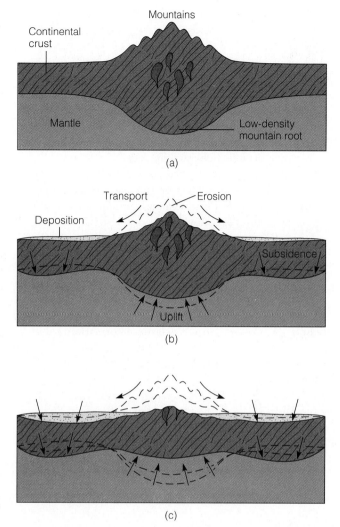

▷ **FIGURE 10-19** A diagrammatic representation showing the isostatic response of the crust to erosion (unloading) and widespread deposition (loading). Notice in (b) and (c) that isostatic rebound occurs as the mountain range is eroded, and subsidence of the crust occurs where deposition occurs.

CHAPTER SUMMARY

1. Contorted and fractured rocks have been deformed by applied forces.
2. Forces are characterized as compression, tension, or shear. Elastic deformation is not permanent, meaning that when the force is removed, the rocks return to their original shape or volume. Plastic deformation and fracture are both permanent types of deformation.
3. The orientation of deformed layers of rock is described by strike and dip.
4. Rock layers that have been buckled into up- and down-arched folds are anticlines and synclines, respectively. They can be identified by the strike and dip of the folded rocks

and by the relative age of the rocks in the center of eroded folds.
5. Domes and basins are the circular to oval equivalents of anticlines and synclines, but are commonly much larger structures.
6. Two types of structures resulting from fracturing are recognized: joints are fractures along which no movement has occurred and faults are fractures along which the blocks on opposite sides of the fracture move parallel to the fracture surface.
7. Joints, which are the commonest geologic structures, form in response to compression, tension, and shear.

8. On dip-slip faults, all movement is in the dip direction of the fault plane. Two varieties of dip-slip faults are recognized: normal faults form in response to tension, while reverse faults are caused by compression.

9. On strike-slip faults, all movement is in the direction of the fault plane's strike. They are characterized as right-lateral or left-lateral depending on the apparent direction of offset of one block relative to the other.

10. Some faults show components of both dip-slip and strike-slip; they are called oblique-slip faults.

11. Mountains can form in a variety of ways, some of which involve little or no folding or faulting. Mountain systems consisting of several mountain ranges result from deformation related to plate movements.

12. A volcanic island arc, deformation, igneous activity, and metamorphism characterize orogenies occurring at oceanic-oceanic plate boundaries. Subduction of oceanic lithosphere at an oceanic-continental plate boundary also results in orogeny.

13. Some mountain systems, such as the Himalayas, are within continents far from a present-day plate boundary. These mountains formed when two continental plates collided and became sutured.

14. According to the principle of isostasy, the Earth's crust is floating in equilibrium with the denser mantle below. Continental crust stands higher than oceanic crust because it is thicker and less dense.

IMPORTANT TERMS

anticline
basin
compression
deformation
dip
dip-slip fault
dome

elastic deformation
fault
footwall block
fracture
hanging wall block
isostatic rebound
joint

microplate
monocline
normal fault
oblique-slip fault
orogeny
plastic deformation
plunging fold

principle of isostasy
reverse fault
shear
strike
strike-slip fault
syncline
tension

REVIEW QUESTIONS

1. Deformation is characterized as _____ if deformed rocks regain their shape when they are no longer subjected to stress.
   a. ____ compression;
   b. ____ elastic;
   c. ____ tensional;
   d. ____ plastic;
   e. ____ shear.

2. An elongated fold in which all the rocks dip in toward the center is a(n):
   a. ____ dome;
   b. ____ monocline;
   c. ____ basin;
   d. ____ syncline;
   e. ____ anticline.

3. An oval to circular fold with all rocks dipping outward from a central point is a(n):
   a. ____ plunging anticline;
   b. ____ dome;
   c. ____ overturned syncline;
   d. ____ recumbent syncline;
   e. ____ basin.

4. A fault on which the hanging wall block appears to have moved down relative to the footwall block is a _____ fault.
   a. ____ thrust;
   b. ____ strike-slip;
   c. ____ normal;
   d. ____ reverse;
   e. ____ joint.

5. Faults on which both dip-slip and strike-slip movement have occurred are referred to as:
   a. ____ plunging;
   b. ____ recumbent;
   c. ____ oblique-slip;
   d. ____ nonplunging;
   e. ____ normal-slip.

6. Strike-slip faults:
   a. ____ are low-angle reverse faults;
   b. ____ have mainly vertical displacement;
   c. ____ have mainly horizontal movement;
   d. ____ are faults on which no movement has yet occurred;

   e. ____ are characterized by uplift of the footwall block.

7. Fractures along which no movement has occurred are:
   a. ____ joints;
   b. ____ monoclines;
   c. ____ transform faults;
   d. ____ axial planes;
   e. ____ fold limbs.

8. In mountain systems that form at continental margins:
   a. ____ the Earth's crust is thicker than average;
   b. ____ most deformation is caused by tensional stresses;
   c. ____ little or no volcanic activity occurs;
   d. ____ stretching and thinning of the continental crust occur;
   e. ____ most deformation results from rifting.

9. According to the principle of isostasy:
   a. ____ more heat escapes from

oceanic crust than from continental crust;

b. ____ the Earth's crust is floating in equilibrium with the denser mantle below;

c. ____ the Earth's crust behaves both as a liquid and a solid;

d. ____ much of the asthenosphere is molten;

e. ____ magnetic anomalies result when the crust is loaded by glacial ice.

10. What types of evidence indicate that deforming forces remain active within the Earth?

11. How do compression, tension, and shear differ from one another?

12. Domes and basins show the same patterns on geologic maps, but differ in two important ways. What are the two criteria for distinguishing between them?

13. Draw a simple cross section showing the displacement on a normal fault.

14. Draw a simple sketch map showing the displacement on a left-lateral strike-slip fault.

15. Cite two examples of mountain systems in which mountain-building processes remain active.

16. Explain why two roughly parallel orogenic belts develop where oceanic lithosphere is subducted beneath continental lithosphere.

17. How do geologists account for mountain systems within continents, such as the Himalayas of Asia?

18. If the continental crust is deeply eroded in one area and loaded by widespread, thick sedimentary deposits in another, how will it respond isostatically at each location?

◆◆ ◆◆

POINTS TO PONDER

1. Over 5 million years, rocks are displaced 6,000 m along a normal fault. What was the average yearly movement on this fault? Is this average likely to represent the actual rate of displacement on this fault?

2. How is it possible for the same kind of rock to behave both elastically and plastically?

Chapter 11

MASS WASTING

OUTLINE

Hong Kong's most destructive landslide occurred on Po Shan road on June 18, 1972. Sixty-seven people were killed when a 68 m wide portion of this steep hillside failed, destroying a four-story building and a 13-story apartment block.

On May 31, 1970, a devastating earthquake occurred about 25 km west of Chimbote, Peru. High in the Peruvian Andes, about 65 km to the east, the violent shaking from the earthquake tore loose a huge block of snow, ice, and rock from the north peak of Nevado Huascarán (6,654 m), setting in motion one of this century's worst landslides. Free-falling for about 1,000 m, this block of material smashed to the ground, displacing thousands of tons of rock and generating a gigantic debris avalanche (▷ Figure 11-1). Hurtling down the mountain's steep glacial valley at speeds up to 320 km per hour, the avalanche, consisting of more than 50,000,000 m^3 of mud, rock, and water, flowed over ridges 140 m high obliterating everything in its path.

About 3 km east of the town of Yungay, part of the debris avalanche overrode the valley walls and within seconds buried Yungay, instantly killing more than 20,000 of its residents (Figure 11-1). The main mass of the flow continued down the valley, where it overwhelmed the town of Ranrahirca and several other villages, burying about 5,000 more people. By the time the flow reached the bottom of the valley, its momentum carried it across the Rio Santa and some 60 m up the opposite bank. In a span of

roughly four minutes from the time of the initial ground shaking, some 25,000 people died, and most of the area's transportation, power, and communication networks were destroyed.

Ironically, the only part of Yungay that was not buried was Cemetery Hill, where 92 people survived by running to its top. As tragic and devastating as this debris avalanche was, it was not the first time a destructive landslide had swept down this valley. In January 1962, another large chunk of snow, ice, and rock broke off from the main glacier and generated a large debris avalanche that buried several villages and killed about 4,000 people.

▷ FIGURE 11-1 An earthquake triggered a landslide on Nevado Huascarán, Peru, that destroyed the towns of Yungay and Ranrahirca and killed more than 25,000 people.

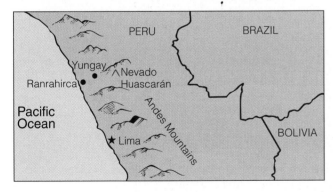

INTRODUCTION

The surface features of the Earth's land areas are the result of the interaction between the Earth's internal processes, the type of rocks exposed at the surface, the effects of weathering, and the erosional agents of water, ice, and wind. The specific type of landscape developed depends, in part, on which agent of erosion is dominant. Landslides (mass movements), which can be very destructive, are part of the normal adjustment of slopes to changing surface conditions.

Geologists use the term *landslide* in a general sense to cover a wide variety of mass movements that may cause loss of life, property damage, or a general disruption of human activities. In 218 B.C., avalanches in the European Alps buried 18,000 people; an earthquake-generated landslide in Hsian, China, killed an estimated 1,000,000 people in 1556; and 7,000 people died when mudflows and avalanches destroyed Huaraz, Peru, in 1941. What makes these mass movements so terrifying, and yet so fascinating, is that they

TABLE 11-1 Selected Landslides, Their Cause, and the Number of People Killed

Date	Location	Type	Deaths
218 B.C.	Alps (European)	Avalanche—destroyed Hannibal's army	18,000
1512	Alps (Biasco)	Landslide—temporary lake burst	>600
1556	China (Hsian)	Landslides—earthquake triggered	1,000,000
1689	Austria (Montaton Valley)	Avalanche	>300
1806	Switzerland (Goldau)	Rock glide	457
1881	Switzerland (Elm)	Rockfall	115
1892	France (Haute-Savoie)	Icefall, mudflow	150
1903	Canada (Frank, Alberta)	Rock glide	70
1920	China (Kansu)	Landslides—earthquake triggered	~200,000
1936	Norway (Loen)	Rockfall into fiord	73
1941	Peru (Huaraz)	Avalanche and mudflow	7,000
1959	USA (Madison Canyon, Montana)	Landslide—earthquake triggered	26
1962	Peru (Mt. Huascarán)	Ice avalanche and mudflow	~4,000
1963	Italy (Vaiont Dam)	Landslide—subsequent flood	~2,000
1966	Brazil (Rio de Janeiro)	Landslides	279
1966	United Kingdom (Aberfan, South Wales)	Debris flow—collapse of mining waste tip	144
1970	Peru (Mt. Huascarán)	Rockfall and debris avalanche—earthquake triggered	25,000
1971	Canada (St. Jean-Vianney, Quebec)	Quick clays	31
1972	USA (West Virginia)	Landslide and mudflow—collapse of mining waste tip	400
1974	Peru (Mayunmarca)	Rock glide and debris flow	430
1978	Japan (Myoko Kogen Machi)	Mudflow	12
1979	Indonesia (Sumatra)	Landslide	80
1980	USA (Washington)	Avalanche and mudflow	63
1981	Indonesia (West Irian)	Landslide—earthquake triggered	261
1981	Indonesia (Java)	Mudflow	252
1983	Iran (Northern area)	Landslide and avalanche	90
1987	El Salvador (San Salvador)	Landslide	1,000
1988	Chile (Tupungatito area)	Mudflow	41
1989	Tadzhikistan	Mudflow—earthquake triggered	274
1989	Indonesia (West Irian)	Landslide—earthquake triggered	120
1991	Guatemala (Santa Maria)	Landslide	33
1994	Colombia (Paez River Valley)	Avalanche—earthquake triggered	>300

SOURCE: Data from J. Whittow, *Disasters: The Anatomy of Environmental Hazards* (Athens, Ga.: University of Georgia Press, 1979) and *Geotimes*.

almost always occur with little or no warning and are over in a very short time, leaving behind a legacy of death and destruction (◉ Table 11-1).

Mass wasting (also called *mass movement*) is defined as the downslope movement of material under the direct influence of gravity. Most types of mass wasting are aided by weathering and usually involve surficial material. The material moves at rates ranging from almost imperceptible, as in the case of creep, to extremely fast as in a rockfall or slide. While water can play an important role, the relentless pull of gravity is the major force behind mass wasting.

Mass wasting is an important geologic process that can occur at any time and almost any place. Though most people associate mass wasting with steep and unstable slopes, it can also occur on near-level land, given the right geologic conditions. Furthermore, while the rapid types of mass wasting, such as avalanches and mudflows, typically get the most publicity, the slow, imperceptible types, such as creep, usually do the greatest amount of property damage.

FACTORS INFLUENCING MASS WASTING

When the gravitational force acting on a slope exceeds its resisting force, slope failure (mass wasting) occurs. The resisting forces helping to maintain slope stability include the slope material's strength and cohesion, the amount of internal friction between grains, and any external support of the slope (▷ Figure 11-2). These factors collectively define a slope's **shear strength.**

Opposing a slope's shear strength is the force of gravity. Gravity operates vertically but has a component acting parallel to the slope, thereby causing instability (Figure 11-2). The greater a slope's angle, the greater the component of force acting parallel to the slope, and the greater the chance for mass wasting. The steepest angle that a slope can maintain without collapsing is its *angle of repose.* At this angle, the shear strength of the slope's material exactly counterbalances the force of gravity. For unconsolidated material, the

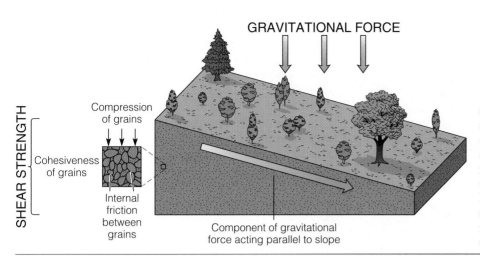

GRAVITATIONAL FORCE

SHEAR STRENGTH

Cohesiveness of grains

Compression of grains

Internal friction between grains

Component of gravitational force acting parallel to slope

▷ **FIGURE 11-2** A slope's shear strength depends on the slope material's strength and cohesiveness, the amount of internal friction between grains, and any external support of the slope. These factors promote slope stability. The force of gravity operates vertically but has a component acting parallel to the slope. When this force, which promotes instability, exceeds a slope's shear strength, slope failure occurs.

angle of repose normally ranges from 25° to 40°. Slopes steeper than 40° usually consist of unweathered rock.

All slopes are in a state of dynamic equilibrium, which means that they are constantly adjusting to new conditions. While we tend to view mass wasting as a disruptive and usually destructive event, it is one of the ways that a slope adjusts to new conditions. Whenever a building or road is constructed on a hillside, the equilibrium of that slope is affected. The slope must then adjust, perhaps by mass wasting, to this new set of conditions.

Many factors can cause mass wasting: a change in slope gradient, weakening of material by weathering, increased water content, changes in the vegetation cover, and overloading. Although most of these are interrelated, we will examine them separately for ease of discussion, but will also show how they individually and collectively affect a slope's equilibrium.

Slope Gradient

Slope gradient is probably the major cause of mass wasting. Generally speaking, the steeper the slope, the less stable it is. Therefore, steep slopes are more likely to experience mass wasting than gentle ones.

A number of processes can oversteepen a slope. One of the most common is undercutting by stream or wave action (▷ Figure 11-3). This removes the slope's base, increases the slope angle, and thereby increases the gravitational force acting parallel to the slope. Wave action, especially during storms, often results in mass movements along the shores of oceans or large lakes.

Excavations for road cuts and hillside building sites are another major cause of slope failure (▷ Figure 11-4). Grading the slope too steeply, or cutting into its side, increases the stress in the rock or soil until it is no longer strong enough to remain at the steeper angle and mass movement ensues. Such action is analogous to undercutting by streams or waves and has the same result, thus explaining why so many mountain roads are plagued by frequent mass movements.

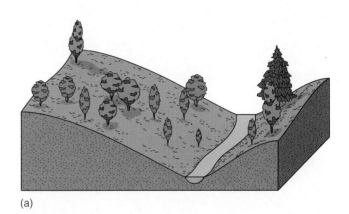

(a)

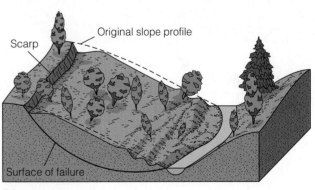

Scarp

Original slope profile

Surface of failure

(b)

▷ **FIGURE 11-3** Undercutting by stream erosion (*a*) removes a slope's base, which increases the slope angle and (*b*) can lead to slope failure.

Weathering and Climate

Mass wasting is more likely to occur in loose or poorly consolidated slope material than in bedrock. As soon as solid rock is exposed at the Earth's surface, weathering begins to disintegrate and decompose it, reducing its shear strength and increasing its susceptibility to mass wasting. The deeper the weathering zone extends, the greater the likelihood of some type of mass movement.

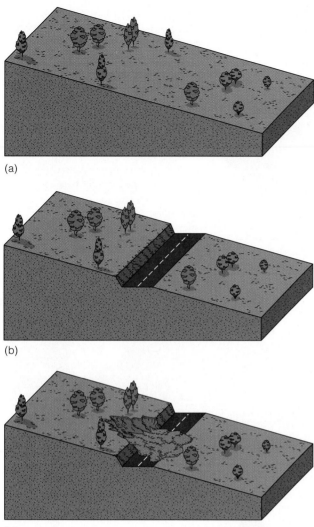

(a)

(b)

(c)

> **FIGURE 11-4** (*a*) Highway excavations disturb the equilibrium of a slope by (*b*) removing a portion of its support as well as oversteepening it at the point of excavation. (*c*) Such action can result in frequent landslides.

Recall from Chapter 6 that some rocks are more susceptible to weathering than others and that climate plays an important role in the rate and type of weathering. In the tropics, where temperatures are high and considerable rainfall occurs, the effects of weathering extend to depths of several tens of meters, and mass movements most commonly occur in the deep weathering zone. In arid and semiarid regions, the weathering zone is usually considerably shallower. Nevertheless, intense, localized cloudbursts can drop large quantities of water on an area in a short time. With little vegetation to absorb this water, runoff is rapid and frequently results in mudflows.

Water Content

The amount of water in rock or soil influences slope stability. Large quantities of water from melting snow or heavy storms greatly increase the likelihood of slope failure. The additional weight that water adds to a slope can be enough to cause mass movement. Furthermore, water percolating through a slope's material helps to decrease friction between grains, contributing to a loss of cohesion. For example, slopes composed of dry clay are usually quite stable, but when wetted, they quickly lose cohesiveness and internal friction and become an unstable slurry. This occurs because clay, which can hold large quantities of water, consists of platy particles that easily slide over each other when wet. For this reason, clay beds are frequently the slippery layer along which overlying rock units slide downslope (see Perspective 11-1).

Vegetation

Vegetation affects slope stability in several ways. By absorbing the water from a rainstorm, vegetation decreases water saturation of a slope's material, and the resultant loss of shear strength frequently leads to mass wasting. On the other hand, the vegetation's root system helps to stabilize a slope

> **FIGURE 11-5** A California Highway Patrol officer stands on top of a 2 m high wall of mud that rolled over a patrol car near the Golden State Freeway on October 23, 1987. Flooding and mudslides also trapped other vehicles and closed the freeway.

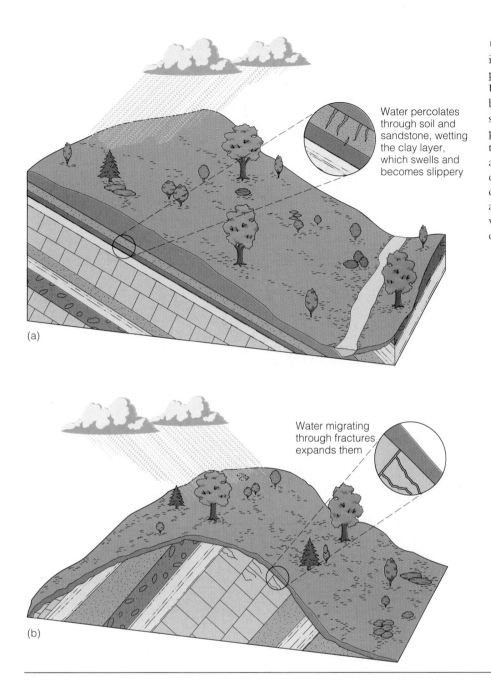

Water percolates through soil and sandstone, wetting the clay layer, which swells and becomes slippery

(a)

Water migrating through fractures expands them

(b)

➤ FIGURE 11-6 (*a*) Rocks dipping in the same direction as a hill's slope are particularly susceptible to mass wasting. Undercutting of the base of the slope by a stream removes support and steepens the slope at the base. Water percolating through the soil and into the underlying rock increases its weight and, if clay layers are present, wets the clay making them slippery. (*b*) Fractures dipping in the same directions as a slope are enlarged by chemical weathering, which can remove enough material to cause mass wasting.

by binding soil particles together and holding the soil to bedrock.

The removal of vegetation by either natural or human activity is a major cause of many mass movements. Summer brush and forest fires in southern California frequently leave the hillsides bare of vegetation. Fall rainstorms saturate the ground causing mudslides that do tremendous damage and cost millions of dollars to clean up (➤ Figure 11-5).

Overloading

Overloading is almost always the result of human activity and typically results from dumping, filling, or piling up of material. Under natural conditions, a material's load is carried by its grain-to-grain contacts, with the friction

between the grains maintaining a slope. The additional weight created by overloading increases the water pressure within the material, which in turn decreases its shear strength, thereby weakening the slope material. If enough material is added, the slope will eventually fail, sometimes with tragic consequences.

Geology and Slope Stability

The relationship between topography and the geology of an area is important in determining slope stability. If the rocks underlying a slope dip in the same direction as the slope, mass wasting is more likely to occur than if the rocks are horizontal or dip in the opposite direction (➤ Figure 11-6). When the rocks dip in the same direction as the slope, water

THE TRAGEDY AT ABERFAN, WALES

The debris brought out of underground coal mines in southern Wales typically consists of a wet mixture of various sedimentary rock fragments. This material is usually dumped along the nearest valley slope where it builds up into large waste piles called tips. A tip is fairly stable as long as the material composing it is relatively dry and its sides are not too steep.

Between 1918 and 1966, seven large tips composed of mine debris were built at various elevations on the valley slopes above the small coal-mining village of Aberfan. Shortly after 9:00 A.M. on October 21, 1966, the 250 m high, rain-soaked Tip No. 7 collapsed, and a black sludge flowed down the valley with the roar of a loud train (▶ Figure 1). Before it came to a halt 800 m from its starting place, the flow had destroyed two farm cottages, crossed a canal, and buried Pantglas Junior School, suffocating virtually all the children of Aberfan. A total of 144 people died in the flow, among them 116 children who had gathered for morning assembly in the school.

After the disaster, everyone asked, "Why did this tragedy occur and could it have been prevented?" The subsequent investigation revealed that no stability studies had ever been made on the tips and that repeated warnings about potential failure of the tips, as well as previous slides, had all been ignored.

In 1939, 8 km to the south, a tip constructed under conditions almost identical to those of Tip No. 7 collapsed. Luckily, no one was injured, but unfortunately, the failure was soon forgotten and the Aberfan tips continued to grow. In 1944 Tip No. 4 failed, and again no one was injured.

In 1958 Tip No. 7 was sited solely on the basis of available space, with no regard to the area's geology. In spite of previous tip failures and warnings of slope failure by tip workers and others, mine debris was being piled onto Tip No. 7 until the day of the disaster.

What exactly caused Tip No. 7 and the others to fail? The official investigation revealed that the foundation of the tips had become saturated with water from the springs over which they were built. In the case of the collapsed tips, pore pressure from the water exceeded the friction between grains, and the entire mass liquefied like a "quicksand." Behaving as a liquid, the mass quickly moved downhill spreading out laterally. As it flowed, water escaped from the mass, and the sedimentary particles regained their cohesion.

Following the inquiry, it was recommended that a National Tip Safety Committee be established to assess the dangers of existing tips and advise on the construction of new tip sites.

---

can percolate along the various bedding planes and decrease the cohesiveness and friction between adjacent rock layers (Figure 11-6a). This is particularly true when clay layers are present because clay becomes very slippery when wet.

Even if the rocks are horizontal or dip in a direction opposite to that of the slope, joints may dip in the same direction as the slope. Water migrating through them weathers the rock and expands these openings until the weight of the overlying rock causes it to fall (Figure 11-6b).

Triggering Mechanisms

While the factors discussed thus far all contribute to slope instability, most—though not all—rapid mass movements are triggered by a force that temporarily disturbs slope equilibrium. The most common triggering mechanisms are strong vibrations from earthquakes and excessive amounts of water from a winter snow melt or a heavy rainstorm.

Volcanic eruptions, explosions, and even loud claps of thunder may also be enough to trigger a landslide if the slope is sufficiently unstable. Many *avalanches,* which are rapid movements of snow and ice down steep mountain slopes, are triggered by the sound of a loud gunshot or, in rare cases, even a person's shout.

TYPES OF MASS WASTING

Geologists recognize a variety of mass movements (◉ Table 11-2). Some are of one distinct type, while others are a combination of different types. It is not uncommon for one type of mass movement to change into another along its course. Even though many slope failures are combinations of different materials and movements, it is still convenient to classify them according to their dominant behavior.

Mass movements are generally classified on the basis of

➤ **FIGURE 1** Location map and aerial view of the Aberfan tip disaster in which 144 people died.

three major criteria (Table 11-2): (1) rate of movement (rapid or slow); (2) type of movement (primarily falling, sliding, or flowing); and (3) type of material involved (rock, soil, or debris).

Rapid mass movements involve a visible movement of material. Such movements usually occur quite suddenly, and the material moves very quickly downslope. Rapid mass movements are potentially dangerous and frequently result in loss of life and property damage. Most rapid mass movements occur on relatively steep slopes and can involve rock, soil, or debris.

Slow mass movements advance at an imperceptible rate and are usually only detectable by the effects of their movement such as tilted trees and power poles or cracked foundations. Although rapid mass movements are more dramatic, slow mass movements are responsible for the downslope transport of a much greater volume of weathered material.

Falls

Rockfalls are a common type of extremely rapid mass movement in which rocks of any size fall through the air (➤ Figure 11-7). Rockfalls occur along steep canyons, cliffs, and road cuts and build up accumulations of loose rocks and rock fragments at their base called *talus* (see Figure 6-5).

Rockfalls result from failure along joints or bedding planes in the bedrock and are commonly triggered by natural or human undercutting of slopes or by earthquakes. Rockfalls range in size from small rocks falling from a cliff to massive falls involving millions of cubic meters of debris that destroy buildings, bury towns, and block highways.

Rockfalls are a particularly common hazard in mountainous areas where roads have been built by blasting and grading through steep hillsides of bedrock (➤ Figure 11-8). Slopes particularly prone to rockfalls are sometimes covered with wire mesh in an effort to prevent dislodged rocks from falling

TABLE 11-2 Classification of Mass Movements and Their Characteristics

Type of Movement	Subdivision	Characteristics	Rate of Movement
Falls	Rockfall	Rocks of any size fall through the air from steep cliffs, canyons, and road cuts	Extremely rapid
Slides	Slump	Movement occurs along a curved surface of rupture; most commonly involves unconsolidated or weakly consolidated material	Extremely slow to moderate
	Rock glide	Movement occurs along a generally planar surface	Rapid to very rapid
Flows	Mudflow	Consists of at least 50% silt- and clay-sized particles and up to 30% water	Very rapid
	Debris flow	Contains larger-sized particles and less water than mudflows	Rapid to very rapid
	Earthflow	Thick, viscous, tongue-shaped mass of wet regolith	Slow to moderate
	Quick clays	Composed of fine silt and clay particles saturated with water; when disturbed by a sudden shock, lose their cohesiveness and flow like a liquid	Rapid to very rapid
	Solifluction	Water-saturated surface sediment	Slow
	Creep	Downslope movement of soil and rock	Extremely slow
Complex movements		Combination of different movement types	Slow to extremely rapid

to the road below. Another tactic is to put up wire mesh fences along the base of the slope to catch or slow down bouncing or rolling rocks.

Slides

A **slide** involves movement of material along one or more surfaces of failure. The type of material may be soil, rock, or

➤ **FIGURE 11-7** Rockfalls result from failure along cracks, fractures, or bedding planes in the bedrock and are common features in areas of steep cliffs.

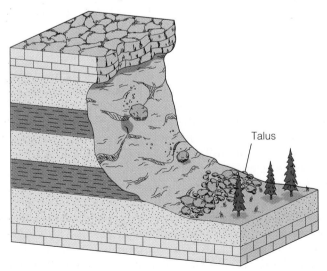

Talus

➤ **FIGURE 11-8** Cutting into the hillside to construct this portion of the Pan American Highway in Mexico resulted in a rockfall that completely blocked the road. (Photo courtesy of R. V. Dietrich.)

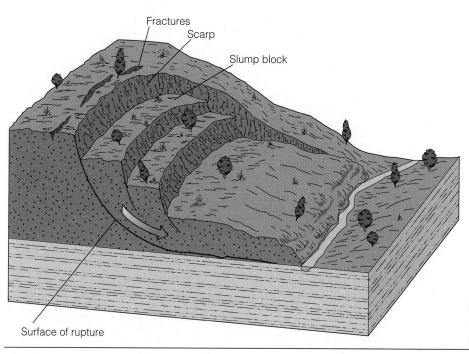

▷ **FIGURE 11-9** In a slump, material moves downward along a curved surface of a rupture, causing the slump block to rotate backward. Most slumps involve unconsolidated or weakly consolidated material and are typically caused by erosion along the slope's base.

a combination of the two, and it may break apart during movement or remain intact. A slide's rate of movement can vary from extremely slow to very rapid (Table 11-2).

Two types of slides are generally recognized: (1) slumps or rotational slides, in which movement occurs along a curved surface; and (2) rock or block glides, which move along a more-or-less planar surface.

A **slump** involves the downward movement of material along a curved surface of rupture and is characterized by the backward rotation of the slump block (▷ Figure 11-9). Slumps occur most commonly in unconsolidated or weakly consolidated material and range in size from small individual sets, such as occur along stream banks, to massive, multiple sets that affect large areas and cause considerable damage.

Slumps can be caused by a variety of factors, but the most common is erosion along the base of a slope, which removes support for the overlying material. This local steepening may be caused naturally by stream erosion along its banks (Figure 11-9) or by wave action at the base of a coastal cliff. Slope oversteepening can also be caused by human activity, such as the construction of highways and housing developments. Slumps are particularly prevalent along highway cuts where they are generally the most frequent type of slope failure observed.

While many slumps are merely a nuisance, large-scale slumps involving populated areas and highways can cause extensive damage. Such is the case in coastal southern California where slumping and sliding have been a constant problem. Many areas along the coast are underlain by poorly to weakly consolidated silts, sands, gravels, and clay layers, some of which are weathered ash falls. In addition, southern California is tectonically active so that many of these deposits are cut by faults and joints, which allow the infrequent rains

to percolate downward rapidly, wetting and lubricating the clay layers.

Southern California lies in a semiarid climate and is dry most of the year. When it does rain, typically between November and March, large amounts of rain can fall in a short time. Thus, the ground quickly becomes saturated, leading to landslides along steep canyon walls as well as along coastal cliffs (▷ Figure 11-10). Most of the slope failures along the southern California coast are the result of slumping.

A **rock** or *block* **glide** occurs when rocks move downslope along a more-or-less planar surface. Most rock glides occur because the local slopes and rock layers dip in the same direction (▷ Figure 11-11), although they can also occur along fractures parallel to a slope. Rock glides are common occurrences along the southern California coast. At Point Fermin, seaward-dipping rocks with interbedded slippery clay layers are undercut by waves causing numerous glides (▷ Figure 11-12).

Flows

Mass movements in which material flows as a viscous fluid or displays plastic movement are termed *flows*. Their rate of movement ranges from extremely slow to extremely rapid (Table 11-2). In many cases, mass movements begin as falls, slumps, or slides and change into flows further downslope.

Of the major mass movement types, **mudflows** are the most fluid and move most rapidly (at speeds up to 80 km per hour). They consist of at least 50% silt- and clay-sized material combined with a significant amount of water (up to 30%). Mudflows are common in arid and semiarid environments where they are triggered by heavy rainstorms that quickly saturate the regolith, turning it into a raging flow of

➤ **FIGURE 11-10** Undercutting of steep sea cliffs by wave action resulted in massive slumping in the Pacific Palisades area of southern California on March 31 and April 3, 1958. Highway 1 was completely blocked. Note the heavy earthmoving equipment for scale.

➤ **FIGURE 11-11** Rock glides occur when material moves downslope along a generally planar surface.

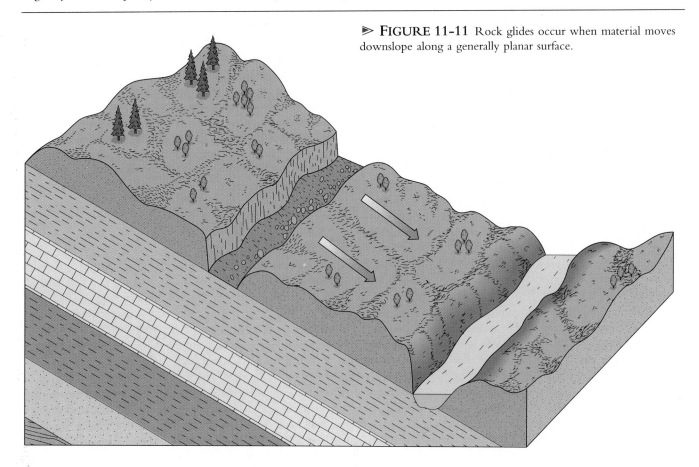

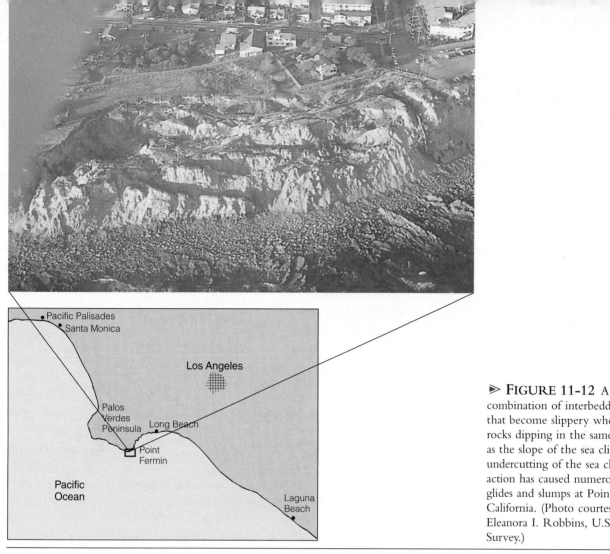

▷ **FIGURE 11-12** A combination of interbedded clay beds that become slippery when wet, rocks dipping in the same direction as the slope of the sea cliffs, and undercutting of the sea cliffs by wave action has caused numerous rock glides and slumps at Point Fermin, California. (Photo courtesy of Eleanora I. Robbins, U.S. Geological Survey.)

mud that engulfs everything in its path. Mudflows can also occur in mountain regions (▷ Figure 11-13) and in areas covered by volcanic ash where they can be particularly destructive (see Chapter 5). Because mudflows are so fluid, they generally follow preexisting channels until the slope decreases or the channel widens, at which point they fan out.

Debris flows are composed of larger-sized particles than those in mudflows and do not contain as much water. Consequently, they are usually more viscous than mudflows, typically do not move as rapidly, and rarely are confined to preexisting channels. Debris flows can be just as damaging, though, because they can transport large objects.

Earthflows move more slowly than either mudflows or debris flows. An earthflow slumps from the upper part of a hillside, leaving a scarp, and flows slowly downslope as a thick, viscous, tongue-shaped mass of wet regolith (▷ Figure 11-14). Like mudflows and debris flows, earthflows can be of any size and are frequently destructive. They occur most commonly in humid climates on grassy soil-covered slopes following heavy rains.

Some clays spontaneously liquefy and flow like water when they are disturbed. Such **quick clays** have caused

serious damage and loss of lives in Sweden, Norway, eastern Canada, and Alaska (Table 11-1). Quick clays are composed of silt and clay particles made by the grinding action of glaciers. Geologists think these fine sediments were originally deposited in a marine environment where their pore space was filled with salt water. The ions in the salt water helped establish strong bonds between the clay particles, thus stabilizing and strengthening the clay. When the clays were subsequently uplifted above sea level, the salt water was flushed out by fresh groundwater, reducing the effectiveness of the ionic bonds between the clay particles and thereby reducing the overall strength and cohesiveness of the clay. Consequently, when the clay is disturbed by a sudden shock or shaking, it essentially turns to a liquid and flows.

An example of the damage that can be done by quick clays occurred in the Turnagain Heights area of Anchorage, Alaska, in 1964 (▷ Figure 11-15). Underlying most of the Anchorage area is the Bootlegger Cove Clay, a massive clay unit of poor permeability. Because the Bootlegger Cove Clay forms a barrier preventing groundwater from flowing through the adjacent glacial deposits to the sea, considerable hydraulic pressure builds up behind the clay. Some of this water has flushed out the salt water in the clay and also has

➤ **FIGURE 11-13** Mudflow near Estes Park, Colorado.

➤ **FIGURE 11-14** (*a*) Earthflows form tongue-shaped masses of wet regolith that move slowly downslope. They occur most commonly in humid climates on grassy soil-covered slopes. (*b*) An earthflow near Baraga, Michigan.

Scarp

(a)

(b)

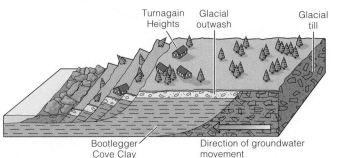

Turnagain Heights   Glacial outwash   Glacial till

Bootlegger Cove Clay   Direction of groundwater movement

(a)

▷ **FIGURE 11-15** (*a*) Groundshaking by the 1964 Alaska earthquake turned parts of the Bootlegger Cove Clay into a quick clay, causing numerous slides (*b*) that destroyed many homes in the Turnagain Heights subdivision of Anchorage.

(b)

saturated the lenses of sand and silt associated with the clay beds. When the 8.5-magnitude Good Friday earthquake struck on March 27, 1964, the shaking turned parts of the Bootlegger Cove Clay into a quick clay and precipitated a series of massive slides in the coastal bluffs that destroyed most of the homes in the Turnagain Heights subdivision (Figure 11-15b).

Solifluction is the slow downslope movement of water–saturated surface sediment. Solifluction can occur in any climate where the ground becomes saturated with water, but is most common in areas of permafrost. *Permafrost* is ground that remains permanently frozen. It covers nearly 20% of the world's land surface (▷ Figure 11-16a). During the warmer season when the upper portion of the permafrost thaws, water and surface sediment form a soggy mass that flows by solifluction and produces a characteristic lobate topography (Figure 11-16b).

Construction of the Alaska pipeline from the oil fields in Prudhoe Bay to the ice-free port of Valdez raised numerous concerns over the effect it might have on the permafrost and the potential for solifluction. Some thought that oil flowing through the pipeline would be warm enough to melt the permafrost, causing the pipeline to sink further into the ground and possibly rupture. After numerous studies were conducted, scientists concluded that the pipeline, completed in 1977, could safely be buried for more than half of its 1,280 km length; where melting of the permafrost might cause structural problems to the pipe, it was insulated and installed above ground.

Creep is the slowest type of flow. It is also the most widespread and significant mass wasting process in terms of the total amount of material moved downslope and the monetary damage caused annually. Creep involves extremely slow downhill movement of soil or rock. Although it can

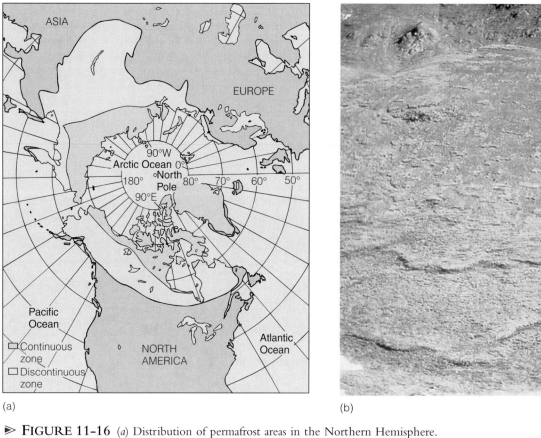

(a)   (b)

▷ **FIGURE 11-16** (*a*) Distribution of permafrost areas in the Northern Hemisphere.
(*b*) Solifluction flows near Suslositna Creek, Alaska, show the typical lobate topography that
is characteristic of solifluction conditions.

occur anywhere and in any climate, it is most effective and significant as a geologic agent in humid regions. In fact, it is the most common form of mass wasting in the southeastern United States and the southern Appalachian Mountains.

Because the rate of movement is essentially imperceptible, we are frequently unaware of creep's existence until we notice its effects: tilted trees and power poles, broken streets and sidewalks, cracked retaining walls or foundations (▷ Figure 11-17). Creep usually involves the whole hillside and probably occurs, to some extent, on any weathered or soil-covered, sloping surface.

Not only is creep difficult to recognize, it is difficult to control. Although engineers can sometimes slow or stabilize creep, many times the only course of action is to simply avoid the area if at all possible or, if the zone of creep is relatively thin, design structures that can be anchored into the solid bedrock.

Complex Movements

Recall that many mass movements are combinations of different movement types. When one type is dominant, the movement can be classified as one of the movements described thus far. If several types are more or less equally involved, it is called a *complex movement*.

The most common type of complex movement is the slide-flow in which there is sliding at the head and then some type of flowage farther along its course. Most slide-flow landslides involve well-defined slumping at the head, followed by a debris flow or earthflow. Any combination of different mass movement types can, however, be classified as a complex movement.

A *debris avalanche* is a complex movement that often occurs in very steep mountain ranges. Debris avalanches typically start out as rockfalls when large quantities of rock, ice, and snow are dislodged from a mountainside, frequently as a result of an earthquake. The material then slides or flows down the mountainside, picking up additional surface material and increasing in speed. The 1970 Peru earthquake set in motion the debris avalanche that destroyed the town of Yungay (see the Prologue).

◉ RECOGNIZING AND MINIMIZING THE EFFECTS OF MASS MOVEMENTS

The most important factor in eliminating or minimizing the damaging effects of mass wasting is a thorough geologic investigation of the region in question. In this way, former landslides and areas susceptible to mass movements can be

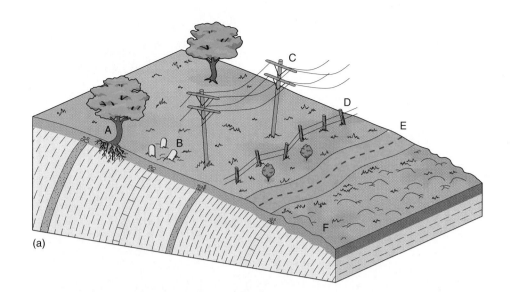

(a)

(b)

▷ **FIGURE 11-17** (*a*) Some evidence of creep: (A) curved tree trunks; (B) displaced monuments; (C) tilted power poles; (D) displaced and tilted fences; (E) roadways moved out of alignment; (F) hummocky surface. (*b*) Creep has bent these sandstone and shale beds of the Haymond Formation near Marathon, Texas.

Identifying areas with a high potential for slope failure is important in any hazard assessment study; these studies include identifying former landslides as well as sites of potential mass movement. Scarps, open fissures, displaced or tilted objects, a hummocky surface, and sudden changes in vegetation are some of the features indicating former landslides or an area susceptible to slope failure. The effects of weathering, erosion, and vegetation may, however, obscure the evidence for previous mass wasting.

Soil and bedrock samples are also studied, both in the field and laboratory, to assess such characteristics as composition, susceptibility to weathering, cohesiveness, and ability to transmit fluids. These studies help geologists and engineers predict slope stability under a variety of conditions.

Although most large mass movements usually cannot be prevented, geologists and engineers can employ various methods to minimize the danger and damage resulting from them. Because water plays such an important role in many landslides, one of the most effective and inexpensive ways to reduce the potential for slope failure and to increase existing slope stability is through surface and subsurface drainage of a hillside. Drainage serves two purposes. It reduces the weight of the material likely to slide and increases the shear strength of the slope material by lowering pore pressure.

Surface waters can be drained and diverted by ditches, gutters, or culverts designed to direct water away from slopes. Drainpipes perforated along one surface and driven into a hillside can help remove subsurface water (▷ Figure 11-18). Finally, planting vegetation on hillsides helps stabilize

identified and perhaps avoided. By assessing the risks of possible mass wasting before construction begins, steps can be taken to eliminate or minimize the effects of such events.

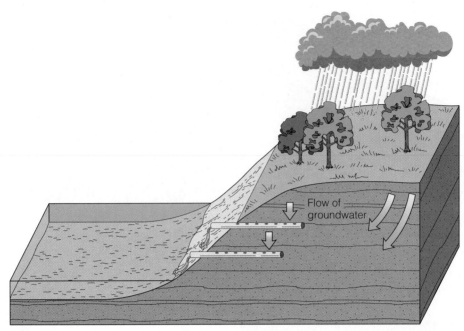

Flow of
groundwater

(a)

(b)

➤ **FIGURE 11-18** (*a*) Driving drainpipes that are perforated on one side into a hillside with the perforated side up can remove some subsurface water and thus help stabilize a hillside. (*b*) A drainpipe driven into the hillside at Point Fermin, California, helps remove subsurface water and stabilize the slope.

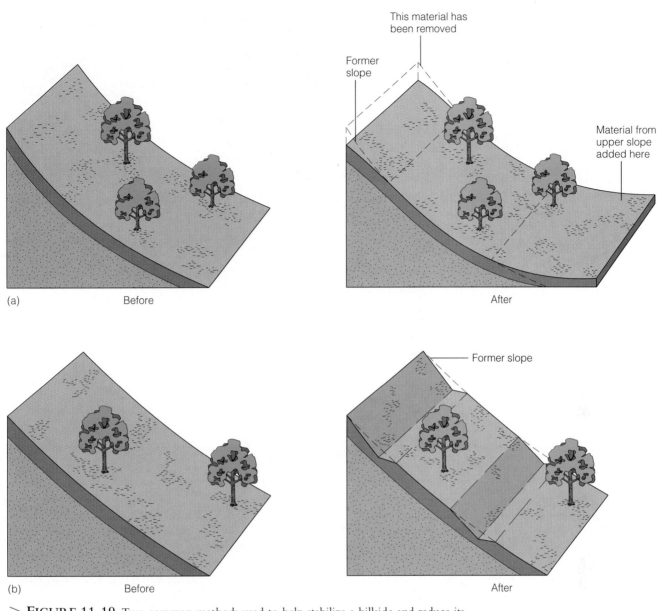

> **FIGURE 11-19** Two common methods used to help stabilize a hillside and reduce its slope. (*a*) In the cut-and-fill method, material from the steeper upper part of the hillside is removed, thereby reducing the slope angle, and is used to fill in the base. This provides some additional support at the base of the slope. (*b*) Benching involves making several cuts along a hillside to reduce its overall slope.

slopes by holding the soil together and reducing the amount of water in the soil.

Another way to help stabilize a hillside is to reduce its slope. Recall that overloading or oversteepening by grading are common causes of slope failure. By reducing the gradient of a hillside, the potential for slope failure is decreased. Two methods are commonly employed to reduce a slope's gradient. In the *cut-and-fill* method, material is removed from the upper part of the slope and used as fill at the base, thus providing a flat surface for construction and reducing the slope (➤ Figure 11-19a). The second method, which is

called *benching,* involves cutting a series of benches or steps into a hillside (Figure 11-19b). This process reduces the overall average slope, and the benches serve as collecting sites for small landslides or rockfalls that might occur. Benching is most commonly used on steep hillsides in conjunction with a system of surface drains to divert runoff.

In some situations, retaining walls can be constructed to provide support for the base of the slope (➤ Figure 11-20). These are usually anchored well into bedrock, backfilled with crushed rock, and provided with drain holes to prevent the buildup of water pressure in the hillside.

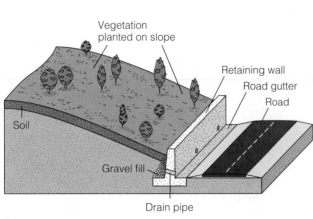

(a)   (b)

➤ **FIGURE 11-20** (*a*) Retaining walls anchored into bedrock, backfilled with gravel, and provided with drainpipes can support a slope's base and reduce landslides. (*b*) Steel retaining wall built to stabilize a slope and keep falling and sliding rocks off of the highway.

CHAPTER SUMMARY

1. Mass wasting is the downslope movement of material under the influence of gravity. It occurs when the gravitational force acting parallel to a slope exceeds the slope's strength.

2. Mass wasting frequently results in loss of life, as well as causing millions of dollars in damage annually.

3. Mass wasting can be caused by many factors including slope gradient, weathering of slope material, water content, overloading, and removal of vegetation. Usually, several of these factors in combination contribute to slope failure.

4. Mass movements are generally classified on the basis of their rate of movement (rapid versus slow), type of movement (falling, sliding, or flowing), and type of material (rock, soil, or debris).

5. Rockfalls are a common mass movement in which rocks free-fall.

6. Two types of slides are recognized. Slumps are rotational slides involving movement along a curved surface; they are most common in poorly consolidated or unconsolidated material. Rock glides occur when movement takes place along a more or less planar surface; they usually involve solid pieces of rock.

7. Several types of flows are recognized on the basis of their rate of movement (rapid vs. slow), type of material (rock, sediment, soil), and amount of water.

8. Mudflows consist of mostly clay- and silt-sized particles and

contain more than 30% water. They are most common in semiarid and arid environments and generally follow pre-existing channels.

9. Debris flows are composed of larger-sized particles and contain less water than mudflows. They are more viscous and do not flow as rapidly as mudflows.

10. Earthflows move more slowly than either debris flows or mudflows; they move downslope as thick, viscous, tongue-shaped masses of wet regolith.

11. Quick clays are clays that spontaneously liquefy and flow like water when they are disturbed.

12. Solifluction is the slow downslope movement of water-saturated surface material and is most common in areas of permafrost.

13. Creep, the slowest type of flow, is the imperceptible downslope movement of soil or rock. Creep is the most widespread of all types of mass wasting.

14. Complex movements are combinations of different types of mass movements in which one type is not dominant. Most complex movements involve sliding and flowing.

15. The most important factor in reducing or eliminating the damaging effects of mass wasting is a thorough geologic investigation of the area to outline areas susceptible to mass movements.

16. Slopes can be stabilized by retaining walls, draining excess water, regrading slopes, and planting vegetation.

creep
debris flow
earthflow
mass wasting

mudflow
quick clay
rapid mass movement
rockfall

rock glide
shear strength
slide

slow mass movement
slump
solifluction

1. Shear strength includes:
   a. ____ the strength and cohesion of material;
   b. ____ the amount of internal friction between grains;
   c. ____ gravity;
   d. ____ all of these;
   e. ____ answers (a) and (b).
2. Which of the following is not a factor influencing mass wasting?
   a. ____ gravity;
   b. ____ weathering;
   c. ____ slope gradient;
   d. ____ water content;
   e. ____ none of these.
3. Which of the following factors can actually enhance slope stability?
   a. ____ water content;
   b. ____ vegetation;
   c. ____ overloading;
   d. ____ rocks dipping in the same direction as the slope;
   e. ____ none of these.
4. Movement of material along a surface or surfaces of failure is a:
   a. ____ slide;
   b. ____ fall;
   c. ____ flow;
   d. ____ solifluction;
   e. ____ none of these.
5. Downslope movement along an essentially planar surface is a(n):
   a. ____ slump;
   b. ____ rockfall;
   c. ____ earthflow;
   d. ____ landslide;
   e. ____ rock glide.
6. The most widespread and costly type of mass wasting in terms of total material moved and monetary damage is:
   a. ____ creep;
   b. ____ solifluction;
   c. ____ mudflow;
   d. ____ debris flow;
   e. ____ slumping.
7. Which of the following features indicate former landslides or areas susceptible to slope failure?
   a. ____ displaced objects;
   b. ____ scarps;
   c. ____ hummocky surfaces;
   d. ____ open fissures;
   e. ____ all of these.
8. What are the forces that help maintain slope stability?
9. What roles do climate and weathering play in mass wasting?
10. Discuss how the relationship between topography and the underlying geology affects slope stability.
11. Where are rockfalls most common and why?
12. Why are slumps particularly common along road cuts and fills?
13. Why are quick clays so dangerous?
14. What precautions must be taken when building in permafrost areas?
15. Why is creep so prevalent, and why does it do so much damage?
16. How can you recognize areas that are susceptible to mass movement?

1. What features would you look for to determine if the site where you wanted to build a house was safe?
2. Do you think it will ever be possible to predict mass wasting events?

Chapter 12

RUNNING WATER

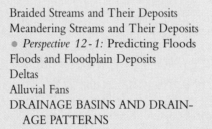

OUTLINE

Cumberland Falls on the big South Fork River in Cumberland State Resort Park, Kentucky. At 38 m wide and 18 m high, it is the second largest waterfall east of the Rocky Mountains and one of the most impressive.

According to one report, the flooding in the midwestern United States from late June to August 1993 caused an estimated $15 to $20 billion in property damage. Fifty people died as a result of the flooding, 70,000 were left homeless, and more than 200 counties in several states including every county in Iowa were declared disaster areas. In addition to Iowa, flooding also occurred in South Dakota, North Dakota, Minnesota, Wisconsin, Nebraska, Missouri, Illinois, and Kansas.

Despite the heroic efforts by thousands of volunteers and the National Guard to stabilize levees with sandbags, levees failed everywhere or were simply overtopped by the rising flood waters. By the end of the first week in July, at least 10 million acres of farmland had been flooded, with more rain expected. On July 16, an additional 17.8 cm of rain fell in North Dakota and Minnesota, and 15.2 cm more fell in South Dakota. By August, 23 million acres had

PROLOGUE

been flooded. Grafton, Illinois, at the juncture of the Mississippi and Illinois rivers was 80% underwater. And much of St. Charles County, Missouri, at the confluence of the Mississippi and Missouri rivers was so extensively flooded that 8,000 people were evacuated from this county alone (▷ Figure 12-1).

Obviously, the direct cause of flooding was too much water for the Mississippi and Missouri rivers and their tributaries to handle. But the reason so much water was present was the unusual weather in the Midwest. Thunderstorms that are usually distributed over a much larger area simply dumped their precipitation in a much smaller region, which received as much as 1½ to 2 times the normal amount of rain.

One important lesson learned from the Flood of '93 is that some floods will occur despite our best efforts at flood control. The 29 dams on the Mississippi and 26 reservoirs

(a)

▷ **FIGURE 12-1** (*a*) Portage des Sioux, St. Charles County, Missouri, on July 16, 1993. The channel of the Mississippi River is at the far right. Only eight homes in the town were not flooded or were only slightly damaged by flooding. Even a 5 m high pedestal of a statue near the Mississippi River was swamped by the flood waters. (*b*) Portage des Sioux on November 7, 1993, after the flood waters had subsided.

(b)

on upstream tributaries, as well as about 5,800 km of levees, had little success at holding the flood waters in check. No doubt debate will now focus on the utility of levees in flood control. Levees are effective in protecting many areas during floods, yet in some cases they actually exacerbate the problem by restricting the flow that would otherwise have spread over a floodplain. They are expensive to build and maintain, and their overall effectiveness has been and will continue to be questioned.

Most criticism was directed at the U.S. Army Corps of Engineers, which has spent about $25 billion in this century to build 500 dams and more than 16,000 km of levees. No one doubts that some of these projects have been successful, at least within the limits of their design. But critics charge that such flood control projects make the problem of flooding worse, particularly because flood-prone areas tend to be developed once the projects are completed even though nothing can be done to prevent some floods. As a consequence, the most destructive and most costly type of natural disaster continues to be flooding.

INTRODUCTION

Among the terrestrial planets, the Earth is unique in having abundant liquid water. Fully 71% of the Earth's surface is covered by water, and a small but important quantity of water vapor is present in its atmosphere.

The volume of water on Earth is estimated at 1.36 billion km^3, most of which (97.2%) is in the oceans. About 2% is frozen in glaciers, and the remaining 0.8% constitutes all the water in streams, lakes, swamps, groundwater, and the atmosphere (● Table 12-1). Only a tiny portion of the total water on Earth is in streams, but running water is nevertheless the most important erosional agent modifying the Earth's surface. Even in most desert regions the effects of running water are manifest, although channels are dry most of the time.

In addition to its significance as a geologic agent, running water is important for many other reasons. It is a source of fresh (nonsaline) water for industry, domestic use, and agriculture, and about 8% of the electricity used in North America is generated by falling water at hydroelectric stations. Streams have been, and continue to be, important avenues of commerce. Much of the interior of North America was first explored by following such large streams as the St. Lawrence, Mississippi, and Missouri rivers.

Much of this discussion of running water is necessarily descriptive, but one should always be aware that streams are dynamic systems that must continually respond to change. For example, paving in urban areas increases surface runoff to streams, while other human actions such as building dams and impounding reservoirs also alter the dynamics of a stream system. Natural changes, too, affect stream dynamics. When more rain falls in a stream's drainage area due to long-term climatic change, more water flows in the stream's channel, and greater energy is available for erosion and transport of sediments. In short, streams continually adjust to change.

THE HYDROLOGIC CYCLE

Water is continually recycled from the oceans, through the atmosphere, to the continents, and back to the oceans. This continual recycling of water is called the **hydrologic cycle** (➤ Figure 12-2). The hydrologic cycle, which is powered by solar radiation, is possible because water changes phases easily under Earth surface conditions. Huge quantities of water evaporate from the oceans each year as the surface waters are heated by solar energy. Approximately 85% of all water that enters the atmosphere is derived from the oceans; the remaining 15% comes from evaporation of water on land.

When water evaporates, the vapor rises into the atmosphere where the complex processes of condensation and cloud formation occur. About 80% of the precipitation falls directly into the oceans, in which case the hydrologic cycle is limited to a three-step process of evaporation, condensation, and precipitation.

About 20% of all precipitation falls on land as rain and snow, and the hydrologic cycle involves more steps: evaporation, condensation, movement of water vapor from the oceans to the continents, precipitation, and runoff and infiltration. Some of the precipitation evaporates as it falls and reenters the hydrologic cycle as vapor; water evaporated from lakes and streams also reenters the cycle as vapor as does moisture evaporated from plants by *transpiration* (Figure 12-2).

Each year about 36,000 km^3 of the precipitation falling on land returns to the oceans by **runoff,** the surface flow of streams. The water returning to the oceans by runoff enters

Location	Volume (km^3)	Percentage of Total
Oceans	1,327,500,000	97.20
Icecaps and glaciers	29,315,000	2.15
Groundwater	8,442,580	.625
Freshwater and saline lakes and inland seas	230,325	.017
Atmosphere at sea level	12,982	.001
Average in stream channels	1,255	.0001

TABLE 12-1 Water on Earth

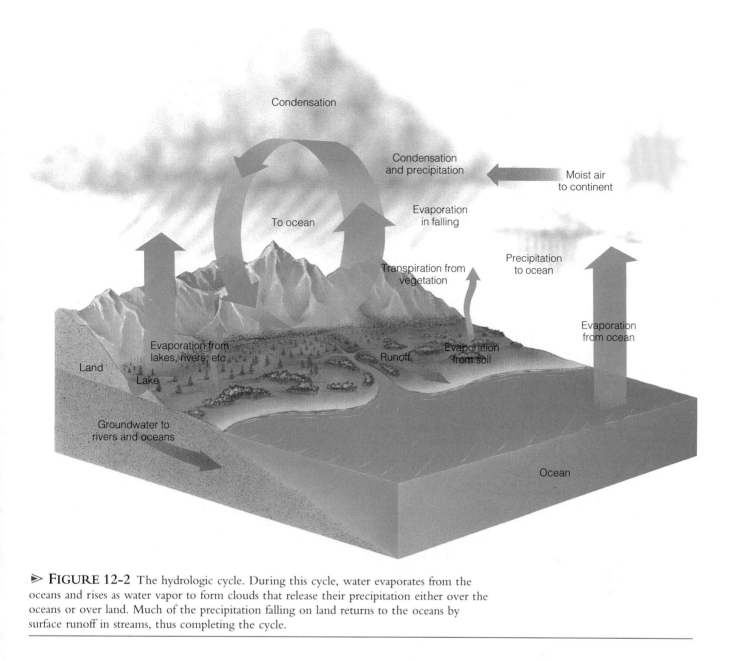

Condensation

Condensation
and precipitation

Moist air
to continent

To ocean

Evaporation
in falling

Transpiration from
vegetation

Precipitation
to ocean

Evaporation
from ocean

Evaporation from
lakes, rivers, etc.

Runoff

Evaporation
from soil

Land

Lake

Groundwater to
rivers and oceans

Ocean

➤ FIGURE 12-2 The hydrologic cycle. During this cycle, water evaporates from the oceans and rises as water vapor to form clouds that release their precipitation either over the oceans or over land. Much of the precipitation falling on land returns to the oceans by surface runoff in streams, thus completing the cycle.

the Earth's ultimate reservoir where it begins the hydrologic cycle again. Some of the precipitation falling on land is temporarily stored in lakes, snow fields, and glaciers or seeps below the surface where it is temporarily stored as groundwater. This water is effectively removed from the system for up to thousands of years, but eventually, glaciers melt, lakes feed streams, and groundwater flows into streams or directly into the oceans (Figure 12-2). Our concern here is with the comparatively small quantity returning to the oceans as runoff, for the energy of running water is responsible for a great many surface features.

RUNNING WATER

The amount of runoff in any area during a rainstorm depends on **infiltration capacity,** the maximum rate that soil or other surface materials can absorb water. Infiltration capacity depends on several factors, including the intensity and duration of rainfall. Loosely packed, dry soils absorb water faster than tightly packed, wet soils.

If rain is absorbed as fast as it falls, no surface runoff occurs. Should the infiltration capacity be exceeded, or should surface materials become saturated, excess water collects on the surface and, if a slope exists, moves downhill. Even on steep slopes flow is initially slow, and hence little or no erosion occurs, but as water continues moving downslope, it accelerates and may move by *sheet flow,* a more-or-less continuous film of water flowing over the surface. Sheet flow is not confined to depressions, and it accounts for *sheet erosion,* a particular problem on some agricultural lands (see Chapter 6).

In *channel flow,* surface runoff is confined to long, trough-like depressions. Channels vary in size from rills containing a trickling stream of water to the Amazon River of South

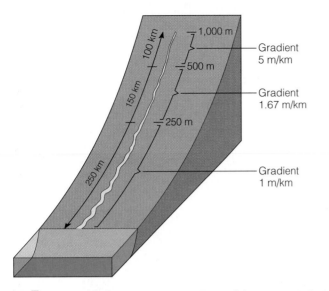

► **FIGURE 12-3** The average gradient of this stream is 2 m/km. Gradient can be calculated for any segment of a stream as shown in this example. Notice that the gradient is steepest in the headwaters area and decreases in a downstream direction.

America, which is 6,450 km long and up to 2.4 km wide and 90 m deep. Channelized flow is described by various terms including rill, brook, creek, stream, and river, most of which are distinguished by size and volume. The term **stream** carries no connotation of size and is used here to refer to all runoff confined to channels regardless of size.

Streams receive water from several sources, including sheet flow and rain falling directly into stream channels. Far more important, though, is the water supplied by soil moisture and groundwater, both of which flow downslope and discharge into streams. In humid areas where groundwater is plentiful, streams may maintain a fairly stable flow year round, even during dry seasons, because they are continuously supplied by groundwater. In contrast, the amount of water in streams of arid and semiarid regions fluctuates widely because these streams depend more on infrequent rainstorms and surface runoff for their water supply.

Stream Gradient, Velocity, and Discharge

Streams flow downhill from a source area to a lower elevation where they empty into another stream, a lake, or the sea.★ The slope over which a stream flows is its **gradient**. If the source (headwaters) of a stream is 1,000 m above sea level and the stream flows 500 km to the sea, it drops 1,000 m vertically over a horizontal distance of 500 km (► Figure 12-3). Its gradient is calculated by dividing the vertical drop by the horizontal distance; in this example, it is 1,000 m/500 km = 2 m/km.

★In certain desert streams, the flow diminishes in a downstream direction by evaporation and infiltration until the streams disappear, and in regions with numerous caverns, some streams may disappear below the ground.

Gradients vary considerably, even along the course of a single stream. Generally, streams are steeper in their upper reaches where their gradients may be tens of meters per kilometer, but in their lower reaches the gradient may be as little as a few centimeters per kilometer.

Stream velocity and discharge are closely related variables. **Velocity** is simply a measure of the downstream distance traveled per unit of time, and is usually expressed in feet per second (ft/sec) or meters per second (m/sec). Variations in flow velocity occur not only with distance along a stream channel but also across a channel's width. Flow velocity is slower and more turbulent near a stream's banks or bed because of friction than it is farther from these boundaries (► Figure 12-4). Other controls on velocity include channel shape and roughness. Broad, shallow channels and narrow, deep channels have proportionally more water in contact with their perimeters than do channels with semicircular cross sections (► Figure 12-5). Consequently the water in semicircular channels flows more rapidly because it encounters less frictional resistance.

Channel roughness is a measure of the frictional resistance within a channel. Frictional resistance to flow is greater in a channel containing large boulders than in one with banks and a bed composed of sand or clay. In channels with abundant vegetation, flow is slower than in barren channels of comparable size.

The most obvious control on velocity is gradient, and one might think that the steeper the gradient, the greater the flow velocity. In fact, the average velocity generally increases in a downstream direction, even though the gradient decreases in the same direction. Three factors contribute to this: First, velocity increases continuously, even as gradient decreases, in response to the acceleration of gravity

---

► **FIGURE 12-4** In a stream, flow velocity varies as a result of friction with the banks and bed. The maximum flow velocity is near the center and top of a stream in a straight channel. The lengths of the arrows in this illustration are proportional to velocity.

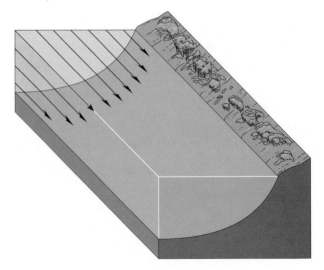

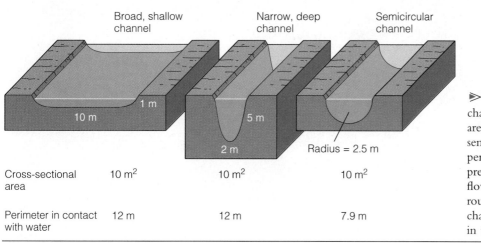

	Broad, shallow channel	Narrow, deep channel	Semicircular channel
Cross-sectional area	10 m²	10 m²	10 m²
Perimeter in contact with water	12 m	12 m	7.9 m

▷ **FIGURE 12-5** All three of these channels have the same cross-sectional area, but each has a different shape. The semicircular channel has the least perimeter in contact with the water and presents the least frictional resistance to flow. If other variables, such as channel roughness, are the same in all of these channels, flow velocity will be greatest in the semicircular channel.

unless other factors retard flow. Secondly, in their upstream reaches, streams commonly have boulder-strewn, broad, shallow channels, so flow resistance is high and velocity is correspondingly slower. Downstream, channels generally become more semicircular, and the bed and banks are usually composed of finer-grained materials, reducing the effects of friction. Thirdly, the number of tributary streams joining a larger stream increases in a downstream direction, so the total volume of water (discharge) increases, and increasing discharge results in increased velocity.

Discharge is the total volume of water in a stream moving past a particular point in a given period of time. To determine discharge, one must know the dimensions of a channel—that is, its cross-sectional area *(A)*—and its flow velocity *(V)*. Discharge *(Q)* can then be calculated by the formula $Q = VA$; it is generally expressed in cubic feet per second (ft³/sec) or cubic meters per second (m³/sec).

STREAM EROSION

Erosion involves the physical removal of dissolved substances and loose particles of soil and rock from a source area. Accordingly, the sediment transported in a stream consists of both dissolved materials and solid particles. Some of the *dissolved load* of a stream is acquired from the stream bed and banks where soluble rocks such as limestone and dolostone are exposed. But much of it is carried into streams by sheet flow and by groundwater.

The solid sediment carried in streams ranges from clay-sized particles to large boulders. Much of this sediment finds its way into streams by mass wasting (▷ Figure 12-6), but some is derived directly from the stream bed and banks. The power of running water, called **hydraulic action,** is sufficient to set particles in motion.

Another process of erosion in streams is **abrasion,** in which exposed rock is worn and scraped by the impact of solid particles. If running water is transporting sand and gravel, the impact of these particles abrades exposed rock surfaces. One obvious manifestation of abrasion is the occurrence of *potholes* in the beds of streams. These circular to oval

holes occur where eddying currents containing sand and gravel swirl around and erode depressions into solid rock.

TRANSPORT OF SEDIMENT LOAD

Streams transport a solid load of sedimentary particles and a **dissolved load** consisting of ions taken into solution by chemical weathering. Sedimentary particles are transported either as suspended load or as bed load. **Suspended load** consists of the smallest particles, such as silt and clay, which are kept suspended by fluid turbulence.

Bed load consists of the coarser particles such as sand and gravel. Fluid turbulence is insufficient to keep large particles suspended, so they move along the stream bed. However, part of the bed load can be suspended temporarily as when an eddying current swirls across a stream bed and lifts sand grains into the water. These particles move forward at approximately the flow velocity, but at the same time they settle toward the stream bed where they come to rest, to be moved again later by the same process. This process of intermittent bouncing and skipping is *saltation.*

Particles too large to be suspended even temporarily are transported by rolling or sliding. Obviously, greater flow velocity is required to move particles of these sizes. The maximum-sized particles that a stream can carry define its *competence,* a factor related to flow velocity. *Capacity* is a measure of the total load a stream can carry. It varies as a function of discharge; with greater discharge, more sediment can be carried. A small, swiftly flowing stream may have the competence to move gravel-sized particles but not to transport a large volume of sediment, so it has a low capacity. A large, slow-flowing stream, on the other hand, has a low competence, but may have a very large suspended load, and hence a large capacity.

STREAM DEPOSITION

Streams can transport sediment a considerable distance from the source area. Some of the sediments deposited in the Gulf of Mexico by the Mississippi River came from such distant

> **FIGURE 12-6** Streams such as the Snake River in Idaho receive some of their sediment load by mass wasting processes, frost wedging in this case. (Photo courtesy of R. V. Dietrich.)

sources as Pennsylvania, Minnesota, and Alberta, Canada. Along the way, deposition may occur in a variety of environments, such as stream channels, the floodplains adjacent to channels, and the points where streams flow into lakes or the seas or flow from mountain valleys onto adjacent lowlands.

Streams do most of their erosion, sediment transport, and deposition when they flood. Consequently, stream deposits, collectively called **alluvium,** do not represent the continuous day-to-day activity of streams, but rather those periodic, large-scale events of sedimentation associated with flooding.

Braided Streams and Their Deposits

Braided streams possess an intricate network of dividing and rejoining channels (▷ Figure 12-7). Braiding develops when a stream is supplied with excessive sediment, which over time is deposited as sand and gravel bars within its channel. During high-water stages, these bars are submerged, but during low-water stages, they are exposed and divide a single channel into multiple channels (Figure 12-7). Braided streams have broad, shallow channels. They are generally characterized as bed load–transport streams, and their deposits are composed mostly of sheets of sand and gravel.

Meandering Streams and Their Deposits

Meandering streams possess a single, sinuous channel with broadly looping curves called *meanders* (▷ Figure 12-8). Such stream channels are semicircular in cross section along straight reaches, but at meanders they are markedly asymmetric, being deepest near the outer bank, which commonly descends vertically into the channel. The outer bank is called the *cut bank* because flow velocity and turbulence are greatest on that side of the channel where it is eroded. In contrast, flow velocity is at a minimum near the inner bank, which slopes gently into the channel (▷ Figure 12-9a).

As a consequence of the unequal distribution of flow velocity across meanders, the cut bank is eroded and deposition occurs along the opposite side of the channel. The deposit formed in this manner is a **point bar;** it consists of cross-bedded sand or, in some cases, gravel (Figure 12-9b).

It is not uncommon for meanders to become so sinuous that the thin neck of land separating adjacent meanders is eventually cut off during a flood. The valley floors of meandering streams are commonly marked by crescent-shaped **oxbow lakes,** which are actually cutoff meanders (Figures 12-8 and ▷ 12-10). These oxbow lakes may persist as lakes for some time, but are eventually filled with organic matter and fine-grained sediment carried by floods. Once filled, oxbow lakes are called *meander scars.*

▷ FIGURE 12-7 A braided stream near Sante Fe, New Mexico.

▷ FIGURE 12-8 Aerial view of a meandering stream. The broad, flat area adjacent to the stream channel is the floodplain. Notice the crescent-shaped lakes—these are cutoff meanders, or what are known as oxbow lakes.

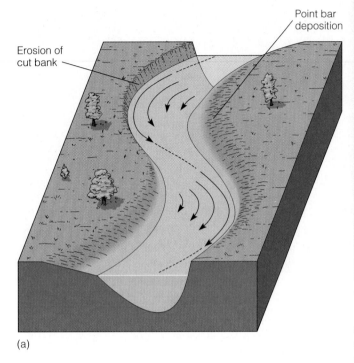

(a)

(b)

▷ **FIGURE 12-9** (*a*) The line of maximum velocity (dashed) switches from side to side in a meandering channel. The arrows show relative velocity at various places in the channel. Because of variations in flow velocity, the outer or cut bank is eroded, and a point bar is deposited on the gently sloping side of the meander. (*b*) Two point bars in a small meandering stream.

▷ **FIGURE 12-10** Four stages in the origin of an oxbow lake. In (*a*) and (*b*), the meander neck becomes narrower. (*c*) The meander neck is cut off, and part of the channel is abandoned. (*d*) When it is completely isolated from the main channel, the abandoned meander is an oxbow lake.

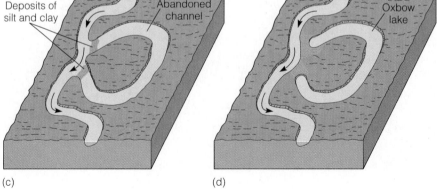

(a)

(b)

(c)

(d)

PREDICTING FLOODS

Occasionally, a stream receives more water than its channel can handle, and it floods, occupying part or all of its floodplain. To monitor stream behavior, the U.S. Geological Survey maintains more than 11,000 stream gauging stations, and various state agencies also monitor streams. Data collected at gauging stations can be used to construct *flood-frequency curves* (▷ Figure 1). To construct such a curve, the peak discharges are first arranged in order of volume; the flood with the greatest discharge has a magnitude rank of 1, the second largest is 2, and so on (◉ Table 1). The *recurrence interval*—the time period during which a flood of a given magnitude or larger can be expected over an average of many years—is determined by the following equation:

$$R = (N + 1)/m$$

where R is the recurrence interval in years, N is the number of years of record, 76 in this case, and m is the magnitude rank. According to this formula, floods with magnitude ranks of 1 and 23 have recurrence intervals of 77.00 and 3.35 years, respectively. Once the recurrence interval has been calculated, it is plotted against discharge, and a line is drawn through the data points (Figure 1).

According to Figure 1, the 10-year flood for the Rio Grande near Lobatos, Colorado, has a discharge of 245 m³/sec. This means that, on average, we can expect one flood of this size or greater to occur within a 10-year interval. One cannot predict that such a flood will occur in any particular year only that it has a probability of 1 in 10 (1/10) of occurring in any year. Furthermore, 10-year floods are not necessarily separated by 10 years. That is, two such events could occur in the same year or in successive years, but over a period of centuries their average occurrence would be once every 10 years.

Unfortunately, stream gauge data in the United States have generally been available for only a few decades, and rarely for more than a century. Accordingly, we have a good idea of stream behavior over short periods, the 2-year and 5-year floods, for example, but our knowledge of long-term behavior is limited by the short period of record keeping. Thus, predictions of 50-year or 100-year floods from Figure 1 are unreliable. In fact, the largest magnitude flood shown in Figure 1 may have been a unique event for this stream that will never be repeated. On the other hand, it may actually turn out to be a magnitude 2 or 3 flood when data for a longer time are available.

Although flood-frequency curves have limited applicability, they are nevertheless helpful in making decisions regarding flood control. Careful mapping of floodplains can identify areas at risk for floods of a given magnitude. For a particular stream, planners must decide what magnitude of flood to protect against because the cost goes up faster than the increasing sizes of floods would indicate.

▷ **FIGURE 1** Flood-frequency curve for the Rio Grande near Lobatos, Colorado.

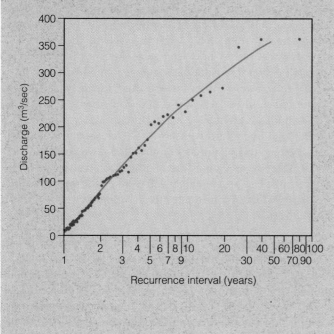

Year	Discharge (m³/sec)	Rank	Recurrence Interval
1900	133	23	3.35
1905	371	1	77.00
1909	211	13	5.92
1975	68	43	1.79

TABLE 1 Some of the Data and Recurrence Intervals for the Rio Grande near Lobatos, Colorado

SOURCE: U.S. Geological Survey Open-File Report 79-681.

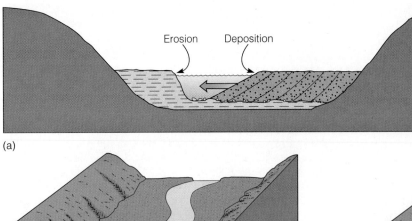

(a)

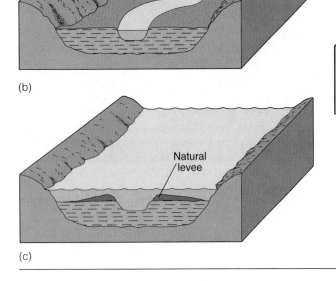

(b)

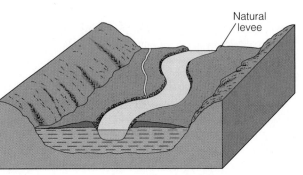

Natural levee

(d)

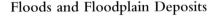

Natural levee

(c)

> **FIGURE 12-11** (*a*) Floodplain deposits forming as a meandering stream migrates laterally, depositing a series of point bars. (*b–d*) Three stages in the formation of deposits on a floodplain. (*b*) Stream at low-water stage. (*c*) Flooding stream and deposition of natural levees. The levees form after many such episodes of flooding. (*d*) After flooding. Notice the tributary stream, which parallels the main stream until it finds a way through the natural levee.

Floods and Floodplain Deposits

Most streams periodically receive more water than their channel can carry, so they spread across low-lying, relatively flat areas called **floodplains** adjacent to their channels (Figure 12-8; see Perspective 12-1). Some floodplains are composed mostly of sand and gravel that were deposited as point bars. When a meandering stream erodes its cut bank and deposits on the opposite bank, it migrates laterally across its floodplain. As lateral migration occurs, a succession of point bars develops (➤ Figure 12-11a).

Many floodplains are dominated by fine-grained sediments, mostly mud. When a stream overflows its banks and floods, the velocity of the water spilling onto the floodplain diminishes rapidly because the water encounters greater frictional resistance to flow as it spreads out as a broad, shallow sheet. In response to the diminished velocity, ridges of sandy alluvium called **natural levees** are deposited along the margins of the stream channel (Figure 12-11b).

The flood waters spilling from a main channel carry large quantities of mud beyond the natural levees and onto the floodplain. During the waning stages of a flood, the flood waters may flow very slowly or not at all, and the suspended silt and clay eventually settle as layers of mud.

Annual property damage from flooding in the United States exceeds $100 million. And in spite of the completion of more and more flood control projects, the amount of property damage is not decreasing. In fact, floodplains are attractive sites for settlement due to the combination of fertile soils, level surfaces for construction, and proximity to water for industry, agriculture, and domestic uses. However, these human activities generally increase the potential for flooding. Urbanization greatly increases surface runoff because concrete and asphalt compact and cover surface materials, thereby reducing their infiltration capacity. Storm drains in urban areas quickly carry water to nearby streams, many of which flood much more commonly than they did in the past.

Deltas

When a stream flows into another body of water, its flow velocity decreases rapidly and deposition occurs. As a result, a **delta** forms, causing the local shoreline to build out, or *prograde* (➤ Figure 12-12). The simplest prograding deltas exhibit a characteristic vertical sequence in which *bottomset*

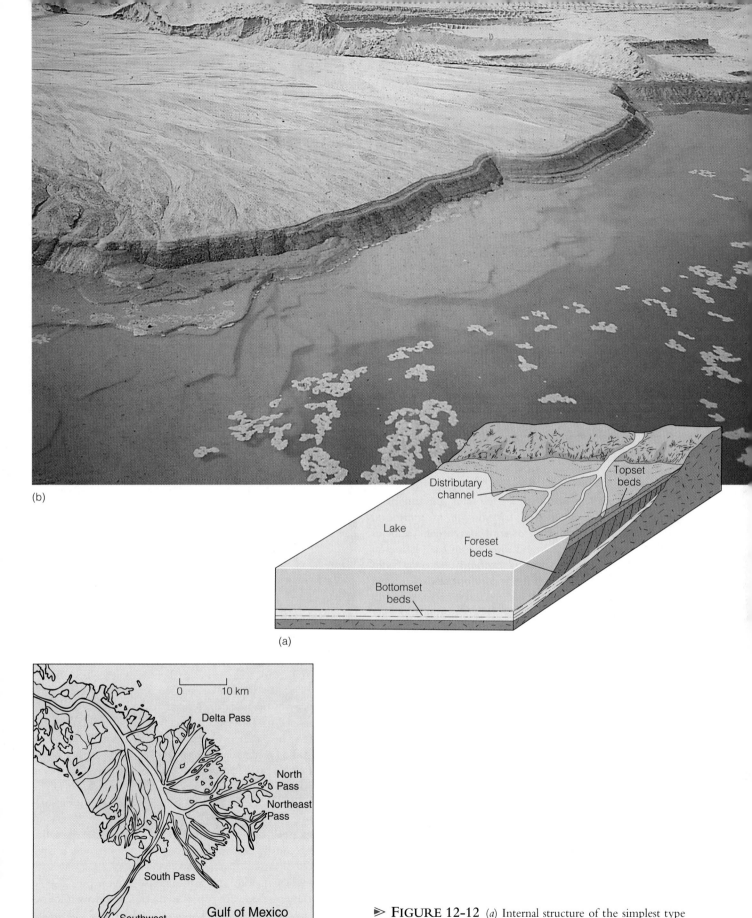

(b)

Distributary
channel

Topset
beds

Lake

Foreset
beds

Bottomset
beds

(a)

0   10 km

Delta Pass

North
Pass

Northeast
Pass

South Pass

Southwest
Pass

Gulf of Mexico

(c)

▷ **FIGURE 12-12** (*a*) Internal structure of the simplest type
of prograding delta. (*b*) A small delta, measuring about 20 m
across, in which bottomset, foreset, and topset beds are visible.
(*c*) The Mississippi River delta of the U.S. Gulf Coast.

➤ **FIGURE 12-13** Alluvial fans form where a stream discharges from a mountain canyon onto an adjacent lowland. This alluvial fan is in Death Valley, California.

beds are successively overlain by *foreset beds* and *topset beds* (Figure 12-12a). This sequence develops when a stream enters another body of water, and the finest sediments are carried some distance beyond the stream's mouth, where they settle from suspension and form bottomset beds. Nearer the stream's mouth, foreset beds are formed as sand and silt are deposited in gently inclined layers. The topset beds consist of coarse-grained sediments deposited in a network of *distributary channels* traversing the top of the delta. In effect, streams lengthen their channels as they extend across prograding deltas (Figure 12-12).

Many small deltas in lakes have the three-part division described above, but large marine deltas are usually much more complex. The Mississippi River delta consists of long fingerlike sand bodies, each deposited in a distributary channel that progrades far seaward (Figure 12-12c). Such deltas are commonly called *bird's-foot deltas* because the projections resemble the toes of a bird.

Progradation of marine deltas is one way that potential reservoirs for oil and gas are formed. Much of the oil and gas production of the Gulf Coast of Texas comes from buried delta deposits, and the present-day deltas of the Niger River in Africa and the Mississippi River are also known to contain reserves of oil and gas. The marshes between distributary channels of deltas are dominated by nonwoody vegetation and are potential areas of coal formation.

Alluvial Fans

Alluvial fans are lobate deposits of alluvium on land (➤ Figure 12-13). They form best on lowlands adjacent to highlands in arid and semiarid regions where little or no vegetation exists to stabilize surface materials. When periodic rainstorms occur, surface materials are quickly saturated and runoff begins. During a particularly heavy rain, all of the surface flow in a drainage area is funneled into a mountain canyon leading to an adjacent lowland. As long as the stream is confined in the mountain canyon, it cannot spread laterally. But when it discharges from the canyon onto the lowland area, it quickly spreads out, its velocity diminishes, and deposition ensues.

The alluvial fans that develop by the process just described are mostly accumulations of sand and gravel, a large proportion of which is deposited by streams. In some cases the water flowing through a mountain canyon picks up so much sediment that it becomes a viscous mudflow. Consequently, mudflow deposits make up a large part of many alluvial fans.

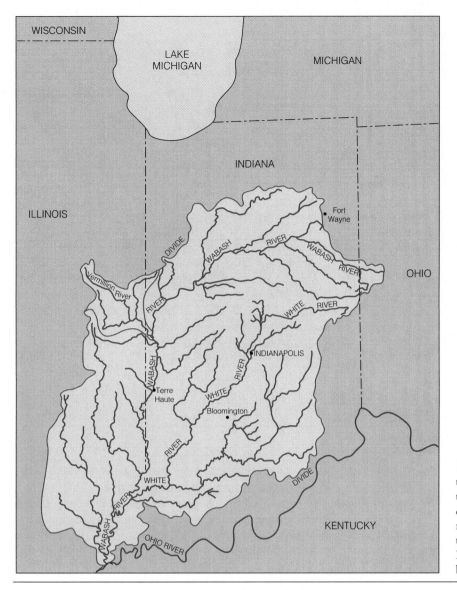

> FIGURE 12-14 Drainage basin of the Wabash River, which is one of the tributaries of the Ohio River. This drainage basin covers about 85,500 km², mostly in Indiana. All of the streams within the drainage basin, such as the Vermillion River, have their own smaller drainage basins. Divides are shown by brown lines.

DRAINAGE BASINS AND DRAINAGE PATTERNS

A stream such as the Mississippi River consists of a main stream and all of the smaller *tributary streams* that supply water to it. The Mississippi and all of its tributaries, or any other drainage system for that matter, carry surface runoff from an area known as the **drainage basin.** Individual drainage basins are separated from adjacent ones by topographically higher areas called **divides** (▷ Figure 12-14).

Various **drainage patterns** are recognized based on the regional arrangement of channels in a drainage system. The most common is *dendritic drainage,* which consists of a network of channels resembling tree branching (▷ Figure 12-15a). Dendritic drainage develops on gently sloping surfaces where the materials respond more or less homogeneously to erosion. Areas of flat-lying sedimentary rocks and

some terrains of igneous or metamorphic rocks usually display a dendritic drainage pattern.

Rectangular drainage is characterized by channels with right angle bends and tributaries that join larger streams at right angles (Figure 12-15b). The positions of the channels are strongly controlled by geologic structures, particularly regional joint systems that intersect at right angles.

In some parts of the eastern United States, such as Virginia and Pennsylvania, erosion of folded sedimentary rocks develops a landscape of alternating parallel ridges and valleys. The ridges consist of more resistant rocks, such as sandstone, whereas the valleys overlie less resistant rocks such as shale. Main streams follow the trends of the valleys. Short tributaries flowing from the adjacent ridges join the main stream at nearly right angles, hence the name *trellis drainage* (Figure 12-15c).

In *radial drainage,* streams flow outward in all directions

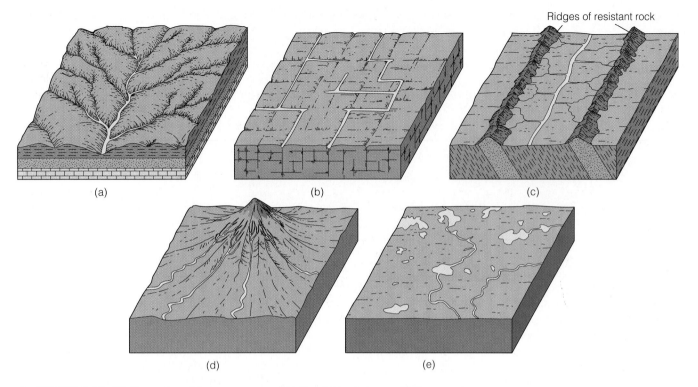

> **FIGURE 12-15** Examples of drainage patterns. (*a*) Dendritic drainage. (*b*) Rectangular drainage. (*c*) Trellis drainage. (*d*) Radial drainage. (*e*) Deranged drainage.

from a central high area (Figure 12-15d). Radial drainage develops on large, isolated volcanic mountains and in areas where the Earth's crust has been arched up by the intrusion of plutons such as laccoliths.

In some areas streams flow in and out of swamps and lakes with irregular flow directions. Drainage patterns character-ized by such irregularity are called *deranged* (Figure 12-15e). The presence of deranged drainage indicates that it devel-oped recently and has not yet formed an organized drainage system. In areas of Minnesota, Wisconsin, and Michigan that were glaciated until about 10,000 years ago, the previously established drainage systems were obliterated by glacial ice.

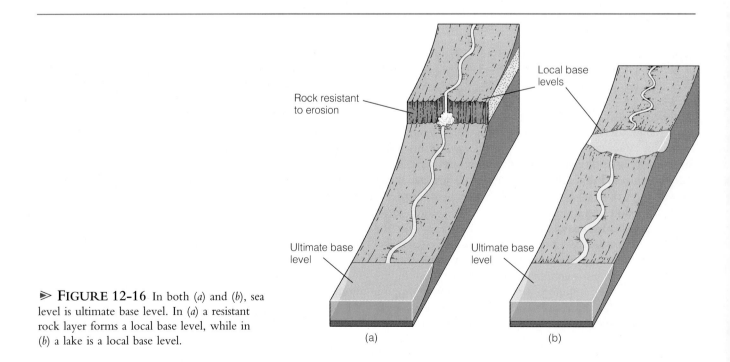

> **FIGURE 12-16** In both (*a*) and (*b*), sea level is ultimate base level. In (*a*) a resistant rock layer forms a local base level, while in (*b*) a lake is a local base level.

Following the final retreat of the glaciers, drainage systems became established, but have not yet become fully organized.

BASE LEVEL

Streams have a lower limit to which they can erode; this limit is called **base level** (▷ Figure 12-16). Theoretically, a stream could erode its entire valley to very near sea level, so sea level is commonly referred to as *ultimate base level*. In reality, though, streams never reach ultimate base level because they must have some gradient in order to maintain flow. Streams flowing into depressions below sea level, such as Death Valley in California, have a base level corresponding to the lowest point of the depression and are not limited by sea level.

In addition to ultimate base level, streams have *local* or *temporary base levels*. For example, a lake or another stream can serve as a local base level for the upstream segment of a stream (Figure 12-16). Likewise, where a stream flows across particularly resistant rock, a waterfall may develop, forming a local base level.

When sea level rises or falls with respect to the land, or the land over which a stream flows is uplifted or subsides, changes in base level occur. During the Pleistocene Epoch when extensive glaciers were present on the Northern Hemisphere continents, sea level was more than 100 m lower than at present. Accordingly, streams deepened their valleys by adjusting to a new, lower base level. Rising sea level at the end of the Pleistocene caused base level to rise, and the streams responded by depositing sediments and backfilling previously formed valleys.

Streams adjust to human intervention, but not always in anticipated or desirable ways. Geologists and engineers are well aware that the process of building a dam and impounding a reservoir creates a local base level (▷ Figure 12-17a). Where a stream enters a reservoir, its flow velocity diminishes rapidly and deposition occurs; consequently, reservoirs are eventually filled with sediment unless they are dredged. Another consequence of building a dam is that the water discharged at the dam is largely sediment free, but it still possesses energy to transport sediment. Commonly, such streams simply acquire a new sediment load by vigorously eroding downstream from the dam.

Draining a lake along a stream's course may seem like a small change that is well worth the time and expense to expose dry land for agriculture or commercial development. However, draining a lake lowers the base level for that part of the stream above the lake, and the stream will very likely respond by rapid downcutting (Figure 12-17b).

THE GRADED STREAM

A stream's *longitudinal profile* shows the elevations of a channel along its length as viewed in cross section (▷ Figure 12-18). The longitudinal profiles of many streams show a number of irregularities such as lakes and waterfalls, which are local base levels (Figure 12-18a). Over time these irregularities tend to be eliminated by stream processes; where the gradient is steep, erosion decreases it, and where the gradient is too low to maintain sufficient flow velocity for sediment transport, deposition occurs, steepening the gradient. In short, streams tend to develop a smooth, concave longitudinal profile of equilibrium, meaning that all parts of the system dynamically adjust to one another (Figure 12-18b).

Streams possessing an equilibrium profile are said to be **graded streams;** that is, a delicate balance exists between gradient, discharge, flow velocity, channel characteristics, and sediment load such that neither significant erosion nor deposition occurs within the channel. Such a delicate bal-

▷ FIGURE 12-17 (*a*) The process of constructing a dam and impounding a reservoir creates a local base level. A stream deposits much of its sediment load where it flows into a reservoir. (*b*) A stream adjusts to a lower base level when a lake is drained.

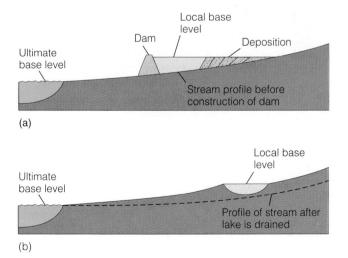

(a)

(b)

▷ FIGURE 12-18 (*a*) An ungraded stream has irregularities in its longitudinal profile. (*b*) Erosion and deposition along the course of a stream eliminate irregularities and cause it to develop the smooth, concave profile typical of a graded stream.

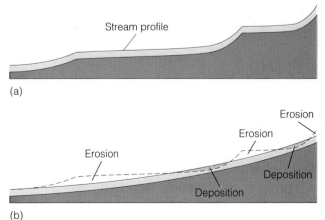

(a)

(b)

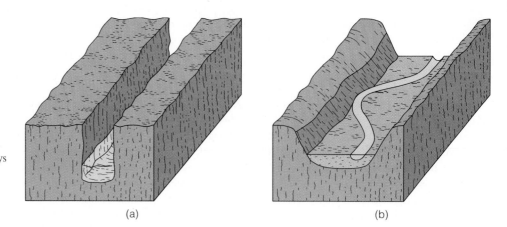

► FIGURE 12-19 Valley development. (a) If valleys formed mostly by downcutting, they would be narrow and steep sided. (b) Valleys are deepened by downcutting, but most of them are also widened by lateral erosion, mass wasting, and sheet wash.

(a)

(b)

ance is rarely attained, so the concept of a graded stream is an ideal. Nevertheless, many streams do indeed approximate the graded condition, although not along their entire courses and usually only temporarily.

Even though the concept of a graded stream is an ideal, we can generally anticipate the responses of a graded stream to changes altering its equilibrium. A change in base level, for instance, would cause a stream to adjust as previously discussed. Increased rainfall in a stream's drainage basin would result in greater discharge and flow velocity. In short, the stream would now possess greater energy—energy that must be dissipated within the stream system by, for example, a change in channel shape. A change from a semicircular to a broad, shallow channel would dissipate more energy by friction. On the other hand, the stream may respond by active downcutting and erode a deeper valley and effectively reduce its gradient until it is once again graded.

⊕ DEVELOPMENT OF STREAM VALLEYS

Valleys are common landforms, and with few exceptions they form and evolve as a consequence of stream erosion, although other processes, especially mass wasting, also contribute. The shapes and sizes of valleys vary considerably; some are small, steep-sided *gullies,* whereas others are broad and have gently sloping valley walls. Some steep-walled, deep valleys of vast proportions are called *canyons,* such as the Grand Canyon of Arizona. Particularly narrow and deep valleys are *gorges.*

A valley may begin where runoff has sufficient energy to dislodge surface materials and excavate a small rill. Once formed, a rill collects more surface runoff and becomes deeper and wider until a full-fledged valley develops (► Figure 12-19). Several processes are involved in the origin and evolution of valleys, including downcutting, lateral erosion, mass wasting, sheet wash, and headward erosion.

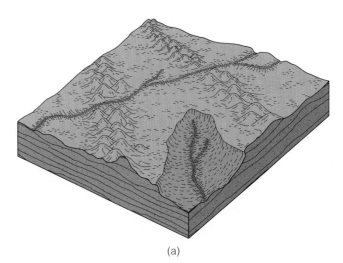

(a)

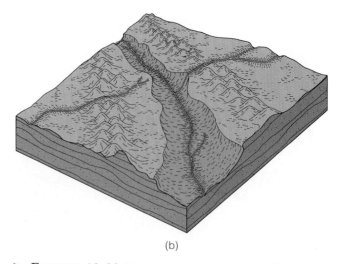

(b)

► FIGURE 12-20 Two stages in stream piracy. (a) In the first stage, the stream at the lower elevation extends its channel by headward erosion. In (b) it has captured some of the drainage of the stream flowing at the higher elevation.

Downcutting occurs when a stream possesses more energy than it requires to transport its sediment load, so some of its excess energy cuts its valley deeper. If downcutting were the only process operating, valleys would be narrow and steep sided, as in Figure 12-19a. In most cases, however, the valley walls are undercut by the stream. Such undermining, termed *lateral erosion,* creates unstable conditions so that part of a bank or valley wall may move downslope by any one or a combination of mass wasting processes (Figure 12-19b). Furthermore, sheet wash and erosion of rill and gully tributaries carry materials from the valley walls into the main stream.

In addition to becoming deeper and wider, stream valleys are commonly lengthened as well. Valleys are lengthened in an upstream direction by *headward erosion* as drainage divides are eroded by entering runoff water (▷ Figure 12-20a). In some cases headward erosion eventually breaches the drainage divide and diverts part of the drainage of another stream by a process called *stream piracy* (Figure 12-20b). Once stream piracy has occurred, both drainage systems must adjust; one now has more water, greater discharge, and greater potential to erode and transport sediment, whereas the other is diminished in all of these aspects.

◉ STEAM TERRACES

Adjacent to many streams are **stream terraces,** erosional remnants of floodplains formed when the streams were flowing at a higher level. These terraces consist of a fairly flat upper surface and a steep slope descending to the level of the lower, present-day floodplain (▷ Figure 12-21). In some cases, a stream has several steplike surfaces above its present-day floodplain, indicating that stream terraces formed several times.

Although all stream terraces result from erosion, they are preceded by an episode of floodplain formation and deposition of sediment. Subsequent erosion causes the stream to cut downward until it is once again graded (Figure 12-21). Once the stream again becomes graded, it begins eroding laterally and establishes a new floodplain at a lower level. Several such episodes account for the multiple terrace levels seen adjacent to some streams.

Renewed erosion and the formation of stream terraces are usually attributed to a change in base level. Either uplift of the land over which a stream flows or lowering of sea level yields a steeper gradient and increased flow velocity, thereby initiating an episode of downcutting. When the stream reaches a level at which it is once again graded, downcutting ceases. Although changes in base level no doubt account for many stream terraces, greater runoff in a stream's drainage basin can also result in the formation of terraces.

◉ INCISED MEANDERS

Some streams are restricted to deep, meandering canyons cut into solid bedrock, where they form features called **incised**

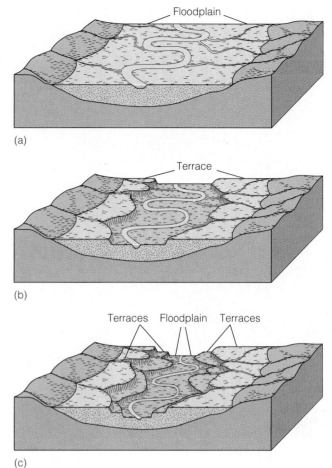

▷ **FIGURE 12-21** Origin of stream terraces. (*a*) A stream has a broad floodplain adjacent to its channel. (*b*) The stream erodes downward and establishes a new floodplain at a lower level. Remnants of its old floodplain are stream terraces. (*c*) Another level of terraces originates as the stream erodes downward again.

meanders. The San Juan River in Utah occupies a meandering canyon more than 390 meters deep (▷ Figure 12-22). Such streams, being restricted by solid rock walls, are generally ineffective in eroding laterally; as a result, they lack a floodplain and occupy the entire width of the canyon floor.

It is not difficult to understand how a stream can cut downward into solid rock, but forming a meandering pattern in bedrock is another matter. Because lateral erosion is inhibited once downcutting begins, one must infer that the meandering course was established when the stream flowed across an area covered by alluvium. Suppose that a stream near base level has established a meandering pattern. If the land over which the stream flows is uplifted, erosion is initiated, and the meanders become incised into the underlying bedrock.

> ▷ FIGURE 12-22 The Goose Necks of the San Juan River, Arizona, are incised meanders.

CHAPTER SUMMARY

1. Water is continuously evaporated from the oceans, rises as water vapor, condenses, and falls as precipitation. About 20% of all precipitation falls on land and eventually returns to the oceans, mostly by surface runoff.

2. Runoff can be characterized as either sheet flow or channel flow. Channels of all sizes are called streams.

3. Gradient generally varies from steep to gentle along the course of a stream, being steep in upper reaches and gentle in lower reaches.

4. Flow velocity and discharge are related. A change in one of these parameters causes the other to change as well.

5. A stream and its tributaries carry runoff from its drainage basin. Drainage basins are separated from one another by divides.

6. Streams erode by hydraulic action, abrasion, and dissolution of soluble rocks.

7. The coarser part of a stream's sediment load is transported as bed load, and the finer part as suspended load. Streams also transport a dissolved load of ions in solution.

8. Braided streams are characterized by a complex of dividing and rejoining channels. Braiding occurs when sediment

transported by the stream is deposited within channels as sand and gravel bars.

9. Meandering streams have a single, sinuous channel with broad looping curves. Meanders migrate laterally as the cut bank is eroded and point bars form on the inner bank. Oxbow lakes are cutoff meanders in which fine-grained sediments and organic matter accumulate.

10. Floodplains are rather flat areas paralleling stream channels. They may be composed mostly of point bar deposits or mud deposited during numerous floods.

11. Deltas are alluvial deposits at a stream's mouth. Many small deltas in lakes conform to the three-part division of bottomset, foreset, and topset beds, but large marine deltas are more complex.

12. Alluvial fans are lobate alluvial deposits on land that consist mostly of sand and gravel. They form best in arid and semiarid regions where erosion rates are high.

13. Sea level is ultimate base level, the lowest level to which streams can erode. Streams, however, commonly have local base levels such as lakes, other streams, or the points where they flow across particularly resistant rocks.

14. Streams tend to eliminate irregularities in their channels so that they develop a smooth, concave profile of equilibrium. Such streams are graded, meaning that a balance exists between gradient, discharge, flow velocity, channel characteristics, and sediment load so that little or no deposition occurs within the channel.

15. Stream valleys develop by a combination of processes including downcutting, lateral erosion, mass wasting, sheet wash, and headward erosion.

16. Renewed downcutting by a stream possessing a floodplain commonly results in the formation of stream terraces, which are remnants of an older floodplain at a higher level.

17. Incised meanders are generally attributed to renewed downcutting by a meandering stream so that it now occupies a deep, meandering valley.

IMPORTANT TERMS

abrasion
alluvial fan
alluvium
base level
bed load
braided stream
delta

discharge
dissolved load
divide
drainage basin
drainage pattern
floodplain
graded stream

gradient
hydraulic action
hydrologic cycle
incised meander
infiltration capacity
meandering stream
natural levee

oxbow lake
point bar
runoff
stream
stream terrace
suspended load
velocity

REVIEW QUESTIONS

1. Mounds of sediment deposited on the margin of a stream are:
   a. _____ natural levees;
   b. _____ oxbow lakes;
   c. _____ bottomset beds;
   d. _____ incised meanders;
   e. _____ alluvial fans.

2. The direct impact of running water is:
   a. _____ bed load;
   b. _____ saltation;
   c. _____ hydraulic action;
   d. _____ meander cutoff;
   e. _____ base level.

3. The vertical drop of a stream in a given horizontal distance is its:
   a. _____ discharge;
   b. _____ gradient;
   c. _____ velocity;
   d. _____ base level;
   e. _____ drainage pattern.

4. A _____ drainage pattern resembles the branching of a tree.
   a. _____ rectangular;
   b. _____ trellis;
   c. _____ dendritic;
   d. _____ deranged;
   e. _____ radial.

5. A meandering stream is one having:
   a. _____ numerous sand and gravel bars in its channel;
   b. _____ a single, sinuous channel;
   c. _____ a broad, shallow channel;
   d. _____ a deep, narrow valley;
   e. _____ long, straight reaches and waterfalls.

6. Erosional remnants of floodplains that are higher than the current level of a stream are:
   a. _____ oxbow lakes;
   b. _____ cut banks;
   c. _____ stream terraces;
   d. _____ incised meanders;
   e. _____ natural bridges.

7. All of the sediment carried by saltation and rolling and sliding along a stream bed is its:
   a. _____ suspended load;
   b. _____ drainage capacity;
   c. _____ stream profile;
   d. _____ bed load;
   e. _____ channel pattern.

8. Infiltration capacity is the:
   a. _____ rate at which a stream erodes;
   b. _____ distance a stream flows from its source to the ocean;
   c. _____ maximum rate that surface materials can absorb water;
   d. _____ vertical distance a stream can erode below sea level;
   e. _____ variation in flow velocity across a stream channel.

9. A stream with a cross-sectional area of 250 m^2 and a flow velocity of 1.5 m/sec has a discharge of _____ m^3/sec.
   a. _____ 500;
   b. _____ 125;
   c. _____ 375;
   d. _____ 1,000;
   e. _____ 200.

10. How do solar radiation, the changing phases of water, and runoff cause the recycling of water from the oceans to the atmosphere and back to the oceans?

11. Explain what infiltration capacity is and why it is important in considering runoff.

12. How do channel shape and roughness control flow velocity?

13. How do streams erode and acquire a sediment load?

14. How is it possible for a meandering stream to erode laterally yet maintain a more or less constant channel width?

15. How do oxbow lakes and meander scars form?

16. Sea level is ultimate base level for most streams. If sea level drops with

respect to the land, how would a stream respond?

17. What is a graded stream, and why are streams rarely graded except temporarily?

18. How do stream terraces form?

POINTS TO PONDER

1. A stream 2,000 m above sea level at its source flows 1,500 km to the sea. What is the stream's gradient? Do you think the gradient you calculated will be correct for all segments of this stream? Explain.

2. What long-term changes may occur in the hydrologic cycle? How might human activities bring about such changes?

GROUNDWATER

OUTLINE

One of the many bathhouses in Bath, England, that were built around hot springs shortly after the Roman Conquest in A.D. 41.

Within the limestone region of western Kentucky lies the largest cave system in the world. In 1941 approximately 51,000 acres were set aside and designated as Mammoth Cave National Park. In 1981 it became a World Heritage Site. Recently, the National Park Service has been considering closing Mammoth Cave because of the health hazard created by raw sewage and contaminated groundwater in the area.

From ground level, the topography of the area is unimposing with numerous sinkholes, lakes, valleys, and disappearing streams. Beneath the surface are more than 230 km of interconnecting passageways whose spectacular geologic features have been enjoyed by numerous cave explorers and tourists alike.

Based on carbon 14 dates from some of the many artifacts found in the cave (such as woven cord and wooden bowls), Mammoth Cave had been explored and used by Native Americans for more than 3,000 years prior to its rediscovery in 1799 by a bear hunter named Robert Houchins. During the War of 1812, approximately 180 metric tons of saltpeter (a potassium nitrate mineral), used in the manufacture of gunpowder, were mined from Mammoth Cave. At the end of the war, the saltpeter market collapsed, and Mammoth Cave was developed as a tourist attraction, easily overshadowing the other caves in the area. Over the next 150 years, the discovery of new passageways and caverns helped establish Mammoth Cave as the world's premier cave and the standard against which all others were measured.

Mammoth Cave formed in much the same way as all other caves. Groundwater flowing through the St. Genevieve Limestone eroded a complex network of openings, passageways, and caverns. Flowing through the various caverns is the Echo River, a system of underground streams that eventually joins the Green River at the surface.

The colorful cave deposits are the primary reason millions of tourists have visited Mammoth Cave (➤ Figure 13-1). Other attractions in Mammoth Cave include the Giant's Coffin, a 15 m collapse block of limestone, and giant rooms such as Mammoth Dome, which is about 58 m high. The cave is also home to more than 200 species of insects and other animals, including about 45 blind species; some of these can be seen on the Echo River Tour, which conveys visitors 5 km along the underground stream.

➤ FIGURE 13-1 Frozen Niagara is a spectacular example of the massive travertine flowstone deposits in Mammoth Cave, Kentucky.

INTRODUCTION

Groundwater—the water stored in the open spaces within underground rocks and unconsolidated material—is a valuable natural resource that is essential to the lives of all people. Its importance to humans is not new. Groundwater rights have always been important in North America and many legal battles have been fought over them. Groundwater also played a crucial role in the development of the U.S. railway system during the nineteenth century when railroads needed a reliable source of water for their steam locomotives. Much of the water used by the locomotives came from groundwater tapped by wells.

Today, the study of groundwater and its movement has become increasingly important as the demand for fresh water by agricultural, industrial, and domestic users has reached an all-time high. More than 65% of the groundwater used in the United States each year goes for irrigation, with industrial use second, followed by domestic needs. Such demands have severely depleted the groundwater supply in many areas and led to such problems as ground subsidence and saltwater contamination. In other areas, pollution from landfills, toxic waste, and agriculture has rendered the groundwater supply unsafe.

As the world's population and industrial development expand, the demand for water, particularly groundwater, will increase. Not only is it important to locate new groundwater sources, but, once found, these sources must be protected from pollution and managed properly to ensure that users do not withdraw more water than can be replenished.

GROUNDWATER AND THE HYDROLOGIC CYCLE

Groundwater is one reservoir of the hydrologic cycle (see Figure 12-2) representing approximately 22% (8.4 million km^3) of the world's supply of fresh water. The major source of groundwater is precipitation that infiltrates the ground and moves through the soil and pore spaces of rocks. Other sources include water infiltrating from lakes and streams, recharge ponds, and wastewater treatment systems. As the groundwater moves through soil, sediment, and rocks, many of its impurities, such as disease-causing microorganisms, are filtered out. Not all soils and rocks are good filters, however, and some serious pollutants are not removed. Groundwater eventually returns to the surface when it enters lakes, streams, or the ocean.

POROSITY AND PERMEABILITY

Porosity and *permeability* are important physical properties of Earth materials and are largely responsible for the amount, availability, and movement of groundwater. Water soaks into the ground because the soil, sediment, or rock has open spaces or pores. **Porosity** is the percentage of a material's total volume that is pore space. While porosity most often consists of the spaces between particles in soil, sediments,

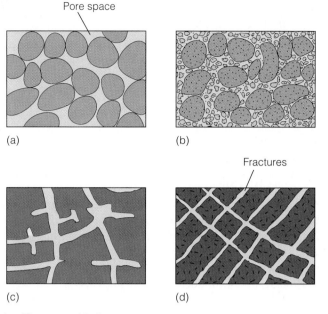

➤ **FIGURE 13-2** A rock's porosity is dependent on the size, shape, and arrangement of the material composing the rock. (*a*) A well-sorted sedimentary rock has high porosity while (*b*) a poorly sorted one has low porosity. (*c*) In soluble rocks such as carbonates, porosity can be increased by solution, while (*d*) crystalline rocks can be rendered porous by fracturing.

Material	Percentage Porosity
TABLE 13-1 Porosity Values for Different Materials	
Unconsolidated sediment	
Soil	55
Gravel	20–40
Sand	25–50
Silt	35–50
Clay	50–70
Rocks	
Sandstone	5–30
Shale	0–10
Solution activity in limestone, dolostone	10–30
Fractured basalt	5–40
Fractured granite	10

SOURCE: U.S. Geological Survey, Water Supply Paper 2220 (1983) and others.

and sedimentary rocks, other types of porosity can include cracks, fractures, faults, and vesicles in volcanic rocks (➤ Figure 13-2).

Porosity varies among different rock types and is dependent on the size, shape, and arrangement of the material composing the rock (● Table 13-1). Most igneous and metamorphic rocks as well as many limestones and dolostones have very low porosity because they are composed of tightly interlocking crystals. Their porosity can be increased, however, if they have been fractured or weathered by groundwater. This is particularly true for limestone and dolostone whose fractures can be enlarged by acidic groundwater.

By contrast, detrital sedimentary rocks composed of well-sorted and well-rounded grains can have very high porosity because any two grains touch only at a single point, leaving relatively large open spaces between the grains (Figure 13-2a). Poorly sorted sedimentary rocks, on the other hand, typically have low porosity because finer grains fill in the space between the larger grains, further reducing porosity (Figure 13-2b). In addition, the amount of cement between grains can also decrease porosity.

Although porosity determines the amount of groundwater a rock can hold, it does not guarantee that the water can be extracted. A material's capacity for transmitting fluids is its **permeability**. Permeability is dependent not only on porosity, but also on the size of the pores or fractures and their interconnections. For example, deposits of silt or clay are typically more porous than sand or gravel, but they have low permeability because the pores between the clay particles are very small, and the molecular attraction between the particles and water is great, thereby preventing movement of the water. In contrast, the pore spaces between grains in sandstone and conglomerate are much larger, and the molecular attraction on the water is therefore low. Chemical and biochemical sedimentary rocks, such as limestone and dolostone, and many igneous and metamorphic rocks that

are highly fractured can also be very permeable provided that the fractures are interconnected.

A permeable layer transporting groundwater is called an **aquifer**, from the Latin *aqua* meaning water. The most effective aquifers are deposits of well-sorted and well-rounded sand and gravel. Limestones in which fractures and bedding planes have been enlarged by solution are also good aquifers. Shales and many igneous and metamorphic rocks make poor aquifers because they are typically impermeable. Rocks such as these and any other materials that prevent the movement of groundwater are called **aquicludes**.

THE WATER TABLE

When precipitation occurs over land, some of it evaporates, some is carried away by runoff in streams, and the remainder seeps into the ground. As this water moves down from the surface, some of it adheres to the material it is moving through and halts its downward progress. This region is the **zone of aeration,** and its water is called *suspended water* (➤ Figure 13-3). The pore spaces in this zone contain both water and air.

Beneath the zone of aeration lies the **zone of saturation** where all of the pore spaces are filled with groundwater (Figure 13-3). The base of the zone of saturation varies from place to place, but usually extends to a depth where an impermeable layer is encountered or to a depth where confining pressure closes all open space. Extending irregularly upward a few centimeters to several meters from the zone of saturation is the *capillary fringe*. Water moves upward

➤ **FIGURE 13-3** The zone of aeration contains both air and water within its open space, while all of the open space in the zone of saturation is filled with groundwater. The water table is the surface separating the zones of aeration and saturation. Within the capillary fringe, water rises upward by surface tension from the zone of saturation into the zone of aeration.

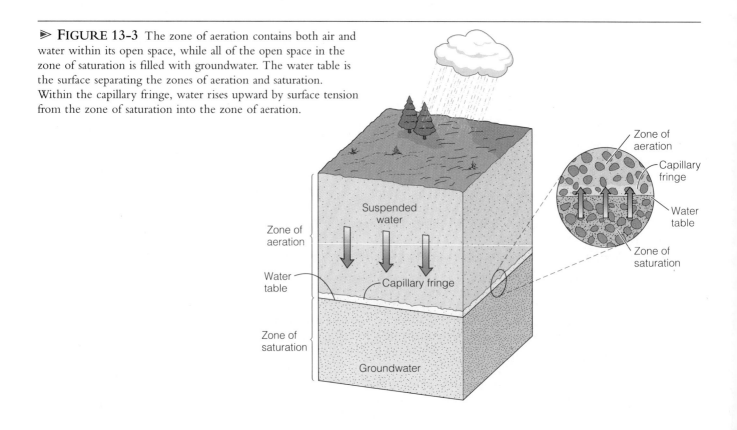

in this region because of surface tension, much as water moves upward through a paper towel.

The surface separating the zone of aeration from the underlying zone of saturation is the **water table** (Figure 13-3). In general, the configuration of the water table is a subdued replica of the overlying land surface; that is, it has its highest elevations beneath hills and its lowest elevations in valleys. In most arid and semiarid regions, the water table is quite flat and is below the level of river valleys.

Several factors contribute to the surface configuration of a region's water table. These include regional differences in the amount of rainfall, permeability, and the rate of groundwater movement. During periods of high rainfall, groundwater tends to rise beneath hills because it cannot flow fast enough into the adjacent valleys to maintain a level surface. During droughts, the water table falls and tends to flatten out because it is not being replenished.

◉ GROUNDWATER MOVEMENT

Groundwater velocity varies greatly and depends on many factors. Velocities range from 250 m per day in some extremely permeable material to less than a few centimeters per year in nearly impermeable material. In most ordinary aquifers, the average velocity of groundwater is a few centimeters per day.

Gravity provides the energy for the downward movement of groundwater. Water entering the ground moves through the zone of aeration to the zone of saturation (Figure 13-3). When water reaches the water table, it continues to move through the zone of saturation from areas where the water table is high toward areas where it is lower, such as at streams, lakes, or swamps. Only some of the water follows the direct route along the slope of the water table. Most of it takes longer curving paths downward and then enters a stream, lake, or swamp from below. This occurs because groundwater moves from areas of high pressure toward areas of lower pressure within the saturated zone.

◉ SPRINGS, WATER WELLS, AND ARTESIAN SYSTEMS

Adding water to the zone of saturation is called *recharge*, and it causes the water table to rise. Water may be added by natural means, such as rainfall or melting snow, or artificially at recharge basins or wastewater treatment plants. If groundwater is discharged without sufficient replenishment, the water table drops. Groundwater discharges naturally whenever the water table intersects the ground surface as at a spring or along a stream, lake, or swamp. Groundwater can also be discharged artificially by pumping water from wells.

Springs

A **spring** is a place where groundwater flows or seeps out of the ground. Springs have always fascinated people because the water flows out of the ground for no apparent reason and from no readily identifiable source. It is not surprising that springs have long been regarded with superstition and revered for their supposed medicinal value and healing powers. Nevertheless, there is nothing mystical or mysterious about springs.

Although springs can occur under a wide variety of geologic conditions, they all form in basically the same way (➤ Figure 13-4). When percolating water reaches the water table or an impermeable layer, it flows laterally, and if this flow intersects the Earth's surface, the water discharges onto the surface as a spring. The Mammoth Cave area in Kentucky is underlain by fractured limestones that have been enlarged into caves by solution activity (see the Prologue). In this geologic environment, springs occur where the fractures and caves intersect the ground surface allowing groundwater to exit onto the surface. Springs most commonly occur along valley walls where streams have cut valleys below the regional water table.

Springs can also develop wherever a perched water table intersects the Earth's surface (➤ Figure 13-5). A *perched*

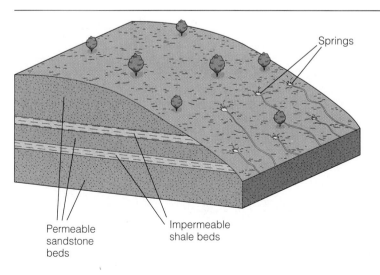

Springs

Permeable sandstone beds

Impermeable shale beds

➤ **FIGURE 13-4** Springs form wherever laterally moving groundwater intersects the Earth's surface. Most commonly, they form when percolating water reaches an impermeable layer and migrates laterally until it seeps out at the surface.

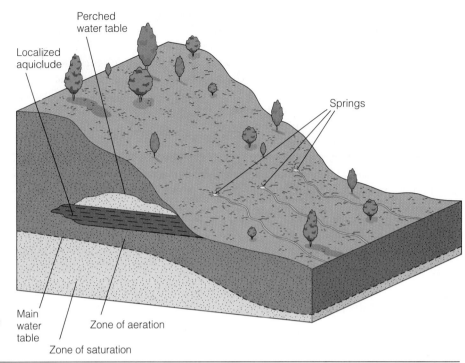

FIGURE 13-5 If a localized aquiclude, such as a shale layer, occurs within an aquifer, a perched water table may result with springs occurring where the perched water table intersects the Earth's surface.

water table may occur wherever a local aquiclude occurs within a larger aquifer, such as a lens of shale within a sandstone. As water migrates through the zone of aeration, it is stopped by the local aquiclude, and a localized zone of saturation "perched" above the main water table is created. Water moving laterally along the perched water table may intersect the Earth's surface to produce a spring.

Water Wells

A **water well** is made by digging or drilling into the zone of saturation. Once the zone of saturation is reached, water percolates into the well and fills it to the level of the water table. Most wells must be pumped to bring the groundwater to the surface.

When a well is pumped, the water table in the area around the well is lowered, because water is removed from the aquifer faster than it can be replenished. A **cone of depression** thus forms around the well, varying in size according to the rate and amount of water being withdrawn (Figure 13-6). If water is pumped out of a well faster than it can be replaced, the cone of depression grows until the well goes dry. This lowering of the water table normally does not pose a problem for the average domestic well provided that the well is drilled sufficiently deep into the zone of saturation. The tremendous amounts of water used by industry and irrigation, however, may create a large cone of depression that lowers the water table sufficiently to cause shallow wells in the immediate area to go dry (Figure 13-6). Such a situation is not uncommon and frequently results in lawsuits by the owners of the shallow dry wells. Furthermore, lowering of the regional water table is becoming a serious problem in many areas, particularly in the southwestern United States where rapid growth has placed tremendous

FIGURE 13-6 A cone of depression forms whenever water is withdrawn from a well. If water is withdrawn faster than it can be replenished, the cone of depression will grow in depth and circumference, lowering the water table in the area and causing nearby shallow wells to go dry.

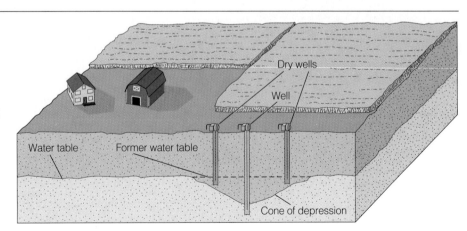

demands on the groundwater system. Unrestricted withdrawal of groundwater cannot continue indefinitely, and the rising costs and decreasing supply of groundwater should soon limit the growth of this region of the United States.

Artesian Systems

The term **artesian system** can be applied to any system in which groundwater is confined and builds up high hydrostatic (fluid) pressure. Water in such a system is able to rise above the level of the aquifer if a well is drilled through the confining layer, thereby reducing the pressure and forcing the water upward. For an artesian system to develop, three geologic conditions must be present (➤ Figure 13-7): (1) the aquifer must be confined above and below by aquicludes to prevent water from escaping; (2) the rock sequence is usually tilted and exposed at the surface, enabling the aquifer to be recharged; and (3) there is sufficient precipitation in the recharge area to keep the aquifer filled.

The elevation of the water table in the recharge area and the distance of the well from the recharge area determine the height to which artesian water rises in a well. The surface defined by the water table in the recharge area, called the *artesian-pressure surface*, is indicated by the sloping dashed line in Figure 13-7. If there were no friction in the aquifer, well water from an artesian aquifer would rise exactly to the elevation of the artesian-pressure surface. Friction, however, slightly reduces the pressure of the aquifer water and consequently the level to which artesian water rises. This is why the pressure surface slopes.

An artesian well will flow freely at the ground surface only if the wellhead is at an elevation below the artesian-pressure surface. In this situation, the water flows out of the well because it rises toward the artesian-pressure surface, which is at a higher elevation than the wellhead. In a nonflowing artesian well, the wellhead is above the artesian-pressure surface, and the water will rise in the well only as high as the artesian-pressure surface.

In addition to artesian wells, many artesian springs also exist. Such springs can occur if a fault or fracture intersects the confined aquifer allowing water to rise above the aquifer. Oases in deserts are commonly artesian springs.

Because the geologic conditions necessary for artesian water can occur in a variety of ways, artesian systems are quite common in many areas of the world underlain by sedimentary rocks. One of the best-known artesian systems in the United States underlies South Dakota and extends southward to central Texas. The majority of the artesian water from this system is used for irrigation. The aquifer of this artesian system, the Dakota Sandstone, is recharged where it is exposed along the margins of the Black Hills of South Dakota. The hydrostatic pressure in this system was

---

➤ FIGURE 13-7 An artesian system must have an aquifer confined above and below by aquicludes, the aquifer must be exposed at the surface, and there must be sufficient precipitation in the recharge area to keep the aquifer filled. The elevation of the water table in the recharge area, which is indicated by the sloping dashed line (the artesian-pressure surface), defines the highest level to which well water can rise. If the elevation of a wellhead is below the elevation of the artesian-pressure surface, the well will be free-flowing because the water will rise toward the artesian-pressure surface, which is at a higher elevation than the wellhead. If the elevation of a wellhead is at or above that of the artesian-pressure surface, the well will be nonflowing.

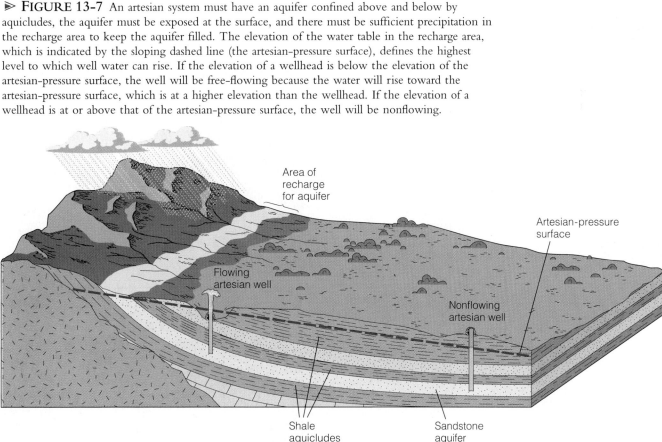

originally great enough to produce free-flowing wells and to operate waterwheels. The extensive use of water for irrigation over the years has reduced the pressure in many of the wells so that they are no longer free-flowing and the water must be pumped.

 GROUNDWATER EROSION AND DEPOSITION

When rainwater begins seeping into the ground, it immediately starts to react with the minerals it contacts, weathering them chemically. In an area underlain by soluble rock, groundwater is the principal agent of erosion and is responsible for the formation of many major features of the landscape.

Limestone, a common sedimentary rock composed primarily of the mineral calcite ($CaCO_3$), underlies large areas of the Earth's surface. Although limestone is practically insoluble in pure water, it readily dissolves if a small amount of acid is present. Carbonic acid (H_2CO_3) is a weak acid that forms when carbon dioxide combines with water ($H_2O + CO_2 \rightarrow H_2CO_3$) (see Chapter 6). Because the atmosphere contains a small amount of carbon dioxide (0.03%), and carbon dioxide is also produced in soil by the decay of organic matter, most groundwater is slightly acidic. When groundwater percolates through the various openings in limestone, the slightly acidic water readily reacts with the calcite to dissolve the rock by forming soluble calcium bicarbonate, which is carried away in solution (see Chapter 6).

Sinkholes and Karst Topography

In regions underlain by soluble rock, the ground surface may be pitted with numerous depressions that vary in size and shape. These depressions, called **sinkholes** or merely *sinks*, mark areas where the underlying rock has been dissolved (➤ Figure 13-8). Sinkholes usually form in one of two ways. The first is when the soluble rock below the soil is dissolved by seeping water. Natural openings in the rock are enlarged and filled in by the overlying soil. As the groundwater continues to dissolve the rock, the soil is eventually removed, leaving depressions that are typically shallow with gently sloping sides.

Sinkholes also form when a cave's roof collapses, usually producing a steep-sided crater. Sinkholes formed in this way are a serious hazard, particularly in populated areas. In regions prone to sinkhole formation, the depth and extent of underlying cave systems must be mapped before any development to ensure that the underlying rocks are thick enough to support planned structures.

A **karst topography** is one that has developed largely by groundwater erosion (➤ Figure 13-9). The name *karst* is derived from the plateau region of the border area of Slovenia, Croatia, and northeastern Italy where this type of topography is well developed. In the United States, regions of karst topography include large areas of southwestern Illinois, southern Indiana, Kentucky, Tennessee, northern Missouri, Alabama, and central and northern Florida.

Karst topography is characterized by numerous caves, springs, sinkholes, solution valleys, and disappearing streams (Figure 13-9). When adjacent sinkholes merge, they form a network of larger, irregular, closed depressions called *solution valleys*. *Disappearing streams* are another feature of areas of karst topography. They are so named because they typically flow only a short distance at the surface and then disappear into a sinkhole. The water continues flowing underground through various fractures or caves until it surfaces again at a spring or other stream.

Karst topography can range from the spectacular high-

➤ **FIGURE 13-8** This sinkhole formed on May 8 and 9, 1981, in Winter Park, Florida, due to a drop in the water table after prior dissolution of the underlying limestone. The sinkhole destroyed a house, numerous cars, and the municipal swimming pool. It has a diameter of 100 m and a depth of 35 m.

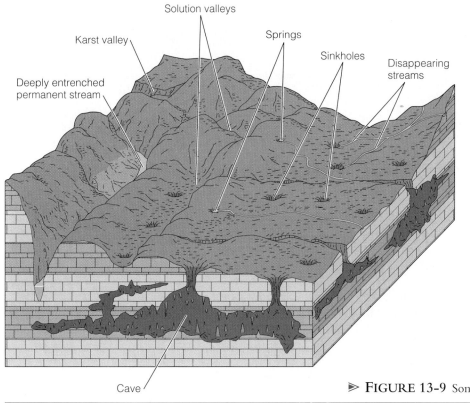

Solution valleys

Karst valley

Springs

Sinkholes

Disappearing streams

Deeply entrenched permanent stream

Cave

≻ **FIGURE 13-9** Some of the features of karst topography.

relief landscapes of China to the subdued and pockmarked landforms of Kentucky (≻ Figure 13-10). Common to all karst topography, though, is the presence of thick-bedded, readily soluble rock at the surface or just below the soil and enough water for solution activity to occur. Karst topography is, therefore, typically restricted to humid and temperate climates.

Caves and Cave Deposits

Caves are some of the most spectacular examples of the combined effects of weathering and erosion by groundwater. As groundwater percolates through carbonate rocks, it dissolves and enlarges original fractures and openings to form a complex interconnecting system of crevices, caves, caverns, and underground streams. A **cave** is usually defined as a

(a)

(b)

≻ **FIGURE 13-10** (a) The Stone Forest, 126 km southeast of Kunming, People's Republic of China, is a high-relief karst landscape formed by the dissolution of carbonate rocks. (b) Solution valleys, sinkholes, and sinkhole lakes dominate the subdued karst topography east of Bowling Green, Kentucky.

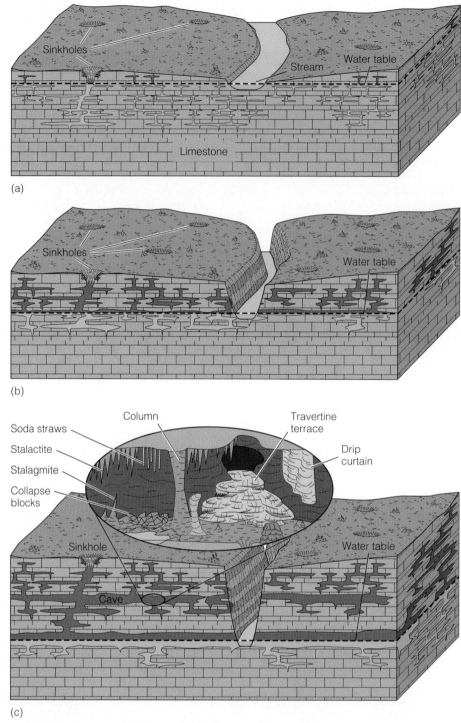

► **FIGURE 13-11** The formation of caves. (*a*) As groundwater percolates through the zone of aeration and flows through the zone of saturation, it dissolves the carbonate rocks and gradually forms a system of passageways. (*b*) As groundwater moves along the surface of the water table, it forms a system of horizontal passageways through which dissolved rock is carried to the surface streams, enlarging the passageways. (*c*) As the surface streams erode deeper valleys, the water table drops, and the abandoned channelways form an interconnecting system of caves and caverns.

naturally formed subsurface opening that is generally connected to the surface and is large enough for a person to enter. A *cavern* is a very large cave or a system of interconnected caves.

More than 17,000 caves are known in the United States. Most of them are small, but some are quite large and spectacular. Some of the more famous caves in the United States are Mammoth Cave, Kentucky (see the Prologue); Carlsbad Caverns, New Mexico; Lewis and Clark Caverns,

Montana; and Meramec Caverns, Missouri, which Jesse James and his outlaw band often used as a hideout. While the United States has many famous caves, the deepest known cave in North America is the 536 m deep Arctomys Cave in Mount Robson Provincial Park, British Columbia, Canada.

Caves and caverns form as a result of the dissolution of carbonate rocks by weakly acidic groundwater (► Figure 13-11). Groundwater percolating through the zone of aeration slowly dissolves the carbonate rock and enlarges its

► FIGURE 13-12 The icicle-shaped structures hanging from the ceiling are stalactites, while the upward-pointing structures on the cave floor are stalagmites. Several columns are present where the stalactites and stalagmites have met in this chamber of Luray Caves, Virginia.

fractures and bedding planes. Upon reaching the water table, the groundwater migrates toward the region's surface streams. As the groundwater moves through the zone of saturation, it continues to dissolve the rock and gradually forms a system of horizontal passageways through which the dissolved rock is carried to the streams. As the surface streams erode deeper valleys, the water table drops in response to the lower elevation of the streams. The water that flowed through the system of horizontal passageways now percolates down to the lower water table where a new system of passageways begins to form. The abandoned channelways now form an interconnecting system of caves and caverns. Caves eventually become unstable and collapse, littering their floors with fallen debris.

When most people think of caves, they think of the seemingly endless variety of colorful and bizarre-shaped deposits found in them. Although a great many different types of cave deposits exist, most form in essentially the same manner and are collectively known as **dripstone**. As water seeps through a cave, some of the dissolved carbon dioxide in the water escapes, and a small amount of calcite is precipitated. In this manner, the various dripstone deposits are formed.

Stalactites are icicle-shaped structures hanging from cave ceilings that form as a result of precipitation from dripping water (► Figure 13-12). With each drop of water, a thin layer of calcite is deposited over the previous layer, forming a cone-shaped projection that grows downward from the ceiling.

The water that drips from a cave's ceiling also precipitates a small amount of calcite when it hits the floor. As additional calcite is deposited, an upward-growing projection called a *stalagmite* forms (Figure 13-12). If a stalactite and stalagmite meet, they form a *column*. Groundwater seeping from a crack in a cave's ceiling may form a vertical sheet of rock called a *drip curtain*, while water flowing across a cave's floor may produce *travertine terraces* (Figure 13-11).

MODIFICATIONS OF THE GROUNDWATER SYSTEM AND THEIR EFFECTS

Groundwater is a valuable natural resource that is rapidly being exploited with little regard to the effects of overuse and misuse. Currently, about 20% of all water used in the United States is groundwater. This percentage is increasing, and unless this resource is used more wisely, sufficient amounts of clean groundwater will not be available in the future. Modifications of the groundwater system may have many consequences including (1) lowering of the water table, causing wells to dry up; (2) loss of hydrostatic pressure, causing once free-flowing wells to require pumping; (3) salt-water encroachment; (4) subsidence; and (5) contamination of the groundwater supply.

Lowering of the Water Table

Withdrawing groundwater at a significantly greater rate than it is replaced by either natural or artificial recharge can have serious effects. For example, the High Plains aquifer is one of the most important aquifers in the United States. Underlying most of Nebraska, large parts of Colorado and Kansas, portions of South Dakota, Wyoming, and New Mexico, as well as the panhandle regions of Oklahoma and Texas, it accounts for approximately 30% of the groundwater used for irrigation in the United States (▷ Figure 13-13). Irrigation from the High Plains aquifer is largely responsible for the regions high agricultural productivity, which includes a significant percentage of the nation's corn, cotton, and wheat and half of U.S. beef cattle. Large areas of land (more than 14 million acres) are currently irrigated with water pumped from the High Plains aquifer. Irrigation is popular because yields from irrigated lands can be triple what they would be without irrigation.

While the High Plains aquifer has contributed to the high productivity of the region, it cannot continue providing the quantities of water that it has in the past. In some parts of the High Plains, from 2 to 100 times more water is being pumped annually than is being recharged. Consequently, water is being removed from the aquifer faster than it is being replenished, causing the water table to drop significantly in many areas (Figure 13-12).

What will happen to this region's economy if long-term withdrawal of water from the High Plains aquifer greatly exceeds its recharge rate so that it can no longer supply the quantities of water necessary for irrigation? Solutions range from going back to farming without irrigation to diverting water from other regions such as the Great Lakes. Farming without irrigation would result in greatly decreased yields and higher costs and prices for agricultural products, while the diversion of water from elsewhere would cost billions of dollars and the price of agricultural products would still rise.

Saltwater Incursion

The excessive pumping of groundwater in coastal areas can result in *saltwater incursion* such as occurred on Long Island, New York, during the 1960s. Along coastlines where permeable rocks or sediments are in contact with the ocean, the fresh groundwater, being less dense than seawater, forms a lens-shaped body above the underlying salt water (▷ Figure 13-14a). The weight of the fresh water exerts pressure on the underlying salt water. As long as rates of recharge equal rates of withdrawal, the contact between the fresh groundwater and the seawater will remain the same. If excessive pumping occurs, a deep cone of depression forms in the fresh groundwater (Figure 13-14b). Because some of the pressure from the overlying fresh water has been removed, salt water forms a *cone of ascension* as it migrates upward to fill the pore space that formerly contained fresh water. When this occurs, wells become contaminated with salt water and remain contaminated until recharge by fresh water restores the former level of the fresh groundwater water table.

Saltwater incursion is a major problem in many rapidly growing coastal communities. As the population in these areas grows, greater demand for groundwater creates an even greater imbalance between recharge and withdrawal.

To counteract the effects of saltwater incursion, recharge wells are often drilled to pump water back into the groundwater system (Figure 13-13c). Recharge ponds that allow large quantities of fresh surface water to infiltrate the groundwater supply may also be constructed.

Subsidence

As excessive amounts of groundwater are withdrawn from poorly consolidated sediments and sedimentary rocks, the water pressure between grains is reduced, and the weight of the overlying materials causes the grains to pack closer together, resulting in subsidence of the ground. As more and more groundwater is pumped to meet the increasing needs of agriculture, industry, and population growth, subsidence is becoming more prevalent.

The San Joaquin Valley of California is a major agricultural region that relies largely on groundwater for irrigation. Between 1925 and 1975, groundwater withdrawals in parts of the valley caused subsidence of almost 9 m. Other

---

▷ **FIGURE 13-13** Areal extent of the High Plains aquifer and changes in the water table, predevelopment to 1980.

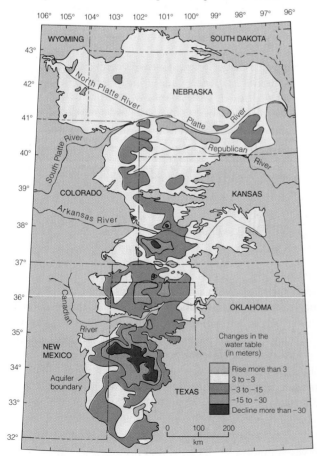

Changes in the water table (in meters)

- Rise more than 3
- 3 to −3
- −3 to −15
- −15 to −30
- Decline more than −30

examples of subsidence in the United States include New Orleans, Louisiana, and Houston, Texas, both of which have subsided more than 2 m, and Las Vegas, Nevada, which has subsided 8.5 m.

Elsewhere in the world, the tilt of the Leaning Tower of Pisa is partly due to groundwater withdrawal (➤ Figure 13-15). The tower started tilting soon after construction began in 1173 because of differential compaction of the foundation. During the 1960s, the city of Pisa withdrew ever larger amounts of groundwater, causing the ground to subside further; as a result, the tilt of the tower increased until it was considered in danger of falling over. Strict control of groundwater withdrawal and stabilization of the

➤ **FIGURE 13-14** Saltwater incursion. (*a*) Because fresh water is not as dense as salt water, it forms a lens-shaped body above the underlying salt water. (*b*) If excessive pumping occurs, a cone of depression develops in the fresh groundwater, and a cone of ascension that forms in the underlying salty groundwater may result in saltwater contamination of the well. (*c*) Pumping water back into the groundwater system through recharge wells can help lower the interface between the fresh groundwater and the salty groundwater and reduce saltwater incursion.

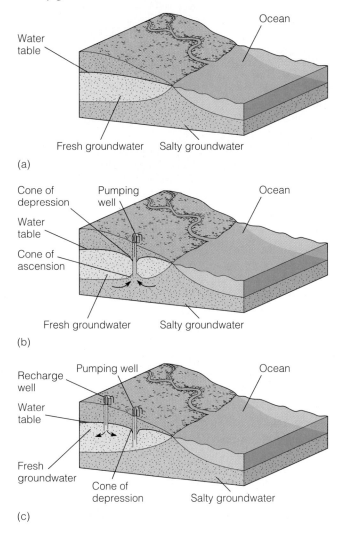

foundation have now reduced the amount of tilting to about 1 mm per year, ensuring that the tower should stand for several more centuries.

The extraction of oil can also cause subsidence. Long Beach, California, has subsided 9 m as a result of 34 years of oil production. More than $100 million of damage was done to the pumping, transportation, and harbor facilities in this area because of subsidence and encroachment of the sea (➤ Figure 13-16). Once water was pumped back into the oil reservoir, subsidence virtually stopped.

Groundwater Contamination

A major problem facing our society is the safe disposal of the numerous pollutant by-products of an industrialized economy. We are becoming increasingly aware that our streams, lakes, and oceans are not unlimited reservoirs for waste, and that we must find new safe ways to dispose of pollutants.

The most common sources of groundwater contamination are sewage, landfills, toxic waste disposal sites (see Perspective 13-1), and agriculture. Once pollutants get into the groundwater system, they will spread wherever groundwater travels, which can make containment of the contamination difficult. Furthermore, because groundwater moves very slowly, it takes a very long time to cleanse a groundwater reservoir once it has become contaminated.

In many areas, septic tanks are the most common way of disposing of sewage. A septic tank slowly releases sewage into the ground where it is decomposed by oxidation and microorganisms and filtered by the sediment as it percolates through the zone of aeration. In most situations, by the time the water from the sewage reaches the zone of saturation, it has been cleansed of any impurities and is safe to use (➤ Figure 13-17a). If the water table is very close to the surface or if the rocks are very permeable, water entering the zone of saturation may still be contaminated and unfit to use.

Landfills are also potential sources of groundwater contamination (Figure 13-17b). Not only does liquid waste seep into the ground, but rainwater also carries dissolved chemicals and other pollutants downward into the groundwater reservoir. Unless the landfill is carefully designed and lined below by an impermeable layer such as clay, many toxic compounds such as paints, solvents, cleansers, pesticides, and battery acid will find their way into the groundwater system.

Toxic waste sites where dangerous chemicals are either buried or pumped underground are an increasing source of groundwater contamination. The United States alone must dispose of several thousand metric tons of hazardous chemical waste per year. Unfortunately, much of this waste has been, and still is being, improperly dumped and is contaminating the surface water, soil, and groundwater.

HOT SPRINGS AND GEYSERS

The subsurface rocks in regions of recent volcanic activity usually stay hot for thousands of years. Groundwater perco-

▷ **FIGURE 13-15** The Leaning Tower of Pisa. The tilting is partly the result of subsidence due to removal of groundwater.

lating through these rocks is heated and, if returned to the surface, forms *hot springs* or *geysers*. Yellowstone National Park in the United States, Rotorua, New Zealand, and Iceland are all famous for their hot springs and geysers. They are all sites of recent volcanism, and consequently their subsurface rocks and groundwater are very hot. The water in some hot springs, however, is circulated deep into the Earth, where it is warmed by the normal increase in temperature, the geothermal gradient (see Chapter 9).

A **hot spring** (also called a *thermal spring* or *warm spring*) is a spring in which the water temperature is warmer than the temperature of the human body (37°C) (▷ Figure 13-18). Some hot springs are much hotter, with temperatures ranging up to the boiling point in many instances. Of the approximately 1,100 known hot springs in the United States, more than 1,000 are in the Far West, while the rest are in the Black Hills of South Dakota, Georgia, the Ouachita region of Arkansas, and the Appalachian region.

Hot springs are also common in other parts of the world. One of the most famous is at Bath, England, where shortly after the Roman conquest of Britain in A.D. 43, numerous bathhouses and a temple were built around the hot springs (see Chapter Opening photo).

Geysers are hot springs that intermittently eject hot water and steam with tremendous force. The word comes from the Icelandic *geysir*, which means to gush or rush forth. One of the most famous geysers in the world is Old Faithful in Yellowstone National Park in Wyoming (▷ Figure 13-19). With a thunderous roar, it erupts a column of hot water and steam every 30 to 90 minutes. Other well-known geyser areas are found in Iceland and New Zealand.

Geysers are the surface expression of an extensive underground system of interconnected fractures within hot igneous rocks (▷ Figure 13-20). Groundwater percolating down into the network of fractures is heated as it comes into contact with the hot rocks. Because the water near the

▷ **FIGURE 13-16** The withdrawal of petroleum from the oil field in Long Beach, California, resulted in up to 9 m of ground subsidence because of sediment compaction. Not until water was pumped back into the reservoir to replace the petroleum did ground subsidence essentially cease. (2 to 29 feet = 0.6 to 8.8 meters)

TOTAL SUBSIDENCE
1928 TO 1968

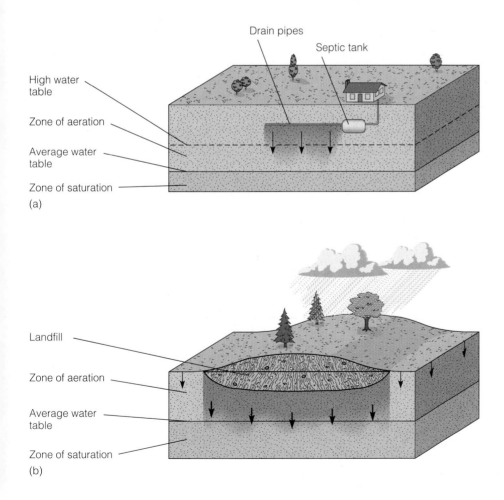

Drain pipes

Septic tank

High water table

Zone of aeration

Average water table

Zone of saturation

(a)

Landfill

Zone of aeration

Average water table

Zone of saturation

(b)

▷ **FIGURE 13-17** (*a*) A septic system slowly releases sewage into the zone of aeration. Oxidation, bacterial degradation, and filtering by the sediments usually remove all of the impurities before they reach the water table. If the rocks are very permeable or the water table is too close to the septic system, contamination of the groundwater can result. (*b*) Unless there is an impermeable barrier between a landfill and the water table, pollutants can be carried into the zone of saturation and contaminate the groundwater supply.

RADIOACTIVE WASTE DISPOSAL

One of the problems of the nuclear age is finding safe storage sites for the radioactive waste from nuclear power plants, the manufacture of nuclear weapons, and the radioactive by-products of nuclear medicine. Radioactive waste can be grouped into two categories: low-level and high-level waste. Low-level wastes are low enough in radioactivity that, when properly handled, they do not pose a significant environmental threat. Most low-level wastes can be safely buried in controlled dump sites where the geology and groundwater system are well known and careful monitoring is provided.

High-level radioactive waste, such as the spent uranium fuel assemblies used in nuclear reactors and the material used in nuclear weapons, is extremely dangerous because of high amounts of radioactivity; it therefore presents a major environmental problem. Currently, more than 15,000 metric tons of spent uranium fuel are awaiting disposal, and the Department of Energy (DOE) estimates that by the year 2000 the nation will have produced almost 50,000 metric tons of highly radioactive waste that must be disposed of safely.

In 1986, Congress chose Yucca Mountain as the only candidate to house the nation's ever increasing amounts of civilian high-level radioactive waste (▷ Figure 1). Congress also authorized the DOE to study the suitability of the site. Such a facility must be able to isolate high-level waste from the environment for at least 10,000 years, which is the minimum time the waste will remain dangerous. The Yucca Mountain site will have a capacity of 70,000 metric tons of waste and will not be completely filled until around the year 2030, at which time its entrance shafts will be sealed and backfilled (▷ Figure 2).

The canisters holding the waste are designed to remain leakproof for at least 300 years, so there is some possibility that leakage could occur over the next 10,000 years. The DOE thinks, however, that the geology of the area will prevent radioactive isotopes from entering the groundwater system. Under an Environmental Protection Agency (EPA) regulation, a radioactive dump site must be located so that the travel time for groundwater from the site to the outside environment is at least 1,000 years.

The radioactive waste at the Yucca Mountain repository will be buried in a volcanic tuff at a depth of about 300 m. The water table in the area will be an additional 200 to 420 m below the dump site. Thus, the canisters will be stored in the zone of aeration, which was one of the reasons Yucca Mountain was selected. Only about 15 cm of rain fall in this area per year, and only a small amount of this percolates into the ground. Most of the water that does seep into the

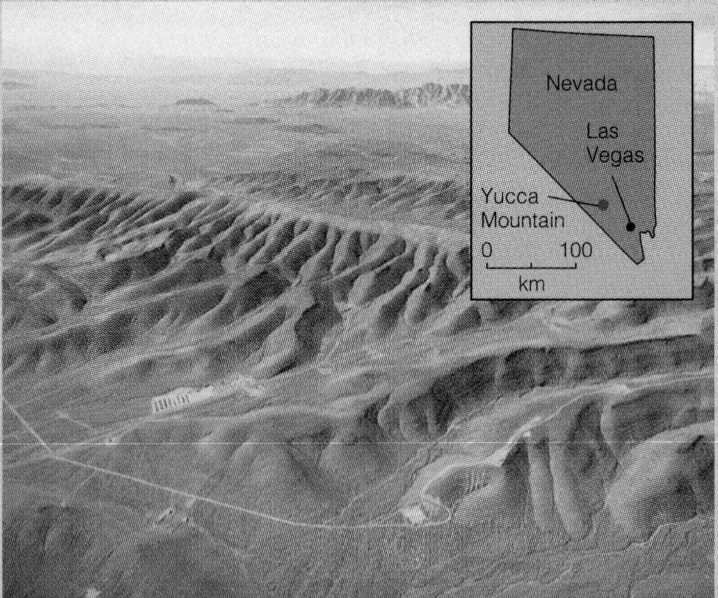

▷ **FIGURE 1** The location and an aerial view of Nevada's Yucca Mountain.

Nevada

Las Vegas

Yucca Mountain

0   100
        km

ground evaporates before it migrates very far, so the rock at the depth the canisters are buried will be very dry, helping prolong the lives of the canisters.

Geologists think that the radioactive waste at Yucca Mountain is most likely to contaminate the environment if it is in liquid form; if liquid, it could seep into the zone of saturation and enter the groundwater supply. But because of the low moisture in the zone of aeration, there is little water to carry the waste downward, and it will take well over 1,000 years to reach the zone of saturation. In fact, the DOE estimates that the waste will take longer than 10,000 years to move from the repository to the water table.

Some geologists are concerned that the climate will change during the next 10,000 years. If the region should become more humid, more water will percolate through the zone of aeration. This will increase the corrosion rate of the canisters and could cause the water table to rise, thereby decreasing the travel time between the repository and the zone of saturation. This area of the country was much more humid during the Ice Age 1.6 million to 10,000 years ago (see Chapter 14).

Another concern is that the area is seismically active. It is, in fact, riddled with faults. At least 27 earthquakes with magnitudes greater than 3 occurred in the area between 1852 and 1991. The most recent occurred on June 29, 1992, with a magnitude of 5.6 and an epicenter only 32 km from Yucca Mountain. Nevertheless, the DOE is convinced that earthquakes pose little danger to the underground repository itself because the disruptive effects of an earthquake are usually confined to the surface.

Finally, some suggest that the DOE has not thoroughly evaluated the economic potential of the area. Exploration is occurring around the Yucca Mountain site, and some Nevada government officials think that there is geologic evidence for various metals and possibly oil and gas in the area. Should human intrusion occur during the thousands of years that the site is supposed to be isolated, dangerous radiation could be released into the environment. Others think the economic potential is being sufficiently evaluated and that the area in question has a low economic potential.

While it appears that Yucca Mountain meets all of the requirements for a safe high-level radioactive waste dump, the site is still controversial, and further studies must be conducted to ensure that the groundwater supply in this area is not rendered unusable by nuclear waste.

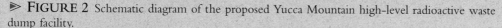

▷ **FIGURE 2** Schematic diagram of the proposed Yucca Mountain high-level radioactive waste dump facility.

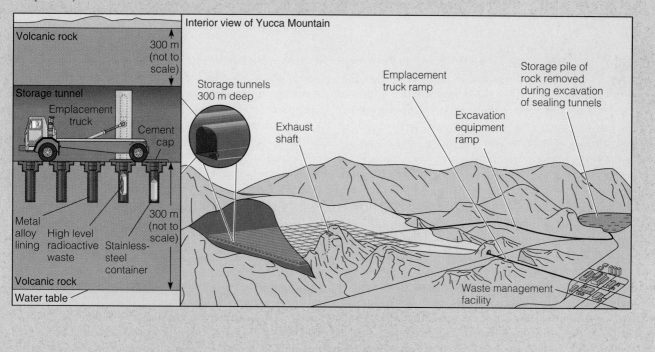

➤ **FIGURE 13-18** Hot springs are springs with a water temperature greater than 37°C. This hot spring is in West Thumb Geyser Basin, Yellowstone National Park, Wyoming.

bottom of the fracture system is under greater pressure than that near the top, it must be heated to a higher temperature before it will boil. Thus, when the deeper water is heated to very near the boiling point, a slight rise in temperature or a drop in pressure, such as from escaping gas, will cause it to instantly change to steam. The expanding steam quickly pushes the water above it out of the ground and into the air, thereby producing a geyser eruption. After the eruption, relatively cool groundwater starts to seep back into the fracture system where it is heated to near its boiling temperature and the eruption cycle begins again. Such a process explains how geysers can erupt with some regularity.

Hot spring and geyser water typically contains large quantities of dissolved minerals because most minerals dissolve more rapidly in warm water than in cold water. Due to this high mineral content, the waters of many hot springs are believed by some to have medicinal properties. Numerous spas and bathhouses have been built throughout the world at hot springs to take advantage of these supposed healing properties.

When the highly mineralized water of hot springs or geysers cools at the surface, some of the material in solution is precipitated, forming various types of deposits. The amount and type of precipitated minerals depend on the solubility and composition of the material through which the groundwater flows. If the groundwater contains dissolved calcium carbonate ($CaCO_3$), then *travertine* or *calcareous tufa* (both of which are varieties of limestone) are precipitated. Spectacular examples of hot spring travertine deposits are found at Mammoth Hot Springs in Yellowstone National Park (➤ Figure 13-21) and at Pamukhale in Turkey.

Geothermal Energy

Energy harnessed from steam and hot water trapped within the Earth's crust is **geothermal energy**. It is a desirable and relatively nonpolluting alternate form of energy. Approximately 1 to 2% of the world's current energy needs could be met by geothermal energy. In those areas where it is plentiful, geothermal energy can supply most, if not all, of the energy needs, sometimes at a fraction of the cost of other types of energy. Some of the countries currently using geothermal energy in one form or another include Iceland,

➤ **FIGURE 13-19** Old Faithful Geyser in Yellowstone National Park, Wyoming, is one of the world's most famous geysers, erupting approximately every 30 to 90 minutes.

➤ **FIGURE 13-20** The formation of a geyser. (*a*) Groundwater percolates downward into a network of interconnected openings and is heated by the hot igneous rocks. The water near the bottom of the fracture system is under greater pressure than that near the top and consequently must be heated to a higher temperature before it will boil. (*b*) Any rise in temperature of the water above its boiling point or a drop in pressure will cause the water to change to steam, which quickly pushes the water above it upward and out of the ground, producing a geyser eruption.

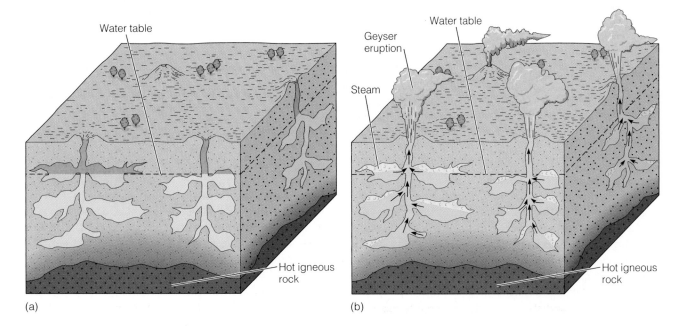

▷ **FIGURE 13-21** Minerva Terrace at Mammoth Hot Springs in Yellowstone National Park, Wyoming, formed when calcium carbonate–rich hot spring water cooled, precipitating travertine deposits.

the United States, Mexico, Italy, New Zealand, Japan, the Philippines, and Indonesia.

The city of Rotorua in New Zealand is world famous for its volcanoes, hot springs, geysers, and geothermal fields. Since the first well was sunk in the 1930s, more than 800 wells have been drilled to tap the hot water and steam below. Geothermal energy in Rotorua is used in a variety of ways, including home, commercial, and greenhouse heating.

In the United States, the first commercial geothermal electrical generating plant was built in 1960 at The Geysers, about 120 km north of San Francisco, California. Here, wells were drilled into the numerous near-vertical fractures underlying the region. As pressure on the rising groundwater decreases, the water changes to steam that is piped directly to electricity-generating turbines.

As oil reserves decline, geothermal energy is becoming an attractive alternative, particularly in parts of the western United States, such as the Salton Sea area of southern California, where geothermal exploration and development have begun. While geothermally generated electricity is a generally clean source of power, it can also be expensive because most geothermal waters are acidic and very corrosive. Consequently, the turbines must either be built of expensive corrosion-resistant alloy metals or frequently replaced. Furthermore, geothermal power is not inexhaustible. The steam and hot water removed for geothermal power cannot be easily replaced, and eventually pressure in the wells drops to the point at which the geothermal field must be abandoned.

CHAPTER SUMMARY

1. The water stored in the pore spaces of subsurface rocks and unconsolidated material is groundwater.
2. Groundwater is part of the hydrologic cycle and represents approximately 22% of the world's supply of fresh water.
3. Porosity is the percentage of a rock, sediment, or soil consisting of pore space. Permeability is the ability of a rock, sediment, or soil to transmit fluids. A material that transmits groundwater is an aquifer, and one that prevents the movement of groundwater is an aquiclude.
4. The water table is the surface separating the zone of aeration (in which pore spaces are filled with both air and

water) from the zone of saturation (in which all pore spaces are filled with water).
5. Groundwater moves slowly through the pore spaces in the zone of aeration and moves through the zone of saturation to outlets such as streams, lakes, and swamps.
6. A spring occurs wherever the water table intersects the Earth's surface. Some springs are the result of a perched water table, that is, a localized aquiclude within an aquifer and above the regional water table.
7. Water wells are made by digging or drilling into the zone of saturation. When water is pumped out of a well, a cone

of depression forms. If water is pumped out faster than it can be recharged, the cone of depression deepens and enlarges and may locally drop to the base of the well, resulting in a dry well.

8. Artesian systems are those in which confined groundwater builds up high hydrostatic pressure. Three conditions must generally be met for an artesian system to form: the aquifer must be confined above and below by aquicludes; the aquifer is usually tilted and exposed at the Earth's surface so it can be recharged; and precipitation must be sufficient to keep the aquifer filled.

9. Karst topography results from groundwater weathering and erosion and is characterized by sinkholes, solution valleys, and disappearing streams.

10. Caves form when groundwater in the zone of saturation weathers and erodes soluble rock such as limestone. Cave deposits, called dripstone, result from the precipitation of calcite.

11. Modifications of the groundwater system can cause serious problems. Excessive withdrawal of groundwater can result in dry wells, loss of hydrostatic pressure, saltwater encroachment, and ground subsidence.

12. Groundwater contamination is becoming a serious problem and can result from sewage, landfills, and toxic waste.

13. Hot springs and geysers may occur where groundwater is heated by hot subsurface volcanic rocks or by the geothermal gradient. Geysers are hot springs that intermittently eject hot water and steam.

14. Geothermal energy comes from the steam and hot water trapped within the Earth's crust. It is a relatively nonpolluting form of energy that is used as a source of heat and for generating electricity.

IMPORTANT TERMS

aquiclude	dripstone	karst topography	water table
aquifer	geothermal energy	permeability	water well
artesian system	geyser	porosity	zone of aeration
cave	groundwater	sinkhole	zone of saturation
cone of depression	hot spring	spring	

REVIEW QUESTIONS

1. The capacity of a material to transmit fluids is:
   a. ____ porosity;
   b. ____ permeability;
   c. ____ solubility;
   d. ____ aeration quotient;
   e. ____ saturation.

2. The water table is a surface separating the:
   a. ____ zone of porosity from the underlying zone of permeability;
   b. ____ capillary fringe from the underlying zone of aeration;
   c. ____ capillary fringe from the underlying zone of saturation;
   d. ____ zone of aeration from the underlying zone of saturation;
   e. ____ zone of saturation from the underlying zone of aeration.

3. Groundwater:
   a. ____ moves slowly through the pore spaces of Earth materials;
   b. ____ moves fastest through the central area of a material's pore space;

   c. ____ can move upward against the force of gravity;
   d. ____ moves from areas of high pressure toward areas of low pressure;
   e. ____ all of these.

4. A perched water table:
   a. ____ occurs wherever there is a localized aquiclude within an aquifer;
   b. ____ is frequently the site of springs;
   c. ____ lacks a zone of aeration;
   d. ____ answers (a) and (b);
   e. ____ answers (b) and (c).

5. An artesian system is one in which:
   a. ____ water is confined;
   b. ____ water can rise above the level of the aquifer when a well is drilled;
   c. ____ there are no aquicludes;
   d. ____ answers (a) and (c);
   e. ____ answers (a) and (b).

6. Which of the following is not an example of groundwater erosion?

   a. ____ karst topography;
   b. ____ stalactites;
   c. ____ sinkholes;
   d. ____ caves;
   e. ____ caverns.

7. Rapid withdrawal of groundwater can result in:
   a. ____ a cone of depression;
   b. ____ ground subsidence;
   c. ____ saltwater incursion;
   d. ____ loss of hydrostatic pressure;
   e. ____ all of these.

8. The water in hot springs and geysers:
   a. ____ is believed to have curative properties;
   b. ____ is noncorrosive;
   c. ____ contains large quantities of dissolved minerals;
   d. ____ answers (a) and (b);
   e. ____ answers (a) and (c).

9. Which of the following is not a cave deposit?
   a. ____ stalagmite;
   b. ____ room;

c. _____ dripstone;

d. _____ stalactite;

e. _____ none of these.

10. Discuss the role of groundwater in the hydrologic cycle.

11. What types of materials make good aquifers and aquicludes?

12. Why is the water table a subdued replica of the surface topography?

13. Where are springs likely to occur?

14. What is a cone of depression and why is it so important?

15. Why are some artesian wells free-flowing while others must be pumped?

16. How does groundwater weather and erode?

17. Discuss the various effects that excessive groundwater removal may have on a region. Give some examples.

18. Discuss the various ways that a groundwater system may become contaminated.

POINTS TO PONDER

1. One of the concerns that geologists have about using Yucca Mountain as a repository for nuclear waste is that the climate may change during the next 10,000 years and become more humid, thus allowing more water to percolate through the zone of aeration. What would the average rate of groundwater movement have to be during the next 10,000 years to reach the canisters containing radioactive waste buried at a depth of 300 m?

2. Why should we be concerned with how fast the groundwater supply is being depleted in some areas?

Chapter 14

GLACIERS AND GLACIATION

OUTLINE

*Glacier on Mount Foresta,
Alaska.*

Following the Great Ice Age, which ended about 10,000 years ago, a general warming trend occurred that was periodically interrupted by short relatively cool periods. One such cool period, from about A.D. 1500 to the mid to late 1800s, was characterized by the expansion of small glaciers in mountain valleys and the persistence of sea ice at high latitudes for longer periods than had occurred previously. This interval of nearly four centuries is known as the *Little Ice Age.*

The climatic changes leading to the Little Ice Age actually began about A.D. 1300. During the preceding centuries, Europe had experienced rather mild temperatures, and the North Atlantic Ocean was warmer and more storm-free than it is now. During this time, the Vikings discovered and settled Iceland, and by A.D. 1200, about 80,000 people resided there. They also sailed to Greenland and North America and established two colonies on the former and one on the latter. As the climate deteriorated, however, the North Atlantic became stormier, and sea ice occurred farther south and persisted longer each year. As a result of poor sea conditions and political problems in Norway, all shipping across the North Atlantic ceased, and the colonies in Greenland and North America eventually disappeared.

During the Little Ice Age, many of the small glaciers in Europe and Iceland expanded and moved far down their valleys, reaching their greatest historic extent by the early 1800s. A small ice cap formed in Iceland where none had existed previously, and glaciers in Alaska and the moun- tains of the western United States and Canada also expanded to their greatest limits during historic time. Although glaciers caused some problems in Europe where they advanced across roadways and pastures, destroying some villages in Scandinavia and threatening villages elsewhere, their overall impact on humans was minimal. Far more important from the human perspective was that during much of the Little Ice Age the summers in northern latitudes were cooler and wetter.

Although worldwide temperatures were a little lower during this time, the change in summer conditions rather than cold winters or glaciers caused most of the problems. Particularly hard hit were Iceland and the Scandinavian countries, but at times much of northern Europe was affected (▷ Figure 14-1). Growing seasons were shorter during many years, resulting in food shortages and a number of famines. Iceland's population declined from its high of 80,000 in 1200 to about 40,000 by 1700. Between 1610 and 1870, sea ice was observed near Iceland for as much as three months a year, and each time the sea ice persisted for long periods, poor growing seasons and food shortages followed.

Exactly when the Little Ice Age ended is debatable. Some authorities put the end at 1880, whereas others think it ended as early as 1850. In any case, during the late 1800s, the sea ice was retreating northward, glaciers were retreating back up their valleys, and summer weather became more stable.

▷ FIGURE 14-1 In this mid-1600s painting by Jan-Abrahamsz Beerstraten titled *The Village of Nieukoop in Winter*, the canals of Holland are shown frozen. These canals rarely freeze today.

INTRODUCTION

Most people have some idea of what a glacier is, but many confuse glaciers with other masses of snow and ice. A **glacier** is a mass of ice composed of compacted and recrystallized snow that flows under its own weight on land. Accordingly, sea ice in the polar regions is not glacial ice, nor are drifting icebergs glaciers even though they may have derived from glaciers that flowed into the sea. Snow fields in high mountains may persist in protected areas for years, but these are not glaciers either because they are not actively moving.

At first glance, glaciers may appear to be static, but like other geologic agents such as streams, glaciers are dynamic systems that are continually adjusting to changes. For instance, just as streams vary their sediment load depending on available energy, increases or decreases in the amount of ice in a glacier alter its ability to erode and transport sediment.

At present, glaciers cover nearly 15 million km^2, or about one-tenth of the Earth's land surface (◉ Table 14-1). Numerous glaciers exist in the mountains of the western United States, especially Alaska, and western Canada and in the Andes in South America, the Alps of Europe, the Himalayas of Asia, and other high mountains. Even Mount Kilimanjaro in Africa, although near the equator, is high enough to support glaciers. In fact, Australia is the only continent lacking glaciers. By far the largest existing glaciers are in Greenland and Antarctica; both areas are almost completely covered by glacial ice (Table 14-1).

Glaciers are particularly effective agents of erosion, transport, and deposition. They deeply scour the land over which they move, producing a number of easily recognized erosional landforms. Eventually, they deposit their sediment load just as do other agents of erosion and transport. Although numerous examples of landscapes that originated from recent glaciation can be found, most glacial landscapes developed during the Pleistocene Epoch, or what is commonly called the Ice Age (1.6 million to 10,000 years ago), a time when glaciers covered much more extensive areas than they do now, particularly on the Northern Hemisphere continents.

GLACIERS AND THE HYDROLOGIC CYCLE

Glaciers contain about 2.15% of the world's water, thereby temporarily removing it from the hydrologic cycle. However, many glaciers at high latitudes, as in Alaska, flow directly into the sea where they melt, or where icebergs break off by a process called *calving* and drift out to sea where they eventually melt. At lower latitudes where they can exist only at high elevations, glaciers flow to lower elevations where they melt and the water returns to the oceans by surface runoff. In areas of low precipitation, as in parts of the western United States, glaciers are important fresh water reservoirs that release water to streams during the dry season.

In addition to melting, glaciers lose water by sublimation, a process in which ice changes directly to water vapor without an intermediate liquid phase. Water vapor so derived rises into the atmosphere where it may condense and fall once again as snow or rain. In any case, it too is eventually returned to the oceans.

THE ORIGIN OF GLACIAL ICE

Ice is a mineral in every sense of the word; it has a crystalline structure and possesses characteristic physical and chemical properties. Accordingly, geologists consider glacial ice to be rock, although it is a type of rock that is easily deformed. It forms in a fairly straightforward manner (▶ Figure 14-2). When an area receives more winter snow than can melt during the spring and summer seasons, a net accumulation occurs. Freshly fallen snow consists of about 80% air and 20% solids, but it compacts as it accumulates, partly thaws, and refreezes; in the process, the original snow layer is converted to a granular type of ice called **firn**. Deeply buried firn is further compacted and is finally converted to **glacial ice**, consisting of about 90% solids (Figure 14-2). When accumulated snow and ice reach a critical thickness of about 40 m, the pressure on the ice at depth is sufficient to cause deformation and flow, even though it remains solid. Once the critical thickness is reached and flow begins, the moving mass of ice becomes a *glacier*.

Plastic flow, which causes permanent deformation, occurs in response to pressure and is the primary way that glaciers move. They may also move by **basal slip,** which occurs when a glacier slides over the underlying surface

TABLE 14-1 Present-Day Ice-Covered Areas	
Antarctica	12,653,000 km^2
Greenland	1,802,600
Northeast Canada	153,200
Central Asian ranges	124,500
Spitsbergen group	58,000
Other Arctic islands	54,000
Alaska	51,500
South American ranges	25,000
West Canadian ranges	24,900
Iceland	11,800
Scandinavia	5,000
Alps	3,600
Caucasus	2,000
New Zealand	1,000
USA (other than Alaska)	650
Others	about 800
	14,971,550
Total volume of present ice: 28 to 35 million km^3	

SOURCE: C. Embleton and C. A. King, *Glacial Geomorphology* (New York: Halsted Press, 1975).

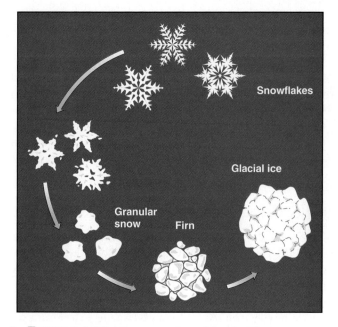

▷ **FIGURE 14-2** The conversion of freshly fallen snow to firn and glacial ice.

(▷ Figure 14-3). Basal slip is facilitated by the presence of meltwater that reduces frictional resistance between the underlying surface and the glacier. If a glacier is frozen to the underlying surface, though, it moves only by plastic flow.

TYPES OF GLACIERS

Geologists generally recognize two basic types of glaciers: *valley* and *continental*. A **valley glacier,** as its name implies, is confined to a mountain valley or perhaps to an interconnected system of mountain valleys (▷ Figure 14-4). Large valley glaciers commonly have several smaller tributary glaciers, much as large streams have tributaries. Valley glaciers flow from higher to lower elevations and are invariably small in comparison to continental glaciers.

Continental glaciers, also called *ice sheets,* cover vast areas (at least 50,000 km²) and are unconfined by topography (▷ Figure 14-5). In contrast to valley glaciers, which flow downhill within the confines of a valley, continental glaciers flow outward in all directions from a central area of accumulation. Currently, only two continental glaciers exist, one in Greenland and the other in Antarctica. Both are more than 3,000 m thick in their central areas, become thinner toward their margins, and cover all but the highest mountains. During the Pleistocene Epoch, glaciers covered large parts of the Northern Hemisphere continents. Many of the erosional and depositional landforms in much of Canada and the northern tier of the United States formed as a consequence of Pleistocene glaciation.

Although valley and continental glaciers are easily differentiated by their size and location, an intermediate variety called an *ice cap* is also recognized. Ice caps are similar to, but

smaller than, continental glaciers and cover less than 50,000 km². Some ice caps form when valley glaciers grow and overtop the divides and passes between adjacent valleys and coalesce to form a continuous ice cap. They also form on fairly flat terrain including some of the islands of the Canadian Arctic and Iceland.

THE GLACIAL BUDGET

Just as a savings account grows and shrinks as funds are deposited and withdrawn, glaciers expand and contract in response to accumulation and wastage. Their behavior can be described in terms of a **glacial budget,** which is essentially a balance sheet of accumulation and wastage. The upper part of a valley glacier is a **zone of accumulation** where additions exceed losses, and the glacier's surface is perennially covered by snow. In contrast, the lower part of the same glacier is a **zone of wastage,** where losses from melting, sublimation, and calving of icebergs exceed the rate of accumulation (▷ Figure 14-6).

At the end of winter, a glacier's surface is usually completely covered with the accumulated seasonal snowfall. During spring and summer, the snow begins to melt, first at lower elevations and then progressively higher up the glacier. The elevation to which snow recedes during a wastage season is called the **firn limit** (Figure 14-6). One can easily identify the zones of accumulation and wastage by noting the position of the firn limit.

Observations of a single glacier reveal that the position of the firn limit usually changes from year to year. If it does not

▷ **FIGURE 14-3** Movement of a glacier by a combination of plastic flow and basal slip. If a glacier is solidly frozen to the underlying surface, it moves only by plastic flow. Notice that the top of the glacier moves farther in a given time than the bottom does.

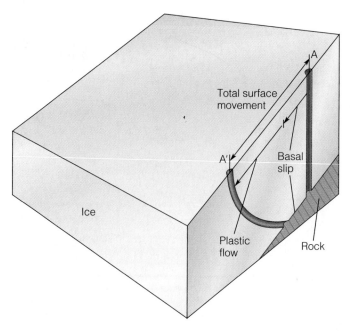

> FIGURE 14-4 A large valley glacier in Alaska. Notice the tributaries to the large glacier.

> FIGURE 14-5 The Antarctic ice sheet, one of two presently existing continental glaciers.

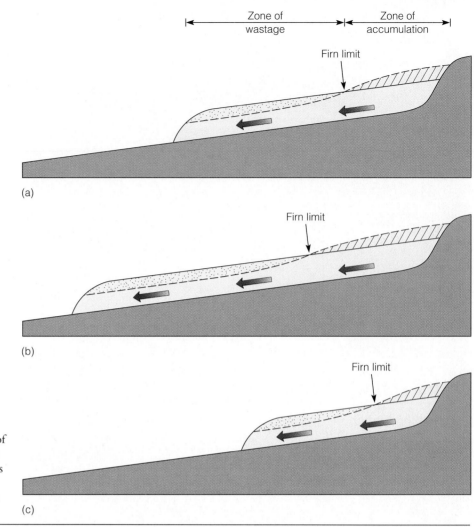

> **FIGURE 14-6** Response of a hypothetical glacier to changes in its budget. (*a*) If the losses in the zone of wastage, shown by stippling, equal additions in the zone of accumulation, shown by crosshatching, the terminus of the glacier remains stationary. (*b*) Gains exceed losses, and the glacier's terminus advances. (*c*) Losses exceed gains, and the glacier's terminus retreats, although the glacier continues to flow.

change or shows only minor fluctuations, the glacier is said to have a balanced budget; that is, additions in the zone of accumulation are exactly balanced by losses in the zone of wastage, and the distal end or *terminus* of the glacier remains stationary (Figure 14-6a). When the firn limit moves down the glacier, the glacier has a positive budget; its additions exceed its losses, and its terminus advances (Figure 14-6b). If the budget is negative, the glacier recedes—its terminus retreats up the glacial valley (Figure 14-6). But even though a glacier's terminus may be receding, the glacial ice continues to move toward the terminus by plastic flow and basal slip. If a negative budget persists long enough, though, a glacier recedes and thins until it no longer flows, thus becoming a *stagnant glacier.*

Although we used a valley glacier as our example, the same budget considerations control the flow of continental glaciers as well. In the case of the Antarctic ice sheet the entire ice sheet is in the zone of accumulation, but wastage occurs where the ice sheet flows into the ocean.

RATES OF GLACIAL MOVEMENT

In general, valley glaciers move more rapidly than continental glaciers, but the rates for both vary, ranging from centimeters to tens of meters per day. Valley glaciers moving down steep slopes flow more rapidly than glaciers of comparable size on gentle slopes, assuming that all other variables are the same. The main glacier in a valley glacier system contains a greater volume of ice and thus has a greater discharge and flow velocity than its tributaries (Figure 14-4). Temperature exerts a seasonal control on valley glaciers because although plastic flow remains rather constant year-round, basal slip is more important during warmer months when meltwater is more abundant.

Flow rates also vary within the ice itself. Flow velocity generally increases in the zone of accumulation until the firn limit is reached; from that point, the velocity becomes progressively slower toward the glacier's terminus. Valley glaciers are similar to streams, in that the valley walls and floor cause frictional resistance to flow, so the ice in contact with the walls and floor moves more slowly than the ice some distance away (▷ Figure 14-7).

Notice in Figure 14-7 that the flow velocity increases upward until the top few tens of meters of ice are reached, but little or no additional increase occurs after that point. This upper ice constitutes the rigid part of the glacier that is moving as a result of basal slip and plastic flow below. The fact that this upper 40 m or so of ice behaves as a brittle solid is clearly demonstrated by large fractures called *crevasses* that develop when a valley glacier flows over a step in its valley floor where the slope increases or where it flows around a corner (▷ Figure 14-8). In either case, the glacial ice is stretched (subjected to tension), and large crevasses develop, but they extend downward only to the zone of plastic flow. In some cases, a valley glacier descends over such a steep precipice that crevasses break up the ice into a jumble of blocks and spires, and an *ice fall* develops (Figure 14-8).

The flow rates of valley glaciers are also complicated by *glacial surges,* which are bulges of ice that move through a glacier at a velocity several times faster than the normal flow. Although surges are best documented in valley glaciers, they occur in ice caps and continental glaciers as well. During a surge, a glacier's terminus may advance several kilometers during a year. In 1993, the terminus of the Bering Glacier in Alaska advanced more than 1.5 km in just three weeks.

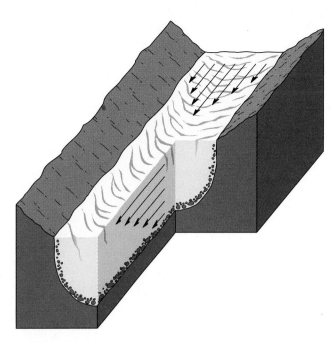

▷ **FIGURE 14-7** Flow velocity in a valley glacier varies both horizontally and vertically. Velocity is greatest at the top-center of the glacier. Friction with the walls and floor of the glacial trough causes the flow to be slower adjacent to these boundaries. The length of the arrows in the figure is proportional to the velocity.

▷ **FIGURE 14-8** Crevasses and an ice fall in a glacier in Alaska.

The causes of surges are not fully understood, but some of them have occurred following a period of unusually heavy precipitation in the zone of accumulation. In any case, rapid changes in the glacial budget occurred.

One reason continental glaciers move comparatively slowly is that they exist at higher latitudes and are frozen to the underlying surface most of the time, which limits the amount of basal slip. Some basal slip does occur even beneath the Antarctic ice sheet, but most of its movement is by plastic flow. Nevertheless, some parts of continental glaciers manage to achieve extremely high flow rates. Near the margins of the Greenland ice sheet, for instance, the ice is forced between mountains in what are called *outlet glaciers*. In some of these outlets, flow velocities exceeding 100 m per day have been recorded.

▷ **FIGURE 14-9** A glacial erratic in Yellowstone National Park, Wyoming.

⊛ GLACIAL EROSION AND TRANSPORT

Glaciers are moving solids that can erode and transport huge quantities of materials, especially unconsolidated sediment and soil. In many areas of Canada and the northern United States, glaciers transported boulders, some of huge proportions, for long distances before depositing them. Such boulders are known as *glacial erratics* (▷ Figure 14-9).

Important erosional processes associated with glaciers include bulldozing, plucking, and abrasion. Bulldozing, although not a formal geologic term, is fairly self-explanatory: a glacier simply shoves or pushes unconsolidated materials in its path. *Plucking,* also called *quarrying,* occurs when glacial ice freezes in the cracks and crevices of a bedrock projection and eventually pulls it loose.

Sediment-laden glacial ice can effectively erode by **abrasion.** Bedrock over which sediment-laden glacial ice has moved commonly develops a *glacial polish,* a smooth surface that glistens in reflected light (▷ Figure 14-10a). Abrasion

also yields *glacial striations,* consisting of rather straight scratches rarely more than a few millimeters deep on rock surfaces (Figure 14-10b). During abrasion, rocks are thoroughly pulverized so that they yield an aggregate of clay- and silt-sized particles having the consistency of flour, hence the name *rock flour.* Rock flour is so common in streams discharging from glaciers that the water generally has a milky appearance.

Continental glaciers can derive sediment from mountains projecting through them, and windblown dust settles on their surfaces. Otherwise, most of their sediment is derived from the surface over which they move and is transported in the lower part of the ice sheet. In contrast, valley glaciers carry sediment in all parts of the ice, but it is concentrated at the base and along the margins. Some of the marginal sediment is derived by abrasion and plucking, but much of it is supplied by mass wasting processes.

▷ **FIGURE 14-10** (*a*) Glacial polish on quartzite near Marquette, Michigan. (*b*) Glacial striations in basalt at Devil's Postpile National Monument, California.

(a) (b)

Erosional Landforms of Valley Glaciers

Some of the world's most inspiring scenery is produced by valley glaciers. Many mountain ranges are scenic to begin with, but when modified by valley glaciers, they take on a unique aspect of jagged, angular peaks and ridges in the midst of broad valleys (▷ Figure 14-11).

U-Shaped Glacial Troughs. A **U-shaped glacial trough** is one of the most distinctive features of valley glaciation (▷ Figure 14-12c). Mountain valleys eroded by running water are typically V-shaped in cross section; that is, they have valley walls that descend to a narrow valley bottom (Figure 14-12a). In contrast, valleys scoured by glaciers are deepened, widened, and straightened so that they possess very steep or vertical walls, but have broad, rather flat valley floors; thus, they exhibit a U-shaped profile (▷ Figure 14-13). Many glacial troughs contain triangular-shaped *truncated spurs,* which are cutoff or truncated ridges that extend into the preglacial valley (Figure 14-12c).

During the Pleistocene, when glaciers were more extensive, sea level was about 130 m lower than at present, so glaciers flowing into the sea eroded their valleys to much greater depths than they do now. When the glaciers melted at the end of the Pleistocene, sea level rose, and the ocean filled the lower ends of the glacial troughs so that now they are long, steep-walled embayments called **fiords** (▷ Figure 14-14).

Lower sea level during the Pleistocene was not responsible for the formation of all fiords. Unlike running water, glaciers can erode a considerable distance below sea level. In fact, a glacier 500 m thick can stay in contact with the sea floor and effectively erode it to a depth of about 450 m before the buoyant effects of water cause the glacial ice to float!

Hanging Valleys. Although waterfalls can form in several ways, some of the world's highest and most spectacular are found in recently glaciated areas. Yosemite Falls in Yosemite National Park, California, plunges from a **hanging valley,** which is a tributary valley whose floor is at a higher level than that of the main valley (▷ Figure 14-15). As Figure 14-12 shows, the large glacier in the main valley vigorously erodes, whereas the smaller glaciers in tributary valleys are less capable of large-scale erosion. When the glaciers disappear, the smaller tributary valleys remain as hanging valleys. Accordingly, streams flowing through hanging valleys plunge over vertical or very steep precipices.

Cirques, Arêtes, and Horns. Perhaps the most spectacular erosional landforms in areas of valley glaciation occur at the upper ends of glacial troughs and along the divides separating adjacent glacial troughs. Valley glaciers form and move out from steep-walled, bowl-shaped depressions called **cirques** at the upper end of their troughs (Figure 14-12c). Cirques are typically steep-walled on three sides, but one side is open and leads into the glacial trough.

Although the details of cirque origin are not fully understood, they apparently form by erosion of a preexisting depression on a mountain side. As snow and ice accumulate in the depression, frost wedging and plucking enlarge it until it takes on the typical cirque shape. Many cirques have a lip

▷ **FIGURE 14-11** The rugged, angular landscape typical of areas eroded by valley glaciers is apparent in Glacier National Park, Montana.

(a)

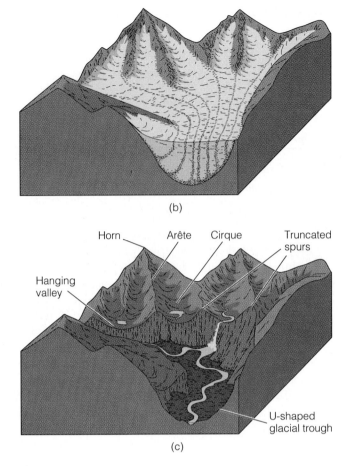
(b)

Horn　Arête　Cirque　Truncated spurs

Hanging valley

U-shaped glacial trough

(c)

 FIGURE 14-12 Erosional landforms produced by valley glaciers. (*a*) A mountain area before glaciation. (*b*) The same area during the maximum extent of the valley glaciers. (*c*) After glaciation.

➤ **FIGURE 14-13** A U-shaped glacial trough in northwestern Montana.

▷ **FIGURE 14-14** Milford Sound, a fiord in New Zealand. (Photo courtesy of George and Linda Lohse.)

or threshold, indicating that the glacial ice not only moves outward but rotates as well, scouring out a depression rimmed by rock.

Cirques become wider and are cut deeper into mountain sides by headward erosion as a consequence of abrasion, plucking, and several mass wasting processes. The largest cirque known is the Walcott Cirque in Antarctica, which is 16 km wide and 3 km deep. The fact that cirques expand laterally and by headward erosion accounts for the origin of two other distinctive erosional features, *arêtes* and *horns*. Arêtes—narrow, serrated ridges—can form in two ways. In many cases, cirques form on opposite sides of a ridge, and headward erosion reduces the ridge until only a thin partition of rock remains. The same effect occurs when erosion in two parallel glacial troughs reduces the intervening ridge to a thin spine of rock (Figure 14-12c).

The most majestic of all mountain peaks are **horns;** these steep-walled, pyramidal peaks are formed by headward erosion of cirques. In order for a horn to form, a mountain peak must have at least three cirques on its flanks, all of which erode headward (Figure 14-12c). Excellent examples of horns include Mount Assiniboine in the Canadian Rockies, the Grand Teton in Wyoming (see Figure 10-1), and the most famous of all, the Matterhorn in Switzerland (▷ Figure 14-16).

Erosional Landforms of Continental Glaciers

Areas eroded by continental glaciers tend to be smooth and rounded because they bevel and abrade high areas that projected into the ice. Rather than yielding the sharp, angular landforms typical of valley glaciation, continental glaciers produce a landscape of rather flat, monotonous topography interrupted by rounded hills.

In a large part of Canada, particularly the vast Canadian Shield region (see Chapter 8), continental glaciation has stripped off the soil and unconsolidated surface sediment, revealing extensive exposures of striated and polished bedrock. Similar though smaller bedrock exposures are also widespread in the northern United States from Maine through Minnesota.

Another result of erosion in these areas is the complete disruption of drainage that has not yet become reestablished. Much of the area is characterized by deranged drainage (see Figure 12-15e), numerous lakes and swamps, low relief, extensive bedrock exposures, and little or no soil. Such areas are generally referred to as *ice-scoured plains*.

◉ GLACIAL DEPOSITS

All sediment deposited by glacial activity is **glacial drift.** Glacial deposits in several upper midwestern states and parts

➤ **FIGURE 14-15** Yosemite Falls in Yosemite National Park, California, plunges 435 m vertically, cascades down a steep slope for another 205 m, and then falls vertically 97 m, for a total descent of 737 m. (Photo courtesy of Sue Monroe.)

of southern Canada are important sources of groundwater and rich soils, and in many areas they are exploited for their sand and gravel.

Geologists generally recognize two distinct types of glacial drift, *till* and *stratified drift*. **Till** consists of sediment deposited directly by glacial ice. It is not sorted or stratified; that is, its particles are not separated by size or density, and it does not exhibit any layering. Till deposited by valley glaciers looks much like the till of continental glaciers except that the latter's deposits are much more extensive and have generally been transported much farther.

Stratified drift is sorted by size and density and, as its name implies, is layered. In fact, most of the sediments recognized as stratified drift are braided stream deposits; the streams in which they were deposited received their water and sediment load directly from melting glacial ice.

Landforms Composed of Till

Landforms composed of till include several types of *moraines* and elongated hills known as *drumlins*.

End Moraines. The terminus of either a valley or a continental glacier may become stabilized in one position for some period of time, perhaps a few years or even decades. Such stabilization of the ice front does not mean that the glacier has ceased flowing, only that it has a balanced budget. When an ice front is stationary, flow within the glacier continues, and the sediment transported within or upon the ice is dumped as a pile of rubble at the glacier's terminus. Such deposits are **end moraines,** which continue to grow as long as the ice front is stabilized (➤ Figure 14-17). End moraines of valley glaciers are commonly crescent-shaped

► **FIGURE 14-16** The Matterhorn in Switzerland is a well-known horn.

► **FIGURE 14-17** (*a*) The origin of an end moraine. (*b*) End moraines are described as terminal moraines or recessional moraines depending on their position relative to the glacier that produced them.

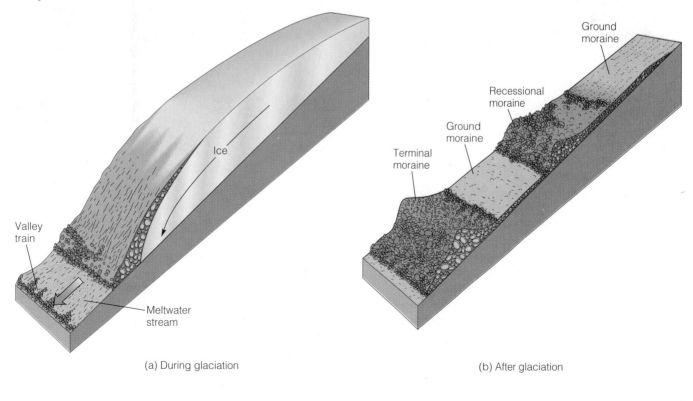

(a) During glaciation

(b) After glaciation

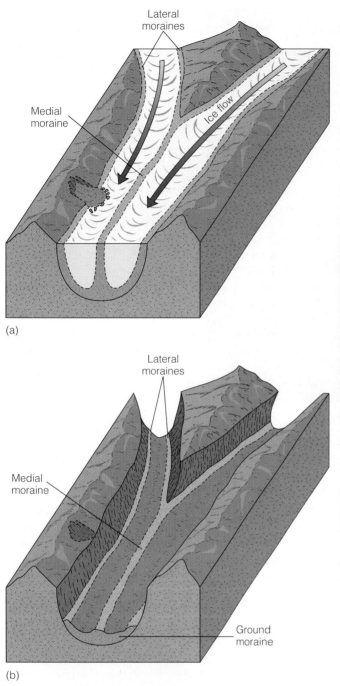

(a)

(b)

(c)

➤ **FIGURE 14-18** (*a*) and (*b*) The material transported and deposited along the margins of a valley glacier is lateral moraine. Where two lateral moraines merge, a medial moraine forms. (*c*) Lateral and medial moraines on a glacier in Alaska.

ridges of till spanning the valley occupied by the glacier. Those of continental glaciers similarly parallel the ice front, but are much more extensive.

Following a period of stabilization, a glacier may advance or retreat, depending on changes in its budget. If it advances, the ice front overrides and modifies its former moraine. Should a negative budget occur, however, the ice front retreats toward the zone of accumulation. As the ice front recedes, till is deposited as it is liberated from the melting ice

and forms a layer of **ground moraine** (Figure 14-17b). Ground moraine has an irregular, rolling topography, whereas end moraine consists of long ridgelike accumulations of sediment.

After a glacier has retreated for some time, its terminus may once again stabilize, and it will deposit another end moraine. Because the ice front has receded, such moraines are called **recessional moraines** (Figure 14-17b). During the Pleistocene, continental glaciers in the mid-continent region extended as far south as southern Ohio, Indiana, and Illinois. Their outermost end moraines, marking the greatest extent of the glaciers, go by the special name **terminal moraine** (valley glaciers also deposit terminal moraines). As the glaciers retreated from the positions where their terminal moraines were deposited, they temporarily ceased retreating numerous times and deposited many recessional moraines.

Lateral and Medial Moraines. Valley glaciers transport considerable sediment along their margins. Much of this sediment is abraded and plucked from the valley walls, but a significant amount falls or slides onto the glacier's surface by mass wasting processes. In any case, when a glacier melts, this sediment is deposited as long ridges of till called **lateral moraines** along the margin of the glacier (▷ Figure 14-18).

Where two lateral moraines merge, as when a tributary glacier flows into a larger glacier, a **medial moraine** forms (Figure 14-18). Although medial moraines are identified by their position on a valley glacier, they are, in fact, formed from the coalescence of two lateral moraines. One can generally determine how many tributaries a valley glacier has by the number of its medial moraines.

Drumlins. In many areas where continental glaciers have deposited till, the till has been reshaped into elongated hills called **drumlins**. Some drumlins measure as much as 50 m high and 1 km long, but most are much smaller. From the side, a drumlin looks like an inverted spoon with the steep end on the side from which the glacial ice advanced, and the gently sloping end pointing in the direction of ice movement (▷ Figure 14-19). Drumlins can therefore be used to determine the direction of ice movement.

Although no one has fully explained the origin of drumlins, it appears that they form in the zone of plastic flow as glacial ice modifies preexisting till into streamlined hills. Drumlins rarely occur as single, isolated hills; instead they occur in *drumlin fields* containing hundreds or thousands of drumlins. Drumlin fields are found in several states and Ontario, Canada, but perhaps the finest example is near Palmyra, New York.

Landforms Composed of Stratified Drift

Stratified drift is associated with both valley and continental glaciers, but as one would expect, it is more extensive in areas of continental glaciation.

Outwash Plains and Valley Trains. Glaciers discharge meltwater laden with sediment most of the time, except perhaps during the coldest months. This meltwater forms a series of braided streams that radiate out from the front of continental glaciers over a wide region (▷ Figure 14-20). So much sediment is supplied to these streams that much of it is deposited within the channels as sand and gravel bars. The vast blankets of sediments so formed are **outwash plains** (Figure 14-20).

Valley glaciers discharge huge amounts of meltwater and, like continental glaciers, have braided streams extending from them. But these streams are confined to the lower parts of glacial troughs, and their long, narrow deposits of stratified drift are known as **valley trains**.

Outwash plains and valley trains commonly contain numerous circular to oval depressions, many of which contain small lakes. These depressions are *kettles;* they form when a retreating ice sheet or valley glacier leaves a block of ice that is subsequently partly or wholly buried (Figure 14-20). When the ice block eventually melts, it leaves a depression; if the depression extends below the water table, it becomes the site of a small lake. So many kettles occur in some outwash plains that they are called *pitted outwash plains.* Although kettles are most common in outwash plains, they can also form in end moraines.

▷ **FIGURE 14-19** These elongated hills in Antrim County, Michigan, are drumlins. (Photo courtesy of B. M. C. Pape.)

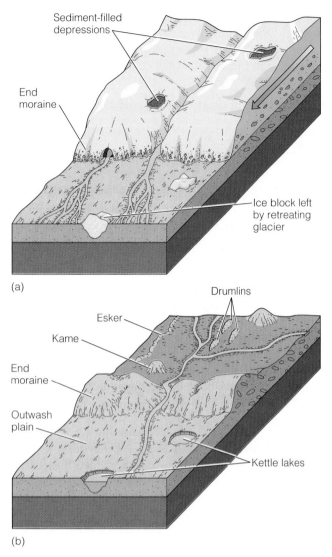

FIGURE 14-20 Two stages in the origin of kettles, kames, eskers, and an outwash plain. (*a*) During glaciation. (*b*) After glaciation.

Kames and Eskers. **Kames** are conical hills as much as 50 m high composed of stratified drift (Figure 14-20). Many kames form when a stream deposits sediment in a depression on a glacier's surface; as the ice melts, the deposit is lowered to the surface. They also form in cavities within or beneath stagnant ice.

Long sinuous ridges of stratified drift, many of which meander and have tributaries, are called **eskers** (Figure 14-20). Some eskers are quite high, as much as 100 m, and can be traced for more than 100 km. Eskers occur most commonly in areas once covered by continental glaciers, but they are also associated with large valley glaciers. The sorting and stratification of the sediments within eskers clearly indicate deposition by running water. The physical properties of ancient eskers and observations of present-day glaciers indicate that they form in tunnels beneath stagnant ice (Figure 14-20).

Glacial Lakes and Their Deposits

During the Pleistocene Epoch, many lakes existed in the western United States. These *pluvial lakes* formed as a result of greater precipitation and overall cooler temperatures that lowered the evaporation rate. Lake Bonneville, which covered 50,000 km², was the largest pluvial lake; the Great Salt Lake in Utah is the remnant of this once much larger lake. Another large pluvial lake existed in Death Valley, California, which is now the hottest, driest place in North America.

In contrast to pluvial lakes, which are far from glaciers, *proglacial lakes* are formed by meltwater accumulating along the margins of glaciers, so at least one shoreline is the ice itself. Lake Agassiz, named in honor of the French naturalist Louis Agassiz, was a large proglacial lake covering about 250,000 km² of North Dakota and Minnesota, and Manitoba, Saskatchewan, and Ontario, Canada. It persisted until the glacial ice along its northern margin melted; then it drained northward into Hudson Bay.

Numerous proglacial lakes existed during the Pleistocene, but most of them eventually disappeared as they drained or were filled with sediments. The most notable exception is the Great Lakes, all of which first formed as proglacial lakes (see Perspective 14-1).

Glacial lakes, like all lakes, are areas of deposition. Sediment may be carried into them and deposited as small deltas, but of special interest are the fine-grained deposits. Mud deposits in glacial lakes are commonly finely laminated, consisting of alternating light and dark layers. Each light-dark couplet is called a *varve* (▶ Figure 14-21). Each varve represents an annual episode of deposition; the light layers form during the spring and summer and consist of silt and clay; the dark layers form during the winter when the smallest particles of clay and organic matter settle from suspension as the lake freezes over. The number of varves indicates how many years a glacial lake has existed.

Dropstones are another distinctive feature of glacial lakes containing varved deposits (Figure 14-21). These are pieces of gravel, some of boulder size, in otherwise very fine-grained deposits. Most of them were probably carried into the lakes by icebergs that eventually melted and released sediment contained in the ice.

CAUSES OF GLACIATION

So far we have examined the effects of glaciation, but have not addressed the central questions of what causes large-scale glaciation and why so few episodes of widespread glaciation have occurred. For more than a century, scientists have been attempting to develop a comprehensive theory explaining all aspects of ice ages, but have not yet been completely successful. One reason for their lack of success is that the climatic changes responsible for glaciation, the cyclic occurrence of glacial-interglacial episodes, and short-term events such as the Little Ice Age operate on vastly different time scales.

Only a few periods of glaciation are recognized in the

> **FIGURE 14-21** Glacial varves with a dropstone. (Photo courtesy of Canadian Geological Survey.)

geologic record, each separated from the others by long intervals of mild climate. Such long-term climatic changes probably result from slow geographic changes related to plate tectonic activity. Moving plates can carry continents to high latitudes where glaciers can exist, provided that they receive enough precipitation as snow. Plate collisions, the subsequent uplift of vast areas far above sea level, and the changing atmospheric and oceanic circulation patterns caused by the changing shapes and positions of plates also contribute to long-term climatic change.

A theory explaining ice ages must also address the fact that during the Pleistocene Ice Age (1.6 million to 10,000 years ago) several intervals of glacial expansion separated by warmer interglacial periods occurred. At least four major episodes of glaciation have been recognized in North America, and six or seven glacial advances and retreats occurred in Europe. These intermediate-term climatic events occur on time scales of tens to hundreds of thousands of years. The cyclic nature of this most recent episode of glaciation has long been a problem in formulating a comprehensive theory of climatic change.

The Milankovitch Theory

A particularly interesting hypothesis for intermediate-term climatic events was put forth by the Yugoslavian astronomer Milutin Milankovitch during the 1920s. He proposed that minor irregularities in the Earth's rotation and orbit are sufficient to alter the amount of solar radiation that the Earth receives at any given latitude and hence can affect climatic changes. Now called the **Milankovitch theory,** it was

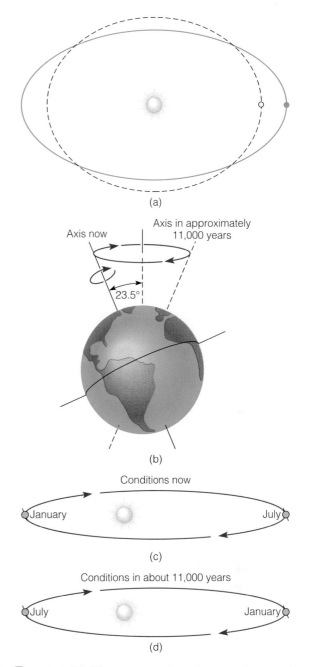

> **FIGURE 14-22** (*a*) The Earth's orbit varies from nearly a circle (dashed line) to an ellipse (solid line) and back again in about 100,000 years. (*b*) The Earth moves around its orbit while spinning about its axis, which is tilted to the plane of the ecliptic at 23.5° and points toward the North Star. The Earth's axis of rotation slowly moves and traces out the path of a cone in space. (*c*) At present, the Earth is closest to the Sun in January when the Northern Hemisphere experiences winter. (*d*) In about 11,000 years, as a result of precession, the Earth will be closer to the Sun in July, when summer occurs in the Northern Hemisphere.

initially ignored, but has received renewed interest during the last 20 years.

Milankovitch attributed the onset of the Pleistocene Ice Age to variations in three parameters of the Earth's orbit (➤ Figure 14-22). The first of these is orbital eccentricity,

A BRIEF HISTORY OF THE GREAT LAKES

Before the Pleistocene, no large lakes existed in the Great Lakes region, which was then an area of generally flat lowlands with broad stream valleys draining to the north (▷ Figure 1). As the glaciers advanced southward, they eroded the stream valleys more deeply, forming what were to become the basins of the Great Lakes. During these glacial advances, the ice front moved forward as a series of lobes, some of which flowed into the preexisting lowlands where the ice became thicker and moved more rapidly. As a consequence, the lowlands were deeply eroded—four of the five Great Lakes basins were eroded below sea level.

At their greatest extent, the glaciers covered the entire Great Lakes region and extended far to the south. As the ice sheet retreated northward during the late Pleistocene, the ice front periodically stabilized, and numerous recessional moraines were deposited. By about 14,000 years ago, parts of the Lake Michigan and Lake Erie basins were ice-free,

and glacial meltwater began forming proglacial lakes (▷ Figure 2). As the retreat of the ice sheet continued—although periodically interrupted by minor readvances of the ice front—the Great Lakes basins were uncovered, and the lakes expanded until they eventually reached their present size and configuration (Figure 2). Currently, the Great Lakes contain nearly 23,000 km³ of water, about 18% of the water in all fresh water lakes.

Although the history of the Great Lakes just presented is generally correct, it is oversimplified. For instance, the areas and depths of the evolving Great Lakes fluctuated widely in response to minor readvances of the ice front. Furthermore, as the lakes filled, they spilled over the lowest parts of their margins, thus cutting outlets that partly drained them. And finally, as the glaciers retreated northward, isostatic rebound raised the southern parts of the Great Lakes region, greatly altering their drainage systems.

▷ **FIGURE 1** Theoretical preglacial drainage in the Great Lakes region. The divide separating the preglacial Mississippi and St. Lawrence drainage basins was probably near its present location. The future sites of the Great Lakes are outlined by dotted lines.

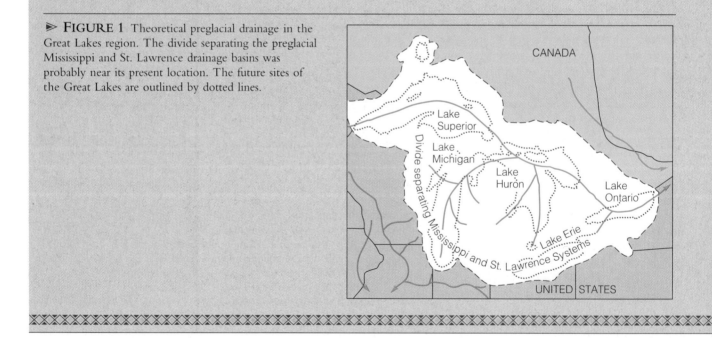

which is the degree that the orbit departs from a perfect circle. Calculations indicate a roughly 100,000-year cycle between times of maximum eccentricity. This corresponds closely to 20 warm-cold climatic cycles that occurred during the Pleistocene. The second parameter is the angle between the Earth's axis and a line perpendicular to the plane of the ecliptic. This angle shifts about 1.5° from its current value of 23.5° during a 41,000-year cycle. The third parameter is the precession of the equinoxes, which causes the position of the

equinoxes and solstices to shift slowly around the Earth's elliptical orbit in a 23,000-year cycle.

Continuous changes in these three parameters cause the amount of solar heat received at any latitude to vary slightly over time. The total heat received by the planet, however, remains little changed. Milankovitch proposed, and now many scientists agree, that the interaction of these three parameters provided the triggering mechanism for the glacial-interglacial episodes during the Pleistocene.

The present-day Great Lakes and their St. Lawrence River drainage constitute one of the great commercial waterways of the world. Oceangoing vessels can sail into the interior of North America as far west as Duluth, Minnesota. To do so, however, they must bypass Niagara Falls between Lake Erie and Lake Ontario via a system of locks. Niagara Falls plunges 51 m over the Niagaran escarpment, which consists of resistant dolostone that was exposed during the last glacial retreat. Erosion of softer shale at the base of the falls is causing Niagara Falls to retreat upstream at a rate of about a meter per year. It is estimated that in about 25,000 years the escarpment between the lakes will have been eliminated! As a result, Lake Erie will become a small lake adjacent to vast swampy areas, and the upper Great Lakes will be considerably smaller than at present.

➤ **FIGURE 2** Four stages in the evolution of the Great Lakes. As the glacial ice retreated northward, the lake basins began filling with meltwater. The dotted lines indicate the present-day shorelines of the lakes.

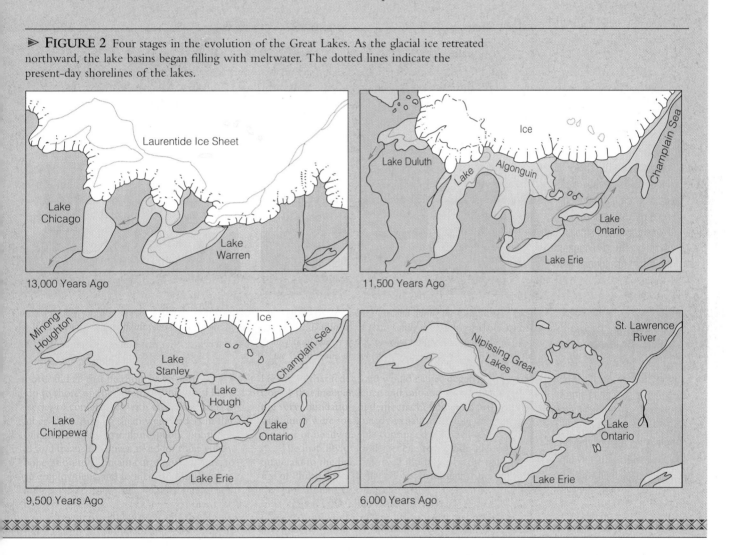

13,000 Years Ago

11,500 Years Ago

9,500 Years Ago

6,000 Years Ago

Short-Term Climatic Events

Climatic events having durations of several centuries, such as the Little Ice Age, are too short to be accounted for by plate tectonics or Milankovitch cycles. Several hypotheses have been proposed, including variations in solar energy and volcanism.

Variations in solar energy could result from changes within the Sun itself or from anything that would reduce the amount of energy the Earth receives from the Sun. The latter could result from the solar system passing through clouds of interstellar dust and gas or from substances in the Earth's atmosphere reflecting solar radiation back into space. Records kept over the past 75 years indicate that during this time the amount of solar radiation has varied only slightly. So although variations in solar energy may influence short-term climatic events, such a correlation has not been demonstrated.

During large volcanic eruptions, tremendous amounts of ash and gases are spewed into the atmosphere where they reflect incoming solar radiation and reduce atmospheric temperatures. Recall from Chapter 5 that small droplets of sulfur gases remain in the atmosphere for years and can have a significant effect on the climate. Several large-scale volcanic events have been recorded, such as the 1815 eruption of Tambora, and are known to have had climatic effects. However, no relationship between periods of volcanic activity and periods of glaciation has yet been established.

CHAPTER SUMMARY

1. Glaciers are masses of ice on land that move by plastic flow and basal slip. Valley glaciers are confined to mountain valleys and flow from higher to lower elevations, whereas continental glaciers cover vast areas and flow outward in all directions from a zone of accumulation.

2. A glacier forms when winter snowfall in an area exceeds summer melt and therefore accumulates year after year. Snow is compacted and converted to glacial ice, and when the ice is about 40 m thick, pressure causes it to flow.

3. The behavior of a glacier depends on its budget, which is the relationship between accumulation and wastage. If a glacier possesses a balanced budget, its terminus remains stationary; a positive or negative budget results in advance or retreat of the terminus, respectively.

4. Glaciers move at varying rates depending on the slope, discharge, and season. Valley glaciers tend to flow more rapidly than continental glaciers.

5. Glaciers are powerful agents of erosion and transport because they are solids in motion. They are particularly effective at eroding soil and unconsolidated sediment, and they can transport any size sediment supplied to them.

6. Continental glaciers transport most of their sediment in the lower part of the ice, whereas valley glaciers may carry sediment in all parts of the ice.

7. Erosion of mountains by valley glaciers yields several sharp, angular landforms including cirques, arêtes, and horns. U-shaped glacial troughs, fiords, and hanging valleys are also products of valley glaciation.

8. Continental glaciers abrade and bevel high areas, producing a smooth, rounded landscape.

9. Depositional landforms include moraines, which are ridge-like accumulations of till. Several types of moraines are recognized, including terminal, recessional, lateral, and medial moraines.

10. Drumlins are composed of till that was apparently reshaped into streamlined hills by continental glaciers.

11. Stratified drift consists of sediments deposited in or by meltwater streams issuing from glaciers; it is found in outwash plains and valley trains. Ridges called eskers and conical hills called kames are also composed of stratified drift.

12. Major glacial intervals separated by tens or hundreds of millions of years probably occur as a result of the changing positions of tectonic plates, which in turn cause changes in oceanic and atmospheric circulation patterns.

13. Currently, the Milankovitch theory is widely accepted as the explanation for glacial-interglacial intervals.

14. The reasons for short-term climatic changes, such as the Little Ice Age, are not understood. Two proposed causes for such events are changes in the amount of solar energy received by the Earth and volcanism.

IMPORTANT TERMS

abrasion
arête
basal slip
cirque
continental glacier
drumlin
end moraine
esker
fiord

firn
firn limit
glacial budget
glacial drift
glacial ice
glacier
ground moraine
hanging valley

horn
kame
lateral moraine
medial moraine
Milankovitch theory
outwash plain
plastic flow
recessional moraine

stratified drift
terminal moraine
till
U-shaped glacial trough
valley glacier
valley train
zone of accumulation
zone of wastage

1. Crevasses in glaciers extend down to:
   a. ____ about 300 m;
   b. ____ the base of the glacier;
   c. ____ the zone of plastic flow;
   d. ____ variable depths depending on how thick the ice is;
   e. ____ the outwash layer.

2. If a glacier has a negative budget:
   a. ____ the terminus will retreat;
   b. ____ its accumulation rate is greater than its wastage rate;
   c. ____ all flow ceases;
   d. ____ the glacier's length increases;
   e. ____ crevasses will no longer form.

3. The bowl-shaped depression at the upper end of a glacial trough is a(n):
   a. ____ inselberg;
   b. ____ cirque;
   c. ____ lateral moraine;
   d. ____ drumlin;
   e. ____ till.

4. Headward erosion of a group of cirques on the flanks of a mountain may produce a:
   a. ____ tarn;
   b. ____ varve;
   c. ____ drumlin;
   d. ____ kettle;
   e. ____ horn.

5. Rocks abraded by glaciers develop a smooth surface that shines in reflected light. Such a surface is called glacial _____.
   a. ____ grooves;
   b. ____ polish;
   c. ____ flour;
   d. ____ striations;
   e. ____ till.

6. Firn is:
   a. ____ freshly fallen snow;
   b. ____ a granular type of ice;
   c. ____ a valley train;
   d. ____ another name for the zone of wastage;
   e. ____ a type of glacial groove.

7. Pressure on ice at depth in a glacier causes it to move by:
   a. ____ rock creep;
   b. ____ fracture;
   c. ____ basal slip;
   d. ____ surging;
   e. ____ plastic flow.

8. Glacial drift is a general term for:
   a. ____ the erosional landforms of continental glaciers;
   b. ____ all the deposits of glaciers;
   c. ____ icebergs floating at sea;
   d. ____ the movement of glaciers by plastic flow and basal slip;
   e. ____ the annual wastage rate of a glacier.

9. The number of medial moraines on a glacier generally indicates the number of its _____ .
   a. ____ tributary glaciers;
   b. ____ terminal moraines;
   c. ____ eskers;
   d. ____ outwash plains;
   e. ____ valley trains.

10. How does glacial ice form, and why is it considered to be a rock?

11. Other than size, how do valley glaciers differ from continental glaciers?

12. What is the relative importance of plastic flow and basal slip for glaciers at high and low latitudes?

13. Explain in terms of the glacial budget how a once active glacier becomes a stagnant glacier.

14. What is a glacial surge and what are the probable causes of surges?

15. Why are glaciers more effective agents of erosion and transport than running water?

16. Describe the processes responsible for the origin of a cirque, U-shaped glacial trough, and hanging valley.

17. Discuss the processes whereby terminal, recessional, and lateral moraines form.

18. How does a medial moraine form, and how can one determine the number of tributaries a valley glacier has by its medial moraines?

1. In a roadside outcrop, you observe a deposit of alternating light and dark laminated mud containing a few large boulders. Explain the sequence of events responsible for deposition of these sediments.

2. In North America, valley glaciers are common in Alaska and western Canada, and small ones are present in the mountains of the western United States; none, however, occur east of the Rocky Mountains. How can you explain this distribution of glaciers?

THE WORK OF WIND AND DESERTS

OUTLINE

A sharp line marks the boundary between pasture and an encroaching dune in Niger, Africa. As the goats eat the remaining bushes, the dune will continue to advance, and more land will be lost to desertification.

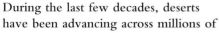
During the last few decades, deserts have been advancing across millions of acres of productive land, destroying rangelands, croplands, and even villages (see the chapter-opening photo). Such expansion, estimated at 70,000 km^2 per year, has exacted a terrible toll in human suffering. Because of the relentless advance of deserts, hundreds of thousands of people have died of starvation or been forced to migrate as "environmental refugees" from their homelands to camps where the majority are severely malnourished. This expansion of deserts into formerly productive lands is *desertification*.

Most regions undergoing desertification lie along the margins of existing deserts. These margins have a delicately balanced ecosystem that serves as a buffer between the desert on one side and a more humid environment on the other. Their potential to adjust to increasing environmental pressures from natural causes or human activity is limited. Ordinarily, desert regions expand and contract gradually in response to natural processes such as climatic change, but much recent desertification has been greatly accelerated by human activities. In many areas, the natural vegetation has been cleared as crop cultivation has expanded into increasingly drier fringes to support the growing population. Because these areas are especially prone to droughts, crop failures are common occurrences, leaving the land bare and susceptible to increased wind and water erosion.

Because grasses constitute the dominant natural vegetation in most fringe areas, raising livestock is a common economic activity. Usually, these areas achieve a natural balance between vegetation and livestock as nomadic herders graze their animals on the available grasses. In many fringe areas, livestock numbers have been greatly increasing in recent years, and they now far exceed the land's capacity to support them. As a result, the vegetation cover that protects the soil has diminished, causing the soil to crumble. This leads to further drying of the soil and accelerated soil erosion by wind and water (see the chapter-opening photo).

Drilling water wells also contributes to desertification because human and livestock activity around a well site strips away the vegetation. With its vegetation gone, the topsoil blows away, and the resultant bare areas merge with the surrounding desert. In addition, the water used for irrigation from these wells sometimes contributes to desertification by increasing the salt content of the soil. As the water evaporates, a small amount of salt is deposited in the soil and is not flushed out as it would be in an area that receives more rain. Over time, the salt concentration becomes so high that plants can no longer grow. Desertification resulting from soil salinization is a major problem in North Africa, the Middle East, southwest Asia, and the western United States.

Collecting firewood for heating and cooking is another major cause of desertification, particularly in many less-developed countries where wood is the major fuel source. In the Sahel of Africa (a belt 300 to 1,100 km wide that lies south of the Sahara), the expanding population has removed all trees and shrubs in the areas surrounding many towns and cities. Journeys of several days on foot to collect firewood are common there. Furthermore, the use of dried animal dung to supplement firewood has exacerbated desertification because important nutrients in the dung are not returned to the soil.

The Sahel averages between 10 and 60 cm of rainfall per year, 90% of which evaporates when it falls. Because drought is common in the Sahel, the region can support only a limited population of livestock and humans. Traditionally, herders and livestock existed in a natural balance with the vegetation, following the rains north during the rainy season and returning south to greener rangeland during the dry seasons. Some areas were alternately planted and left fallow to help regenerate the soil. During fallow periods, livestock fed off the stubble of the previous year's planting, and their dung helped fertilize the soil.

With the emergence of new nations and increased foreign aid to the Sahel during the 1950s and 1960s, nomads and their herds were restricted, and large areas of grazing land were converted to cash crops such as peanuts and cotton that have a short growing season. Expanding human and animal populations and more intensive agriculture put increasing demands on the land. These factors, combined with the worst drought of the century (1968–1973), brought untold misery to the people of the Sahel. Without rains, the crops failed and the livestock denuded the land of what little vegetation remained. As a result, nearly 250,000 people and 3.5 million cattle died of starvation, and the adjacent Sahara expanded southward as much as 150 km.

The tragedy of the Sahel and prolonged droughts in other desert fringe areas serve to remind us of the delicate equilibrium of ecosystems in such regions. Once the fragile soil cover has been removed by erosion, it takes centuries for new soil to form (see Chapter 6).

INTRODUCTION

Most people associate the work of wind with deserts. Wind is an effective geologic agent in desert regions, but it also plays an important role wherever loose sediment can be eroded, transported, and deposited, such as along shorelines or on the plains. Therefore, we will first consider the work of wind in general and then discuss the distribution, characteristics, and landforms of deserts.

⬡ SEDIMENT TRANSPORT BY WIND

Wind is a turbulent fluid and therefore transports sediment in much the same way as running water. Although wind typically flows at a greater velocity than water, it has a lower density and, thus, can carry only clay- and silt-size particles as *suspended load*. Sand and larger particles are moved along the ground as *bed load*.

Bed Load

Sediments too large or heavy to be carried in suspension by water or wind are moved as bed load either by *saltation* or by rolling and sliding. As we discussed in Chapter 12, saltation is the process by which a portion of the bed load moves by intermittent bouncing along a stream bed. Saltation also occurs on land. The wind starts sand grains rolling and lifts and carries some grains short distances before they fall back to the surface. As the descending sand grains hit the surface, they strike other grains causing them to bounce along by saltation (➤ Figure 15-1). Wind tunnel experiments have shown that once sand grains begin moving, they will continue to move, even if the wind drops below the speed necessary to start them moving! This happens because once saltation begins, it sets off a chain reaction of collisions between grains that keeps the sand grains in constant motion.

Saltating sand usually moves near the surface, and even when winds are strong, grains are rarely lifted higher than about a meter. If the winds are very strong, these wind-whipped grains can cause extensive abrasion. A car's paint can be removed by sandblasting in a short time, and its windshield will become completely frosted and translucent from pitting.

Suspended Load

Silt- and clay-sized particles constitute most of a wind's suspended load. Even though these particles are much smaller and lighter than sand-sized particles, wind usually starts the latter moving first. The reason for this phenom-

➤ FIGURE 15-1 Most sand is moved near the ground surface by saltation. Sand grains are picked up by the wind and carried a short distance before falling back to the ground where they usually hit other grains, causing them to bounce and move in the direction of the wind.

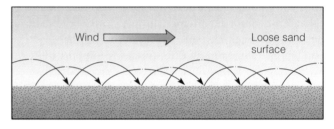

➤ FIGURE 15-2 (*a*) A ventifact forms when wind-borne particles (1) abrade the surface of a rock (2) forming a flat surface. If the rock is moved, (3) additional flat surfaces are formed. (*b*) Large ventifacts lying on desert pavement in Death Valley National Monument, California.

(b)

enon is that a very thin layer of motionless air lies next to the ground where the small silt and clay particles remain undisturbed. The larger sand grains, however, stick up into the turbulent air zone where they can be moved. Unless the stationary air layer is disrupted, the silt and clay particles remain on the ground providing a smooth surface. This phenomenon can be observed on a dirt road on a windy day. Unless a vehicle travels over the road, little dust is raised even though it is windy. When a vehicle moves over the road, it breaks the calm boundary layer of air and disturbs the smooth layer of dust, which is picked up by the wind and forms a dust cloud in the vehicle's wake.

In a similar manner, when a sediment layer is disturbed, silt- and clay-sized particles are easily picked up and carried in suspension by the wind, creating clouds of dust or even dust storms. Once these fine particles are lifted into the atmosphere, they may be carried thousands of kilometers from their source. For example, large quantities of fine dust from the southwestern United States were blown eastward and fell on New England during the Dust Bowl of the 1930s (see Figure 6-16).

✺ WIND EROSION

Although wind action produces many distinctive erosional features and is an extremely efficient sorting agent, running water is responsible for most erosional landforms in deserts, even though stream channels are typically dry. Wind erodes material in two ways: *abrasion* and *deflation*.

Abrasion

Abrasion involves the impact of saltating sand grains on an object and is analogous to sandblasting. The effects of abrasion are usually minor because sand, the most common agent of abrasion, is rarely carried more than 1 m above the surface. Rather than creating major erosional features, wind abrasion merely modifies existing features by etching, pitting, smoothing, or polishing. Thus, wind abrasion is most effective on soft sedimentary rocks.

Ventifacts are a common product of wind abrasion; these are stones whose surfaces have been polished, pitted, grooved, or faceted by the wind (► Figure 15-2). If the wind blows from different directions, or if the stone is moved, the ventifact will have multiple facets. Ventifacts are most common in deserts, yet they can also form wherever stones are exposed to saltating sand grains, as on beaches in humid regions and some outwash plains in New England.

Yardangs are larger features than ventifacts and also result from wind abrasion (► Figure 15-3). They are elongated and streamlined ridges that look like an overturned ship's hull. They are typically found grouped in clusters aligned parallel to the prevailing winds. They probably form by differential erosion in which depressions, parallel to the direction of wind, are carved out of a rock body, leaving sharp, elongated ridges. These ridges may then be further modified by wind abrasion into their characteristic shape.

Deflation

Another important mechanism of wind erosion is **deflation,** which is the removal of loose surface sediment by the wind. Among the characteristic features of deflation in many arid and semiarid regions are *deflation hollows* or *blowouts* (► Figure 15-4). These shallow depressions of variable dimensions result from differential erosion of surface materials. Ranging in size from several kilometers in diameter and tens of meters deep to small depressions only a few meters wide and less

► **FIGURE 15-3** Profile view of a streamlined yardang in the Roman playa deposits of the Kharga Depression, Egypt. (Photo courtesy of Marion A. Whitney.)

▷ **FIGURE 15-4** A deflation hollow in Death Valley, California.

than a meter deep, deflation hollows are common in the southern Great Plains region of the United States.

In many dry regions, the removal of sand-sized and smaller particles by wind leaves a surface of pebbles, cobbles, and boulders. As the wind removes the fine-grained material from the surface, the effects of gravity and occasional floodwaters rearrange the remaining coarse particles into a mosaic of close-fitting rocks called **desert pavement**

(Figures 15-2b and ▷ 15-5). Once a desert pavement forms, it protects the underlying material from further deflation.

⊚ WIND DEPOSITS

Although wind is of minor importance as an erosional agent, wind can form impressive deposits. These deposits are primarily of two types. The first, called *dunes,* occur in

▷ **FIGURE 15-5** (*a*) Desert pavement forms when deflation removes fine-grained material from the ground surface leaving larger-sized particles. (*b*) As deflation continues and more material is removed, the larger particles are concentrated and form a desert pavement, which protects the underlying material from additional deflation.

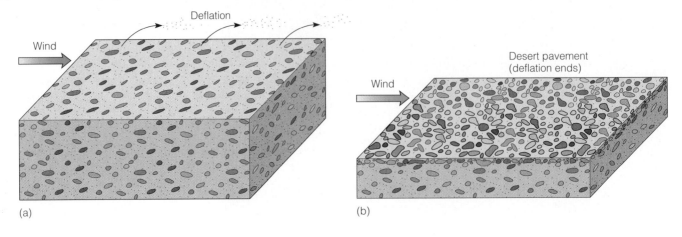

several distinctive types, all of which consist of sand-sized particles that are usually deposited near their source. The second is *loess,* which consists of layers of windblown silt and clay that are deposited over large areas downwind and commonly far from their source.

The Formation and Migration of Dunes

The most characteristic features associated with sand-covered regions are **dunes,** which are mounds or ridges of wind-deposited sand. Dunes form when wind flows over and around an obstruction. Most dunes have an asymmetrical profile, with a gentle windward slope and a steeper downwind or leeward slope that is inclined in the direction of the prevailing wind (➤ Figure 15-6a). Sand grains move up the gentle windward slope by saltation and accumulate on the leeward side forming an angle between 30° and 34° from the horizontal, which is the angle of repose of dry sand. When this angle is exceeded by accumulating sand, the slope collapses, and the sand slides down the leeward slope, coming to rest at its base. As sand moves from a dune's windward side and periodically slides down its leeward slope, the dune slowly migrates in the direction of the prevailing wind (Figure 15-6b).

Dune Types

Four major dune types are generally recognized (*barchan, longitudinal, transverse,* and *parabolic*), although intermediate forms between the major types also exist. The size, shape, and arrangement of dunes result from the interaction of such factors as sand supply, the direction and velocity of the prevailing wind, and the amount of vegetation. While dunes are usually found in deserts, they can also occur wherever

there is an abundance of sand such as along the upper parts of many beaches.

Barchan dunes are crescent-shaped dunes whose tips point downwind (➤ Figure 15-7). They form in areas where there is a generally flat, dry surface with little vegetation, a limited supply of sand, and a nearly constant

➤ **FIGURE 15-6** (*a*) Profile view of a sand dune. (*b*) Dunes migrate when sand moves up the windward side and slides down the leeward slope. Such movement of the sand grains produces a series of cross-beds that slope in the direction of wind movement.

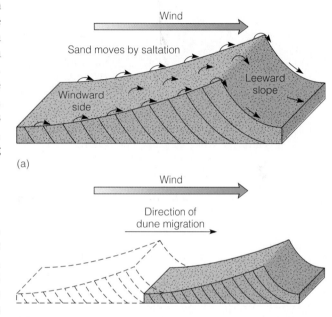
(a)

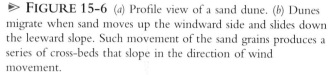
(b)

(a)

➤ **FIGURE 15-7** (*a*) Barchan dunes form where there is a limited amount of sand, a nearly constant wind direction, and a generally flat, dry surface with little vegetation. The tips of barchan dunes point downwind. (*b*) Several barchan dunes west of the Salton Sea, California.

(b)

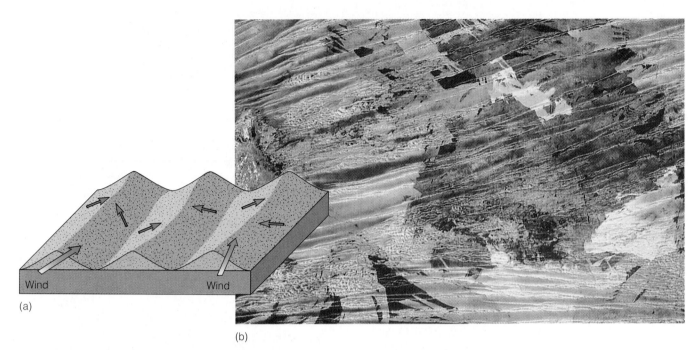

► FIGURE 15-8 (*a*) Longitudinal dunes form long, parallel ridges of sand aligned roughly parallel to the prevailing wind direction. They typically form where sand supplies are limited. (*b*) These longitudinal dunes in the Gibson Desert in west central Australia are 15 m high. The bright blue areas between the dunes are shallow pools of rainwater, while the darkest patches are areas where the Aborigines have set fires to encourage the growth of spring grasses.

wind direction. Most barchans are small, with the largest reaching about 30 m in height. Barchans are the most mobile of the major dune types, moving at rates that can exceed 10 m per year.

Longitudinal dunes (also called *seif dunes*) are long, parallel ridges of sand aligned generally parallel to the direction of the prevailing winds; they form where the sand supply is somewhat limited (► Figure 15-8). Longitudinal dunes result when winds converge from slightly different directions to produce the prevailing wind. They range in size from about 3 m to more than 100 m high, and some stretch for more than 100 km. These dunes are especially well developed in central Australia, where they cover nearly one-fourth of the continent. They also cover extensive areas in Saudi Arabia, Egypt, and Iran.

Transverse dunes form long ridges perpendicular to the prevailing wind direction in areas where abundant sand is available and little or no vegetation exists (► Figure 15-9). When viewed from the air, transverse dunes have a wavelike appearance, and the areas they cover are therefore sometimes called sand seas. The crests of transverse dunes can be as high as 200 m, and the dunes may be as much as 3 km wide.

Parabolic dunes are most common in coastal areas with abundant sand, strong onshore winds, and a partial cover of vegetation (► Figure 15-10). Although parabolic dunes have a crescent shape like barchan dunes, their tips point upwind. Parabolic dunes form when the vegetation cover is broken and deflation produces a deflation hollow or blow-

out. As the wind transports the sand out of the depression, it builds up on the convex downwind dune crest. The central part of the dune is excavated by the wind, while vegetation holds the ends and sides fairly well in place.

Loess

Windblown silt and clay deposits composed of angular quartz grains, feldspar, micas, and calcite are known as **loess.** The distribution of loess shows that it is derived from three main sources: deserts, Pleistocene glacial outwash deposits, and the floodplains of rivers in semiarid regions. It must be stabilized by moisture and vegetation in order to accumulate. Consequently, loess is not found in deserts, even though they provide much of its material. Because of its unconsolidated nature, loess is easily eroded, and as a result, eroded loess areas are characterized by steep cliffs and rapid lateral and headward stream erosion.

At present, loess deposits cover approximately 10% of the Earth's land surface and 30% of the United States. The most extensive and thickest loess deposits occur in northeast China where accumulations greater than 30 m are common. The extensive deserts in central Asia are the source for this loess. Other important loess deposits are on the North European Plain from Belgium eastward to Ukraine, central Asia, and the Pampas of Argentina. In the United States, they occur in the Great Plains, the Midwest, the Mississippi River Valley, and eastern Washington.

(a)

(b)

➤ FIGURE 15-9 (*a*) Transverse dunes form long ridges of sand that are perpendicular to the prevailing wind direction in areas of little or no vegetation and abundant sand. (*b*) Aerial view of transverse dunes, Great Sand Dunes National Monument, Colorado.

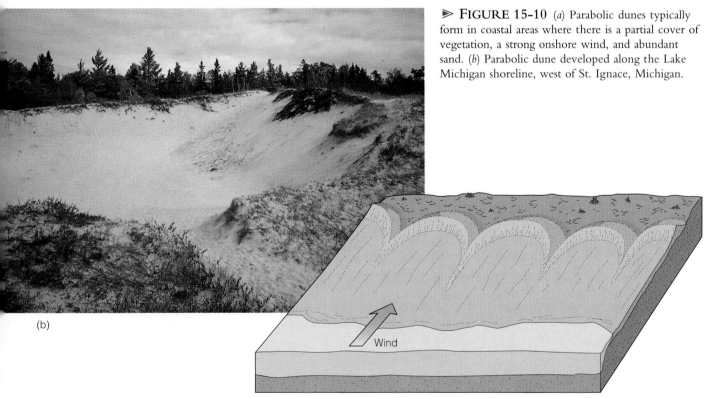

➤ FIGURE 15-10 (*a*) Parabolic dunes typically form in coastal areas where there is a partial cover of vegetation, a strong onshore wind, and abundant sand. (*b*) Parabolic dune developed along the Lake Michigan shoreline, west of St. Ignace, Michigan.

(b)

(a)

Loess-derived soils are some of the world's most fertile. It is therefore not surprising that the world's major grain-producing regions correspond to the distribution of large loess deposits such as the North European Plain, Ukraine, and the Great Plains of North America.

◉ AIR PRESSURE BELTS AND GLOBAL WIND PATTERNS

To understand the work of wind and the distribution of deserts, we need to consider the global pattern of air pressure belts and winds, which are responsible for the Earth's atmospheric circulation patterns. Air pressure is the density of air exerted on its surroundings (that is, its weight). When air is heated, it expands and rises, reducing its mass for a given volume and causing a decrease in air pressure. Conversely, when air is cooled, it contracts and air pressure increases. Therefore, those areas of the Earth's surface that receive the most solar radiation, such as the equatorial regions, have low air pressure, while the colder areas, such as the polar regions, have high air pressure.

Air flows from high-pressure zones to low-pressure zones. If the Earth did not rotate, winds would move in a straight line from one zone to another. Because the Earth rotates, winds are deflected to the right of their direction of motion (clockwise) in the Northern Hemisphere and to the left of their direction of motion (counterclockwise) in the Southern Hemisphere. Such deflection of air between latitudinal zones resulting from the Earth's rotation is known as the **Coriolis effect.** Therefore, the combination of latitudinal pressure differences and the Coriolis effect produces a worldwide pattern of east-west–oriented wind belts (▶ Figure 15-11).

The Earth's equatorial zone receives the most solar energy, which heats the surface air, causing it to rise. As the air rises, it cools and releases moisture that falls as rain in the equatorial region (Figure 15-11). The rising air is now much drier as it moves northward and southward toward each pole. By the time it reaches 20° to 30° north and south latitude, the air has become cooler and denser and begins to descend. Compression of the atmosphere warms the descending air mass, producing a dry, high-pressure area, providing the perfect conditions for the formation of the low-latitude deserts of the Northern and Southern hemispheres.

◉ THE DISTRIBUTION OF DESERTS

Dry climates occur in the low and middle latitudes, where the potential loss of water by evaporation exceeds the yearly precipitation. Dry climates cover 30% of the Earth's land surface and are subdivided into *semiarid* and *arid* regions. Semiarid regions receive more precipitation than arid regions, yet are moderately dry. Their soils are usually well developed and fertile and support a natural grass cover. Arid regions, generally described as **deserts,** are very dry; they receive less than 25 cm of rain per year on average, typically have poorly developed soils, and are mostly or completely devoid of vegetation.

▶ **FIGURE 15-11** The general circulation of the Earth's atmosphere.

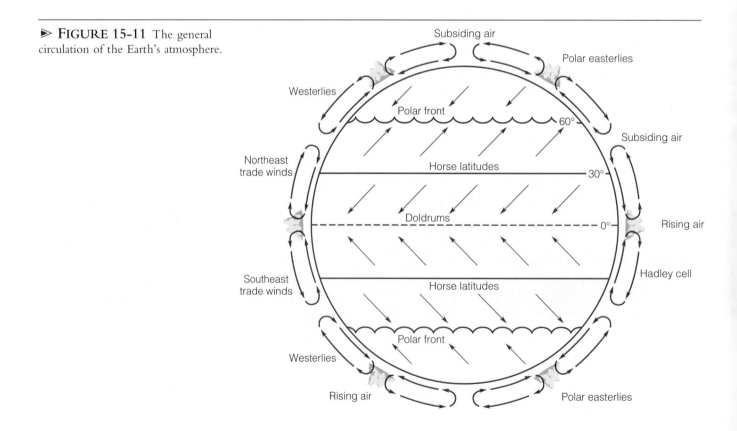

The majority of the world's deserts are found in the dry climates of the low and middle latitudes. In North America, most of the southwestern United States and northern Mexico are characterized by this hot, dry climate, while in South America this climate is primarily restricted to the Atacama Desert of coastal Chile and Peru. The Sahara in northern Africa and the Arabian Desert in the Middle East, along with the majority of Pakistan and western India, form the largest essentially unbroken desert environment in the Northern Hemisphere. More than 40% of Australia is desert, and most of the rest of it is semiarid.

The remaining dry climates of the world are found in the middle and high latitudes, mostly within continental interiors in the Northern Hemisphere. Many of these areas are dry because of their remoteness from moist maritime air and the presence of mountain ranges that produce a **rainshadow desert** (▷ Figure 15-12). When moist marine air moves inland and meets a mountain range, it is forced upward. As it rises, it cools, forming clouds and producing precipitation that falls on the windward side of the mountains. The air that descends on the leeward side of the mountain range is much warmer and drier, producing a rainshadow desert.

Three widely separated areas are included within the mid-latitude dry climate zone. The largest of these is the central part of Eurasia extending from just north of the Black Sea eastward to north-central China. The Gobi Desert in China is the largest desert in this region. The Great Basin area of North America is the second largest mid-latitude dry climate zone and results from the rainshadow produced by the Sierra Nevada (see Perspective 15-1). This region adjoins the southwestern deserts of the United States that formed as a result of the low-latitude subtropical high-pressure zone. The smallest of the mid-latitude dry climate areas is the Patagonian region of southern and western Argentina. Its dryness results from the rainshadow effect of the Andes.

The remainder of the world's deserts are found in the cold but dry high latitudes, such as Antarctica.

CHARACTERISTICS OF DESERTS

To people who live in humid regions, deserts may seem stark and inhospitable. Instead of a landscape of rolling hills and gentle slopes with an almost continuous vegetation cover, deserts are dry, have little vegetation, and consist of nearly continuous rock exposures, desert pavement, or sand dunes. And yet despite the great contrast between deserts and more humid areas, the same geologic processes are at work, only operating under different climatic conditions.

Temperature and Vegetation

The heat and dryness of deserts are well known. Many of the deserts of the low latitudes have average summer temperatures ranging between 32° and 38°C. It is not uncommon for some low-elevation inland deserts to record daytime highs of 46° to 50°C for weeks at a time. The highest temperature ever recorded was 58°C in El Azizia, Libya, on September 13, 1922.

During the winter months when the angle of the Sun is lower and there are fewer daylight hours, daytime temperatures average between 10° and 18°C. Winter nighttime lows can be quite cold, with frost and freezing temperatures common in the more poleward deserts.

Deserts display a wide variety of vegetation. While the driest deserts, or those with large areas of shifting sand, are almost devoid of vegetation, most deserts support at least a sparse plant cover. Compared to humid areas, desert vegetation may appear monotonous. A closer examination, however, reveals an amazing diversity of plants that have evolved the ability to live in the near-absence of water.

Desert plants are widely spaced, typically small, and grow slowly. Their stems and leaves are usually hard and waxy to minimize water loss by evaporation and protect the plant from sand erosion. Most plants have a widespread shallow root system to absorb the dew that forms each morning in all

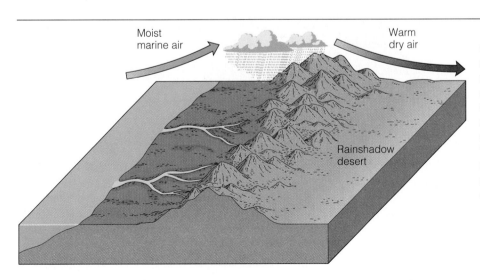

▷ **FIGURE 15-12** Many deserts in the middle and high latitudes are rainshadow deserts, so named because they form on the leeward sides of mountain ranges. When moist marine air moving inland meets a mountain range, it is forced upward where it cools and forms clouds that produce rain. This rain falls on the windward side of the mountains. The air descending on the leeward side is much warmer and drier, producing a rainshadow desert.

DEATH VALLEY NATIONAL MONUMENT

Death Valley National Monument was established in 1933 and encompasses 7,700 km² of southeastern California and part of western Nevada (➤ Figure 1). The hottest, driest, and lowest of the U.S. National Monuments and Parks, it receives less than 5 cm of rain per year and features normal daytime summer temperatures above 42°C. The highest temperature ever recorded was 57°C in the shade! The topographic relief in Death Valley is impressive. Telescope Peak near the southwestern border is 3,368 m high, while the lowest point in the Western Hemisphere—86 m below sea level—is less than 32 km to the east at Badwater.

Within Death Valley and its bordering mountains are excellent examples of a wide variety of desert landforms and economically valuable evaporite deposits. In addition, numerous folds, faults, landslides, and considerable evidence of volcanic activity can be seen.

The geologic history of Death Valley is complex and still being worked out, but rocks from every geologic era can be found in the valley or the surrounding mountains.

➤ **FIGURE 1** Death Valley National Monument, California, encompasses 7,700 km² of southeastern California and part of western Nevada.

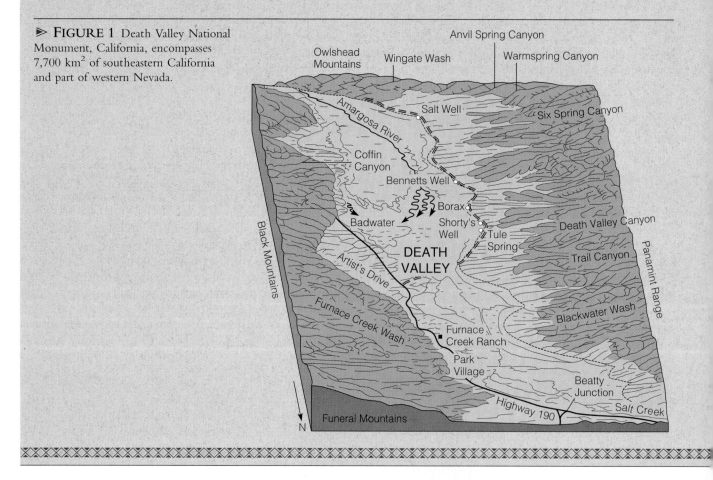

but the driest deserts and to help anchor the plant in what little soil there may be. In extreme cases, many plants lie dormant during particularly dry years and spring to life after the first rain shower with a beautiful profusion of flowers.

Weathering and Soils

Mechanical weathering is dominant in desert regions. Daily temperature fluctuations and frost wedging are the primary forms of mechanical weathering (see Chapter 6). The breakdown of rocks by roots and from salt crystal growth is of minor importance. Some chemical weathering does occur, but its rate is greatly reduced by aridity and the scarcity of organic acids produced by the sparse vegetation. Most chemical weathering takes place during the winter months when more precipitation occurs, particularly in the mid-latitude deserts.

An interesting feature seen in many deserts is a thin, red,

Although the geologic history of the region extends back to the Precambrian, Death Valley itself formed less than 4 million years ago.

Death Valley formed during the Pliocene Epoch, when the Earth's crust was stretched and rifted, forming various horsts and grabens. Death Valley continues to subside along normal faults and is sinking most rapidly along its western side. This movement has been so great that more than 3,000 m of sediments are buried beneath the present valley floor.

During the Pleistocene Epoch, when the climate of this region was more humid than it is today, numerous lakes spread over the valley. Lake Manly, the largest of these lakes (145 km long and 178 m deep), dried up about 10,000 years ago, when the climate became arid.

Volcanic activity has also been occurring during the last several thousand years. The most famous volcanic feature in Death Valley is Ubehebe Crater, an explosion crater that formed approximately 2,000 years ago.

In addition to the usual desert features, Death Valley also includes some unusual ones such as the Devil's Golf Course, a bed of solid rock salt displaying polygonal ridges and pinnacles that are almost impossible to traverse (➤ Figure 2). The Harmony Borax Works was home to the famous 20-mule teams that hauled out countless wagons of borax. The borax, used for ceramic glazes, fertilizers, glass, solder, and pharmaceuticals, was leached from volcanic ash by hot groundwater and then accumulated in layers of lake sediment.

Besides the numerous geologic features that have made Death Valley famous, it is also home to more than 600 species of plants as well as numerous animal species.

➤ **FIGURE 2** Devil's Golf Course consists of a layer of solid rock salt that has formed a network of polygonal ridges and pinnacles making it very difficult to traverse.

brown, or black shiny coating on the surface of many rocks. This coating, called *rock varnish,* is composed of iron and manganese oxides (➤ Figure 15-13). Because many of the varnished rocks contain little or no iron and manganese oxides, the varnish is thought to result from either wind-blown iron and manganese dust that settles on the ground or from the precipitated waste of microorganisms.

Desert soils, if developed, are usually thin and patchy because the limited rainfall and the resultant scarcity of vegetation reduce the efficiency of chemical weathering and hence soil formation. Furthermore, the sparseness of the vegetative cover enhances wind and water erosion of what little soil actually forms.

Mass Wasting, Streams, and Groundwater

When traveling through a desert, most people are impressed by such wind-formed features as moving sand, sand dunes,

▷ **FIGURE 15-13** The shiny black coating on this rock exposed at Castle Valley, Utah, is rock varnish. It is composed of iron and manganese oxides.

gullies and overflow their banks. During these times, a tremendous amount of sediment is rapidly transported and deposited far downstream.

While water is the major erosive agent in deserts today, it was even more important during the Pleistocene Epoch when these regions were more humid. During that time, many of the major topographic features of deserts were forming. Today that topography is being modified by wind and infrequently flowing streams.

Most desert streams are poorly integrated and flow only intermittently. Many of them never reach the sea because the water table is usually far deeper than the channels of most streams, so they cannot draw upon groundwater to replace water lost to evaporation and absorption into the ground. This type of drainage in which a stream's load is deposited within the desert is called *internal drainage* and is common in most arid regions.

While the majority of deserts have internal drainage, some deserts have permanent through-flowing streams such as the Nile and Niger rivers in Africa, the Rio Grande and Colorado River in the southwestern United States, and the Indus River in Asia. These streams are able to flow through desert regions because their headwaters are well outside the desert and water is plentiful enough to offset losses resulting from evaporation and infiltration. Demands for greater amounts of water for agriculture and domestic use from the Colorado River, however, are leading to increased salt concentrations in its lower reaches and causing political problems between the United States and Mexico.

and sand and dust storms. They may also notice the dry washes and dry stream beds. Because of the lack of running water, most people would conclude that wind is the most important erosional geologic agent in deserts. They would be wrong. Running water, even though it occurs infrequently, causes most of the erosion in deserts. The dry conditions and sparse vegetation characteristic of deserts enhance water erosion. If you look closely, you will see the evidence of erosion and transportation by running water nearly everywhere except in areas covered by sand dunes.

Most of a desert's average annual rainfall of 25 cm or less comes in brief, heavy, localized cloudbursts. During these times, considerable erosion occurs because the ground cannot absorb all of the rainwater. With so little vegetation to hinder its flow, runoff is rapid, especially on moderately to steeply sloping surfaces, resulting in flash floods and sheetflows. Dry stream channels quickly fill with raging torrents of muddy water and mudflows, which carve out steep-sided

Wind

Although running water does most of the erosional work in deserts, wind can also be an effective geologic agent capable of producing a variety of distinctive erosional and depositional features. It is very effective in transporting and depositing unconsolidated sand-, silt-, and dust-sized particles. Contrary to popular belief, most deserts are not sand-covered wastelands, but rather consist of vast areas of rock

▷ **FIGURE 15-14** Racetrack Playa, Death Valley, California. The Inyo Mountains can be seen in the background.

▷ FIGURE 15-15 Aerial view of an alluvial fan, Death Valley, California.

exposures and desert pavement. Sand-covered regions, or sandy deserts, constitute less than 25% of the world's deserts. The sand in these areas has accumulated primarily by the action of wind.

DESERT LANDFORMS

Because of differences in temperature, precipitation, and wind, as well as the underlying rocks and recent tectonic events, landforms in arid regions vary considerably.

After an infrequent and particularly intense rainstorm, excess water not absorbed by the ground may accumulate in low areas and form *playa lakes.* These lakes are very temporary, lasting from a few hours to several months. Most of them are very shallow and have rapidly shifting boundaries as water flows in or leaves by evaporation and seepage into the ground. The water is often very saline.

When a playa lake evaporates, the dry lake bed is called a **playa** or *salt pan* and is characterized by mudcracks and precipitated salt crystals (▷ Figure 15-14). Salts in some playas are thick enough to be mined commercially. For example, borates have been mined in Death Valley, California, for more than a hundred years (see Perspective 15-1).

Other common features of deserts, particularly in the Basin and Range region of the United States, are *alluvial fans* and *bajadas.* **Alluvial fans** form when sediment-laden streams flowing out from generally straight, steep mountain fronts deposit their load on the desert floor. Once beyond the mountain front where no valley walls contain streams, the

sediment spreads out laterally, forming a gently sloping and poorly sorted fan-shaped sedimentary deposit (▷ Figure 15-15). Alluvial fans are similar in origin and shape to deltas (see Chapter 12) but are formed entirely on land. Alluvial fans may coalesce to form a *bajada.* This broad alluvial apron typically has an undulating surface resulting from the overlap of adjacent fans (▷ Figure 15-16).

Large alluvial fans and bajadas are frequently important sources of groundwater for domestic and agricultural use. Their outer portions are typically composed of fine-grained sediments suitable for cultivation, and their gentle slopes allow good drainage of water. Many alluvial fans and bajadas are also the sites of large towns and cities, such as San Bernardino, California, Salt Lake City, Utah, and Teheran, Iran.

Most mountains in desert regions, including those of the Basin and Range region, rise abruptly from gently sloping surfaces called pediments. **Pediments** are erosional bedrock surfaces of low relief that slope gently away from mountain bases (▷ Figure 15-17). Most pediments are covered by a thin layer of debris or by alluvial fans or bajadas.

Rising conspicuously above the flat plains of many deserts are isolated steep-sided erosional remnants called **inselbergs,** a German word meaning "island mountain" (▷ Figure 15-18). Inselbergs have survived longer than other mountains because of their greater resistance to weathering.

Other easily recognized erosional remnants common to arid and semiarid regions are mesas and buttes (▷ Figure 15-19). A **mesa** is a broad, flat-topped erosional remnant

➤ **FIGURE 15-16** Coalescing alluvial fans forming a bajada at the base of the Black Mountains, Death Valley, California.

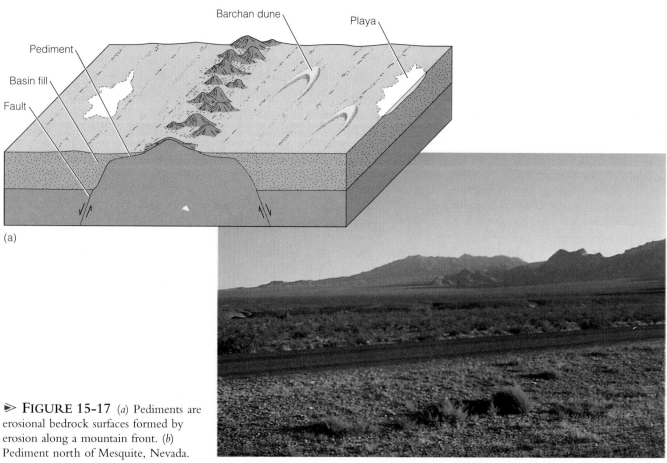

➤ **FIGURE 15-17** (*a*) Pediments are erosional bedrock surfaces formed by erosion along a mountain front. (*b*) Pediment north of Mesquite, Nevada.

(a)

(b)

➤ **FIGURE 15-18** Ayers Rock in central Australia is a good example of an inselberg. This photo shows Ayers Rock at sunset.

(a)

(b)

➤ **FIGURE 15-19** (*a*) Mesas southeast of Zuni Pueblo, New Mexico. (*b*) Butte in Monument Valley, Arizona.

bounded on all sides by steep slopes. Continued weathering and stream erosion form isolated pillarlike structures known as **buttes**. Buttes and mesas consist of relatively easily weathered sedimentary rocks capped by nearly horizontal, resistant rocks such as sandstone, limestone, or basalt. They form when the resistant rock layer is breached, allowing rapid erosion of the less resistant underlying sediment. One of the best-known areas of mesas and buttes in the United States is Monument Valley on the Arizona-Utah border.

CHAPTER SUMMARY

1. Wind can transport sediment in suspension or as bed load, which involves saltation and surface creep.
2. Wind erodes material either by abrasion or deflation. Abrasion is a near-surface effect caused by the impact of saltating sand grains. Ventifacts are common wind-abraded features.
3. Deflation is the removal of loose surface material by the wind. Deflation hollows resulting from differential erosion of surface material are common features of many deserts, as is desert pavement, which effectively protects the underlying surface from additional deflation.
4. The two major wind deposits are dunes and loess. Dunes are mounds or ridges of wind-deposited sand, whereas loess is wind-deposited silt and clay.
5. The four major dune types are barchan, longitudinal, transverse, and parabolic. The amount of sand available, the prevailing wind direction, the wind velocity, and the amount of vegetation present determine which type will form.
6. Loess is derived from deserts, Pleistocene glacial outwash deposits, and river floodplains in semiarid regions. Loess covers approximately 10% of the Earth's land surface and weathers to a rich and productive soil.
7. Deserts are very dry (averaging less than 25 cm rain/year), have poorly developed soils, and are mostly or completely devoid of vegetation.
8. The winds of the major east-west–oriented air pressure belts resulting from rising and cooling air are deflected by the Coriolis effect. These belts help control the world's climate.
9. Dry climates are located in the low and middle latitudes where the potential loss of water by evaporation exceeds the yearly precipitation. Dry climates cover 30% of the Earth's surface and are subdivided into semiarid and arid regions.
10. The majority of the world's deserts are in the low-latitude dry climate zone between 20° and 30° north and south latitudes. Their dry climate results from a high-pressure belt of descending dry air. The remaining deserts are in the middle latitudes where their distribution is related to the rainshadow effect and in the dry polar regions.
11. Deserts are characterized by lack of precipitation and high evaporation rates. Furthermore, rainfall is unpredictable and, when it does occur, tends to be very intense and of short duration. As a consequence of such aridity, desert vegetation and animals are scarce.
12. Mechanical weathering is the dominant form of weathering in deserts. The sparse precipitation and slow rates of chemical weathering result in poorly developed soils.
13. Running water is the dominant agent of erosion in deserts and was even more important during the Pleistocene Epoch when wetter climates resulted in humid conditions.
14. Wind is an erosional agent in deserts and is very effective in transporting and depositing unconsolidated fine-grained sediments.
15. Important desert landforms include playas, which are dry lake beds; when temporarily filled with water, they form playa lakes. Alluvial fans are poorly-sorted, fan-shaped sedimentary deposits that may coalesce to form bajadas. Pediments are low-relief erosional bedrock surfaces that gently slope away from mountain bases. Inselbergs are isolated steep-sided erosional remnants that rise above the surrounding desert plains. Buttes and mesas are, respectively, pinnacle-like and flat-topped erosional remnants with steep sides.

IMPORTANT TERMS

abrasion
alluvial fan
barchan dune
butte
Coriolis effect

deflation
desert
desert pavement
desertification
dune

inselberg
loess
longitudinal dune
mesa
parabolic dune

pediment
playa
rainshadow desert
transverse dune
ventifact

1. Deserts:
   a. ____ can be found in the low, middle, and high latitudes;
   b. ____ receive more than 25 cm of rain per year;
   c. ____ are mostly or completely devoid of vegetation;
   d. ____ answers (a) and (c);
   e. ____ answers (b) and (c).
2. The primary process by which bed load is transported is:
   a. ____ suspension;
   b. ____ abrasion;
   c. ____ saltation;
   d. ____ precipitation;
   e. ____ answers (a) and (c).
3. Which particle size constitutes most of a wind's suspended load?
   a. ____ sand;
   b. ____ silt;
   c. ____ clay;
   d. ____ answers (a) and (b);
   e. ____ answers (b) and (c).
4. Which of the following is a feature produced by wind erosion?
   a. ____ playa;
   b. ____ loess;
   c. ____ dune;
   d. ____ yardang;
   e. ____ none of these.
5. Which of the following is a crescent-shaped dune whose tips point downwind?
   a. ____ barchan;
   b. ____ longitudinal;
   c. ____ parabolic;
   d. ____ transverse;
   e. ____ none of these.
6. Where are the thickest and most extensive loess deposits in the world?
   a. ____ United States;
   b. ____ Pampas of Argentina;
   c. ____ Belgium;
   d. ____ Ukraine;
   e. ____ northeast China.
7. The dominant form of weathering in deserts is _____ , desert vegetation is _____ , and soils are _____ :
   a. ____ mechanical, limited, thick;
   b. ____ mechanical, diverse, thin;
   c. ____ mechanical, limited, thin;
   d. ____ chemical, diverse, thick;
   e. ____ chemical, diverse, thin.
8. The dry lake beds in many deserts are:
   a. ____ playas;
   b. ____ bajadas;
   c. ____ inselbergs;
   d. ____ pediments;
   e. ____ mesas.
9. What are some of the problems associated with desertification?
10. Describe how the global distribution of air pressure belts and winds operates.
11. What are the two ways that sediments are transported by wind?
12. Why is desert pavement important in a desert environment?
13. How do sand dunes migrate?
14. Describe the four major dune types and the conditions necessary for their formation.
15. What is loess and why is it important?
16. Why are most of the world's deserts located in the low latitudes?
17. How are temperature, precipitation, and vegetation interrelated in desert environments?
18. What is the dominant form of weathering in desert regions, and why is it so effective?

1. Because much of the recent desertification has been greatly accelerated by human activities, is there anything we can do to slow the process or restore some type of equilibrium or buffer zone between encroaching deserts and adjacent productive lands?
2. If deserts are very dry regions in which mechanical weathering predominates, why are so many of their distinctive landforms the result of running water and not wind?

Chapter 16

THE SEA FLOOR, SHORELINES, AND SHORELINE PROCESSES

OUTLINE

Beach in a small cove along California's Pacific Ocean shoreline. (Photo courtesy of Sue Monroe.)

In 1979, researchers aboard the submersible *Alvin* descended about 2,500 m to the Galapagos Rift in the eastern Pacific Ocean basin and observed hydrothermal vents on the sea floor (➤ Figure 16-1). Such vents occur near spreading ridges where seawater seeps down into the oceanic crust through cracks and fissures, is heated by the hot rocks, and then rises and is discharged onto the sea floor as hot springs. During the 1960s, hot metal-rich brines apparently derived from hydrothermal vents were detected and sampled in the Red Sea. These dense brines were concentrated in pools along the axis of the sea; beneath them thick deposits of metal-rich sediments were found. During the early 1970s, researchers observed hydrothermal vents on the Mid-Atlantic Ridge about 2,900 km east of Miami, Florida, and in 1978 moundlike mineral deposits were sampled from the East Pacific Rise just south of the Gulf of California.

When the submersible *Alvin* descended to the Galapagos Rift in 1979, mounds of metal-rich sediments were observed. Near these mounds the researchers saw what they called black smokers (chimneylike vents) discharging plumes of hot, black water (Figure 16-1). Since 1979 similar vents have been observed at or near spreading ridges in several other areas.

Submarine hydrothermal vents are interesting for several reasons. Near the vents live communities of organisms, including bacteria, crabs, mussels, starfish, and tubeworms, many of which had never been seen before (Figure 16-1). In most biological communities, photosynthesizing organisms form the base of the food chain and provide nutrients for the herbivores and carnivores. In vent communities, however, no sunlight is available for photosynthesis, and the base of the food chain consists of bacteria that practice chemosynthesis; they oxidize sulfur compounds from the hot vent waters, thus providing their own nutrients and nutrients for other members of the food chain.

Another interesting aspect of these submarine hydrothermal vents is their economic potential. When seawater circulates downward through the oceanic crust, it is heated as high as 400°C. The hot water reacts with the crust and is transformed into a metal-bearing solution. As the hot solution rises and discharges onto the sea floor, it cools, precipitating iron, copper, and zinc sulfides, and other minerals that accumulate to form a chimneylike vent (Figure 16-1). These vents are ephemeral; one observed in 1979 was inactive six months later. When their activity ceases, the vents eventually collapse and are incorporated into a moundlike mineral deposit.

The economic potential of hydrothermal vent deposits is tremendous. The deposits in the Atlantis II Deep of the Red Sea contain an estimated 100 million tons of metals, including iron, copper, zinc, silver, and gold. Many of these deposits now on land are thought to have formed on the sea floor by hydrothermal vent activity. Hydrothermal vent deposits have formed throughout geologic time. None are currently being mined, but the technology to exploit them exists.

➤ FIGURE 16-1 The submersible *Alvin* sheds light on hydrothermal vents at the Galapagos Rift, a branch of the East Pacific Rise. Seawater seeps down through the oceanic crust, becomes heated, and then rises and builds chimneys on the sea floor. The plume of "black smoke" is simply heated water saturated with dissolved minerals. Communities of organisms, including tubeworms, giant clams, crabs, and several types of fish, live near the vents.

INTRODUCTION

During most of historic time, people knew little of the oceans and, until fairly recently, still believed that the sea floor was flat and featureless. The ancient Greeks determined the size of the Earth rather accurately, but Western Europeans were not aware of the vastness of the oceans until the fifteenth and sixteenth centuries when various explorers sought new trade routes to the Indies. When Christopher Columbus set sail on August 4, 1492, in an attempt to find a route to the Indies, he greatly underestimated the width of the Atlantic Ocean. Contrary to popular belief, Columbus was not attempting to demonstrate that the Earth is spherical—the Earth's spherical shape was well accepted by then.

We now know that about 71% of the Earth's surface is covered by an interconnected body of salt water we refer to as *oceans* and *seas*. Four very large areas in this body of water are distinct enough to be designated as oceans: the Pacific, the Atlantic, the Indian, and the Arctic (Table 16-1). Seas are simply marginal parts of oceans that occupy an indentation into a continent, for example, the Caribbean Sea and the Sea of Japan.

Research during the last 200 years, and particularly during the last few decades, has shown that the sea floor possesses varied topography including such features as oceanic trenches, submarine canyons and ridges, and broad plateaus, hills, and plains. Furthermore, scientists have come to more fully appreciate the dynamic nature of shorelines and the continuous interplay between their materials and the energy of waves, tides, and nearshore currents.

EXPLORING THE SEA FLOOR

Oceanic depths were first measured by lowering a weighted line to the sea floor and measuring its length. Now an instrument called an *echo sounder* is used. Depth is calculated by knowing the velocity of sound waves in water and determining how long sound waves from a ship take to reach the sea floor and return to the ship. *Seismic profiling* uses waves that penetrate the layers beneath the sea floor and are reflected back to the source (Figure 16-2). This technique is particularly useful in determining the structure of the oceanic crust beneath sea-floor sediments.

Submersibles, both remotely controlled and carrying scientists, are also used to explore the sea floor. In 1985, the *Argo,* towed by a surface vessel and equipped with sonar and television systems, provided the first views of the British ocean liner R.M.S. *Titanic* since it sank in 1912. The U.S. Geological Survey is using a towed device that uses sonar to produce images resembling aerial photographs. Researchers aboard the submersible *Alvin* have observed various parts of the sea floor (see the Prologue).

Deep-sea drilling has also provided a wealth of information about the sea floor. Ships such as the *Glomar Challenger* (now retired) and the JOIDES★ *Resolution* are equipped to drill in water more than 6,000 m deep and recover long cores of sea-floor sediment and oceanic crust. Much of what we know of the composition and structure of the upper oceanic crust was obtained by deep-sea drilling.

★JOIDES is an acronym for Joint Oceanographic Institutions for Deep Earth Sampling.

 FIGURE 16-2 Diagram showing how seismic profiling is used to detect buried layers at sea. Some of the energy generated at the energy source is reflected from various horizons back to the surface where it is detected by hydrophones.

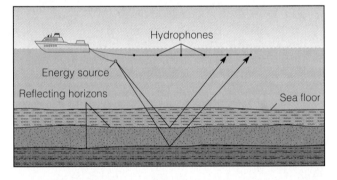

	TABLE 16-1 Numerical Data for the Oceans			
Ocean★	Surface Area (Million km²)	Water Volume (Million km³)	Average Depth (km)	Maximum Depth (km)
Pacific	180	700	4.0	11.0
Atlantic	93	335	3.6	9.2
Indian	77	285	3.7	7.5
Arctic	15	17	1.1	5.2

SOURCE: Pinet, P. R. 1992. *Oceanography.*

★Excludes adjacent seas.

SEA-FLOOR TOPOGRAPHY

The ocean floor is not monotonous and flat as once believed. Indeed, it is as varied as the surface of the continents, but for centuries people were unaware of this fact. In 1959, Maurice Ewing, Bruce Heezen, and Marie Tharp published a spectacular three-dimensional map of the North Atlantic showing vast plains and conical seamounts, as well as the Mid-Atlantic Ridge with its central rift valley. As more of the world's ocean floors were explored, this original map was expanded to reveal numerous other ocean floor features (▷ Figure 16-3).

Continental Margins

The zone separating each continent above sea level from the deep-sea floor is the **continental margin.** It consists of a gently sloping *continental shelf,* a more steeply inclined *continental slope,* and, in some cases, a gently inclined *continental rise* (▷ Figure 16-4). At its outer limit, the continental margin merges with the deep-sea floor or descends into an oceanic trench.

The width of the continental shelf varies from only a few tens of meters to more than 1,000 km. It outer edge is the shelf-slope break, the point at which the inclination of the sea floor increases rather abruptly to several degrees. Rarely is the inclination of the shelf even 1°, whereas the continental slope's inclination is anywhere from several degrees up to 25°. In many places the slope merges with the more gently sloping continental rise, but in some areas it descends directly into an oceanic trench, and a rise is absent (Figure 16-4).

The shelf-slope break, which occurs at an average depth of 135 m, serves as an important control on the processes operating on the sea floor. Landward of the break, the shelf is affected by waves and tidal currents, but seaward of the break, the sea floor is unaffected by surface processes. Sediment transport onto the slope and rise is therefore controlled by gravity-driven processes, especially turbidity currents, which are sediment-water mixtures. Because they are denser than seawater, turbidity currents flow downslope through *submarine canyons* to the deep-sea floor where they typically deposit sequences of graded beds (see Figure 7-12). In fact, where a continental rise is present, it is composed mostly of *submarine fans,* which are simply accumulations of turbidity current deposits (▷ Figure 16-5b).

The *submarine canyons* through which turbidity currents move are deep and steep walled and are best developed on the continental slope. Some of these canyons can be traced across the shelf to streams on land and apparently formed as stream valleys when sea level was lower during the Ice Age.

▷ **FIGURE 16-3** This map of the sea floor resulted from the work of Maurice Ewing, Bruce Heezen, and Marie Tharp.

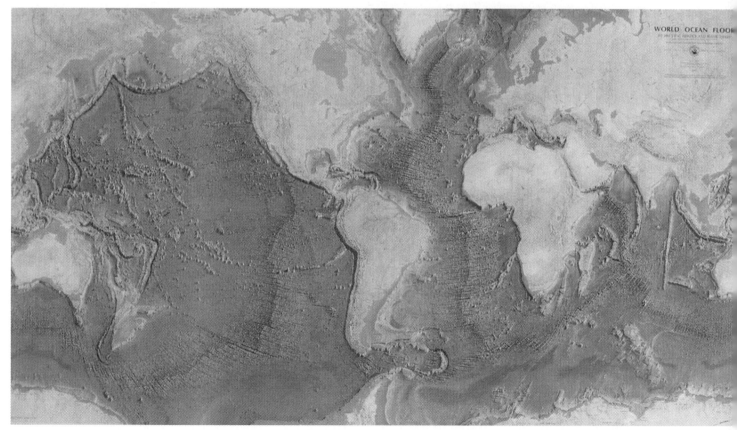

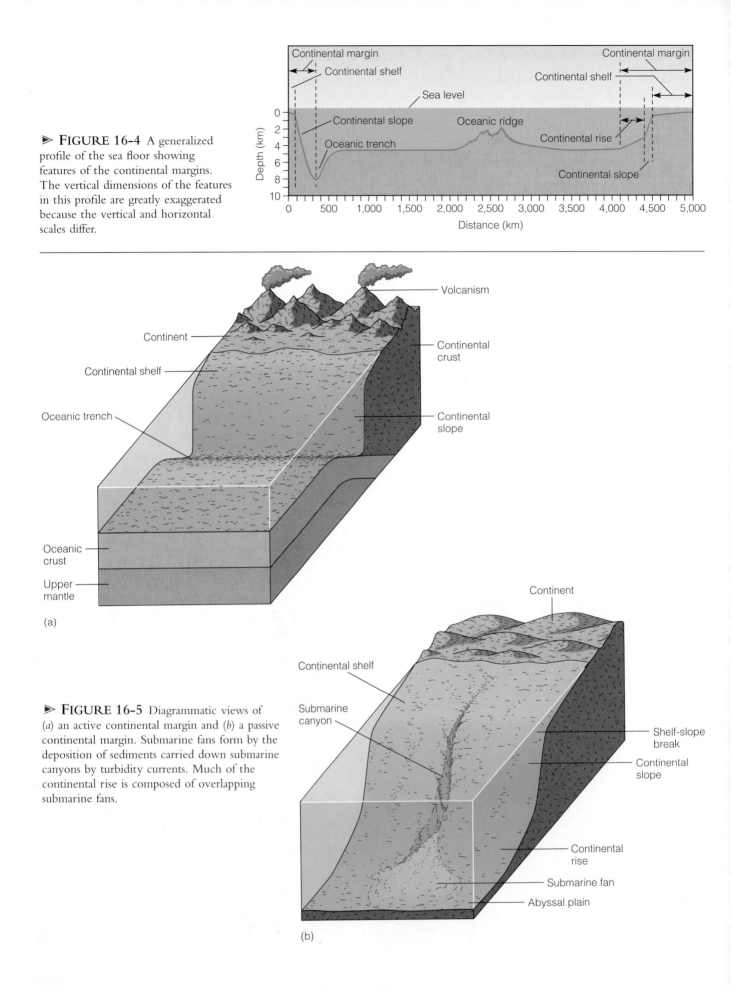

➤ **FIGURE 16-4** A generalized profile of the sea floor showing features of the continental margins. The vertical dimensions of the features in this profile are greatly exaggerated because the vertical and horizontal scales differ.

➤ **FIGURE 16-5** Diagrammatic views of (*a*) an active continental margin and (*b*) a passive continental margin. Submarine fans form by the deposition of sediments carried down submarine canyons by turbidity currents. Much of the continental rise is composed of overlapping submarine fans.

Many have no such association, however, and are thought to have originated through erosion by turbidity currents.

Two types of continental margins are generally recognized, *active* and *passive*. An **active continental margin** develops at the leading edge of a continental plate where oceanic lithosphere is subducted (Figure 16-5a). This kind of continental margin is characterized by andesitic volcanism, an inclined seismic zone, and a geologically young mountain range. Furthermore, the continental shelf is quite narrow, and the continental slope descends directly into an oceanic trench.

A **passive continental margin** occurs at the trailing edge of a continental plate (Figure 16-5b). It possesses a broad shelf and a continental rise that commonly merges with vast, flat *abyssal plains* (Figure 16-5b). No volcanism and little seismicity occur on this type of continental margin.

The proximity of the oceanic trench to the continent explains why the shelf is so narrow along an active continental margin. Sediment derived from the continent is simply transported down the slope into the trench. On a passive continental margin, which lacks a trench, the entire continental margin is much wider because land-derived sedimentary deposits build outward into the ocean.

The Deep-Ocean Basin

Submersibles have carried scientists into the oceanic depths, so some of the sea floor has been observed directly. Nevertheless, most of our knowledge of the deep-sea floor comes from deep-sea drilling, echo sounding, seismic profiling, and data gathered by remote devices that descend in excess of 11,000 m.

As noted previously, the deep-sea floor is not completely flat and featureless as once believed (Figure 16-3). It possesses a variety of features and has a topography as diverse as that on the continents. Nevertheless, there are vast flat areas known as **abyssal plains** adjacent to the rises of passive continental margins (Figure 16-5b). The oceanic crust beneath these areas is rugged, but it has been buried by land-derived sediments, thus yielding a very flat surface. Abyssal plains are not found along active continental margins where sediments from the land are trapped in an oceanic trench (Figure16-5a).

Although **oceanic trenches** constitute only a small part of the sea floor, they are very important for it is here that lithospheric plates are consumed by subduction (see Figure 2-16). Oceanic trenches are long, narrow features restricted to active continental margins, so they are common around the margins of the Pacific Ocean basin. The Peru-Chile Trench, for instance, lies just off the west coast of South America; it is 5,900 km long and only about 100 km wide, but more than 8,000 m deep. The greatest oceanic depth of more than 11,000 m is in an oceanic trench.

Oceanic ridges are part of a worldwide system of mostly submarine ridges that in some areas have a central rift (▶ Figure 16-6). The Mid-Atlantic Ridge, part of this ridge system, is more than 2,000 km wide and rises about 2.5 km above the sea floor adjacent to it. This oceanic ridge system extends for about 65,000 km, surpassing the length of mountain ranges on land. Oceanic ridges, however, are composed mostly of basalt and have features produced by tensional forces, whereas mountains on land are composed of granitic, metamorphic, and sedimentary rocks that have been deformed by compression. Recall from Chapter 2 that oceanic ridges are the sites of plate divergence and are also referred to as *spreading ridges* (see Figure 2-14).

Oceanic ridges abruptly terminate where they are offset along large fractures in the sea floor. These fractures, better known as *transform faults* (Figure 16-6a), extend for hundreds of kilometers, and many geologists are convinced that some of them continue into continents. Offsets along these fractures yield nearly vertical escarpments more than 3 km high on the sea floor (Figure 16-6a).

The sea floor also has numerous seamounts and guyots. Both are of volcanic origin and rise more than 1 km above the sea floor, but **seamounts** are conical whereas **guyots** are flat topped. Guyots are simply volcanoes that once extended above sea level near a spreading ridge. But as the plate upon which they were situated continued to move away from the ridge and into deeper water, what was an island was eroded by waves as it sank beneath the sea (Figure 16-6b). Other features known as *volcanic hills* are similar to seamounts but average only about 250 km high.

Long, linear ridges and broad plateaus rising above the sea floor are also common in the oceanic basins. They are known as *aseismic ridges* because they lack seismic activity. A few of these features, called *microcontinents,* are thought to be small fragments separated from continents during rifting. Most aseismic ridges form as a linear succession of seamounts and/or guyots extending from an oceanic ridge.

⬢ DEEP-SEA SEDIMENTATION

Deep-sea sediments consist mostly of fine-grained wind-blown dust and volcanic ash from the continents and oceanic islands and the shells of microscopic organisms living in the oceans' near-surface waters. Sand and gravel deposits beyond the continental margins or fringes of oceanic islands are rare because the only mechanism that can transport such large particles far into the ocean basins is ice rafting, which is effective only adjacent to Greenland and Antarctica.

Most of the sediments on the sea floor are *pelagic,* meaning that they settled from suspension far from land. Two types of pelagic sediments are generally recognized. **Pelagic clay** covers most of the deeper parts of the ocean basins. It is composed largely of clay-sized particles derived from the continents and oceanic islands. **Ooze** is composed of shells of microscopic marine plants and animals. If it contains mostly calcium carbonate ($CaCO_3$) shells, it is *calcareous ooze,* and if it is composed predominantly of silica (SiO_2) shells, it is *siliceous ooze.*

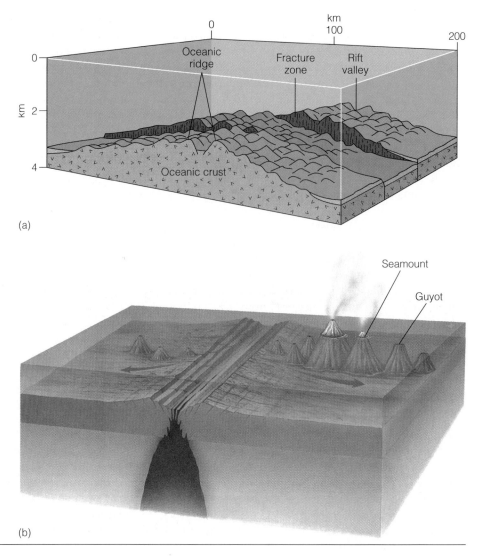

▷ **FIGURE 16-6** (*a*) Diagrammatic view of a fracture or transform fault offsetting a ridge. Earthquakes occur only in the segments between offset ridge crests. (*b*) Submarine volcanoes may build up above sea level to form seamounts. As the plate upon which these volcanoes rest moves away from a spreading ridge, the volcanoes sink beneath sea level and become guyots.

REEFS

Reefs are moundlike, wave-resistant structures composed of the skeletons of various organisms such as corals, clams, sponges, and algae. Reefs grow to a depth of 45 or 50 m and are restricted to shallow tropical seas where the water is clear, and the temperature does not fall below about 20°C.

Three types of reefs are recognized: fringing, barrier, and atoll (▷ Figure 16-7). *Fringing reefs* are solidly attached to the margins of an island or continent. They have a rough, tablelike surface and are up to 1 km wide. *Barrier reefs* are similar except that they are separated from the mainland by a lagoon. The 2,000 km long Great Barrier Reef of Australia is a good example of a barier reef.

An *atoll* is a circular to oval reef surrounding a lagoon (Figure 16-7). Atolls first form as fringing reefs around volcanic islands that are carried below sea level on a moving plate. As the island is carried into progressively deeper water, the reef grows upward so that the living part of the reef remains in shallow water, leaving a circular lagoon surrounded by a more-or-less continuous reef (Figure 16-7). As

Figure 16-7 shows, a fringing reef around an island can evolve into a barrier reef and then an atoll as the island subsides below sea level.

SHORELINES AND SHORELINE PROCESSES

Shorelines are the areas between low tide and the highest level on land affected by storm waves. Here we are concerned mostly with ocean shorelines where processes such as waves, tides, and nearshore currents continually modify existing shoreline features. Waves and nearshore currents are also effective geologic agents in large lakes, where the shorelines exhibit many of the same features present along seashores. The most notable differences are that waves and nearshore currents are more energetic on seashores, and even the largest lakes lack appreciable tides.

The continents possess more than 400,000 km of shorelines. They vary from rocky, steep shorelines, such as those in Maine and much of the western United States and Canada, to those with broad sandy beaches as in eastern North America from New Jersey southward. Whatever their

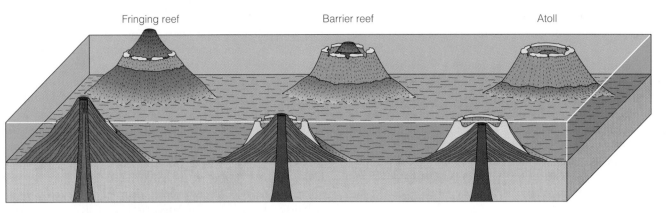

(a)

(b)

➤ FIGURE 16-7 (*a*)Three-stage development of an atoll. In the first stage, a fringing reef forms, but as the island sinks, a barrier reef becomes separated from the island by a lagoon. As the island disappears beneath the sea, the barrier reef continues to grow upward, thus forming an atoll. An oceanic island carried into deeper water by plate movement can account for this sequence. (*b*) View of an atoll in the Pacific Ocean.

Fringing reef   Barrier reef   Atoll

type, all shorelines exhibit continual interplay between the energy levels of shoreline processes and the shoreline materials.

Scientists from several disciplines have contributed to our understanding of shorelines as dynamic systems. Elected officials and city planners of coastal communities must become familiar with shoreline processes so they can develop policies that serve the public as well as protect the fragile shoreline environment. In short, the study of shorelines is not only interesting, but has many practical applications.

Tides

On seacoasts the surface of the ocean rises and falls once or twice daily in response to the gravitational attraction of the Moon and Sun. Such regular fluctuations in the sea's surface are **tides.** Two high tides and two low tides occur daily in most areas as sea level rises and falls anywhere from a few centimeters to more than 15 m. During rising or *flood tide,* more and more of the nearshore area is flooded until high tide is reached. *Ebb tide* occurs when currents flow seaward during a decrease in the height of the tide.

Both the Moon and the Sun have sufficient gravitational attraction to exert tide-generating forces strong enough to deform the solid body of the Earth, but they have a much greater influence on the oceans. The Sun is 27 million times more massive than the Moon, but it is 390 times as far from the Earth, and its tide-generating force is only 46% as strong

as that of the Moon. Accordingly, the tides are dominated by the Moon, but the Sun plays an important role as well.

If we consider only the Moon acting on a spherical, water-covered Earth, the tide-generating forces produce two bulges on the ocean surface (➤ Figure 16-8). One bulge extends toward the Moon because it is on the side of the Earth where the Moon's gravitational attraction is greatest. The other bulge occurs on the opposite side of the Earth, where the Moon's gravitational attraction is least. These two bulges always point toward and away from the Moon (Figure 16-8a), so as the Earth rotates and the Moon's position changes, an observer at a particular shoreline location experiences the rhythmic rise and fall of tides twice daily. The heights of two successive high tides may vary depending on the Moon's inclination with respect to the equator.

The Moon revolves around the Earth every 28 days, so its position with respect to any latitude changes slightly each day. That is, as the Moon moves in its orbit and the Earth rotates on its axis, it takes the Moon 50 minutes longer each day to return to the same position it was in the previous day. Thus, an observer would experience a high tide at 1:00 P.M. on one day, for example, and at 1:50 P.M. on the following day.

Tides are also complicated by the combined effects of the Moon and the Sun. Even though the Sun's tide-generating force is weaker than the Moon's, when the Moon and Sun are aligned every two weeks, their forces are added together and generate *spring tides,* which are about 20% higher than

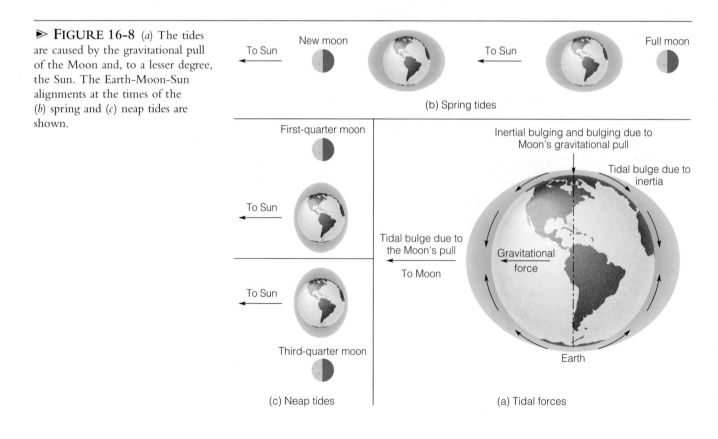

➤ **FIGURE 16-8** (*a*) The tides are caused by the gravitational pull of the Moon and, to a lesser degree, the Sun. The Earth-Moon-Sun alignments at the times of the (*b*) spring and (*c*) neap tides are shown.

average tides (Figure 16-8b). When the Moon and Sun are at right angles to one another, also at two-week intervals, the Sun's tide-generating force cancels some of that of the Moon, and *neap tides* about 20% lower than average occur (Figure 16-8c).

Tidal ranges are also affected by shoreline configuration. Broad, gently sloping continental shelves as in the Gulf of Mexico have low tidal ranges, whereas steep, irregular shorelines experience a much greater rise and fall of tides. Tidal ranges are greatest in some narrow, funnel-shaped bays and inlets. In the Bay of Fundy in Nova Scotia, a tidal range of 16.5 m has been recorded, and ranges greater than 10 m occur in several other areas.

Tides have an important impact on shorelines because the area of wave attack constantly shifts onshore and offshore as the tides rise and fall. Tidal currents themselves, however, have little modifying effect on shorelines, except in narrow passages where tidal current velocity is great enough to erode and transport sediment. Indeed, if it were not for strong tidal currents, some passageways would be blocked by sediments deposited by nearshore currents.

Waves

Waves, or oscillations of a water surface, occur on all bodies of water, but are most significant in large lakes and the oceans where they serve as agents of erosion, transport, and deposition. The highest part of a wave is its *crest,* whereas the low point between crests is the *trough. Wave length* is the distance between successive wave crests (or troughs), and *wave height* is the vertical distance from trough to crest (▶ Figure 16-9). As waves move across a water surface, the water "particles" rotate in circular orbits, with little or no net movement in the direction of wave travel (Figure 16-9). They do, however, transfer energy in the direction of wave advance.

The diameters of the orbits followed by water particles in waves diminish rapidly with depth, and at a depth of about one-half wave length, called **wave base,** they are essentially zero (Figure 16-9). Thus, at depths exceeding wave base, the water and sea floor, or lake floor, are unaffected by surface waves. The significance of wave base will be explored more fully in later sections.

Wave Generation. Waves can be generated by several processes including displacement of water by landslides, displacement of the sea floor by faulting, and volcanic explosions. But most of the geologic work done on shorelines occurs from wind-generated waves. When wind blows over water, some of its energy is transferred to the water, causing the water surface to oscillate. The mechanism that transfers energy from wind to water is related to the frictional drag resulting from one fluid (air) moving over another (water).

As one would expect, the harder and longer the wind blows, the larger the waves generated. Wind velocity and duration, however, are not the only factors controlling the

▶ **FIGURE 16-9** Waves and the terminology applied to them. The water in waves moves in circular orbits that decrease in size with depth. At wave base, which is at a depth of one-half wave length, water is not disturbed by surface waves. As deep water-waves move toward shore, the orbital motion of water within them is disrupted when they enter water shallower than wave base. Wave length decreases while wave height increases, causing the waves to oversteepen and plunge forward as breakers.

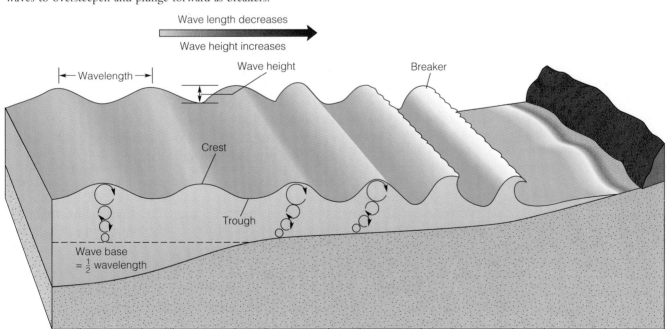

size of waves. High-velocity wind blowing over a small pond will never generate large waves regardless of how long it blows. In fact, waves occur on ponds and most lakes only while the wind is blowing; once the wind stops, the water quickly smooths out. In contrast, the surface of the ocean is always in motion, and waves with heights of 34 m have been recorded during storms.

The reason for the disparity between the wave sizes on ponds and lakes and on the oceans is the **fetch,** which is the distance the wind blows over a continuous water surface. The greater the fetch, the greater the size of the waves. Fetch is limited by the available water surface, so on ponds and lakes it corresponds to their length or width, depending on wind direction.

Shallow-Water Waves and Breakers. Waves moving out from the area of generation form swells and lose only a small amount of energy as they travel across the ocean. In deep-water swells, the water surface oscillates and water particles move in orbital paths, with little net displacement of water occurring in the direction of wave advance (Figure 16-9).

When deep-water waves enter shallow water, they are transformed from broad, undulating swells into sharp-crested waves. This transformation begins at wave base; that is, it begins where wave base intersects the sea floor. At this point, the waves "feel" the bottom, and the orbital motions of water particles within waves are disrupted (Figure 16-9). As they move further shoreward, the speed of wave advance and wave length decrease, and wave height increases. In effect, as waves enter shallower water, they become over-steepened; the wave crest advances faster than the wave form, until eventually the crest plunges forward as a **breaker** (Figure 16-9). Breakers are commonly several times higher than deep-water waves, and when they plunge forward, their kinetic energy is expended on the shoreline.

Nearshore Currents

It is convenient to identify the *nearshore zone* as the area extending seaward from the shoreline to just beyond the area where waves break. The width of the nearshore zone varies depending on the wave length of the approaching waves, because long waves break at a greater depth, and thus farther offshore, than do short waves. Two types of currents are important in the nearshore zone, *longshore currents* and *rip currents.*

Wave Refraction and Longshore Current. Deep-water waves are characterized by long, continuous crests, but rarely are their crests parallel with the shoreline (▷ Figure 16-10). One part of a wave enters shallow water where it encounters wave base and begins breaking before other parts of the same wave. As a wave begins breaking, its velocity diminishes, but the part of the wave still in deep water races ahead until it too encounters wave base. The net effect of this oblique approach is that the waves bend so that they more nearly

▷ **FIGURE 16-10** Wave refraction. These oblique waves are refracted and more nearly parallel the shoreline as they enter progressively shallower water.

parallel the shoreline (Figure 16-10). Such a phenomenon is called **wave refraction.**

Even though waves are refracted, they still usually strike the shoreline at some angle, causing the water between the breaker zone and the beach to flow parallel to the shoreline. These **longshore currents,** as they are called, are long and narrow and flow in the same general direction as the approaching waves. These currents are particularly important agents of transport and deposition in the nearshore zone.

Rip Currents. Waves carry water into the nearshore zone, so there must be a mechanism for mass transfer of water back out to sea. One way that water moves seaward from the nearshore zone is in **rip currents,** which are narrow surface currents that flow out to sea through the breaker zone (▷ Figure 16-11). Surfers commonly take advantage of rip currents for an easy ride out beyond the breaker zone, but such currents pose a danger to inexperienced swimmers. Some rip currents flow at several kilometers per hour, so if a swimmer is caught in one, it is useless to try to swim directly back to shore. Instead, because rip currents are narrow and usually nearly perpendicular to the shore, one can swim parallel to the shoreline for a short distance and then turn shoreward with no difficulty.

SHORELINE DEPOSITION

Depositional features of shorelines include *beaches, spits, baymouth bars,* and *barrier islands.* The characteristics of these deposits are determined largely by wave energy and longshore currents. Rip currents play only a minor role in the configuration of shorelines.

Beaches

Beaches are the most familiar of all coastal landforms, attracting millions of visitors each year and providing the

➤ FIGURE 16-11 Suspended sediment, indicated by discolored water, being carried seaward by a rip current.

(a)   (b)

➤ FIGURE 16-12 (*a*) A pocket beach in California. (*b*) The Grand Strand of South Carolina, shown here at Myrtle Beach, is 100 km of nearly continuous beach.

economic base for many communities. By definition a **beach** is a deposit of unconsolidated sediment extending landward from low tide to a change in topography such as a line of sand dunes, a sea cliff, or the point where permanent vegetation begins. Depending on shoreline configuration and wave intensity, beaches may be discontinuous, existing only as *pocket beaches* in protected areas such as embayments,

or they may be continuous for long distances (➤ Figure 16-12).

Some of the sediment on beaches is derived from weathering and wave erosion of the shoreline, but most of it is transported to the coast by streams and redistributed along the shoreline by longshore currents. **Longshore drift** is the phenomenon by which sand is transported along a shoreline

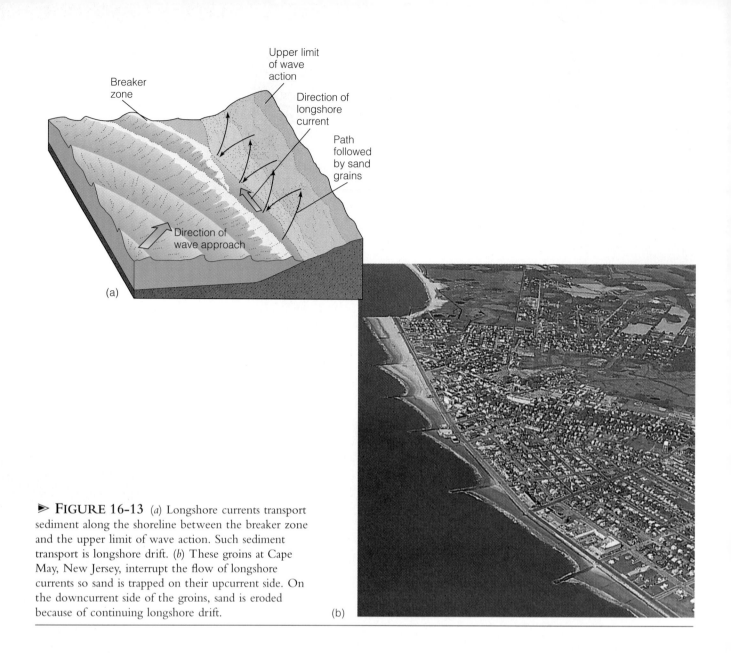

Breaker zone

Upper limit of wave action

Direction of longshore current

Path followed by sand grains

Direction of wave approach

(a)

➤ **FIGURE 16-13** (*a*) Longshore currents transport sediment along the shoreline between the breaker zone and the upper limit of wave action. Such sediment transport is longshore drift. (*b*) These groins at Cape May, New Jersey, interrupt the flow of longshore currents so sand is trapped on their upcurrent side. On the downcurrent side of the groins, sand is eroded because of continuing longshore drift.

(b)

by longshore currents (➤ Figure 16-13a). As previously noted, waves usually strike beaches at some angle, causing the sand grains to move up the beach face at a similar angle; as the sand grains are carried seaward in the backwash, however, they move perpendicular to the long axis of the beach. Thus, individual sand grains move in a zigzag pattern in the direction of longshore currents. This movement is not restricted to the beach; it extends seaward to the outer edge of the breaker zone (Figure 16-13a).

In an attempt to widen a beach or prevent erosion, shoreline residents often build *groins*, structures that project seaward at right angles from the shoreline (Figure 16-13b). Groins interrupt the flow of longshore currents, causing sand to be deposited on their upcurrent side and widening the beach at that location. However, erosion inevitably occurs on the downcurrent side of a groin.

A beach is an area where wave energy is dissipated, so the loose grains composing the beach are constantly affected by

wave motion. But the overall configuration of a beach remains unchanged as long as equilibrium conditions persist. The beach profile can be thought of as a profile of equilibrium; that is, all parts of the beach are adjusted to the prevailing conditions of wave intensity and nearshore currents.

Tides and longshore currents affect the configuration of beaches to some degree, but storm waves are by far the most important agent modifying their equilibrium profile. In many areas, beach profiles change with the seasons, so, we recognize *summer beaches* and *winter beaches,* each of which is adjusted to the conditions prevailing at these times. Summer beaches are generally wide and covered with sand and are characterized by a gently sloping, smooth offshore profile. Winter beaches, on the other hand, tend to be narrower, coarser grained and steeper; their offshore profiles reveal sand bars paralleling the shoreline.

Seasonal changes in beach profiles are related to changing wave intensity. During the winter, energetic storm waves erode the sand from the beach and transport it offshore where it is stored in sand bars. The same sand that was eroded from the beach during the winter returns the next summer when it is driven onshore by the more gentle swells that occur during that season. The volume of sand in the system remains more or less constant; it simply moves farther offshore or onshore depending on the energy of waves.

Spits, Baymouth Bars, and Barrier Islands

Other than the beach itself, some of the most common depositional landforms on shorelines are *spits* and *baymouth bars,* both of which are variations of the same feature. A **spit** is simply a continuation of a beach forming a point, or "free end," that projects into a body of water, commonly a bay. A **baymouth bar** is a spit that has grown until it completely closes off a bay from the open sea (▷ Figure 16-14).

Both spits and baymouth bars form and grow as a result of longshore drift. Where currents are weak, as in the deeper water at the opening to a bay, longshore current velocity diminishes, and sediment is deposited, forming a sand bar. The free ends of many spits are curved by wave refraction or waves approaching from a different direction. Such spits are called *hooks* or *recurved spits* (Figure 16-14a). A rarer type of spit, called a *tombolo,* extends out into the sea and connects

▷ **FIGURE 16-15** This chain of barrier islands comprises the outer banks of North Carolina. Cape Hatteras, near the center of the photo, juts the furthest out into the Atlantic.

an island to the mainland. Tombolos develop on the shoreward sides of islands (Figure 16-14b). Wave refraction around an island causes converging currents that turn seaward and deposit a sand bar connecting the shore with the island.

Barrier Islands are long, narrow islands composed of sand and separated from the mainland by a lagoon (▷ Figure 16-15). The origin of barrier islands has been long debated and is still not completely resolved. It is known that they form on gently sloping continental shelves with abundant sand in areas where both tidal fluctuations and wave energy levels are low. According to one model, barrier islands formed as spits that became detached from the land, while another model proposes that they formed as beach ridges on coasts that were subsequently partly submerged when sea level rose.

Because sea level is currently rising, most barrier islands are migrating in a landward direction. Such migration is a natural consequence of the evolution of these islands, but it is a problem for the island residents and communities. Barrier islands generally migrate rather slowly, but the rates for many are rapid enough to cause shoreline problems (see Perspective 16-1).

▷ **FIGURE 16-14** (*a*) Spits form where longshore currents deposit sand in deeper water as at the entrance to a bay. A baymouth bar is simply a spit that has grown until it extends across the mouth of a bay. (*b*) Origin of a tombolo. Wave refraction around an island causes longshore currents to converge and deposit a sand bar that joins the island with the mainland.

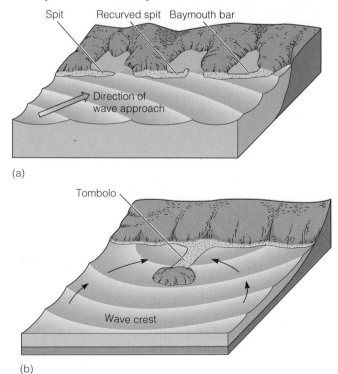

RISING SEA LEVEL AND COASTAL MANAGEMENT

Shorelines in most of the United States and much of Canada are eroding as sea level rises. According to one study, 54% of U.S. shorelines are eroding at rates ranging from millimeters per year to more than 10 m in a few areas. Many other areas of the world are experiencing shoreline problems as well.

During the last century, sea level rose about 12 cm worldwide, and all indications are that it will continue to rise. The absolute rate of sea level rise in a particular shoreline region depends on two factors. The first is the volume of water in the ocean basins, which is increasing as a result of the melting of glacial ice and the thermal expansion of near-surface seawater. Many scientists think that sea level will continue to rise as a consequence of global warming resulting from concentrations of greenhouse gases in the atmosphere.

The second factor controlling sea level is the rate of uplift or subsidence of a coastal area. In some areas, uplift is occurring so fast that sea level is actually falling with respect to the land. In other areas, sea level is rising while the coastal region is simultaneously subsiding, resulting in a net change in sea level of as much as 30 cm per century. Perhaps such a "slow" rate of sea level change seems insignificant; after all it amounts to only a few millimeters per year. But, in gently sloping coastal areas, as in the eastern United States from New Jersey southward, even a slight rise in sea level would eventually have widespread effects.

Many of the nearly 300 barrier islands along the east and Gulf coasts of the United States are migrating landward as a result of rising sea level. During storms, their beaches are eroded by large waves that carry the beach sand over the islands and into their lagoons. Thus, sand is removed from the seaward sides of barrier islands and deposited on their landward sides, resulting in a gradual landward shift of the entire island complex (⊳ Figure 1). During the last 120 years, Hatteras Island, North Carolina, has migrated nearly 500 m landward so that Cape Hatteras lighthouse, which was 460 m from the shoreline when it was built in 1870, now stands on a promontory in the Atlantic Ocean. Landward migration of barrier islands would pose few problems if it were not for the numerous communities, resorts, and vacation homes located on them.

Rising sea level also directly threatens many beaches communities depend on for revenue. The beach at Miami Beach, Florida, was disappearing at an alarming rate until the Army Corps of Engineers began replacing the eroded beach sand. The problem is even more serious in other countries. While a rise in sea level of only 2 m would inundate large areas of the east and Gulf coasts, it would cover 20% of the entire country of Bangladesh.

Because nothing can be done to prevent sea level from rising, engineers and scientists must examine what can be done to prevent or minimize the effects of shoreline erosion. At present, only a few viable options exist. One is to put strict controls on coastal development. North Carolina permits large structures to be sited no closer to the shoreline than 60 times the annual erosion rate. Although a growing awareness of shoreline processes has resulted in similar legislation elsewhere, some states have virtually no restrictions on coastal development.

Regulating coastal development is commendable, but it has no impact on existing structures and coastal communities. A general retreat from the shoreline may be possible,

⊳ **FIGURE 1** Rising sea level and the landward migration of barrier islands. (*a*) Barrier island before landward migration in response to rising sea level. (*b*) Landward movement occurs when storm waves wash sand from the seaward side of the islands and deposit it in the lagoon. (*c*) Over time, the entire complex migrates landward.

(b)

(a) Lagoon   Barrier island

(c) Lagoon   Migrating barrier island   Original barrier island position   Sea level rise   Barrier island movement

but expensive, for individual dwellings and small communities, but it is impractical for large population centers. Such communities as Atlantic City, New Jersey, Miami Beach, Florida, and Galveston, Texas, have adopted one of two strategies to combat coastal erosion. One is to build protective barriers such as seawalls. Seawalls can be effective, but they are tremendously expensive to construct and maintain. More than $50 million has been spent to replenish the beach and build a seawall at Ocean City, Maryland, in just the last five years. Furthermore, they retard erosion only in the area directly behind the seawall.

Another option, adopted by both Atlantic City, New Jersey, and Miami Beach, Florida, is to pump sand onto the beaches to replace that lost to erosion. This, too, is expensive as the sand must be replenished periodically because erosion is a continuing process. In many areas, groins are constructed to preserve beaches, but unless additional sand is artificially supplied to the beaches, longshore currents invariably erode sand from the downcurrent sides of the groins.

Rising sea level has already had a significant economic impact, and all options for dealing with this phenomenon are expensive. Fortifying the shoreline with seawalls, groins, and other structures is initially expensive, requires constant maintenance, and in the long run will be ineffective if sea level continues to rise. A general retreat from the shoreline is simply impractical for most coastal communities. Perhaps the best option is to replace sand lost to erosion by pumping it from elsewhere, usually farther offshore. In some areas, however, little can be done to offset the effects of rising sea level.

In addition to rising sea level, many coastal communities are threatened by storm waves, especially from hurricanes, which are vast storms with winds that may exceed 300 km/hr. When these storms sweep across coastal areas, the intense wind can cause considerable damage, but most of the damage and about 90% of all hurricane fatalities are caused by coastal flooding. In fact, of the nearly $2 billion paid out by the federal government's National Flood Insurance Program since 1974, most has gone to owners of beachfront homes.

Flooding during hurricanes is caused by large storm-generated waves being driven onshore and by intense rainfall, in some cases more than 60 cm in 24 hours. In addition, as the storm moves over the ocean, low atmospheric pressure beneath the eye of the storm causes the ocean surface to bulge upward as much as 0.5 m. When the eye of the storm reaches the shoreline, the bulge, coupled with wind-driven waves, piles up in a storm surge that can rise several meters above normal high tide and inundate areas several kilometers inland.

Several coastal areas in the United States have been devastated by storm surges, including Galveston, Texas, in 1900 and Charleston, South Carolina, in 1989 (▷ Figure 2). One of the greatest natural disasters of the twentieth century occurred in 1970 when a storm surge estimated at 8 to 10 m high flooded the low-lying coastal areas of Bangladesh, drowning 300,000 people. Since 1970, coastal Bangladesh has been flooded several more times, the most recent and most tragic being on April 30, 1991, when more than 100,000 people drowned.

▷ **FIGURE 2** Charleston, South Carolina, was flooded by a storm surge produced by Hurricane Hugo on September 22, 1989.

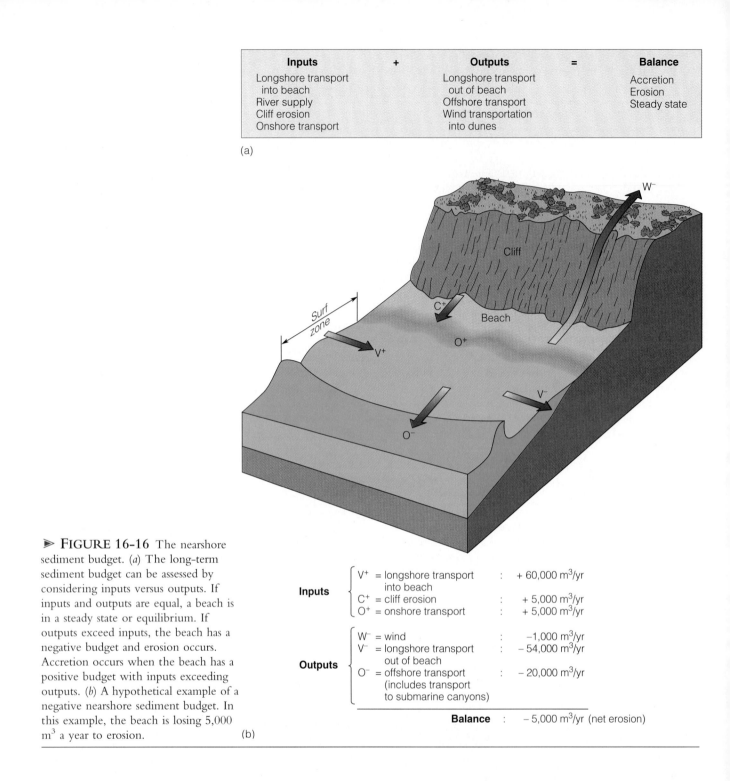

Inputs	+	Outputs	=	Balance
Longshore transport into beach		Longshore transport out of beach		Accretion
River supply		Offshore transport		Erosion
Cliff erosion		Wind transportation into dunes		Steady state
Onshore transport				

(a)

Inputs
V^+ = longshore transport into beach : + 60,000 m³/yr
C^+ = cliff erosion : + 5,000 m³/yr
O^+ = onshore transport : + 5,000 m³/yr

Outputs
W^- = wind : – 1,000 m³/yr
V^- = longshore transport out of beach : – 54,000 m³/yr
O^- = offshore transport (includes transport to submarine canyons) : – 20,000 m³/yr

Balance : – 5,000 m³/yr (net erosion)

➤ **FIGURE 16-16** The nearshore sediment budget. (*a*) The long-term sediment budget can be assessed by considering inputs versus outputs. If inputs and outputs are equal, a beach is in a steady state or equilibrium. If outputs exceed inputs, the beach has a negative budget and erosion occurs. Accretion occurs when the beach has a positive budget with inputs exceeding outputs. (*b*) A hypothetical example of a negative nearshore sediment budget. In this example, the beach is losing 5,000 m³ a year to erosion. (b)

The Nearshore Sediment Budget

We can think of the gains and losses of sediment in the nearshore zone in terms of a **nearshore sediment budget** (➤ Figure 16-16). If a nearshore system has a balanced budget, sediment is supplied to it as fast as it is removed, and the volume of sediment remains more or less constant, although sand may shift offshore and onshore with the changing seasons. A positive budget means gains exceed losses, whereas a negative budget results when losses exceed gains. If a negative budget prevails long enough, a nearshore system is depleted and beaches may disappear.

Erosion of sea cliffs provides some sediment to beaches, but in most areas probably no more than 5 to 10% of the total sediment supply is derived from this source. Most of the sediment on typical beaches is transported to the shoreline by streams and then redistributed along the shoreline by longshore drift. So, longshore drift also plays a role in the nearshore sediment budget because it continually moves sediment into and away from beach systems.

The primary ways in which a nearshore system loses sediment include longshore drift, offshore transport, wind, and deposition in submarine canyons (Figure 16-16). Offshore transport mostly involves fine-grained sediment that is carried seaward where it eventually settles in deeper water. Wind is an important process because it removes sand from beaches and blows it inland where it commonly piles up as sand dunes.

If the heads of submarine canyons are nearshore, huge quantities of sand are funneled into them and deposited in deeper water. La Jolla and Scripps submarine canyons off the coast of southern California funnel off an estimated 2 million m³ of sand each year. In most areas, submarine canyons are too far offshore to interrupt the flow of sand in the nearshore zone.

It should be apparent from the preceding discussion that if a nearshore system is in equilibrium, its incoming supply of sediment exactly offsets its losses. Such a delicate balance tends to continue unless the system is somehow disrupted. One common change that affects this balance is the construction of dams across the streams supplying sand. The sediment contribution from a stream is proportional to its drainage area, but once dams have been built, all sediment from the upper reaches of the drainage systems is trapped in reservoirs and thus cannot reach the shoreline.

⊛ Shoreline Erosion

Along seacoasts where erosion rather than deposition predominates, beaches are lacking or poorly developed, and a sea cliff commonly develops (⊳ Figure 16-17). Sea cliffs are erosional features frequently pounded by waves, especially during storms: the cliff face retreats landward as a result of *corrosion, hydraulic action,* and *abrasion. Corrosion* is an erosional process involving the wearing away of rock by chemical processes, especially the solvent action of seawater. The force of the water itself, called *hydraulic action,* is a particularly effective erosional process. Waves exert tremendous pressure on shorelines by direct impact, but are most effective on sea cliffs composed of unconsolidated sediment or rocks that are highly fractured. *Abrasion* is an erosional process involving the grinding action of rocks and sand carried by waves.

Wave-Cut Platforms and Associated Landforms

Sea cliffs erode in a landward direction mostly as a result of hydraulic action and abrasion at their bases. As a sea cliff is undercut by erosion, the upper part is left unsupported and susceptible to mass wasting processes. Sea cliffs retreat little by little, and as they do, they leave a beveled surface called a

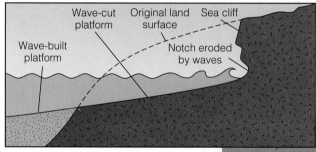

(a)

(b)

⊳ **FIGURE 16-17** (*a*) Wave erosion of a sea cliff produces a gently sloping surface called a wave-cut platform. Deposition at the seaward margin of the wave-cut platform forms a wave-built platform. (*b*) Sea cliffs and a wave-cut platform.

(a)

(b)

(c)

▷ **FIGURE 16-18** (*a*) Erosion of a headland and the origin of sea caves, sea arches, and sea stacks. (*b*) This sea stack in Australia has an arch developed in it. (*c*) Sea stacks along Australia's south coast.

wave-cut platform that slopes gently in a seaward direction (Figure 16-17). Broad wave-cut platforms exist in many areas, but invariably the water over them is shallow because the abrasive planing action of waves is only effective to a depth of about 10 m. The sediment eroded from sea cliffs is transported seaward and deposited to form a *wave-built platform,* a seaward extension of the wave-cut platform (Figure 16-17a).

Sea cliffs do not retreat uniformly because some of the materials of which they are composed are more resistant to erosion than others. *Headlands* are seaward-projecting parts of the shoreline that are eroded on both sides due to wave refraction (▷ Figure 16-18). *Sea caves* may form on opposite sides of a headland, and if these join, they form a *sea arch* (Figure 16-18a and b). Continued erosion generally causes the span of an arch to collapse, yielding isolated *sea stacks* on wave-cut platforms (Figure 16-18c). In the long run, shoreline processes tend to straighten an initially irregular shoreline. They do so because wave refraction causes more wave energy to be expended on headlands and less on embayments. As the headlands become eroded, some of the sediment yielded by erosion is deposited in the embayments. The net effect of these processes is to straighten the shoreline.

TYPES OF COASTS

Depositional coasts, such as the U.S. Gulf Coast, are characterized by an abundance of detrital sediment, long sandy beaches, and the presence of such depositional landforms as deltas and barrier islands. Erosional coasts are generally steep and irregular and typically lack well-developed beaches except in protected areas. They are further characterized by erosional features such as sea cliffs, wave-cut platforms, and sea stacks. Many of the beaches along the west coast of North America fall into this category.

The following section examines coasts in terms of their changing relationships to sea level. But note that while some coasts, such as those in southern California, are described as emergent (uplifted), these same coasts may be erosional as well. In other words, coasts commonly possess features allowing them to be classified in several ways.

If sea level rises with respect to the land or the land subsides, coastal regions are flooded and said to be **submergent** or *drowned* (▷ Figure 16-19). Much of the east coast of North America from Maine southward through South Carolina was flooded during the post-Pleistocene rise in sea level, so that it is now an extremely irregular coast. Recall that during the expansion of glaciers during the Pleistocene, sea level was as much as 130 m lower than at present, and that streams eroded their valleys more deeply as they adjusted to a lower base level. When sea level rose, the lower ends of these valleys were drowned, forming *estuaries* such as Delaware and Chesapeake bays (Figure 16-19). Estuaries are the seaward ends of river valleys where seawater and fresh water mix.

Emergent coasts are found where the land has risen with respect to sea level. Emergence can occur when water is withdrawn from the oceans, as occurred during the Pleistocene expansion of glaciers. At present, however, coasts are emerging as a result of isostasy or tectonism. In northeastern Canada and the Scandinavian countries, the coasts are irregular because isostatic rebound is elevating formerly glaciated terrain from beneath the sea.

Coasts rising in response to tectonism, tend to be rather straight because the sea-floor topography being exposed as uplift proceeds is smooth. The west coasts of North and South America are rising as a consequence of plate tectonics. Distinctive features of such coasts are **marine terraces** (▷ Figure 16-20), which are old wave-cut platforms now elevated above sea level. Uplift in such areas appears to be episodic rather than continuous, as indicated by the multiple levels of terraces in some areas. In southern California, several terrace levels are present, each of which probably represents a period of tectonic stability followed by uplift. The highest of these terraces is now about 425 m above sea level.

RESOURCES FROM THE SEA

Seawater contains many elements in solution, some of which are extracted for various industrial and domestic uses. In

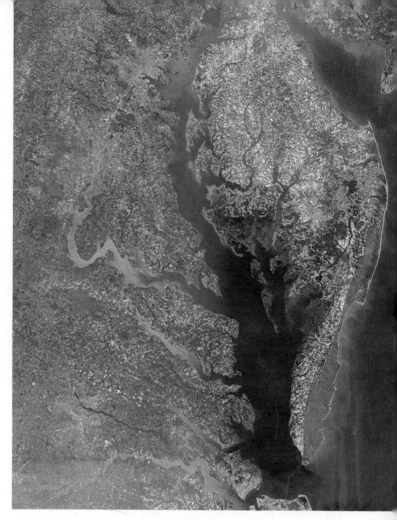

▷ **FIGURE 16-19** Chesapeake Bay is a large estuary. It formed when the east coast of the United States was flooded as sea level rose following the Pleistocene Epoch.

many places sodium chloride (table salt) is produced by the evaporation of seawater, and a large proportion of the world's magnesium is produced from seawater. Numerous other elements and compounds can be extracted from seawater, but for many, such as gold, the cost is prohibitive.

In addition to substances in seawater, deposits on the sea floor or within sea-floor sediments are becoming increasingly important. Many of these potential resources lie well beyond the margins of the continents, so their ownership is a political and legal problem that has not yet been resolved. Most nations bordering the ocean claim those resources occurring within their adjacent continental margin. The United States, by a presidential proclamation issued on March 10, 1983, claims sovereign rights over an area designated as the **Exclusive Economic Zone (EEZ).** The EEZ extends seaward 200 nautical miles (371 km) from the coast, giving the United States jurisdiction over an area about 1.7 times larger than its land area. Also included within the EEZ are the areas adjacent to U.S. territories, such as Guam, American Samoa, Wake Island, and Puerto Rico. In short, the United States claims a huge area of the sea floor and any resources on or beneath it.

➤ FIGURE 16-20 This gently sloping surface in Ireland is a marine terrace.

Numerous resources occur within the EEZ, some of which have been exploited for many years. Sand and gravel for construction are mined from the continental shelf in several areas. About 17% of U.S. oil and natural gas production comes from wells on the continental shelf. Ancient shelf deposits in the Persian Gulf region contain the world's largest reserves of oil (see Perspective 7-1).

Other resources of interest include deposits that form by submarine hydrothermal activity at spreading ridges (see the Prologue). Such deposits containing iron, copper, zinc, and other metals have been identified within the EEZ at the Gorda Ridge off the coasts of California and Oregon; similar deposits occur at the Juan de Fuca Ridge within the Canadian EEZ.

Other potential resources include irregular to spherical masses known as *manganese nodules* that are fairly common in all the ocean basins. These nodules result from chemical reactions in seawater and, in addition to manganese, contain iron oxides, copper, nickel, and cobalt. The United States is interested in this potential resource because it is heavily dependent on imports of manganese and cobalt.

Within the EEZ, manganese nodules occur near Johnston Island in the Pacific Ocean and on the Blake Plateau off the east coast of South Carolina and Georgia. In addition, seamounts and seamount chains within the EEZ in the Pacific are known to have metalliferous oxide crusts several centimeters thick from which cobalt and manganese could be mined.

CHAPTER SUMMARY

1. Continental margins separate continents above sea level from the deep-sea floor. They consist of a continental shelf, a continental slope, and in some places a continental rise.
2. Submarine canyons are best developed on the continental slope, but some extend well up onto shelves and lie offshore from streams. Many submarine canyons were probably eroded by turbidity currents that transport sediments to the rise.
3. Active continental margins are characterized by seismicity,

volcanism, a narrow shelf, and a slope that descends into an oceanic trench. Passive continental margins have little seismic activity, no volcanism, and possess broad shelves and a rise that commonly merges with abyssal plains.
4. Oceanic trenches are long, narrow, deep features where oceanic lithosphere is subducted. Oceanic ridges, also known as spreading ridges, nearly encircle the globe, but they are interrupted and offset by large fractures in the sea floor.

5. Other important features of the sea floor include seamounts, flat-topped seamounts known as guyots, volcanic hills, and aseismic ridges.

6. Deep-sea sediments consist of small particles derived from continents and oceanic islands and the shells of microscopic marine organisms. These sediments are characterized as pelagic clay and ooze.

7. Reefs are wave-resistant structures composed of animal skeletons, particularly those of corals. There are three types of reefs: fringing, barrier, and atoll.

8. Shorelines are continually modified by the energy of nearshore currents and waves, which are oscillations on water surfaces that transmit energy in the direction of wave movement. Surface waves affect the sea floor only to wave base, a depth equal to one-half of the wave length.

9. Breakers form where waves enter shallow water and the orbital motion of water particles is disrupted. The waves become oversteepened and plunge forward onto the shoreline, thus expending their energy.

10. Longshore currents are generated by waves approaching a shoreline at an angle. These currents are capable of considerable erosion, transport, and deposition.

11. Rip currents are narrow surface currents that carry water from the nearshore zone seaward through the breaker zone.

12. Beaches are the most common shoreline depositional features. They are continually modified by nearshore processes, and their profiles generally exhibit seasonal changes.

13. Spits, baymouth bars, and tombolos all form and grow as a result of longshore current transport and deposition. Barrier islands, which are separated from the mainland by a lagoon, are nearshore sediment deposits of uncertain origin.

14. The volume of sediment in a nearshore system remains rather constant unless the system is somehow disrupted as when dams are built across the streams that supply sand to the system.

15. Shorelines characterized by erosion have sea cliffs, wave-cut platforms with sea stacks, and discontinuous beaches, whereas depositional shorelines exhibit deltas, long sandy beaches, and barrier islands.

16. Submergent coasts and emergent coasts are defined on the basis of their relationships to change in sea level. The former have been inundated by rising sea level or subsidence of the coast, and the latter have risen with respect to sea level.

IMPORTANT TERMS

abyssal plain	Exclusive Economic Zone (EEZ)	oceanic ridge	shoreline
active continental margin	fetch	oceanic trench	spit
barrier island	guyot	ooze	submergent coast
baymouth bar	longshore current	passive continental margin	tide
beach	longshore drift	pelagic clay	wave
breaker	marine terrace	reef	wave base
continental margin	nearshore sediment budget	rip current	wave-cut platform
emergent coast		seamount	wave refraction

REVIEW QUESTIONS

1. Submarine canyons are best developed in:
   a. ____ rift valleys;
   b. ____ fractures in the sea floor;
   c. ____ abyssal plains;
   d. ____ continental slopes;
   e. ____ submarine fans.

2. Continental shelves:
   a. ____ are composed of pelagic sediments;
   b. ____ lie between continental slopes and rises;
   c. ____ descend to an average depth of 1,500 m;
   d. ____ slope gently from the shoreline to the shelf-slope break;
   e. ____ are widest along active continental margins.

3. The flattest, most featureless areas on Earth are:
   a. ____ oceanic ridges;
   b. ____ abyssal plains;
   c. ____ aseismic ridges;
   d. ____ continental margins;
   e. ____ oceanic trenches.

4. Which of the following statements is correct?
   a. ____ North America's east coast is a passive continental margin;
   b. ____ oceanic ridges are composed largely of granite and deformed sedimentary rocks;
   c. ____ the deposits of turbidity currents consist of calcareous ooze;

d. ____ most of the Earth's volcanism occurs at aseismic ridges;

e. ____ the greatest oceanic depths are at continental rises.

5. Wave base is:

a. ____ the distance offshore that waves break;

b. ____ the width of a longshore current;

c. ____ the depth at which the orbital motion in surface waves dies out;

d. ____ the distance wind blows over a water surface;

e. ____ the height of storm waves.

6. Waves approaching a shoreline at an angle generate:

a. ____ flood tides;

b. ____ longshore currents;

c. ____ tidal currents;

d. ____ submergent coasts;

e. ____ marine terraces.

7. The distance the wind blows over a water surface is the:

a. ____ fetch;

b. ____ spit;

c. ____ barrier island;

d. ____ wave period;

e. ____ rip current.

8. The bending of waves so that they more nearly parallel the shoreline is wave _____:

a. ____ translation;

b. ____ oscillation;

c. ____ deflection;

d. ____ inversion;

e. ____ refraction.

9. Erosional remnants of a shoreline now rising above a wave-cut platform are:

a. ____ barrier islands;

b. ____ sea stacks;

c. ____ beaches;

d. ____ marine terraces;

e. ____ spits.

10. What is wave base, and how does it affect waves as they enter shallow water?

11. Make a sketch showing how a tombolo develops.

12. Explain the concept of a nearshore sediment budget.

13. How does a wave-cut platform develop?

14. Why does an observer at a shoreline experience two high and two low tides each day?

15. What are the characteristics of a passive continental margin? How does such a margin develop?

16. Describe the continental rise, and explain why a rise occurs at some continental margins and not at others.

17. What are the characteristics of oceanic ridges, and how do they differ from mountain ranges on land?

18. What is the Exclusive Economic Zone? What types of resources occur within it?

POINTS TO PONDER

1. Why are long, broad sandy beaches more common in eastern North America than western North America?

2. An initially straight shoreline is composed of granite flanked on both sides by glacial drift. Diagram and explain this shoreline's probable response to erosion.

Chapter 17

GEOLOGIC TIME:
CONCEPTS AND PRINCIPLES

OUTLINE

Three Sisters, in New South Wales, Australia. The Three Sisters are distinctive erosional remnants that formed over a long period of geologic time. They are a popular tourist attraction in Australia's Blue Mountains.

What is time? We seem obsessed with it, and organize our lives around it with the help of clocks, calendars, and appointment books. Yet most of us feel we don't have enough of it—we are always running "behind" or "out of time." According to biologists and psychologists, children less than two years old and animals have no conscious concept of time and exist in a "timeless present," where there is no past or future. Some scientists think that our early ancestors may also have lived in a state of timelessness with little or no perception of a past or future. According to Buddhist, Taoist, and Mayan beliefs, time is circular, and like a circle, all things are destined to return to where they once were. Thus, in these belief systems, there is no beginning or end, but rather a cyclicity to everything. For most people, though, time is linear and moves like a flowing stream.

In some respects, time is defined by the methods used to measure it. Many prehistoric monuments are oriented to detect the summer solstice. Sundials were used to divide the day into measurable units. As civilization advanced, mechanical devices were invented to measure time, the earliest being the water clock, first used by the ancient Egyptians and further developed by the Greeks and Romans. The pendulum clock was invented in the seventeenth century and provided the most accurate timekeeping for the next two and a half centuries. For most people though, time is linear and moves like a flowing stream.

Today the quartz watch is the most popular timepiece. Powered by a battery, a quartz crystal vibrates approximately 100,000 times per second. An integrated circuit counts these vibrations and converts them into a digital or dial reading on your watch face. An inexpensive quartz watch today is more accurate than the best mechanical watch, and precision-manufactured quartz clocks are accurate to within one second per 10 years.

Precise timekeeping is important in our technological world. Ships and aircraft plot their locations by satellite, relying on an extremely accurate time signal. Deep-space probes such as the *Voyagers* require radio commands timed to billionths of a second, while physicists exploring the motion inside the nucleus of an atom deal in trillionths of a second as easily as we talk about minutes.

To achieve such accuracy, scientists use atomic clocks. First developed in the 1940s, these clocks rely on an atom's oscillating electrons, a rhythm so regular that they are accurate to within a few thousandths of a second per day. Recently, an atomic clock accurate to within one second per three million years was installed at the National Institute of Standards and Technology.

While physicists deal with incredibly short intervals of time, astronomers and geologists are concerned with geologic time measured in millions or billions of years. When astronomers look at a distant galaxy, they are seeing what it looked like billions of years ago. When geologists investigate rocks in the walls of the Grand Canyon, they are deciphering events that occurred over an interval of 2 billion years. Geologists can measure decay rates of such radioactive elements as uranium, thorium, and rubidium to determine how long ago an igneous rock formed. Furthermore, geologists know that the Earth's rotational velocity has been slowing down a few thousandths of a second per century as a result of the frictional effects of tides, ocean currents, and varying thicknesses of polar ice. Five hundred million years ago a day was only 20 hours long, and at the current rate of slowing, 200 million years from now a day will be 25 hours long.

Time is a fascinating topic that has been the subject of numerous essays and books. And though we can comprehend concepts like milliseconds and understand how a quartz watch works, deep time, or geologic time, is still very difficult for most people to comprehend.

INTRODUCTION

Time is what sets geology apart from most of the other sciences, and an appreciation of the immensity of geologic time is fundamental to understanding both the physical and biological history of our planet. Most people have difficulty comprehending geologic time because they tend to view time from the perspective of their own existence. Ancient history is what occurred hundreds or even thousands of years ago, but when geologists talk in terms of ancient geologic history, they are referring to events that happened millions or even billions of years ago!

Geologists use two different frames of reference when speaking of geologic time. **Relative dating** involves placing geologic events in a sequential order as determined from their position in the geologic record. Relative dating will not tell us how long ago a particular event occurred, only that one event preceded another. The various principles used to determine relative dating were discovered hundreds of years ago and since then have been used to construct the *relative geologic time scale* (▶ Figure 17-1). These principles are still widely used today.

Absolute dating results in specific dates for rock units or events expressed in years before the present. Radiometric dating is the most common method of obtaining absolute age dates. Such dates are calculated from the natural rates of decay of various radioactive elements occurring in trace amounts in some rocks. It was not until the discovery of radioactivity near the end of the nineteenth century that

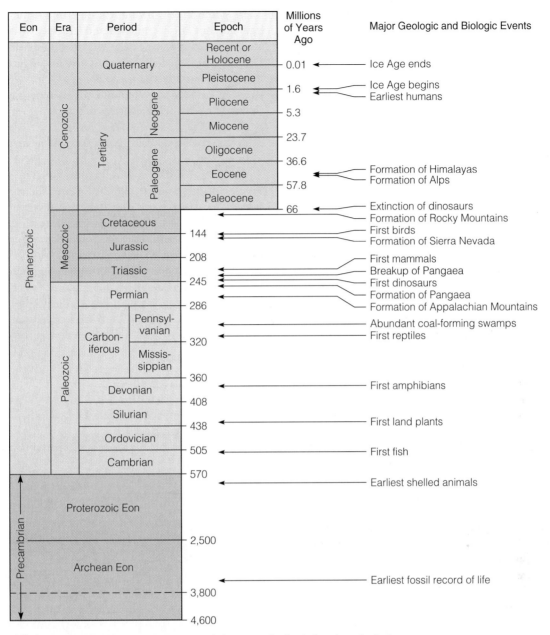

Eon	Era	Period		Epoch	Millions of Years Ago	Major Geologic and Biologic Events
Phanerozoic	Cenozoic	Quaternary		Recent or Holocene	0.01	Ice Age ends
				Pleistocene	1.6	Ice Age begins / Earliest humans
		Tertiary	Neogene	Pliocene	5.3	
				Miocene	23.7	
			Paleogene	Oligocene	36.6	
				Eocene	57.8	Formation of Himalayas / Formation of Alps
				Paleocene	66	
	Mesozoic	Cretaceous				Extinction of dinosaurs / Formation of Rocky Mountains
		Jurassic			144	First birds / Formation of Sierra Nevada
		Triassic			208	First mammals
					245	Breakup of Pangaea / First dinosaurs
	Paleozoic	Permian			286	Formation of Pangaea / Formation of Appalachian Mountains
		Carboniferous	Pennsylvanian		320	Abundant coal-forming swamps / First reptiles
			Mississippian		360	
		Devonian			408	First amphibians
		Silurian			438	First land plants
		Ordovician			505	First fish
		Cambrian			570	Earliest shelled animals
Precambrian		Proterozoic Eon			2,500	
		Archean Eon			3,800	Earliest fossil record of life
					4,600	

▷ **FIGURE 17-1** The geologic time scale. Some of the major biological and geological events are indicated along the right-hand margin.

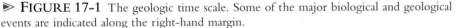

absolute ages could be accurately applied to the relative geologic time scale. Today the geologic time scale is really a dual scale: a relative scale based on rock sequences with radiometric dates expressed as years before the present added to it (Figure 17-1).

EARLY CONCEPTS OF GEOLOGIC TIME AND THE AGE OF THE EARTH

The concept of geologic time and its measurement have changed through human history. Many early Christian scholars and clerics tried to establish the date of creation by analyzing historical records and the genealogies found in Scripture. Based on their analyses, they generally believed that the Earth and all of its features were no more than about 6,000 years old. The idea of a very young Earth provided the basis for most Western chronologies of Earth history prior to the eighteenth century.

During the eighteenth and nineteenth centuries, several attempts were made to determine the age of the Earth on the basis of scientific evidence rather than revelation. The French zoologist Georges Louis de Buffon (1707–1788) assumed that the Earth gradually cooled to its present condition from a molten beginning. To simulate this history, he melted iron balls of various diameters and allowed them to cool to the surrounding temperature. By extrapolating their cooling rate to a ball the size of the Earth, he determined that the Earth was at least 75,000 years old. While this age was much older than that derived from Scripture, it was still vastly younger than we now know the Earth to be.

Other scholars were equally ingenious in attempting to calculate the Earth's age. For example, if deposition rates could be determined for various sediments, geologists reasoned that they could calculate how long it would take to deposit any rock layer. They could then extrapolate how old the Earth was from the total thickness of sedimentary rock in the Earth's crust. Rates of deposition vary, however, even for the same type of rock. Furthermore, it is impossible to estimate how much rock has been removed by erosion, or how much a rock sequence has been reduced by compaction. As a result of these variables, estimates ranged from less than a million years to more than a billion years.

In addition to trying to determine the Earth's age, naturalists of the eighteenth and nineteenth centuries were also formulating some of the fundamental geologic principles that are used in deciphering Earth history. From the evidence preserved in the geologic record, it was clear to them that the Earth is very old and that geologic processes have operated over long periods of time.

JAMES HUTTON AND THE RECOGNITION OF GEOLOGIC TIME

The Scottish geologist James Hutton (1726–1797) is considered by many to be the father of modern geology. His detailed studies and observations of rock exposures and present-day geological processes were instrumental in establishing the **principle of uniformitarianism** (see Chapter 1), the concept that the same processes have operated over vast

➢ FIGURE 17-2 The Grand Canyon of Arizona illustrates three of the six fundamental principles of relative dating. The sedimentary rocks of the Grand Canyon were originally deposited horizontally in a variety of marine and continental environments (principle of original horizontality). The oldest rocks are therefore at the bottom of the canyon, and the youngest rocks are at the top, forming the rim (principle of superposition). The exposed rock layers extend laterally for some distance (principle of lateral continuity).

amounts of time. Because Hutton relied on known processes to account for Earth history, he concluded that the Earth must be very old and wrote that "we find no vestige of a beginning, and no prospect of an end."

Unfortunately, Hutton was not a particularly good writer, so his ideas were not widely disseminated or accepted. In 1830, Charles Lyell published a landmark book, *Principles of Geology,* in which he championed Hutton's concept of uniformitarianism. Instead of relying on catastrophic events to explain various features of the Earth, Lyell recognized that imperceptible changes brought about by present-day processes could, over long periods of time, have tremendous cumulative effects. Through his writings, Lyell firmly established uniformitarianism as the guiding philosophy of geology.

⊛ RELATIVE DATING METHODS

Before the development of radiometric dating techniques, geologists had no reliable means of absolute age dating and therefore depended solely on relative dating methods. These methods only allow events to be placed in sequential order and do not tell us how long ago an event took place. Though the principles of relative dating may now seem self-evident, their discovery was an important scientific achievement because they provided geologists with a means to interpret geologic history and develop a relative geologic time scale.

Six fundamental geologic principles are used in relative dating: *superposition, original horizontality, lateral continuity, cross-cutting relationships, inclusions,* and *fossil succession.*

Fundamental Principles of Relative Dating

The seventeenth century was an important time in the development of geology as a science because of the widely circulated writings of the Danish anatomist, Nicolas Steno (1638–1686). Steno observed that during flooding, streams spread out across their floodplains and deposit layers of sediment that bury organisms dwelling on the floodplain. Subsequent flooding events produce new layers of sediments that are deposited or superposed over previous deposits. When lithified, these layers of sediment become sedimentary rock. Thus, in an undisturbed succession of sedimentary rock layers, the oldest layer is at the bottom and the youngest layer is at the top. This **principle of superposition** is the basis for relative age determinations of strata and their contained fossils (➤ Figure 17-2).

Steno also observed that because sedimentary particles settle from water under the influence of gravity, sediment is deposited in essentially horizontal layers, illustrating the **principle of original horizontality** (Figure 17-2). Therefore, a sequence of sedimentary rock layers that is steeply inclined from the horizontal must have been tilted after deposition and lithification.

Steno's third principle, the **principle of lateral continuity**, states that a layer of sediment extends laterally in all

directions until it thins and pinches out or terminates against the edge of the depositional basin (Figure 17-2).

James Hutton is credited with discovering the **principle of cross-cutting relationships**. Based on his detailed studies and observations of rock exposures in Scotland, Hutton recognized that an igneous intrusion or fault must be younger than the rocks it intrudes or displaces (➤ Figure 17-3).

Another way to determine relative ages is by using the **principle of inclusions**. This principle holds that inclusions, or fragments of one rock contained within a layer of another, are older than the rock layer itself. The batholith shown in ➤ Figure 17-4a contains sandstone inclusions, and the sandstone unit shows the effects of baking. Accordingly, we conclude that the sandstone is older than the batholith. In Figure 17-4b, however, the sandstone contains granite rock

---

➤ **FIGURE 17-3** The principle of cross-cutting relationships. (*a*) A dark-colored dike has been intruded into older light-colored granite, north shore of Lake Superior, Ontario, Canada. (*b*) A small fault displacing tilted beds along Templin Highway, Castaic, California.

(a)

(b)

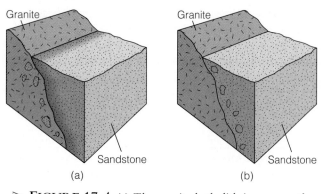

(a)

(b)

> **FIGURE 17-4** (*a*) The granite batholith is younger than the sandstone because the sandstone has been baked at its contact with the granite and the granite contains sandstone inclusions. (*b*) Granite inclusions in the sandstone indicate that the batholith was a source of the sandstone and therefore is older.

fragments, indicating that the batholith was the source rock for the inclusions and is therefore older than the sandstone.

Fossils have been known for centuries, yet their utility in relative dating and geologic mapping was not fully appreci-

ated until the early nineteenth century. William Smith (1769–1839), an English civil engineer involved in surveying and building canals in southern England, independently recognized the principle of superposition by reasoning that the fossils at the bottom of a sequence of strata are older than those at the top of the sequence. This recognition served as the basis for the **principle of fossil succession** or the *principle of faunal and floral succession* as it is sometimes called (> Figure 17-5). According to this principle, fossil assemblages succeed one another through time in a regular and predictable order.

Unconformities

Our discussion so far has been concerned with conformable sequences of strata, sequences in which no depositional breaks of any consequence occur. A sharp bedding plane (see Figure 7-11) separating strata may represent a depositional break of minutes, hours, years, or even tens of years, but it is inconsequential when considered in the context of geologic time.

Surfaces of discontinuity that encompass significant amounts of geologic time are **unconformities**, and any

> **FIGURE 17-5** This generalized diagram shows how William Smith used fossils to identify strata of the same age in different areas (principle of fossil succession). The composite section on the right shows the relative ages of all strata in this area.

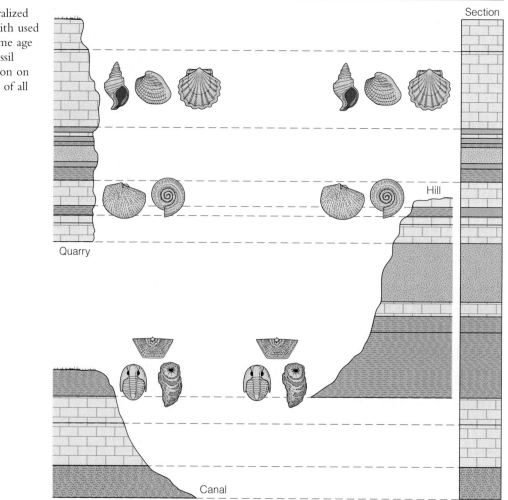

interval of geologic time not represented by strata in a particular area is a *hiatus* (▷ Figure 17-6). Thus, an unconformity is a surface of nondeposition or erosion that separates younger strata from older rocks. As such, it represents a break in our record of geologic time. The famous 12-minute gap in the Watergate tapes of Richard Nixon's presidency is somewhat analogous. Just as we have no record of the conversations that were occurring during this period of time, we have no record of the events that occurred during a hiatus.

Three types of unconformities are recognized. A **disconformity** is a surface of erosion or nondeposition between younger and older beds that are parallel with one another (▷ Figure 17-7). Unless a well-defined erosional surface separates the older from the younger parallel beds, the disconformity frequently resembles an ordinary bedding plane. Accordingly, many disconformities are difficult to recognize and must be identified on the basis of fossil assemblages.

An **angular unconformity** is an erosional surface on tilted or folded strata over which younger strata have been deposited (▷ Figure 17-8). Both younger and older strata may dip, but if their dip angles are different (generally the older strata dip more steeply), an angular unconformity is present.

The angular unconformity illustrated in Figure 17-8b is probably the most famous in the world. It was here at Siccar Point, Scotland, that James Hutton realized that severe upheavals had tilted the lower rocks and formed mountains that were then worn away and covered by younger, flat-lying rocks. The erosional surface between the older tilted rocks and the younger flat-lying strata meant that there was a significant gap in the rock record. Although Hutton did not use the term unconformity, he was the first to understand and explain the significance of such discontinuities in the geologic record.

The third type of unconformity is a **nonconformity**. Here an erosion surface cut into metamorphic or igneous rocks is covered by sedimentary rocks (▷ Figure 17-9). This type of unconformity closely resembles an intrusive igneous contact with sedimentary rocks. The principle of inclusions is helpful in determining whether the relationship between igneous rocks and overlying sedimentary rocks is the result of an intrusion or erosion. In the case of an intrusion, the igneous rocks are younger, but in the case of erosion, the sedimentary rocks are younger. Being able to distinguish between a nonconformity and an intrusive contact is very important because they represent different sequences of events.

⊕ CORRELATION

If geologists are to decipher Earth history, they must demonstrate the time equivalency of rock units in different areas. This process is known as **correlation**.

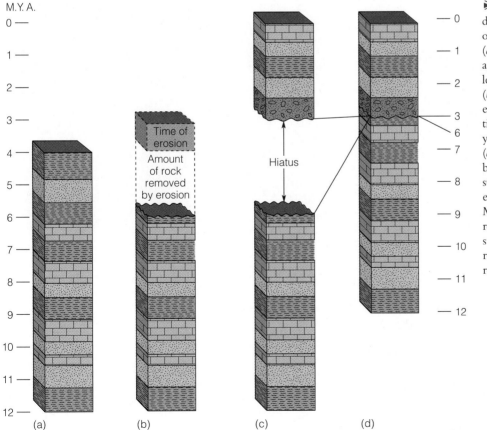

▷ **FIGURE 17-6** Simplified diagram showing the development of an unconformity and a hiatus. (*a*) Deposition began 12 million years ago (M.Y.A.) and continued more or less uninterrupted until 4 M.Y.A. (*b*) A 1-million-year episode of erosion occurred, and during that time rocks representing 2 million years of geologic time were eroded. (*c*) A hiatus of 3 million years exists between the older strata and the strata that formed during a renewed episode of deposition that began 3 M.Y.A. (*d*) The actual stratigraphic record. The unconformity is the surface separating the strata and represents a major break in our record of geologic time.

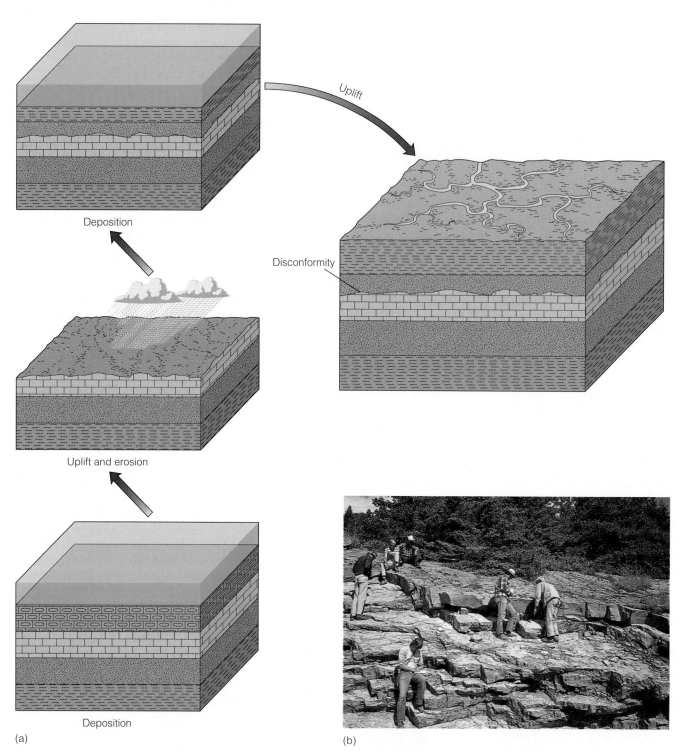

(a)   (b)

➤ FIGURE 17-7 (*a*) Formation of a disconformity. (*b*) Disconformity between Mississippian and Jurassic rocks in Montana. The geologist at the upper left is sitting on Jurassic rocks, and his right foot is resting upon Mississippian rocks.

If exposures are adequate, units may simply be traced laterally (principle of lateral continuity), even if occasional gaps exist (➤ Figure 17-10). Other criteria used to correlate units are similarity of rock type, position in a sequence, and *key beds*. Key beds are units, such as coal beds or volcanic ash layers, that are sufficiently distinctive to allow identification of the same unit in different areas (Figure 17-10). In addition to surface correlation, geologists frequently use well logs, cores, or cuttings to correlate subsurface rock units when exploring for minerals, coal, and petroleum.

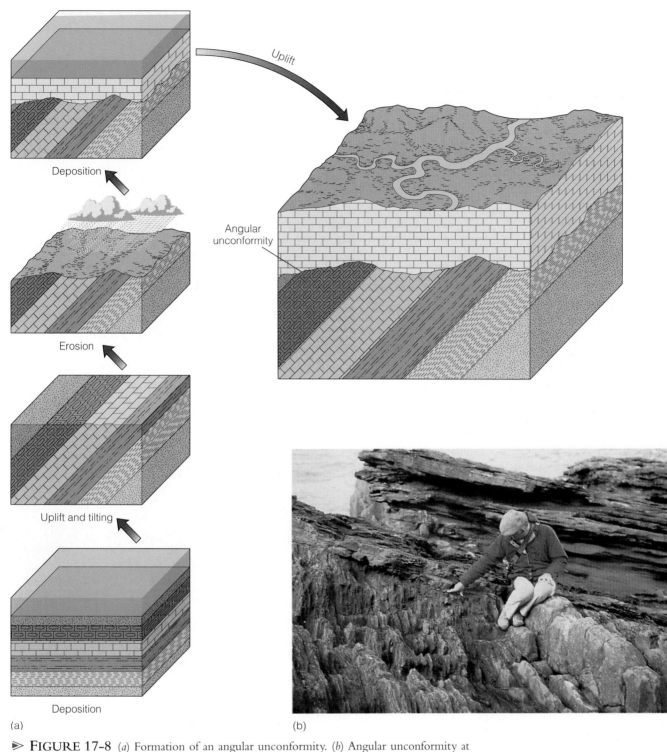

Uplift

Deposition

Erosion

Uplift and tilting

Deposition

Angular
unconformity

(a)

(b)

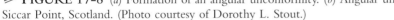

▷ FIGURE 17-8 (*a*) Formation of an angular unconformity. (*b*) Angular unconformity at Siccar Point, Scotland. (Photo courtesy of Dorothy L. Stout.)

Generally, no single location in a region has a geologic record of all the events that occurred during its history; therefore geologists must correlate from one area to another in order to decipher the complete geologic history of the region. An excellent example is the history of the Colorado Plateau. A record of events occurring over approximately 2 billion years is present in this region. Because of the forces of erosion, the entire record is not preserved at any single location. Within the walls of the Grand Canyon are rocks of the Precambrian and Paleozoic Era, while Paleozoic and

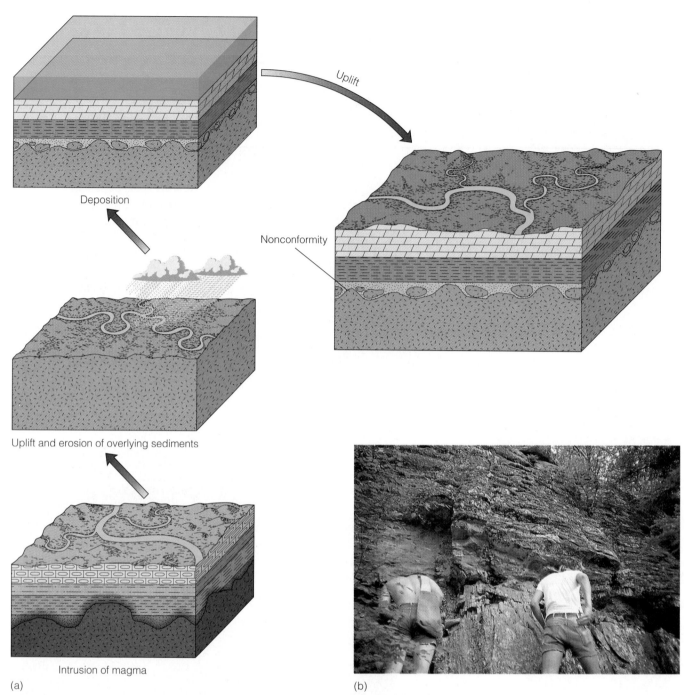

Deposition

Uplift

Nonconformity

Uplift and erosion of overlying sediments

Intrusion of magma

(a)

(b)

▷ **FIGURE 17-9** (*a*) Formation of a nonconformity. (*b*) Nonconformity between Precambrian granite and the overlying Cambrian-age Deadwood Formation, Wyoming.

Mesozoic Era rocks are found in Zion National Park, and Mesozoic and Cenozoic Era rocks are exposed in Bryce Canyon. By correlating the uppermost rocks at one location with the lowermost equivalent rocks of another area, the history of the entire region can be deciphered.

Although geologists can match up rocks on the basis of similar rock type and superposition, correlation of this type can only be done in a limited area where beds can be traced from one site to another. In order to correlate rock units over a large area or to correlate age-equivalent units of different composition, fossils and the principle of fossil succession must be used.

Fossils are useful as time indicators because they are the remains of organisms that lived for a certain length of time during the geologic past. Fossils that are easily identified, are geographically widespread, and existed for a rather short

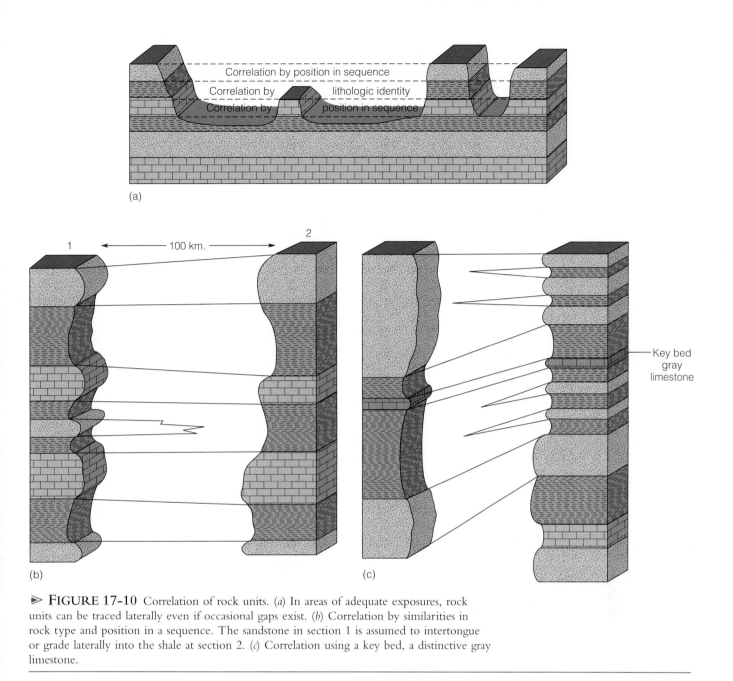

Correlation by position in sequence

Correlation by   lithologic identity

Correlation by   position in sequence

(a)

1 ← 100 km. → 2

(b)

(c)

Key bed
gray
limestone

➤ FIGURE 17-10 Correlation of rock units. (*a*) In areas of adequate exposures, rock units can be traced laterally even if occasional gaps exist. (*b*) Correlation by similarities in rock type and position in a sequence. The sandstone in section 1 is assumed to intertongue or grade laterally into the shale at section 2. (*c*) Correlation using a key bed, a distinctive gray limestone.

geologic time are particularly useful. Such fossils are called **guide fossils** or *index fossils* (➤ Figure 17-11). The trilobite *Isotelus* and the clam *Inoceramus* meet all of these criteria and are therefore good guide fossils. In contrast, the brachiopod *Lingula* is easily identified and widespread, but its geologic range of Ordovician to Recent makes it of little use in correlation.

⬡ ABSOLUTE DATING METHODS

Although most of the isotopes of the 92 naturally occurring elements are stable, some are radioactive and spontaneously decay to other more stable isotopes of elements, releasing energy in the process. The discovery, in 1903 by Pierre and

Marie Curie, that radioactive decay produces heat as a by-product meant that geologists had a mechanism for explaining the internal heat of the Earth that did not rely on residual cooling from a molten origin. Furthermore, geologists had a powerful tool to date geologic events accurately and verify the long time periods postulated by Hutton and Lyell.

Atoms, Elements, and Isotopes

As we discussed in Chapter 3, all matter is made up of chemical elements, each of which is composed of extremely small particles called *atoms*. The nucleus of an atom is composed of *protons* and *neutrons* with *electrons* encircling it

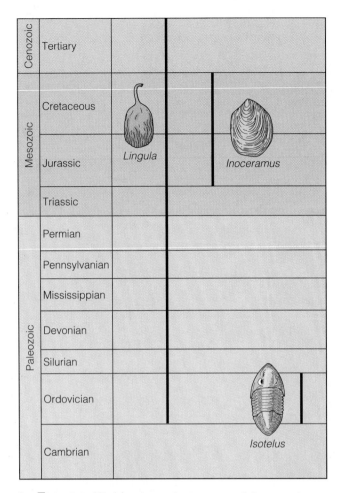

	Cenozoic	Tertiary
	Mesozoic	Cretaceous
		Jurassic
		Triassic
	Paleozoic	Permian
		Pennsylvanian
		Mississippian
		Devonian
		Silurian
		Ordovician
		Cambrian

Lingula

Inoceramus

Isotelus

▶ **FIGURE 17-11** The geologic ranges of three marine invertebrates. The brachiopod *Lingula* is of little use in correlation because of its long geologic range. The trilobite *Isotelus* and the bivalve *Inoceramus* are good guide fossils because they are geographically widespread, are easily identified, and have short geologic ranges.

(see Figure 3-2). The number of protons defines an element's *atomic number* and helps determine its properties and characteristics. The combined number of protons and neutrons in an atom is its *atomic mass number*. However, not all atoms of the same element have the same number of neutrons in their nuclei. These variable forms of the same element are called *isotopes* (see Figure 3-3). Most isotopes are stable, but some are unstable and spontaneously decay to a more stable form. It is the decay rate of unstable isotopes that geologists measure to determine the absolute age of rocks.

Radioactive Decay and Half-Lives

Radioactive decay is the process whereby an unstable atomic nucleus is spontaneously transformed into an atomic nucleus of a different element. Three types of radioactive decay are recognized, all of which result in a change of atomic structure (▶ Figure 17-12). In **alpha decay**, two protons and two neutrons are emitted from the nucleus,

resulting in a loss of two atomic numbers and four atomic mass numbers. In **beta decay**, a fast-moving electron is emitted from a neutron in the nucleus, changing that neutron to a proton and consequently increasing the atomic number by one, with no resultant atomic mass number change. **Electron capture** results when a proton captures an electron from an electron shell and thereby converts to a neutron, resulting in a loss of one atomic number and no change in the atomic mass number.

Some elements undergo only one decay step in the conversion from an unstable form to a stable form. For example, rubidium 87 decays to strontium 87 by a single beta emission, and potassium 40 decays to argon 40 by a single electron capture. Other radioactive elements undergo several decay steps (see Perspective 17-1). Uranium 235 decays to lead 207 by seven alpha and six beta steps, while uranium 238 decays to lead 206 by eight alpha and six beta steps (▶ Figure 17-13).

When discussing decay rates, it is convenient to refer to them in terms of half-lives. The **half-life** of a radioactive element is the time it takes for one-half of the atoms of the original unstable **parent element** to decay to atoms of a new, more stable **daughter element**. The half-life of a given radioactive element is constant and can be precisely measured in the laboratory. Half-lives of various radioactive elements range from less than a billionth of a second to 49 billion years.

Radioactive decay occurs at a geometric rate rather than a linear rate. Therefore, a graph of the decay rate produces a curve rather than a straight line (▶ Figure 17-14). For example, an element with *1,000,000* parent atoms will have *500,000* parent atoms and 500,000 daughter atoms after one half-life. After two half-lives, it will have *250,000* parent atoms (one-half of the previous parent atoms, which is equivalent to one-fourth of the original parent atoms) and 750,000 daughter atoms. After three half-lives, it will have *125,000* parent atoms (one-half of the previous parent atoms or one-eighth of the original parent atoms) and 875,000 daughter atoms, and so on until the number of parent atoms remaining is so few that they cannot be accurately measured by present-day instruments.

By measuring the parent-daughter ratio and knowing the half-life of the parent (determined in the laboratory), geologists can calculate the age of a sample containing the radioactive element. The parent-daughter ratio is usually determined by a *mass spectrometer,* an instrument that measures the proportions of elements of different masses.

Sources of Uncertainty. The most accurate radiometric dates are obtained from igneous rocks. As a magma cools and begins to crystallize, radioactive parent atoms are separated from previously formed daughter atoms. Because they are the right size, some radioactive parent atoms are incorporated into the crystal structure of certain minerals. The stable daughter atoms, however, are a different size than the radioactive parent atoms and consequently cannot fit into the crystal structure of the same mineral as the parent atoms.

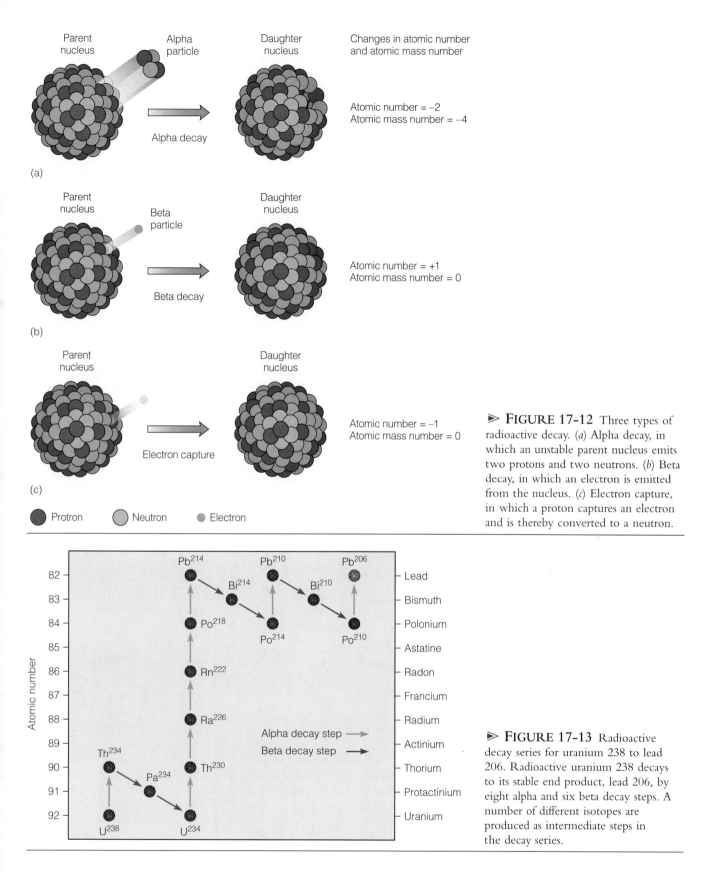

FIGURE 17-12 Three types of radioactive decay. (*a*) Alpha decay, in which an unstable parent nucleus emits two protons and two neutrons. (*b*) Beta decay, in which an electron is emitted from the nucleus. (*c*) Electron capture, in which a proton captures an electron and is thereby converted to a neutron.

FIGURE 17-13 Radioactive decay series for uranium 238 to lead 206. Radioactive uranium 238 decays to its stable end product, lead 206, by eight alpha and six beta decay steps. A number of different isotopes are produced as intermediate steps in the decay series.

Therefore a mineral crystallizing in a cooling magma will contain radioactive parent atoms but no stable daughter atoms (➤ Figure 17-15). Thus, the time that is being measured is the time of crystallization of the mineral containing the radioactive atoms, and not the time of formation of the radioactive atoms.

To obtain accurate radiometric dates, geologists must be sure that they are dealing with a closed system, meaning that

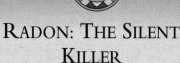

RADON: THE SILENT KILLER

Radon is a colorless, odorless, naturally occurring radioactive gas that has a 3.8-day half-life. It is part of the uranium 238–lead 206 radioactive decay series (Figure 17-13) and occurs in any rock or soil that contains uranium 238. Radon concentrations are reported in picocuries per liter (pCi/L) of air (a curie is the standard measure of radiation, and a picocurie is one-trillionth of a curie, or the equivalent of the decay of about two radioactive atoms per minute). Outdoors, radon escapes into the atmosphere where it is diluted and dissipates to harmless levels (0.2 pCi/L is the ambient outdoor level of radon). Radon levels for indoor air range from less than 1 pCi/L to about 3,000 pCi/L, but average about 1.5 pCi/L. The Environmental Protection Agency (EPA) considers radon levels exceeding 4 pCi/L to be unhealthy and recommends remedial action be taken to lower them. Continued exposure to elevated levels of radon over an extended period of time is thought by many researchers to increase the risk of lung cancer.

Radon is one of the natural decay products of uranium 238. It rapidly decays by the emission of an alpha particle, producing two short-lived radioactive isotopes—polonium 218 and polonium 214 (Figure 17-13). Both of these isotopes are solid and can become trapped in your lungs every time you breathe. When polonium decays, it emits alpha and beta particles that can damage lung cells and cause lung cancer.

Your chances of being adversely affected by radon depend on numerous interrelated factors such as geographic location, the geology of the area, the climate, how buildings are constructed, and the amount of time spent in them. Because radon is a naturally occurring gas, contact with it is unavoidable, but atmospheric concentrations of it are probably harmless. Only when concentrations of radon build up in poorly ventilated structures does it become a potential health risk.

Concern about the health risks posed by radon first arose during the 1960s when the news media revealed that some homes in the West had been built with uranium mine tailings. Since then, geologists have found that high indoor radon levels can be caused by natural uranium in minerals of the rock and soil on which buildings are constructed. In response to the high cost of energy during the 1970s and 1980s, old buildings were insulated, and new buildings were constructed to be as energy efficient and airtight as possible. Ironically, these energy-saving measures also sealed in radon.

Radon enters buildings through dirt floors, cracks in the floor or walls, joints between floors and walls, floor drains, sumps, and utility pipes as well as any cracks or pores in hollow block walls (▷ Figure 1). Radon can also be released into a building whenever the water is turned on, particularly if the water comes from a private well. Municipal water is generally safe because it has usually been aerated before it gets to your home.

To find out if your home has a radon problem, you must test for it with commercially available, relatively inexpensive, simple home testing devices. If radon readings are above the recommended EPA levels of 4 pCi/L, several remedial measures can be taken to reduce your risk. These include sealing up all cracks in the foundation, pouring a concrete slab over a dirt floor, increasing the circulation of air throughout the house, especially in the basement and crawl space, providing filters for drains and other utility openings, and limiting time spent in areas with higher concentrations of radon.

It is important to remember that although the radon hazard covers most of the country, some areas are more likely to have higher natural concentrations of radon than others (▷ Figure 2). Such rocks as uranium-bearing granites, metamorphic rocks of granitic composition, and black shales (high carbon content) are quite likely to cause indoor

▷ **FIGURE 1** Some of the common points where radon can enter a house.

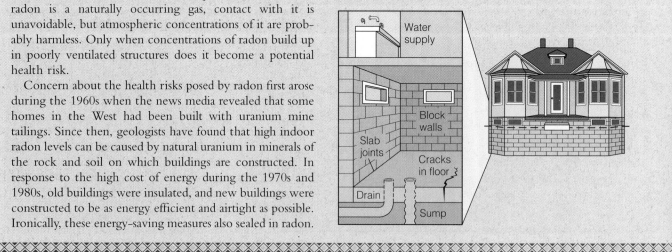

neither parent nor daughter atoms have been added or removed from the system since crystallization and that the ratio between them results only from radioactive decay. Otherwise, an inaccurate date will result. If daughter atoms have leaked out of the mineral being analyzed, the calculated age will be too young; if parent atoms have been removed, the calculated age will be too great.

Leakage may occur if the rock is heated as occurs during

radon problems. Other rocks such as marine quartz sandstone, noncarbonaceous shales and siltstones, most volcanic rocks, and igneous and metamorphic rocks rich in iron and magnesium typically do not cause radon problems. The permeability of the soil overlying the rock can also affect the indoor levels of radon gas. Some soils are more permeable than others and allow more radon to escape into the overlying structures.

The climate and type of construction affect not only how much radon enters a structure, but how much escapes. Concentrations of radon are highest during the winter in northern climates because buildings are sealed as tightly as possible. Homes with basements are likely to have higher radon levels than those built on concrete slabs. While research continues into the sources of indoor radon and ways of controlling it, the most important thing people can do is to test their home, school, or business for radon.

There is currently a heated debate among scientists concerning the large-scale health hazards resulting from radon exposure and how much money should be spent for its remediation. On the one hand, the EPA and Surgeon General estimate that exposure to high levels of indoor radon cause between 5,000 and 20,000 lung cancer deaths each year. Other scientists dispute these figures because of the difficulty of attributing mortality rates for lung cancer directly to radon, particularly when so many other factors, such as smoking, are involved. Central to this debate are two questions: What concentration levels of indoor radon are acceptable, and exactly how serious is the risk from radon at those levels? Unfortunately, the data for making these determinations are simply not available at this time.

➤ **FIGURE 2** Areas in the United States where granite, phosphate-bearing rocks, carbonaceous shales, and uranium occur. These rocks are all potential sources of radon gas.

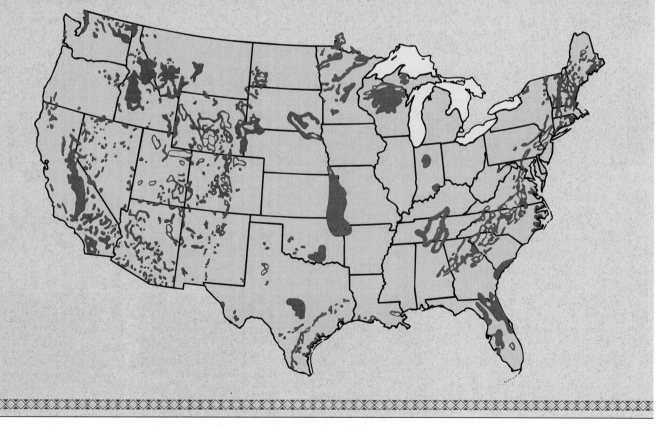

metamorphism. If this happens, some of the parent or daughter atoms may be driven from the mineral being analyzed, resulting in an inaccurate age determination. If the daughter product was completely removed, then one would

be measuring the time since metamorphism (a useful measurement itself), and not the time since crystallization of the mineral (➤ Figure 17-16). Because heat affects the parent-daughter ratio, metamorphic rocks are difficult to age-date

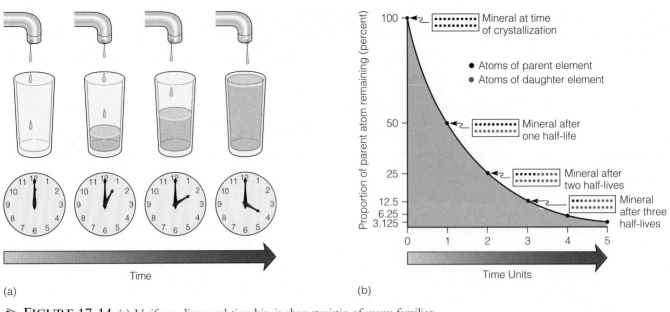

(a)   (b)

➤ **FIGURE 17-14** (*a*) Uniform, linear relationship is characteristic of many familiar processes. (*b*) Geometric radioactive decay curve, in which each time unit represents one half-life, and each half-life is the time it takes for one-half of the parent element to decay to the daughter element.

accurately. Remember that while the parent-daughter ratio may be affected by heat, the decay rate of the parent element remains constant, regardless of any physical or chemical changes.

Long-Lived Radioactive Isotope Pairs

◉ Table 17-1 shows the five common, long-lived parent-daughter isotope pairs used in radiometric dating. Long-lived pairs have half-lives of millions or billions of years. All of these were present when the Earth formed and are still present in measurable quantities. Other shorter-lived radioactive isotope pairs have decayed to the point that only small quantities near the limit of detection remain.

The most commonly used isotope pairs are the uranium-lead and thorium-lead series, which are used principally to date ancient igneous intrusives, lunar samples, and some meteorites. The rubidium-strontium pair is also used for very old samples and has been effective in dating the oldest rocks on Earth as well as meteorites. The potassium-argon method is typically used for dating fine-grained volcanic rocks from which individual crystals cannot be separated; hence the whole rock is analyzed. Argon is a gas, however, so great care must be taken to assure that the sample has not been subjected to heat, which would allow argon to escape; such a sample would yield an age that is too young. Other long-lived radioactive isotope pairs exist, but they are rather rare and are used only in special situations.

➤ **FIGURE 17-15** (*a*) A magma contains both radioactive and stable atoms. (*b*) As the magma cools and begins to crystallize, some radioactive atoms are incorporated into certain minerals because they are the right size and can fit into the crystal structure. Therefore, at the time of crystallization, the mineral will contain 100% radioactive parent atoms and 0% stable daughter atoms. (*c*) After one half-life, 50% of the radioactive parent atoms will have decayed to stable daughter atoms.

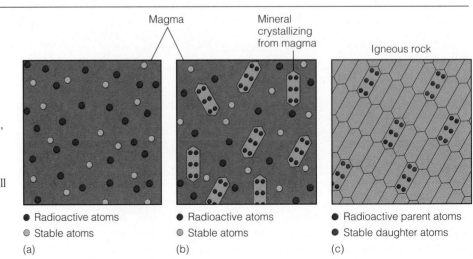

● Radioactive atoms
● Stable atoms
(a)

● Radioactive atoms
● Stable atoms
(b)

● Radioactive parent atoms
● Stable daughter atoms
(c)

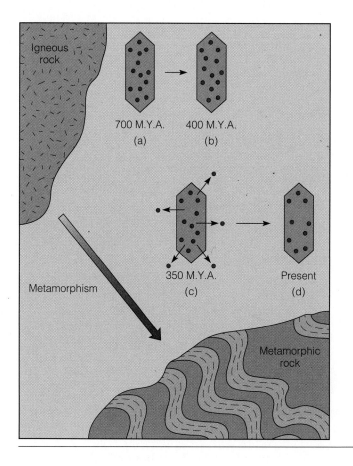

► FIGURE 17-16 The effect of metamorphism in driving out daughter atoms from a mineral that crystallized 700 million years ago (M.Y.A.). The mineral is shown immediately after crystallization (*a*), then at 400 million years (*b*), when some of the parent atoms had decayed to daughter atoms. Metamorphism at 350 M.Y.A. (*c*) drives the daughter atoms out of the mineral into the surrounding rock. (*d*) Assuming the rock has remained a closed chemical system throughtout its history, dating the mineral today yields the time of metamorphism, while dating the whole rock provides the time of its crystallization, 700 M.Y.A.

Carbon 14 Dating Method

Carbon is an important element in nature and is one of the basic elements found in all forms of life. It has three isotopes; two of these, carbon 12 and 13, are stable, whereas carbon 14 is radioactive (see Figure 3-3). Carbon 14 has a half-life of 5,730 years plus or minus 30 years. The **carbon 14 dating technique** is based on the ratio of carbon 14 to carbon 12 and is generally used to date once-living material.

The short half-life of carbon 14 makes this dating technique practical only for specimens younger than about 70,000 years. Consequently, the carbon 14 dating method is

TABLE 17-1 Five of the Principal Long-Lived Radioactive Isotope Pairs Used in Radiometric Dating

Isotopes		Half-Life of Parent (Years)	Effective Dating Range (Years)	Minerals and Rocks That Can Be Dated	
Parent	*Daughter*				
Uranium 238	Lead 206	4.5 billion	10 million to 4.6 billion	Zircon Uraninite	
Uranium 235	Lead 207	704 million			
Thorium 232	Lead 208	14 billion			
Rubidium 87	Strontium 87	48.8 billion	10 million to 4.6 billion	Muscovite Biotite Potassium feldspar Whole metamorphic or igneous rock	
Potassium 40	Argon 40	1.3 billion	100,000 to 4.6 billion	Glauconite Muscovite Biotite	Hornblende Whole volcanic rock

especially useful in archaeology and has greatly aided in unraveling the events of the latter portion of the Pleistocene Epoch.

Carbon 14 is constantly formed when cosmic rays, which are high-energy particles (mostly protons), strike the atoms of upper-atmospheric gases, splitting their nuclei into protons and neutrons. When a neutron strikes the nucleus of a nitrogen atom (atomic number 7, atomic mass number 14), it may be absorbed into the nucleus and a proton emitted. Thus, the atomic number of the atom decreases by one, while the atomic mass number stays the same. Because the atomic number has changed, a new element, carbon 14 (atomic number 6, atomic mass number 14), is formed. The newly formed carbon 14 is rapidly assimilated into the carbon cycle and, along with carbon 12 and 13, is absorbed in a nearly constant ratio by all living organisms ($\triangleright$ Figure 17-17). When an organism dies, carbon 14 is not replenished, and the ratio of carbon 14 to carbon 12 decreases as carbon 14 decays back to nitrogen by a single beta decay step (Figure 17-17).

Currently, the ratio of carbon 14 to carbon 12 is remarkably constant in both the atmosphere and living organisms. There is good evidence, however, that the production of carbon 14, and thus the ratio of carbon 14 to carbon 12, has varied somewhat over the past several thousand years. This was determined by comparing ages established by carbon 14 dating of wood samples against those established by counting annual tree-rings in the same samples. As a result, carbon 14 ages have been corrected to reflect such variations in the past.

DEVELOPMENT OF THE GEOLOGIC TIME SCALE

The geologic time scale is a hierarchical scale in which the 4.6-billion-year history of the Earth is divided into time units of varying duration (Figure 17-1). It was not developed by any one individual, but rather evolved, primarily during the nineteenth century, through the efforts of many people. By applying relative dating methods to rock outcrops, geologists in England and western Europe defined the major geologic time units without the benefit of radiometric dating techniques. Using the principles of superposition and fossil succession, they were able to correlate various rock exposures and piece together a composite geologic section. This composite section is, in effect, a relative time scale because the rocks are arranged in their correct sequential order.

By the beginning of the twentieth century, geologists had developed a relative geologic time scale, but did not yet have any absolute dates for the various time unit boundaries. Following the discovery of radioactivity near the end of the last century, radiometric dates were added to the relative geologic time scale.

Because sedimentary rocks, with rare exceptions, cannot be radiometrically dated, geologists have had to rely on interbedded volcanic rocks and igneous intrusions to apply

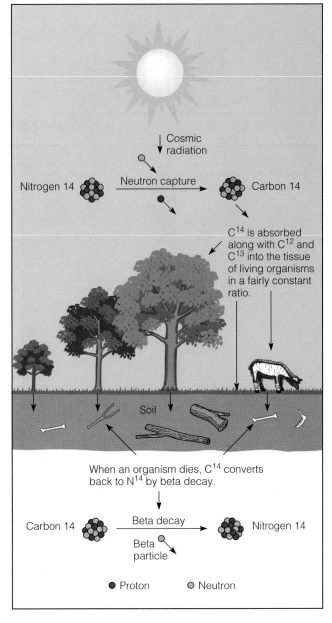

$\triangleright$ **FIGURE 17-17** The carbon cycle showing the formation, dispersal, and decay of carbon 14.

absolute dates to the boundaries of the various subdivisions of the geologic time scale ($\triangleright$ Figure 17-18). An ash fall or lava flow provides an excellent marker bed that is a time-equivalent surface, supplying a minimum age for the sedimentary rocks below and a maximum age for the rocks above. Ash falls are particularly useful because they may fall over both marine and nonmarine sedimentary environments and can provide a connection between these different environments.

Thousands of absolute ages are now known for sedimentary rocks of known relative ages, and these absolute dates have been added to the relative time scale. In this way, geologists have been able to determine the absolute ages of the various geologic periods and to determine their durations (Figure 17-1).

➤ FIGURE 17-18 Absolute ages of sedimentary rocks can be determined by dating associated igneous rocks. In (a) and (b), sedimentary rocks are bracketed by rock bodies for which absolute ages have been determined.

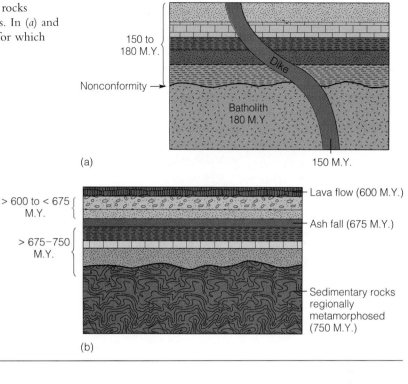

150 to
180 M.Y.

Nonconformity →

Dike

Batholith
180 M.Y.

(a)

150 M.Y.

> 600 to < 675
M.Y.

> 675–750
M.Y.

Lava flow (600 M.Y.)

Ash fall (675 M.Y.)

Sedimentary rocks
regionally
metamorphosed
(750 M.Y.)

(b)

CHAPTER SUMMARY

1. Relative dating involves placing geologic events in a sequential order as determined from their position in the geologic record. Absolute dating results in specific dates for events, expressed in years before the present.

2. During the eighteenth and nineteenth centuries, attempts were made to determine the age of the Earth based on scientific evidence rather than revelation. Though some attempts were quite ingenious, they yielded a variety of ages that now are known to be much too young.

3. James Hutton thought that present-day processes operating over long periods of time could explain all the geologic features of the Earth. His observations were instrumental in establishing the basis for the principle of uniformitarianism.

4. Uniformitarianism, as articulated by Charles Lyell, soon became the guiding principle of geology. It holds that the laws of nature have been constant through time and that the same processes operating today have operated in the past, although not necessarily at the same rates.

5. In addition to uniformitarianism, the principles of superposition, original horizontality, lateral continuity, cross-cutting relationships, inclusions, and fossil succession are basic for determining relative geologic ages and for interpreting the geologic history of the Earth.

6. Surfaces of discontinuity that encompass significant amounts of geologic time are common in the geologic record. Such surfaces are unconformities and result from times of nondeposition, erosion, or both.

7. Correlation is the practice of demonstrating equivalency of units in different areas. Time equivalence is most commonly demonstrated by correlating strata containing similar fossils.

8. Radioactivity was discovered during the late nineteenth century, and soon thereafter radiometric dating techniques allowed geologists to determine absolute ages for geologic events.

9. Absolute age dates for rock samples are usually obtained by determining how many half-lives of a radioactive parent element have elapsed since the sample originally crystallized. A half-life is the time it takes for one-half of the radioactive parent element to decay to a stable daughter element.

10. The most accurate radiometric dates are obtained from long-lived radioactive isotope pairs in igneous rocks.

11. Carbon 14 dating can be used only for organic matter such as wood, bones, and shells and is effective back to about 70,000 years ago. Unlike the long-lived isotopic pairs, the carbon 14 dating technique determines age by the ratio of radioactive carbon 14 to stable carbon 12.

12. Through the efforts of many geologists applying the principles of relative dating, a relative geologic time scale was established.

13. Most absolute ages of sedimentary rocks and their contained fossils are obtained indirectly by dating igneous rocks.

IMPORTANT TERMS

absolute dating	electron capture	principle of fossil succession	principle of superposition
alpha decay	guide fossil	principle of inclusions	principle of
angular unconformity	half-life	principle of lateral	uniformitarianism
beta decay	nonconformity	continuity	radioactive decay
carbon 14 dating technique	parent element	principle of original	relative dating
correlation	principle of cross-cutting	horizontality	unconformity
daughter element	relationships		
disconformity			

REVIEW QUESTIONS

1. In which type of unconformity are the beds parallel to each other?
   a. ____ nonconformity;
   b. ____ angular unconformity;
   c. ____ disconformity;
   d. ____ hiatus;
   e. ____ none of these.

2. Placing geologic events in sequential order as determined by their position in the geologic record is called:
   a. ____ absolute dating;
   b. ____ uniformitarianism;
   c. ____ relative dating;
   d. ____ correlation;
   e. ____ historical dating.

3. If a rock is heated during metamorphism and the daughter atoms migrate out of a mineral that is subsequently radiometrically dated, an inaccurate date will be obtained. This date will be _____ the actual date.
   a. ____ younger than;
   b. ____ older than;
   c. ____ the same as;
   d. ____ it cannot be determined;
   e. ____ none of these.

4. Which of the following methods can be used to demonstrate age equivalency of rock units?
   a. ____ lateral tracing;
   b. ____ radiometric dating;
   c. ____ guide fossils;
   d. ____ position in a sequence;
   e. ____ all of these.

5. Which fundamental geologic principle states that the oldest layer is on the bottom of a succession of sedimentary rocks and the youngest is on top?
   a. ____ lateral continuity;
   b. ____ fossil succession;
   c. ____ original horizontality;
   d. ____ superposition;
   e. ____ cross-cutting relationships.

6. In which type of radioactive decay are two protons and two neutrons emitted from the nucleus?
   a. ____ alpha;
   b. ____ beta;
   c. ____ electron capture;
   d. ____ fission track;
   e. ____ radiocarbon.

7. Which of the following is not a long-lived radioactive isotope pair?
   a. ____ uranium–lead;
   b. ____ thorium–lead;
   c. ____ potassium–argon;
   d. ____ carbon–nitrogen;
   e. ____ none of these.

8. What is being measured in radiometric dating?
   a. ____ the time when a radioactive isotope formed;
   b. ____ the time of crystallization of a mineral containing an isotope;
   c. ____ the amount of the parent isotope only;
   d. ____ when the dated mineral became part of a sedimentary rock;
   e. ____ when the stable daughter isotope was formed.

9. If a radioactive element has a half-life of 4 million years, the amount of parent material remaining after 12 million years of decay will be what fraction of the original amount?
   a. ____ $\frac{1}{32}$;
   b. ____ $\frac{1}{16}$;
   c. ____ $\frac{1}{8}$;
   d. ____ $\frac{1}{4}$;
   e. ____ $\frac{1}{2}$.

10. In carbon 14 dating, which ratio is being measured?
    a. ____ the parent to daughter isotope;
    b. ____ C^{14}/N^{14};
    c. ____ C^{12}/C^{13};
    d. ____ C^{12}/N^{14};
    e. ____ C^{12}/C^{14}.

11. How many half-lives are required to yield a mineral with 625 atoms of U^{238} and 19,375 atoms of Pb^{206}?
    a. ____ 4;
    b. ____ 5;
    c. ____ 6;
    d. ____ 8;
    e. ____ 10.

12. What is the difference between relative and absolute dating of geologic events?

13. Explain how a geologist would determine the relative ages of a granite batholith and an overlying sandstone formation.

14. Why is the principle of uniformitarianism important to geologists?

15. Are volcanic eruptions, earthquakes, and storm deposits geologic events encompassed by uniformitarianism?

16. Assume a hypothetical radioactive isotope with an atomic number of 150 and an atomic mass number of 300 emits five alpha decay particles and three beta decay particles and undergoes two electron capture steps. What is the atomic number of

the resulting stable daughter product? The atomic mass number?

17. Why is it difficult to date sedimentary and metamorphic rocks radiometrically?

18. How does the carbon 14 dating technique differ from the uranium-lead dating method?

POINTS TO PONDER

1. The controversies surrounding the health hazards posed by radon and asbestos (see Perspective 8-1) are similar in that massive amounts of money are being spent to remedy what may not be as serious an environmental hazard as policy makers have been led to believe. Using these two highly publicized environmental hazards as examples, discuss how you might objectively assess their potential danger to the public and how you would determine what constitutes an acceptable risk for them.

2. A zircon mineral in an igneous rock contains both uranium 235 and uranium 238. In measuring the parent-daughter ratio for both uranium isotopes in the zircon, it was discovered that uranium 238 had undergone 2 half-lives in decaying to lead 206, while uranium 235 had undergone 10 half-lives during its decay to lead 207. Using Table 17-1, determine the age of the igneous rock based on each long-lived radioactive isotope pair. Are they the same age? If not, why? What explanations can you give for the difference in geologic age between the two samples?

Chapter 18

EARTH HISTORY

OUTLINE

The Late Proterozoic
Jacobsville Sandstone
exposed along the shore
of Lake Superior at
Marquette, Michigan.

Imagine a barren, lifeless, waterless, hot planet with a poisonous atmosphere. Volcanoes erupt nearly continuously, meteorites and comets flash through the atmosphere, and cosmic radiation is intense. The planet's crust, which is composed entirely of dark-colored igneous rock, is thin and unstable. Storms form in the turbulent atmosphere, and lightning discharges are common, but no rain falls. And because the atmosphere contains no oxygen, nothing burns. Rivers and pools of molten rock emit a continuous reddish glow.

This may sound like a science fiction novel, but it is probably a reasonably accurate description of the Earth shortly after it formed (▷ Figure 18-1). We emphasize "probably" because no record exists for the earliest chapter of Earth history, the interval from 4.6 to 3.8 billion years ago, although one area of rocks 3.96 billion years old is now known in Canada. We can only speculate about what the Earth was like during this time, based on our knowledge of how planets form and our information about other Earth-like planets.

When the Earth formed, it had a tremendous reservoir of primordial heat. This heat was generated by colliding particles as the Earth accreted, by gravitational compression, and by the decay of short-lived radioactive elements. Many geologists think that the early Earth was so hot that it was partly or perhaps almost entirely molten. No one knows what the Earth's surface temperature was during its earliest history, but it was almost certainly too hot for liq-

uid water to exist or for any known organism to survive. Volcanism must have been ubiquitous and nearly continuous. Molten rock later solidified to form a thin, discontinuous, dark-colored crust, only to be disrupted by upwelling magmas. Assuming that visitors to the early Earth could tolerate the high temperatures, they would also have had to contend with other factors. The atmosphere would be unbreathable by any of today's inhabitants. It probably contained considerable carbon dioxide and water vapor, but little or no oxygen. No ozone layer existed in the upper atmosphere, so our hypothetical visitors, unless protected, would receive a lethal dose of ultraviolet radiation and would be threatened constantly by comet and meteorite impacts. Their view of the Moon would have been spectacular because it was much closer to the Earth. Its gravitational attraction, however, would have caused massive Earth tides. And finally, our visitors would experience a much shorter day because the Earth rotated on its axis in as little as 10 hours.

Eventually, much of the Earth's primordial heat was dissipated into space, and its surface cooled. As the Earth cooled, water vapor began to condense, rain fell, and surface water began to accumulate. The bombardment by comets and meteorites slowed. By 3.8 billion years ago, a few small areas of continental crust existed. The atmosphere still lacked oxygen and an ozone layer, but by as much as 3.5 billion years ago, life appeared.

▷ FIGURE 18-1 The Earth as it is thought to have appeared about 4.6 billion years ago.

INTRODUCTION

In this chapter we will examine the geologic history of the Earth beginning with the origin of continents sometime during the Archean Eon. It was during the Archean and Proterozoic eons that the earliest continents and ocean basins formed and plate movement began. At the beginning of the Phanerozoic Eon, six major continental landmasses existed, four of which straddled the paleoequator. Plate movements during the Phanerozoic created a changing panorama of continents and ocean basins whose positions affected atmospheric and oceanic circulation patterns and created new environments for habitation by the rapidly evolving biota.

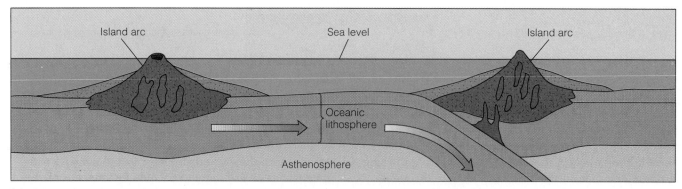

(a)

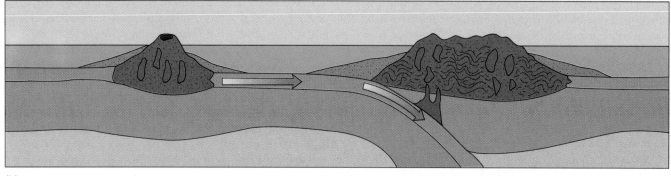

(b)

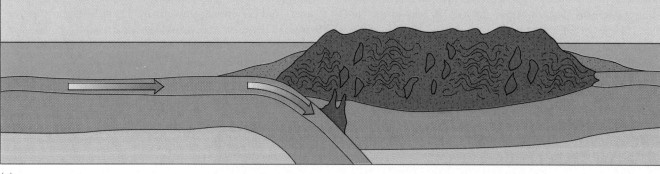

(c)

➤ FIGURE 18-2 Three stages in the origin of granitic continental crust. Andesitic island arcs formed by the partial melting of basaltic oceanic crust are intruded by granitic magmas. As a result of plate movements, island arcs collide and form larger units or cratons. (*a*) Two island arcs on separate plates move toward one another. (*b*) The island arcs shown in (*a*) collide, forming a small craton, and another island arc approaches this craton. (*c*) The island arc shown in (*b*) collides with the craton.

THE ORIGIN AND EVOLUTION OF CONTINENTS

Rocks, 3.8 billion years old, thought to represent continental crust are known from several areas, including Minnesota, Greenland, and South Africa. Most geologists agree that even older continental crust probably existed, and, in fact, rocks dated 3.96 billion years were recently discovered in Canada.

According to one model for the origin of continents, the earliest crust was thin, unstable, and composed of ultramafic igneous rock. This early ultramafic crust was disrupted by upwelling basaltic magmas at ridges and was consumed at subduction zones. Ultramafic crust would have been destroyed because it was dense enough to make recycling by subduction very likely. Apparently, only crust of a more granitic composition, which has a lower density, is resistant to destruction by subduction.

A second stage in crustal evolution began when partial melting of earlier-formed basaltic crust resulted in the formation of andesitic island arcs, and partial melting of lower crustal andesites yielded granitic magmas that were emplaced in the earlier-formed crust. As plutons were emplaced in these island arcs, they became more like continental crust. By 3.96 to 3.8 billion years ago, plate motions accompanied by subduction and collisions of island arcs had formed several granitic continental nuclei (▷ Figure 18-2).

Shields, Cratons, and the Evolution of Continents

Each continent is characterized by one or more areas of exposed ancient rocks called a *shield* (see Figure 8-2). Extending outward from these shields are broad platforms of ancient rocks buried beneath younger sediments and sedimentary rocks. The shields and buried platforms are collectively called **cratons**, so shields are simply the exposed parts of cratons. Cratons are considered to be the stable interior parts of continents.

In North America, the *Canadian Shield* includes much of Canada; a large part of Greenland; parts of the Lake Superior region in Minnesota, Wisconsin, and Michigan; and parts of the Adirondack Mountains of New York (▷ Figure 18-3). In general, the Canadian Shield is a vast area of subdued topography, numerous lakes, and exposed ancient metamorphic, volcanic, plutonic, and sedimentary rocks.

Each continent evolved by accretion along the margins of ancient cratons. To this extent, all continents developed similarly, but the details of each continent's history differ. Several cratons that would eventually become part of the North American craton had formed by 2.5 billion years ago, but these were independent minicontinents that were later assembled into a larger craton. The Slave, Hearn, and Superior cratons in ▷ Figure 18-4, for instance, were separate landmasses until a larger craton formed.

A major episode in the Precambrian evolution of North America took place between 2.0 and 1.8 billion years ago when several major *orogens* developed. Orogens are zones of complexly deformed rocks, many of which have been metamorphosed and intruded by plutons and thus represent areas of ancient mountain building. Smaller cratons were sutured along these orogenic belts, so by 1.8 billion years ago, much of what is now Greenland, central Canada, and the north-central United States formed a large craton (Figure 18-4a).

Following this initial stage of North America's evolution, continental accretion occurred along the southern and eastern margins of the craton (Figures 18-4b and c). By about 1 billion years ago, the size and shape of North America were approximately as shown in Figure 18-4c. No further episode of continental accretion occurred until the Paleozoic.

✶ PALEOZOIC PALEOGEOGRAPHY

By the beginning of the Paleozoic Era, six major continents were present. In addition to these large landmasses, geolo-gists have also identified numerous small microcontinents and island arcs associated with various microplates that were present during the Paleozoic. The six major Paleozoic continents are *Baltica* (Russia west of the Ural Mountains and the major part of northern Europe), *China* (a complex area consisting of at least three Paleozoic continents that were not widely separated and are here considered to include China, Indochina, and the Malay Peninsula), *Gondwana* (Africa, Antarctica, Australia, Florida, India, Madagascar, and parts of the Middle East and southern Europe), *Kazakhstania* (a triangular continent centered on Kazakhstan, but considered by some to be an extension of the Paleozoic Siberian continent), *Laurentia* (most of present North America, Greenland, northwestern Ireland, Scotland, and part of eastern Russia), and *Siberia* (Russia east of the Ural Mountains and Asia north of Kazakhstan and south of Mongolia).

In contrast to today's global geography, the Cambrian world consisted of these six continents dispersed around the globe at low tropical latitudes (▷ Figure 18-5a). Water circulated freely among ocean basins, and the polar regions were apparently ice-free. By the Late Cambrian, shallow seas had covered large areas of Laurentia, Baltica, Siberia, Kazakhstania, and China, while major highlands were present in northeastern Gondwana, eastern Siberia, and central Kazakhstania.

During the Ordovician, Gondwana moved southward and began to cross the South Pole as indicated by Upper Ordovician glacial deposits found today in the Sahara (Figure 18-5b). In contrast to the passive continental margin Laurentia exhibited during the Cambrian, an active convergent plate boundary formed along its eastern margin during the Ordovician as indicated by the Late Ordovician *Taconic orogeny* that occurred in New England. During the Silurian, Baltica moved northwestward relative to Laurentia and collided with it to form the larger continent of *Laurasia*. This collision, which closed the northern *Iapetus Ocean*, is marked by the *Caledonian orogeny*. Following this orogeny, the southern part of the Iapetus Ocean still remained open between Laurentia and Gondwana. Siberia and Kazakhstania moved from a southern equatorial position during the Cambrian to north temperate latitudes by the end of the Silurian Period.

During the Devonian, as the southern Iapetus Ocean narrowed between Laurasia and Gondwana, mountain building continued along the eastern margin of Laurasia with the *Acadian orogeny* (▷ Figure 18-6a). The erosion of the resulting highlands provided vast amounts of reddish fluvial sediments that covered large areas of northern Europe and eastern North America. Other Devonian tectonic events, probably related to the collision of Laurentia and Baltica, include the Cordilleran *Antler orogeny* and the change from a passive continental margin to an active convergent plate boundary in the Uralian mobile belt of eastern Baltica. The distribution of reefs, evaporites, and red beds, as well as the existence of similar floras throughout the world, suggests a rather uniform global climate during the Devonian Period.

During the Carboniferous Period, southern Gondwana

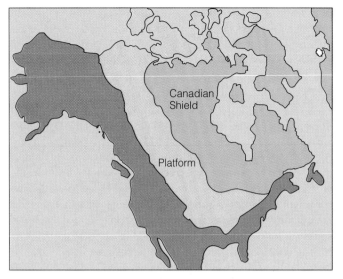

► **FIGURE 18-3 (above)** The North American craton. The Canadian Shield is a large area of exposed Precambrian-aged rocks. Extending from the shield are platforms of buried Precambrian rocks. The shield and platforms collectively make up the craton.

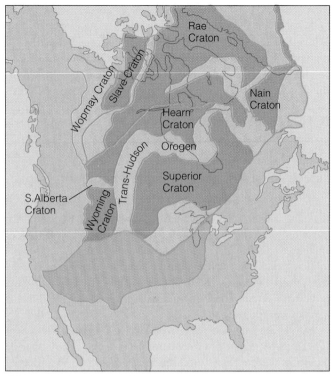

(b)

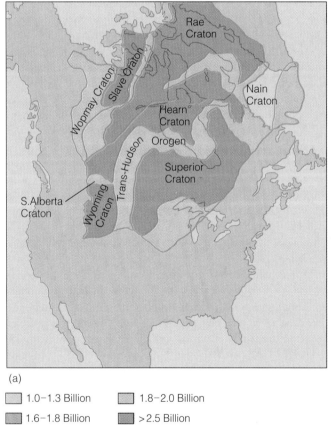

(a)

▢ 1.0–1.3 Billion	▢ 1.8–2.0 Billion
▢ 1.6–1.8 Billion	▢ >2.5 Billion

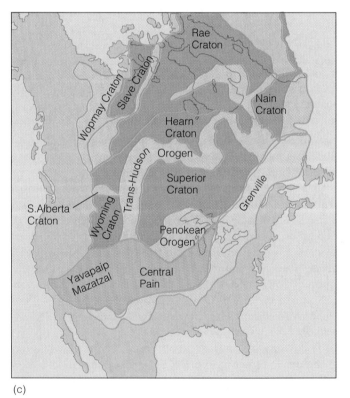

(c)

► **FIGURE 18-4** Three stages in the early evolution of the North American craton. (*a*) By about 1.8 billion years ago, North America consisted of the elements shown here. The various orogens formed when older cratons collided to form a larger craton. (*b*) and (*c*) Continental accretion along the southern and eastern margins of North America. By about 1 billion years ago, North America had the size and shape shown diagrammatically in (*c*).

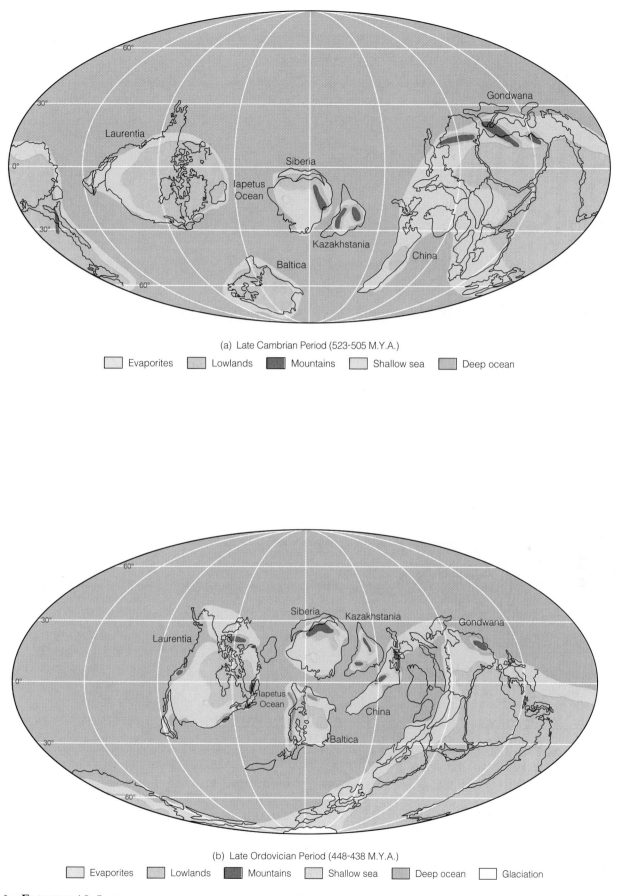

(a) Late Cambrian Period (523-505 M.Y.A.)

☐ Evaporites   ☐ Lowlands   ■ Mountains   ☐ Shallow sea   ☐ Deep ocean

(b) Late Ordovician Period (448-438 M.Y.A.)

☐ Evaporites   ☐ Lowlands   ■ Mountains   ☐ Shallow sea   ☐ Deep ocean   ☐ Glaciation

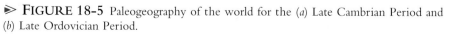

➤ FIGURE 18-5 Paleogeography of the world for the (*a*) Late Cambrian Period and
(*b*) Late Ordovician Period.

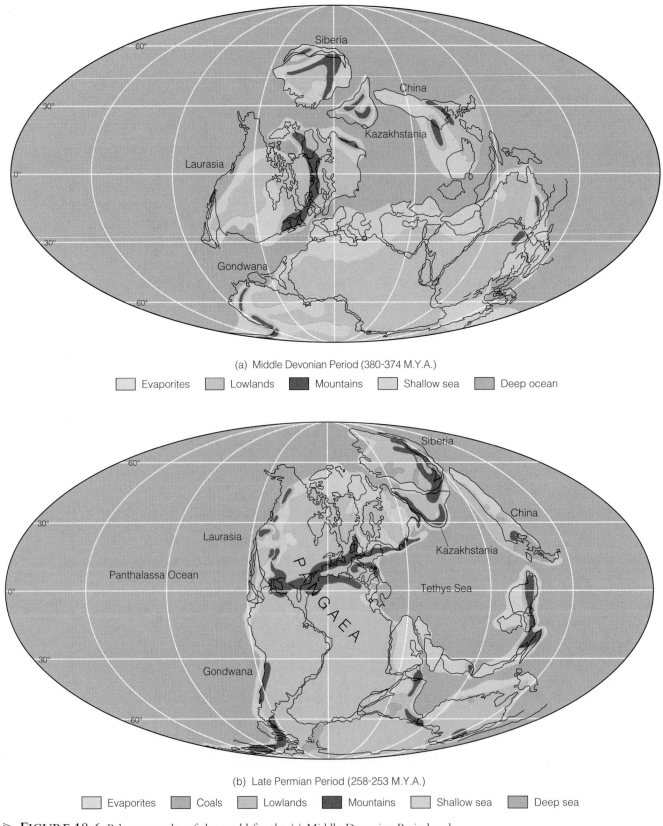

(a) Middle Devonian Period (380-374 M.Y.A.)

Evaporites Lowlands Mountains Shallow sea Deep ocean

(b) Late Permian Period (258-253 M.Y.A.)

Evaporites Coals Lowlands Mountains Shallow sea Deep sea

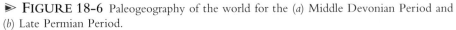

➤ **FIGURE 18-6** Paleogeography of the world for the (*a*) Middle Devonian Period and
(*b*) Late Permian Period.

moved over the South Pole, resulting in extensive continental glaciation. The advance and retreat of these glaciers produced global changes in sea level that affected sedimentation patterns on the cratons. As Gondwana continued moving northward, it first collided with Laurasia during the Early Carboniferous and continued suturing with it throughout the rest of the Carboniferous. The final phase of collision between Gondwana and Laurasia is indicated by the Ouachita Mountains of Oklahoma, which were formed by thrusting during the Late Carboniferous and Early Permian. Elsewhere, Siberia collided with Kazakhstania and moved toward the Uralian margin of Laurasia (Baltica), colliding with it during the Early Permian.

The assemblage of Pangaea was essentially concluded during the Permian with the completion of many of the continental collisions that began during the Carboniferous (Figure 18-6b). An enormous single ocean, *Panthalassa*, surrounded the supercontinent, spanning the Earth from pole to pole. Waters of this ocean probably circulated more freely than at present, resulting in more equable water temperatures.

The formation of a single landmass had climatic consequences for the terrestrial environment as well. Terrestrial Permian sediments indicate that arid and semiarid conditions were widespread over Pangaea. The mountain ranges produced by the *Hercynian, Alleghenian,* and *Ouachita orogenies* were high enough to create rainshadows that blocked the moist, subtropical, easterly winds—much as the southern Andes Mountains do in western South America today. This produced very dry conditions in North America and Europe, as is evident from the extensive Permian evaporites found in western North America, central Europe, and parts of Russia.

⊕ PALEOZOIC EVOLUTION OF NORTH AMERICA

It is convenient to divide the Paleozoic history of the North American craton into two parts, the first dealing with the relatively stable continental interior over which shallow seas transgressed and regressed, and the second with the mobile belts where mountain building occurred.

In 1963 the American geologist Laurence L. Sloss proposed that the sedimentary-rock record of North America could be subdivided into six cratonic sequences. A *cratonic sequence* is a large-scale rock unit representing a major transgressive-regressive cycle bounded by cratonwide unconformities (▷ Figure 18-7). The transgressive phase, which is usually covered by younger sediments, commonly is well preserved, while the regressive phase of each sequence is marked by an unconformity. Where rocks of the appropriate age are preserved, each of the six unconformities can be shown to extend across the various sedimentary basins of the North American craton and into the mobile belts along the cratonic margin.

Geologists have also recognized major unconformity-bound sequences in cratonic areas outside North America.

Such global transgressive and regressive cycles are caused by changes in sea level and are thought to result from major tectonic and glacial events.

The Sauk Sequence

Rocks of the **Sauk sequence** record the first major transgression onto the North American craton. During the Late Proterozoic and Early Cambrian, deposition of marine sediments was limited to the passive shelf areas of the Appalachian and Cordilleran borders of the craton. The craton itself was above sea level and experiencing weathering and erosion. Because North America was located in a tropical climate at this time (Figure 18-5a) and there is no evidence of any terrestrial vegetation, weathering and erosion of the exposed Precambrian rocks must have proceeded at a very rapid rate. During the Middle Cambrian, the transgressive phase of the Sauk began with shallow seas encroaching over the craton. By the Late Cambrian, the Sauk Sea had covered most of North America, leaving only a portion of the Canadian Shield and a few large islands above sea level (▷ Figure 18-8). These islands, collectively referred to as the *Transcontinental Arch*, extended from New Mexico to Minnesota and the Lake Superior region.

The Tippecanoe Sequence

As the Sauk Sea regressed from the craton during the Early Ordovician, it revealed a landscape of low relief. The exposed rocks were predominantly limestones and dolostones that were deeply eroded because North America was still in a tropical environment (Figure 18-5b). The resulting cratonwide unconformity marks the boundary between the Sauk and Tippecanoe sequences.

Like the Sauk sequence, deposition of the **Tippecanoe sequence** began with a major transgression onto the craton. This transgressing sea deposited clean quartz sands over most of the craton. The Tippecanoe basal sandstones were followed by widespread carbonate deposition. The limestones were generally the result of deposition by calcium carbonate–secreting organisms. In addition to the limestones, there were also many dolostones.

As the Tippecanoe Sea gradually regressed from the craton during the Late Silurian, evaporite minerals were precipitated in the Appalachian, Ohio, and Michigan basins (▷ Figure 18-9). In the Michigan Basin alone, approximately 1,500 m of sediments were deposited, nearly half of which are halite and anhydrite.

By the Early Devonian, the regressing Tippecanoe Sea had retreated to the craton margin exposing an extensive lowland topography. During this regression, marine deposition was initially restricted to a few interconnected cratonic basins and finally, by the end of the Tippecanoe, to only the margins surrounding the craton.

During the Early Devonian as the Tippecanoe Sea regressed, the craton experienced mild deformation resulting in the formation of many domes, arches, and basins. These

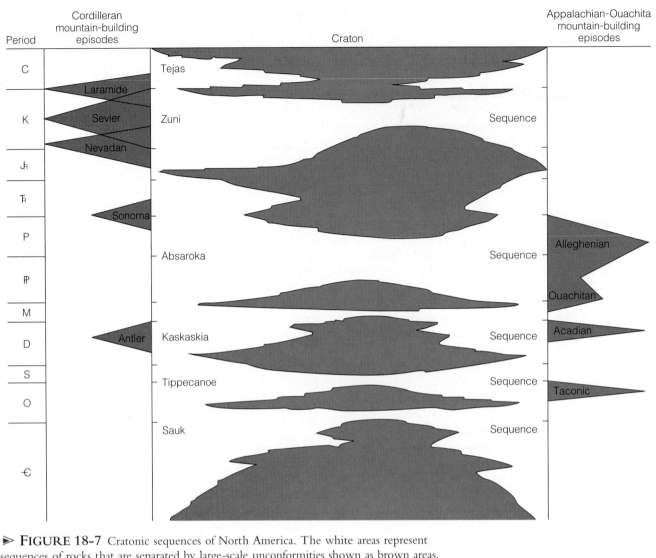

FIGURE 18-7 Cratonic sequences of North America. The white areas represent sequences of rocks that are separated by large-scale unconformities shown as brown areas. The major Cordilleran orogenies are shown on the left side of the figure, and the major Appalachian orogenies are shown on the right side.

structures were mostly eroded during the time the craton was exposed so that they were eventually covered by deposits from the encroaching Kaskaskia Sea.

The Kaskaskia Sequence

The boundary between the Tippecanoe sequence and the overlying **Kaskaskia sequence** is marked by a major unconformity. As the Kaskaskia Sea transgressed over the low-relief landscape of the craton, most of the basal beds consisted of clean, well-sorted, quartz sandstones.

Except for widespread Late Devonian and Early Mississippian black shales, the majority of Kaskaskian rocks are carbonates, including reefs, and associated evaporite deposits. In many other parts of the world, such as southern England, Belgium, central Europe, Australia, and Russia, the Middle

and early Late Devonian epochs were times of major reef building.

During the Late Mississippian regression of the Kaskaskia Sea from the craton, carbonate deposition was replaced by vast quantities of detrital sediments. The resulting sandstones, particularly in the Illinois Basin, have been studied in great detail because they are excellent petroleum reservoirs. Prior to the end of the Mississippian, the Kaskaskia Sea had retreated to the craton margin, once again exposing the craton to widespread weathering and erosion that resulted in a cratonwide unconformity.

The Absaroka Sequence

The extensive unconformity separating the Kaskaskia and Absaroka sequences essentially divides the rocks into the

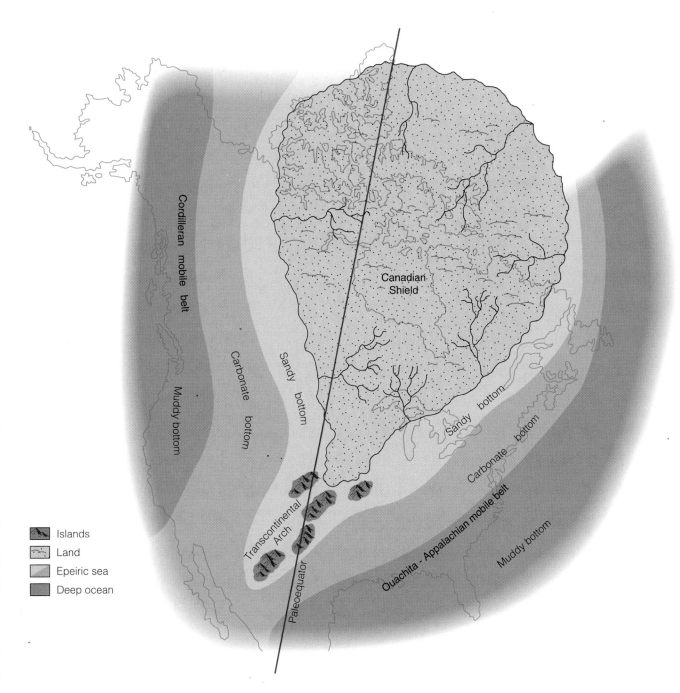

Islands
Land
Epeiric sea
Deep ocean

Cordilleran mobile belt

Muddy bottom

Carbonate bottom

Sandy bottom

Canadian Shield

Sandy bottom

Carbonate bottom

Ouachita - Appalachian mobile belt

Muddy bottom

Transcontinental Arch

Paleoequator

➤ **FIGURE 18-8** Paleogeography of North America during the Cambrian Period. Note the position of the Cambrian paleoequator. During this time North America straddled the equator as indicated in Figure 18-5a.

North American Mississippian and Pennsylvanian systems. The rocks of the **Absaroka sequence** not only differ from those of the Kaskaskia sequence, but they also result from quite different tectonic regimes affecting the North American craton.

One of the characteristic features of Pennsylvanian rocks is their cyclical pattern of alternating marine and nonmarine strata. Such rhythmically repetitive sedimentary sequences are known as **cyclothems** (➤ Figure 18-10). They result from repeated alternations of marine and nonmarine deposition, usually in areas of low relief. Though seemingly simple, cyclothems reflect a delicate interplay between nonmarine deltaic and shallow marine interdeltaic and shelf environments.

Cyclothems represent transgressive and regressive sequences with an erosional surface separating one cyclothem

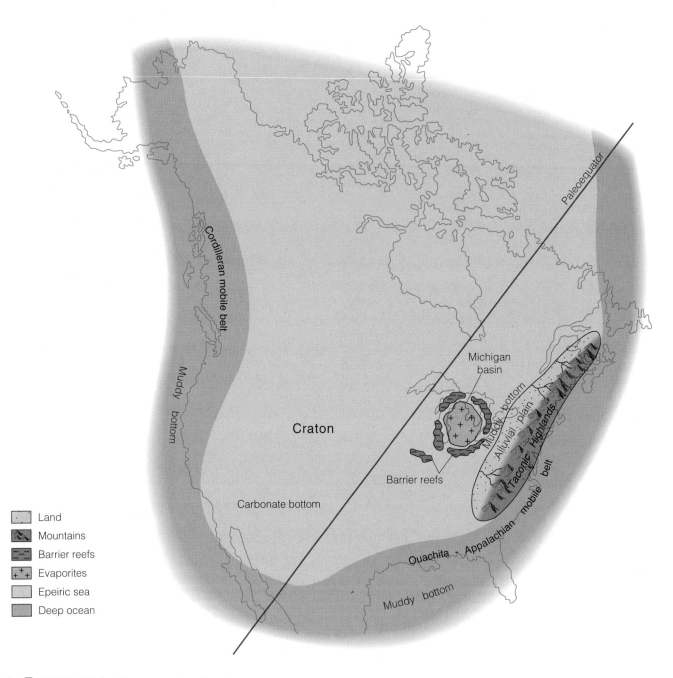

Land
Mountains
Barrier reefs
Evaporites
Epeiric sea
Deep ocean

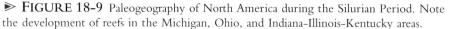

➤ **FIGURE 18-9** Paleogeography of North America during the Silurian Period. Note the development of reefs in the Michigan, Ohio, and Indiana-Illinois-Kentucky areas.

from another. An idealized cyclothem, therefore, passes upward from fluvial-deltaic deposits, through coals, to detrital shallow-water marine sediments, and finally to limestones typical of an open marine environment (Figure 18-10a).

Such regularity and cyclicity in sedimentation over a large area requires an explanation. The hypothesis currently favored by most geologists is a rise and fall of sea level related to advances and retreats of Gondwanan continental glaciers. When the Gondwanan ice sheets advanced, sea level dropped, and when they melted, sea level rose. Late Paleo-

zoic cyclothem activity on all of the cratons closely corresponds to Gondwanan glacial-interglacial cycles.

During the Pennsylvanian, the area of greatest deformation occurred in the southwestern part of the North American craton where a series of fault-bounded uplifted blocks formed the *Ancestral Rockies* (➤ Figure 18-11). These mountain ranges had diverse geologic histories and were not elevated at the same time. Uplift of these mountains along near-vertical faults resulted in the erosion of the overlying Paleozoic sediments and exposure of the Precambrian igne-

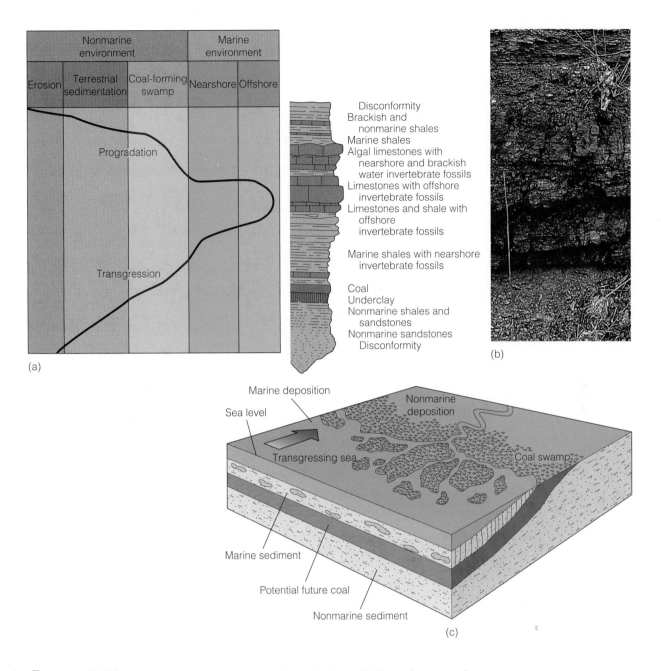

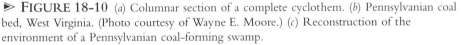

➤ FIGURE 18-10 (*a*) Columnar section of a complete cyclothem. (*b*) Pennsylvanian coal bed, West Virginia. (Photo courtesy of Wayne E. Moore.) (*c*) Reconstruction of the environment of a Pennsylvanian coal-forming swamp.

ous and metamorphic basement rocks. As the mountains eroded, tremendous quantities of coarse-grained sediment were deposited in the surrounding basins (Figure 18-11).

While the various intracratonic basins were filling with sediment during the Late Pennsylvanian, the Absaroka Sea slowly began retreating from the craton. During the Middle and Late Permian, the Absaroka Sea was restricted to west Texas and southern New Mexico, forming an interrelated complex of lagoonal, reef, and open-shelf environments.

Massive reefs grew around the basin margins while limestones, evaporites, and red beds were deposited in the lagoonal areas behind the reefs.

Spectacular deposits representing the geologic history of this region can be seen today in the Guadalupe Mountains of Texas and New Mexico where the Capitan Limestone forms the caprock of these mountains. These reefs have been extensively studied because of the tremendous oil production that comes from this region.

Legend:
- Land
- Mountains
- Evaporites
- Coal swamps
- Volcanoes
- Epeiric sea
- Deep ocean

Labels on map: Volcanic Island arc, Cordilleran mobile belt, Muddy bottom, Sandy to muddy bottom, Antler Highlands, Ancestral Rockies and basins, Carbonate bottom, Paleoequator, Ouachita Mountains, Muddy bottom, Lowlands, Uplands, Coal swamps, Appalachian Mountains

➤ **FIGURE 18-11** Paleogeography of North America during the Pennsylvanian Period.

By the end of the Permian Period, the Absaroka Sea had retreated from the craton, and continental red beds were deposited over most of the southwestern and eastern region.

HISTORY OF THE PALEOZOIC MOBILE BELTS

Having examined the Paleozoic history of the craton, we now turn to the orogenic activity in the **mobile belts** (elongated areas of mountain-building activity occurring along the margins of continents). The mountain building

that occurred during this time had a profound influence on the climate and sedimentary history of the craton. In addition, it was part of the global tectonic regime that sutured the continents together, forming Pangaea by the end of the Paleozoic Era.

Appalachian Mobile Belt

Throughout Sauk time, the Appalachian region was a broad, passive continental margin. Sedimentation was closely balanced by subsidence as thick, shallow marine sands were succeeded by extensive carbonate deposits. During this time,

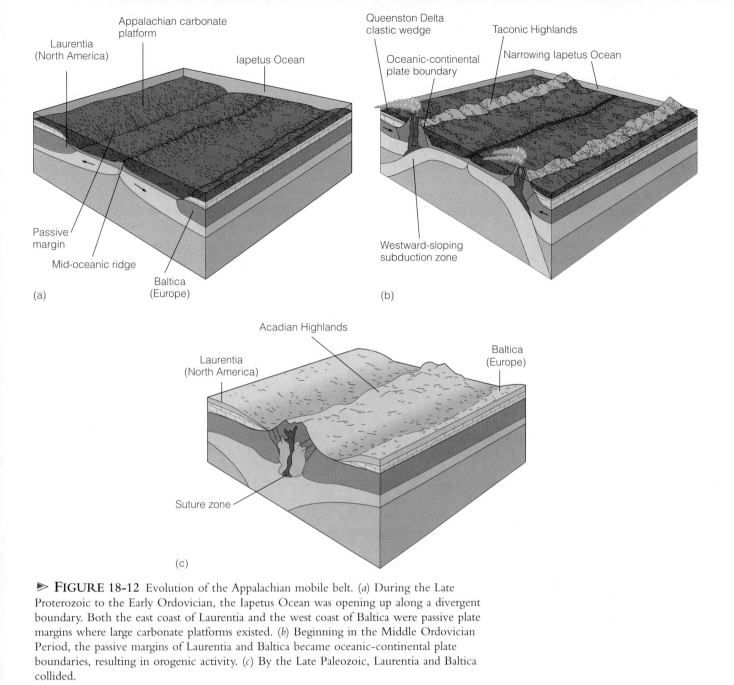

(a)

(b)

(c)

▷ **FIGURE 18-12** Evolution of the Appalachian mobile belt. (*a*) During the Late Proterozoic to the Early Ordovician, the Iapetus Ocean was opening up along a divergent boundary. Both the east coast of Laurentia and the west coast of Baltica were passive plate margins where large carbonate platforms existed. (*b*) Beginning in the Middle Ordovician Period, the passive margins of Laurentia and Baltica became oceanic-continental plate boundaries, resulting in orogenic activity. (*c*) By the Late Paleozoic, Laurentia and Baltica collided.

the Iapetus Ocean was widening as a result of movement along a divergent plate boundary (▷ Figure 18-12a).

Beginning with the subduction of the Iapetus plate beneath Laurentia (an oceanic-continental convergent plate boundary), the Appalachian mobile belt was born (Figure 18-12b). The resulting **Taconic orogeny**, named after the present-day Taconic Mountains of eastern New York, central Massachusetts, and Vermont, was the first of several orogenies to affect the Appalachian region.

A large *clastic* wedge (an extensive accumulation of mostly detrital sediments deposited adjacent to an uplifted area) formed in the shallow seas to the west of the Taconic orogeny. These deposits are thickest and coarsest nearest the highland area and become thinner and finer grained away from the source area, eventually grading into carbonates on

the craton. The clastic wedge resulting from the erosion of the Taconic Highlands is referred to as the *Queenston Delta*.

The second Paleozoic orogeny to affect Laurentia began during the Late Silurian and concluded at the end of the Devonian Period. The **Acadian orogeny** affected the Appalachian mobile belt from Newfoundland to Pennsylvania as sedimentary rocks were folded and thrust against the craton.

Like the preceding Taconic orogeny, the Acadian orogeny occurred along an oceanic-continental convergent plate boundary. As the northern Iapetus Ocean continued to close during the Devonian, the plate carrying Baltica finally collided with Laurentia, forming a continental-continental convergent plate boundary along the zone of collision (Figure 18-12c). Weathering and erosion of the Acadian Highlands produced another thick clastic wedge called the

Catskill Delta, named for the Catskill Mountains in northern New York where it is well exposed.

Geologists think that the Taconic and Acadian orogenies were part of the same major orogenic event related to the closing of the Iapetus Ocean (Figure 18-12). This event began with an oceanic-continental convergent plate boundary during the Taconic orogeny and culminated with a continental-continental convergent plate boundary during the Acadian orogeny as Laurentia and Baltica became sutured. Following this, the Hercynian-Alleghenian orogeny began, followed by orogenic activity in the Ouachita mobile belt.

The Hercynian mobile belt of southern Europe and the Appalachian and Ouachita mobile belts of North America mark the zone along which Europe (part of Laurasia) collided with Gondwana. While Gondwana and southern Laurasia collided during the Pennsylvanian and Permian, eastern Laurasia (Europe and southeastern North America) joined together with Gondwana (Africa) as part of the **Hercynian-Alleghenian orogeny** (Figure 18-6b).

Cordilleran Mobile Belt

During the Late Proterozoic and Early Paleozoic, the Cordilleran area was a passive continental margin along which extensive continental shelf sediments were deposited. Beginning in the Middle Paleozoic, an island arc formed on the western margin of the craton. This eastward-moving island arc collided with the western border of the craton during the Late Devonian and Early Mississippian resulting in a highland area. This orogenic event, called the *Antler orogeny*, was caused by subduction. Erosion of the Antler Highlands produced large quantities of sediment that were deposited in the shallow sea covering the craton to the east and in the deep sea to the west (Figure 18-11).

Ouachita Mobile Belt

The Ouachita mobile belt extends for approximately 2, 100 km from the subsurface of Mississippi to the Marathon region of Texas. Approximately 80% of the former mobile belt is buried beneath a Mesozoic and Cenozoic sedimentary cover. The two major exposed areas in this region are the Ouachita Mountains of Oklahoma and Arkansas and the Marathon Mountains of Texas.

During the Late Proterozoic to Early Mississippian, shallow-water detrital and carbonate sediments were slowly deposited on a broad continental shelf, while in the deeper-water portion of the adjoining mobile belt, bedded cherts and shales were also slowly accumulating. Beginning in the Mississippian Period, the rate of sedimentation increased dramatically as the region changed from a passive continental margin to an active convergent plate boundary. Rapid deposition of sediments continued into the Pennsylvanian with the formation of a clastic wedge that thickened to the south. The formation of a clastic wedge marks the beginning of uplift of the area and the formation of a mountain range

during the **Ouachita orogeny**. Thrusting of sediments continued throughout the Pennsylvanian and Early Permian as a result of the compressive forces generated along the zone of subduction as Gondwana collided with Laurasia (Figure 18-6b). The collision of Gondwana and Laurasia is marked by the formation of a large mountain range, most of which was eroded during the Mesozoic Era.

THE BREAKUP OF PANGAEA

Just as the formation of Pangaea influenced geologic and biologic events during the Paleozoic, the breakup of this supercontinent profoundly affected geologic and biologic events during the Mesozoic. The movement of continents affected global climatic and oceanic regimes as well as the climates of individual continents.

Geologic, paleontologic, and paleomagnetic data indicate that the breakup of Pangaea occurred in four general stages. The first stage involved rifting between Laurasia and Gondwana during the Late Triassic. By the end of the Triassic, the expanding Atlantic Ocean separated North America from Africa (▶ Figure 18-13a). This was followed by the rifting of North America from South America sometime during the Late Triassic and Early Jurassic.

The second stage in Pangaea's breakup involved rifting and movement of the various Gondwana continents during the Late Triassic and Jurassic periods. As early as the Late Triassic, Antarctica and Australia, which remained sutured together, separated from South America and Africa, while India split away from all four Gondwana continents and began moving northward (Figures 18-13a and b).

The third stage of breakup began during the Late Jurassic, when South America and Africa began separating (Figure 18-13b). During this stage, the eastern end of the Tethys Sea began closing as a result of the clockwise rotation of Laurasia and the northward movement of Africa. This narrow Late Jurassic and Cretaceous seaway between Africa and Europe was the forerunner of the present Mediterranean Sea.

By the end of the Cretaceous, Australia and Antarctica had separated, India had nearly reached the equator, South America and Africa were widely separated, and the eastern side of what is now Greenland had begun separating from Europe (Figure 18-13c).

The final stage in Pangaea's breakup occurred during the Cenozoic. During this stage, Australia continued moving northward, and Greenland completely separated from Europe and rifted from North America to form a separate landmass.

THE MESOZOIC HISTORY OF NORTH AMERICA

In terms of mountain building and sedimentation, the beginning of the Mesozoic Era was essentially the same as the preceding Permian Period in North America. Terrestrial sedimentation continued over much of the craton, while block-faulting and igneous activity began in the Appalachian

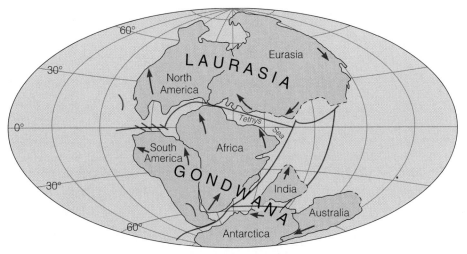

(a) Triassic Period (245–208 M.Y.A)

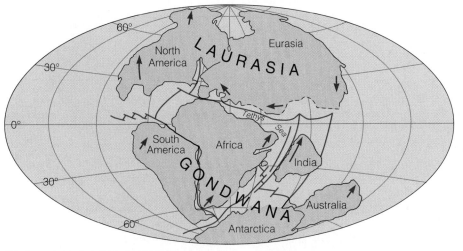

(b) Jurassic Period (208–144 M.Y.A)

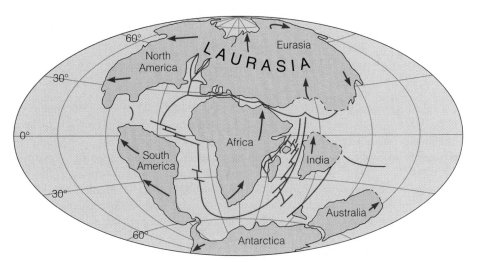

(c) Cretaceous Period (144–66 M.Y.A)

➤ FIGURE 18-13
Paleogeography of the world during the Mesozoic. Blue arrows show the direction of movement for the continents. (*a*) The Triassic Period, (*b*) the Jurassic Period, and (*c*) the Cretaceous Period.

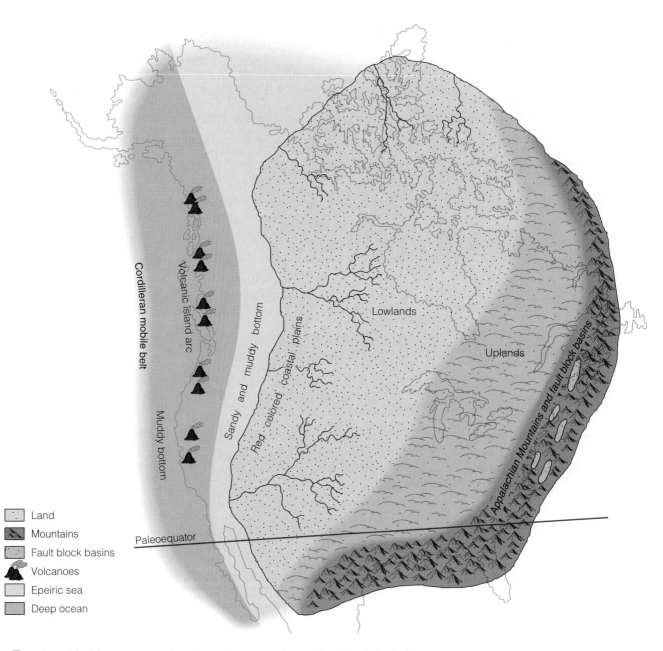

Land

Mountains

Fault block basins

Volcanoes

Epeiric sea

Deep ocean

Cordilleran mobile belt

Volcanic island arc

Muddy bottom

Sandy and muddy bottom

Red colored coastal plains

Lowlands

Uplands

Appalachian Mountains and fault block basins

Paleoequator

▷ **FIGURE 18-14** Paleogeography of North America during the Triassic Period.

region as North America and Africa began separating (▷ Figure 18-14). The newly forming Gulf of Mexico was the site of extensive evaporite deposition during the Late Triassic and Jurassic as North America separated from South America (▷ Figure 18-15).

A global rise in sea level during the Cretaceous resulted in worldwide transgressions onto the continents (▷ Figure 18-16). These transgressions were caused by higher heat flow along the oceanic ridges due to increased rifting and the consequent expansion of oceanic crust. By the Middle Cretaceous, sea level probably was as high as at any time since the Ordovician, and approximately one-third of the present land area was covered by shallow seas.

Marine deposition was continuous over much of western

North America. A volcanic island arc system that formed off the western edge of the craton during the Permian was sutured to North America sometime during the Permian or Triassic. During the Jurassic, the entire Cordilleran area was involved in a series of major mountain-building episodes that resulted in the formation of the Sierra Nevada, the Rocky Mountains, and other lesser mountain ranges.

Eastern Coastal Region

During the Early and Middle Triassic, coarse detrital sediments derived from the erosion of the recently uplifted Appalachians (Alleghenian orogeny) filled the various intermontane basins and spread over the surrounding areas. As

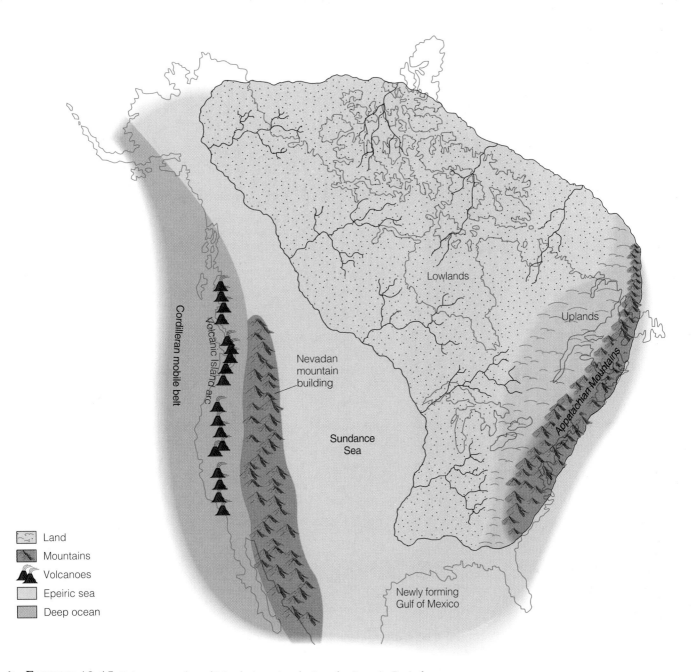

➤ **FIGURE 18-15** Paleogeography of North America during the Jurassic Period.

Land

Mountains

Volcanoes

Epeiric sea

Deep ocean

Cordilleran mobile belt

Volcanic Island arc

Nevadan mountain building

Sundance Sea

Lowlands

Uplands

Appalachian Mountains

Newly forming Gulf of Mexico

---

erosion continued during the Mesozoic, this once lofty mountain system was reduced to a low-lying plain. During the Late Triassic, the first stage in the breakup of Pangaea began as North America separated from Africa and the Atlantic Ocean began to form (Figure 18-13a). Fault-block basins developed in response to upwelling magma beneath Pangaea in a zone stretching from present-day Nova Scotia to North Carolina (Figure 18-14). Erosion of the adjacent fault-block mountains filled these basins with great quantities (up to 6,000 m) of poorly sorted, red, nonmarine detrital sediments.

As the Atlantic Ocean grew, rifting ceased along the eastern margin of North America, and this once active plate margin became a passive, trailing continental margin. The fault-block mountains that were produced by this rifting continued to erode during the Jurassic and Early Cretaceous until only a broad, low-lying erosional surface remained. The sediments resulting from erosion contributed to the growing eastern continental shelf. During the Cretaceous Period, the Appalachian region was reelevated and once again shed sediments onto the continental shelf.

Gulf Coastal Region

The Gulf Coastal region was above sea level until the Late Triassic. As North America separated from South America

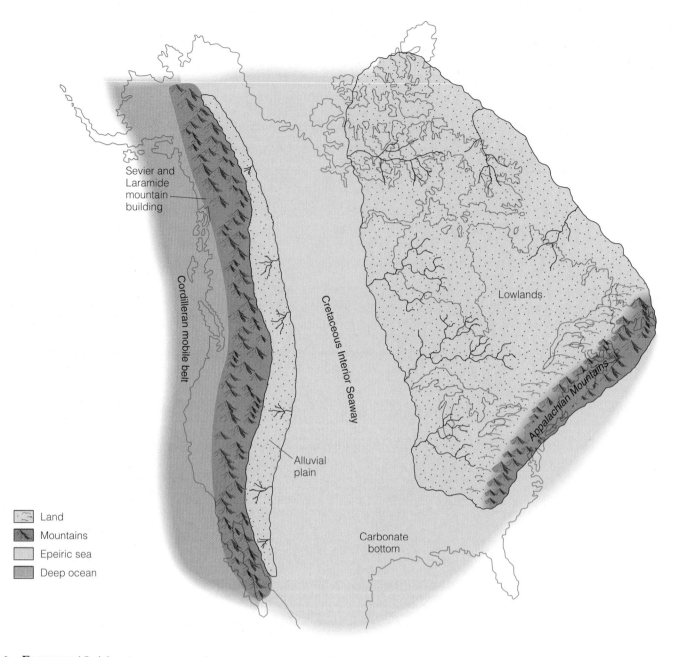

Land
Mountains
Epeiric sea
Deep ocean

Sevier and Laramide mountain building

Cordilleran mobile belt

Cretaceous Interior Seaway

Alluvial plain

Lowlands

Appalachian Mountains

Carbonate bottom

➤ **FIGURE 18-16** Paleogeography of North America during the Cretaceous Period.

during the Late Triassic, the Gulf of Mexico began to form (Figure 18-15). With oceanic waters flowing into this newly formed, shallow, restricted basin, conditions were ideal for evaporite formation. More than 1,000 m of evaporites were precipitated at this time, and most geologists think that these Jurassic evaporites are the source for the Tertiary salt domes found today in the Gulf of Mexico and southern Louisiana.

By the Late Jurassic, circulation in the Gulf of Mexico was less restricted, and evaporite deposition ended. Normal marine conditions with alternating transgressing and regressing seas returned to the area, resulting in the deposition of sandstones, shales, and limestones. These sedimentary rocks were later covered and deeply buried by great thicknesses of Cretaceous and Cenozoic sediments.

During the Cretaceous, the Gulf Coastal region, like the rest of the continental margin, was inundated by northward-transgressing seas as a wide seaway extended from the Arctic Ocean to the Gulf of Mexico (Figure 18-16).

Western Region

During the Permian, an island arc and ocean basin formed off the western North American craton (Figure 18-14). This was followed by subduction of an oceanic plate beneath the island arc and the thrusting of oceanic and island arc rocks eastward against the craton margin. This event initiated the *Sonoma orogeny* at or near the Permian-Triassic boundary.

Following the Late Paleozoic–Early Mesozoic destruction of the volcanic island arc during the Sonoma orogeny, the western margin of North America became an oceanic-continental convergent plate boundary. During the Late Triassic, a steeply dipping subduction zone developed along the western margin of North America in response to the westward movement of North America over the Pacific plate. This newly created oceanic-continental plate boundary controlled Cordilleran tectonics for the rest of the Mesozoic Era and for most of the Cenozoic Era.

The general term *Cordilleran orogeny* is applied to the mountain-building activity that began during the Jurassic and continued into the Cenozoic (Figure 18-7). The Cordilleran orogeny consisted of a series of individual mountain-building events that occurred in different regions at different times. Most of this Cordilleran orogenic activity is related to the continued westward movement of the North American plate.

The first phase of the Cordilleran orogeny, the **Nevadan orogeny** (Figure 18-7), began during the Late Jurassic and continued into the Cretaceous as large volumes of granitic magma were generated at depth beneath the western edge of North America. These granitic masses ascended as huge batholiths that are now recognized as the Sierra Nevada, southern California, Idaho, and Coast Range batholiths.

The second phase of the Cordilleran orogeny, the **Sevier orogeny**, was mostly a Cretaceous event (Figure 18-7). As subduction of the Pacific plate beneath the North American plate continued, compressive forces generated along this convergent plate boundary were transmitted eastward, resulting in numerous overlapping, low-angle thrust faults. This thrusting produced generally north–south–trending mountain ranges consisting of blocks of Paleozoic shelf and slope strata.

During the Late Cretaceous to Early Cenozoic, the final pulse of the Cordilleran orogeny occurred (Figure 18-7). The **Laramide orogeny** developed east of the Sevier orogenic belt in the present-day Rocky Mountain areas of New Mexico, Colorado, and Wyoming. Most of the features of the present-day Rocky Mountains resulted from the Cenozoic phase of the Laramide orogeny, and for that reason, it will be discussed later in this chapter.

Concurrent with the tectonism occurring in the Cordilleran mobile belt, Early Triassic sedimentation occurred on the western continental shelf in a shallow-water marine environment. During the Middle and Late Triassic, the shallow western seas regressed further west, exposing large areas of former sea floor to erosion. Marginal marine and nonmarine Triassic rocks, particularly red beds, contribute to the spectacular and colorful scenery of the region.

These rocks represent a variety of depositional environments, including fluvial, deltaic, floodplain, and desert dunes. The Upper Triassic Chinle Formation, for example, is widely exposed over the Colorado Plateau and is probably most famous for its petrified wood in Petrified Forest National Park, Arizona (see Perspective 18-1). This formation, like other Triassic formations in the Southwest, also contains the fossilized remains and tracks of amphibians and reptiles.

The Early Jurassic deposits in a large part of the western region consist mostly of clean, cross-bedded sandstones indicative of windblown deposits. The thickest and most prominent of these is the Navajo Sandstone, a widespread cross-bedded sandstone that accumulated in a coastal dune environment along the southwestern margin of the craton. The sandstone's most distinctive feature is its large-scale cross-beds, some of which are more than 25 m high (➤ Figure 18-17).

Marine conditions returned to the area during the Middle Jurassic when a wide seaway called the *Sundance Sea* twice flooded the interior of western North America (Figure 18-15). The resulting deposits were largely derived from tectonic highlands to the west that paralleled the shoreline and were the result of intrusive igneous activity and associated volcanism that began during the Triassic.

During the Late Jurassic, the folding and thrust faulting that began as part of the Nevadan orogeny in Nevada, Utah, and Idaho formed a large mountain chain paralleling the coastline. As the mountain chain grew and shed sediments eastward, the Sundance Sea retreated northward. A large part of the area formerly occupied by the Sundance Sea was then covered by multicolored detrital sediments that comprise the Morrison Formation, which contains the world's richest assemblage of Jurassic dinosaur remains (➤ Figure 18-18).

Shortly before the end of the Early Cretaceous, Arctic waters spread southward over the craton. By the beginning of the Late Cretaceous, this incursion joined the northward-transgressing waters from the Gulf area to create an enormous *Cretaceous Interior Seaway* that occupied the area east of the Sevier orogenic belt (Figure 18-16). Extending from the Gulf of Mexico to the Arctic Ocean, and more than 1,500 km wide at its maximum extent, this seaway effectively divided North America into two large landmasses until just before the end of the Late Cretaceous.

As the Mesozoic Era ended, the Cretaceous Interior Seaway withdrew from the craton. During this regression, marine waters retreated to the north and south, and marginal marine and continental deposition formed widespread coal-bearing deposits on the coastal plain.

🏵 CENOZOIC PLATE TECTONICS AND OROGENY

The Late Triassic fragmentation of the supercontinent Pangaea (Figure 18-13a) began an episode of plate motions that continues even now. As a consequence of these plate motions, Cenozoic orogenic activity was largely concentrated in two major zones or belts, the *Alpine-Himalayan belt* and the *circum-Pacific belt* (➤ Figure 18-19). The Alpine-Himalayan belt includes the mountainous regions of southern Europe and North Africa and extends eastward through the Middle East and India and into southeast Asia, whereas the circum-Pacific belt, as its name implies, nearly encircles the Pacific Ocean basin.

The Alpine and Himalayan orogens developed in the Alpine-Himalayan orogenic belt. The *Alpine orogeny* began

▶ **FIGURE 18-17** Large cross-beds of the Jurassic Navajo Sandstone in Zion National Park, Utah.

during the Mesozoic, but major deformation also occurred from the Eocene to Late Miocene as the African and Arabian plates moved northward against Eurasia. Deformation resulting from plate convergence formed the Pyrenees Mountains between Spain and France, the Alps of mainland Europe, the Apennines of Italy, and the Atlas Mountains of North Africa (Figure 18-19). Active volcanoes in Italy and seismic activity in much of southern Europe and the Middle East indicate that this orogenic belt remains geologically active.

Farther east in the Alpine-Himalayan orogenic belt, the *Himalayan orogen* resulted from the collision of India with Asia (see Figure 10-16). The exact time of this collision is uncertain, but sometime during the Eocene, India's northward drift rate decreased abruptly, indicating the probable time of collision. In any event, as the two continental plates became sutured, an orogeny resulted, accounting for the present-day Himalayas being far inland rather than at a continental margin.

Plate subduction in the circum-Pacific orogenic belt occurred throughout the Cenozoic, giving rise to orogenies in the Aleutians, the Philippines, and Japan and along the west coasts of North, Central, and South America. The Andes Mountains in western South America, for example, formed as a result of convergence of the Nazca and South American plates (see Figure 10-15). Spreading at the East Pacific Rise and subduction of the Cocos and Nazca plates beneath Central and South America, respectively, account for continuing orogenic activity in these regions.

The North American Cordillera

The *North American Cordillera* is a complex mountainous region in western North America extending from Alaska into central Mexico (▶ Figure 18-20). It has a long, complex geologic history involving accretion of island arcs along the continental margin, orogeny at an oceanic-continental boundary, vast outpourings of basaltic lavas, and block-faulting. Although the Cordillera has a long history of deformation, the most recent episode of large-scale deformation was the Laramide orogeny, which began 85 to 90 million years ago. Like many other orogenies, it occurred along an oceanic-continental boundary. The main Laramide orogeny was centered in the Rocky Mountains of present-day Colorado and Wyoming, but deformation occurred far

PETRIFIED FOREST NATIONAL PARK

Petrified Forest National Park is located in eastern Arizona about 42 km east of Holbrook. The park consists of two sections: the Painted Desert, which is north of Interstate 40, and the Petrified Forest, which is south of the interstate.

The Painted Desert is a brilliantly colored landscape where colors and hues change constantly throughout the day. The multicolored rocks of the Triassic Chinle Formation have been weathered and eroded to form a badlands topography of numerous gullies, valleys, ridges, mounds, and mesas. The Chinle Formation is composed predominantly of various-colored shale beds that are easily weathered and eroded.

The Petrified Forest was originally set aside as a national monument to protect the large number of petrified logs that lay exposed in what is now the southern part of the park (▷ Figure 1). When the transcontinental railroad constructed a coaling and watering stop in Adamana, Arizona, passengers were encouraged to take excursions to "Chalcedony Park," as the area was then called, to see the petrified forests. In a short time, collectors and souvenir hunters hauled off tons of petrified wood, quartz crystals, and Indian relics. It was not until a huge rock crusher was built to crush the logs for the manufacture of abrasives that the area was declared a national monument and the petrified forests preserved and protected.

During the Triassic Period, the climate of the area was much wetter than today. Numerous fossils of seedless vascular plants such as rushes and ferns as well as gymnosperms such as cycads and conifers are preserved in the Chinle Formation. Most of the logs are conifers, and some were more than 60 m tall and up to 4 m in diameter. Burial of the logs was rapid, and groundwater saturated with silica from the ash of nearby volcanic eruptions quickly preserved the trees.

Deposition continued in the Colorado Plateau region during the Jurassic and Cretaceous, further burying the Chinle Formation. During the Laramide orogeny, the Colorado Plateau area was uplifted and eroded, exposing the Chinle Formation. Because the Chinle is mostly shales, it was easily eroded, leaving the more resistant petrified logs and log fragments exposed on the surface—much as we see them today (Figure 1).

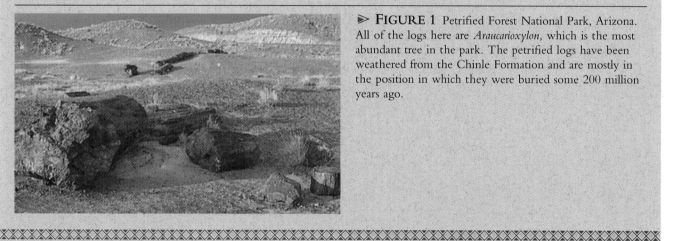

▷ FIGURE 1 Petrified Forest National Park, Arizona. All of the logs here are *Araucarioxylon,* which is the most abundant tree in the park. The petrified logs have been weathered from the Chinle Formation and are mostly in the position in which they were buried some 200 million years ago.

to the north in Canada and Alaska and as far south as Mexico City. The Lewis overthrust of Montana resulted from Laramide compression, and in the Canadian Rockies, thrust sheets piled one upon another (▷ Figure 18-21).

The Laramide orogeny ceased about 40 million years ago, but since that time the Rocky Mountains have continued to evolve. The mountain ranges formed during the orogeny were eroded, and the valleys between the ranges filled with sediments. Many of the ranges were nearly buried in their own erosional debris, and their present-day elevations are the result of renewed uplift, which is continuing in such areas as the Teton Range of Wyoming.

In other parts of the Cordillera, the Colorado Plateau was uplifted far above sea level, but the rocks were little deformed. In the Basin and Range Province, block-faulting began during the Middle Cenozoic and continues to the present. At its

► **FIGURE 18-18** The north wall of the visitors' center at Dinosaur National Monument, showing dinosaur bones in bas relief, just as they were deposited 140 million years ago in the Morrison Formation.

The present-day elements of the Pacific coast section of the Cordillera developed as a result of the westward drift of North America, the partial consumption of the oceanic Farallon plate, and the collision of North America with the Pacific-Farallon Ridge. During the Early Cenozoic, the entire Pacific coast was bounded by a subduction zone that stretched from Mexico to Alaska (► Figure 18-22). Most of the Farallon plate was consumed at this subduction zone, and now only two small remnants exist—the Juan de Fuca and Cocos plates (Figure 18-22). The continuing subduction of these small plates accounts for seismicity and volcanism in the Cascade Range of the Pacific Northwest and Central America, respectively. Westward drift of the North American plate also resulted in its collision with the Pacific-Farallon Ridge and the origin of the Queen Charlotte and San Andreas transform faults (Figure 18-22).

The Gulf Coastal Plain

The Gulf Coast sedimentation pattern was established during the Jurassic and persisted through the Cenozoic. Much of the Gulf Coastal Plain was dominated by detrital sediment deposition during the Cenozoic. In the Florida section of the coastal plain and the Gulf Coast of Mexico, however, significant carbonate deposition occurred. A carbonate platform was established in Florida during the Cretaceous, and shallow-water carbonate deposition continued through the Early Tertiary. Carbonate deposition continues at the present, but now it occurs only in Florida and the Florida Keys.

western edge, the province is bounded by a large escarpment that forms the east face of the Sierra Nevada. This escarpment resulted from movement on a normal fault that has elevated the Sierra Nevada 3,000 m above the basins to the east.

In the Pacific Northwest, an area of about 200,000 km² is covered by the Cenozoic Columbia River basalts (see Figure 5-16). Issuing from long fissures, these flows overlapped to produce an aggregate thickness of about 1,000 m.

► **FIGURE 18-19** Most of the Earth's geologically recent and current orogenic activity is concentrated in the circum-Pacific and Alpine-Himalayan orogenic belts.

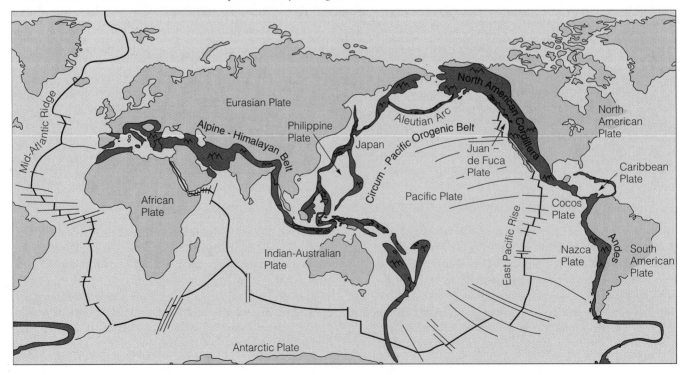

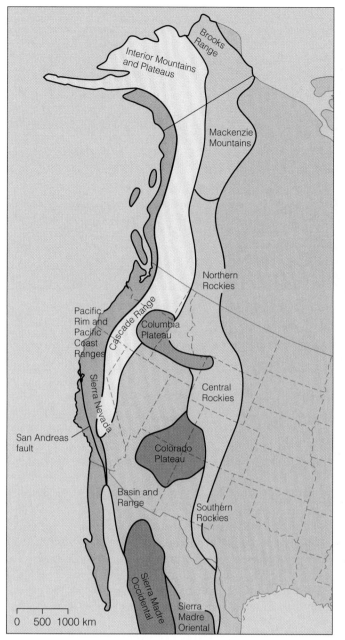

> **FIGURE 18-20** The North American Cordillera is a complex mountainous region extending from Alaska into central Mexico. It consists of a number of elements as shown here.

Eastern North America

The eastern seaboard has been a passive continental margin since Late Triassic rifting separated North America from North Africa and Europe. The present distinctive topography of the Appalachian Mountains is the product of Cenozoic uplift and erosion. By the end of the Mesozoic, the Appalachian Mountains had been eroded to a plain. Cenozoic uplift rejuvenated the streams, which responded by renewed downcutting. As the streams eroded downward, they were superposed on resistant strata and cut large canyons across these strata. The distinctive topography of the Valley and Ridge Province is the product of Cenozoic erosion and preexisting geologic structures. It consists of northeast-southwest–trending ridges of resistant upturned strata and intervening valleys eroded into less resistant strata.

⊚ PLEISTOCENE GLACIATION

We know today that the last Ice Age began about 1.6 million years ago and consisted of several intervals of glacial expansion separated by warmer interglacial periods. It appears that the present interglacial period began about 10,000 years ago, but geologists do not know whether we are still in an interglacial period or are entering another colder glacial interval.

The climatic effects responsible for Pleistocene glaciation were, as one would expect, worldwide. Nevertheless, the world was not as frigid as it is commonly portrayed in cartoons and movies, nor was the onset of glacial conditions as rapid as many people think. In fact, evidence from various lines of research indicates that the world's climate gradually cooled from the beginning of the Eocene through the Pleistocene, and that during the past 2 million years, at least 20 major warm-cold cycles occurred.

From such glacial features as the distribution of moraines, erratic boulders, and drumlins, it seems that at their greatest extent, Pleistocene glaciers covered about three times as much of the Earth's surface as they do now and were up to 3 km thick (▷ Figure 18-23). Geologists have determined that during the Pleistocene of North America, at least four major glacial stages occurred, with each followed by an interglacial stage. These four stages, the *Wisconsin, Illinoian,*

> **FIGURE 18-21** Cross section of part of the Canadian Rocky Mountains. The folding and thrust faulting occurred during the Laramide orogeny.

Canadian Rocky Mountains

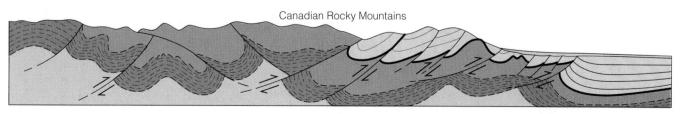

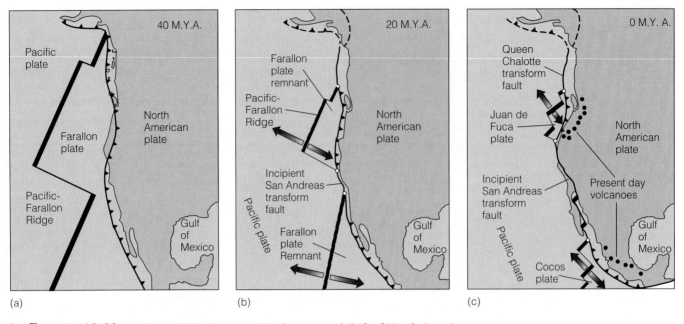

► **FIGURE 18-22** (*a*), (*b*), and (*c*) Three stages in the westward drift of North America and its collision with the Pacific-Farallon Ridge. As the North American plate overrode the ridge, its margin became bounded by transform faults rather than a subduction zone.

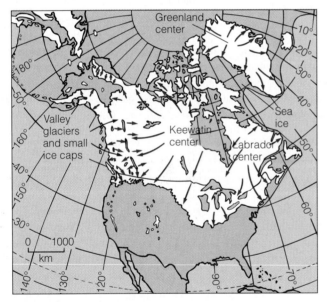

► **FIGURE 18-23** Centers of ice accumulation and maximum extent of Pleistocene glaciation in North America.

Kansan, and *Nebraskan,* are named for the states in which the most southerly glacial deposits are well exposed. The three interglacial stages, the *Sangamon, Yarmouth,* and *Aftonian,* are named for localities of well-exposed interglacial soils and other deposits. In Europe, six or seven major glacial advances and retreats are recognized.

Recent studies indicate that there were an as yet undetermined number of pre-Illinoian glacial events and that the history of glacial advances and retreats in North America is more complex than previously thought. In view of these data, the traditional four-part subdivision of the Pleistocene of North America must be modified.

1. A craton is the stable core of a continent. Broad areas where the cratons of continents are exposed are called shields; each continent has at least one shield area.

2. Cratons formed as a result of accretion, a process involving the addition of eroded continental material, igneous rocks, and island arcs to the margin of a craton during orogenesis.

3. In North America, the craton evolved during Precambrian time by collision of smaller cratons along belts of deformation known as orogens and by accretion along its southern and eastern margins. Since then, orogenies have resulted in continental accretion along the craton's margin.

4. Six major continents existed at the beginning of the Paleozoic Era.

5. During the Early Paleozoic (Cambrian-Silurian), Laurentia was moving northward, and Gondwana moved to a south polar location, as indicated by glacial deposits.

6. During the Late Paleozoic, Baltica and Laurentia collided, forming Laurasia. Siberia and Kazakhstania collided and finally were sutured to Laurasia. Gondwana moved over the South Pole and experienced several glacial-interglacial periods, resulting in global changes in sea level and transgressions and regressions along the low-lying craton margins.

7. Laurasia and Gondwana underwent a series of collisions beginning in the Carboniferous. During the Permian, the formation of Pangaea was completed.

8. The Sauk Sea was the first of several major transgressions onto the North American craton. At its peak it covered the craton except for a series of large, northeast-southwest–trending islands called the Transcontinental Arch.

9. The Tippecanoe sequence began with deposition of an extensive sandstone over the exposed and eroded Sauk landscape. During Tippecanoe time, extensive carbonate deposition occurred.

10. Except for a persistent and widespread black shale during the Late Devonian and Early Mississippian, most of the Kaskaskia sequence was dominated by carbonates and evaporites.

11. Transgressions and regressions over the low-lying craton resulted in cyclothems and the formation of coals during the Pennsylvanian Period of the Absaroka sequence. Cratonic mountain building also occurred during the Pennsylvanian, with thick nonmarine detrital rocks and evaporites accumulating in the intervening basins. The Absaroka Sea occupied a narrow zone of the south-central craton by the Early Permian and completely retreated from the craton by the end of the Permian.

12. During Tippecanoe time, an oceanic-continental convergent plate boundary formed, resulting in the Taconic orogeny, the first of several orogenies to affect the Appalachian mobile belt. The newly formed Taconic Highlands shed sediments into the shallow western sea, producing a clastic wedge called the Queenston Delta.

13. The Taconic orogeny was followed by the Acadian orogeny, which began as an oceanic-continental convergent plate boundary until Baltica collided with Laurentia; it then formed a continental-continental convergent plate boundary. The Hercynian-Alleghenian orogeny marks the suturing of Laurasia and Gondwana and the formation of Pangaea.

14. The breakup of Pangaea can be divided into four stages:
    a. The first stage involved the separation of North America from Africa during the Late Triassic, followed by the separation of North America from South America.
    b. The second stage involved the separation of Antarctica, India, and Australia from South America and Africa during the Jurassic. During this stage, India broke away from the still-united Antarctica and Australia landmass.
    c. During the third stage, South America separated from Africa, while Europe and Africa began to converge.
    d. In the last stage, Greenland separated from North America and Europe.

15. The eastern coastal region was the initial site of the separation of North America from Africa that began during the Late Triassic.

16. The Gulf Coastal region was the site of major evaporite accumulation during the Jurassic as North America rifted from South America. During the Cretaceous, it was inundated by a transgressing sea, which, at its maximum, connected with a sea transgressing from the north to create the Cretaceous Interior Seaway.

17. Mesozoic rocks of the western region of North America were deposited in a variety of continental and marine environments. In addition, this region was affected by four interrelated orogenies: the Sonoma, Nevadan, Sevier, and Laramide. Each involved batholithic intrusions as well as eastward thrust faulting and folding.

18. Cenozoic orogenic activity occurred mostly in two major belts—the Alpine-Himalayan orogenic belt and the circum-Pacific orogenic belt. Each belt is composed of smaller units called orogens.

19. The North American Cordillera is a complex mountainous region extending from Alaska into Mexico. Its Cenozoic evolution included deformation during the Laramide orogeny, extensional tectonics that formed the Basin and Range structures, intrusive and extrusive volcanism, and uplift and erosion.

20. The westward drift of North America resulted in its collision with the Pacific-Farallon Ridge. Subduction ceased, and the continental margin became bounded by major transform faults, except where the Juan de Fuca plate continues to collide with North America.

21. Cenozoic uplift and erosion are responsible for the present topography of the Appalachian Mountains.

22. About 20 warm-cold Pleistocene climatic cycles are recognized, while several intervals of widespread glaciation, separated by interglacial periods, occurred in North America.

Absaroka sequence
Acadian orogeny
craton
cyclothem

Hercynian-Alleghenian
    orogeny
Kaskaskia sequence
Laramide orogeny

mobile belt
Nevadan orogeny
Ouachita orogeny
Sauk sequence

Sevier orogeny
Taconic orogeny
Tippecanoe sequence

REVIEW QUESTIONS

1. The largest exposed area of the North American craton is the:
   a. ____ Wyoming craton;
   b. ____ American platform;
   c. ____ Canadian Shield;
   d. ____ Appalachian Mountains;
   e. ____ Grand Canyon.

2. An elongated area marking the site of former mountain building is a:
   a. ____ craton;
   b. ____ platform;
   c. ____ shield;
   d. ____ shallow sea;
   e. ____ mobile belt.

3. Which of the following was not a Paleozoic continent?
   a. ____ Gondwana;
   b. ____ Baltica;
   c. ____ Kazakhstania;
   d. ____ Eurasia;
   e. ____ Laurentia.

4. Which was the first major transgressive sequence onto the North American craton?
   a. ____ Sauk;
   b. ____ Tippecanoe;
   c. ____ Kaskaskia;
   d. ____ Absaroka;
   e. ____ Zuni.

5. Rhythmically repetitive sedimentary sequences are:
   a. ____ tillites;
   b. ____ cyclothems;
   c. ____ orogenies;
   d. ____ reefs;
   e. ____ evaporites.

6. A major transgressive-regressive cycle bounded by cratonwide unconformities is (a)n:
   a. ____ cratonic sequence;
   b. ____ shallow sea;
   c. ____ orogeny;
   d. ____ biostratigraphic unit;
   e. ____ cyclothem.

7. Weathering of which highlands or mountains produced the Catskill Delta?
   a. ____ Acadian;
   b. ____ Alleghenian;
   c. ____ Antler;
   d. ____ Appalachian;
   e. ____ Caledonian.

8. The first exclusively Mesozoic orogeny in the Cordilleran region was the:
   a. ____ Antler;
   b. ____ Laramide;
   c. ____ Nevadan;
   d. ____ Sevier;
   e. ____ Sonoma.

9. The breakup of Pangaea began with initial Triassic rifting between which two continental landmasses?
   a. ____ South America and Africa;
   b. ____ Laurasia and Gondwana;
   c. ____ North America and Eurasia;
   d. ____ Antarctica and India;
   e. ____ India and Australia.

10. During which geologic period did the greatest post-Paleozoic inundation of the craton occur?
    a. ____ Triassic;
    b. ____ Jurassic;
    c. ____ Cretaceous;
    d. ____ Paleogene;
    e. ____ Neogene.

11. The Cenozoic history of the Appalachians involved mostly:
    a. ____ uplift and erosion;
    b. ____ subduction and island arc collision;
    c. ____ tension and block-faulting;
    d. ____ compression and folding;
    e. ____ carbonate deposition and the origin of salt domes.

12. How many years ago did the Pleistocene Epoch begin?
    a. ____ 6.6 million;
    b. ____ 2 million;
    c. ____ 1.6 million;
    d. ____ 1 million;
    e. ____ 10,000.

13. Briefly discuss the Paleozoic geologic history of the world.

14. What are cyclothems? How do they form? Why are they economically important?

15. Compare the Taconic and Acadian orogenies in terms of the tectonic forces that caused them and the sedimentary features that resulted.

16. Provide a general global history of the breakup of Pangaea.

17. Discuss the tectonics of the Cordilleran mobile belt during the Mesozoic Era.

18. Briefly discuss the Cenozoic geologic history of the North American Cordillera.

1. Why are cratonic sequences a convenient way to study the geologic history of the Paleozoic Era?

2. How did the breakup of Pangaea affect oceanic and climatic circulation patterns?

Chapter 19

LIFE HISTORY

OUTLINE

Diorama of the environment and biota of the Phyllopod bed of the Burgess Shale, British Columbia, Canada. In the background is the vertical wall of a submarine escarpment with algae growing on it. The large cylindrical ribbed organisms on the muddy bottom in the foreground are sponges.

On August 30 and 31, 1909, near the end of the summer field season, Charles D. Walcott, geologist and head of the Smithsonian Institution, was searching for fossils along a trail on Burgess Ridge between Mount Field and Mount Wapta, near Field, British Columbia, Canada. On the west slope of the ridge, he discovered the first soft-bodied fossils from the Burgess Shale, a discovery of immense importance in deciphering the early history of life. During the following week, Walcott and his collecting party split open numerous blocks of shale, many of which yielded the impressions of a number of soft-bodied organisms beautifully preserved on bedding planes (⊳ Figure 19-1). Returning the next summer, Walcott quarried the site and shipped back thousands of fossil specimens to the U.S. National Museum of Natural History for later cataloging and study.

Walcott had discovered more than just another collection of well-preserved Cambrian fossils; his find provided a rare glimpse into a world previously almost unknown—that of the soft-bodied animal community that lived some 530 million years ago. The beautifully preserved fossils from the Burgess Shale present a much more complete picture of a Middle Cambrian community than deposits containing only fossils with hard parts. Specifically, the Burgess Shale contains species of trilobites, sponges, brachiopods, mollusks, and echinoderms, all of which have hard parts and are characteristic of Cambrian faunas throughout the world. In addition to the diverse skeletonized fauna, a large and varied fossil assemblage of soft-bodied animals is also present. In all, more than 100 genera of animals, at least 60 of which were soft-bodied and preserved as impressions, have been recovered from the Burgess Shale.

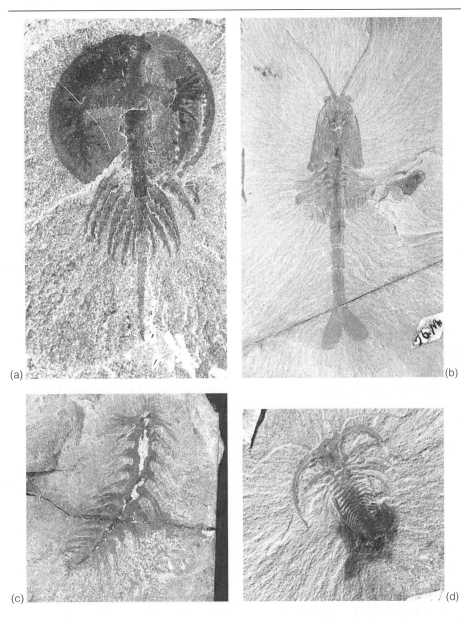

⊳ **FIGURE 19-1** Some of the fossil animals preserved in the Burgess Shale. (a) *Burgessia bella*, an arthropod. (b) *Waptia fieldensis*, another arthropod. (c) *Burgessochaeta setigera*, an annelid worm. (d) *Marrella splendens*, an arthropod that is the most abundant fossil animal in the Phyllopod bed of the Burgess Shale.

The animals whose exquisitely preserved fossil remains are found in the Burgess Shale lived in and on mud banks that formed the top of a steep escarpment. Periodically, this area would slide down the escarpment carrying the mud and animals to the base where they were deposited in a deep-water anaerobic environment devoid of life. Here, bacterial degradation did not destroy the buried animals, and they were compressed by the weight of the overlying sediments, eventually resulting in their preservation as carbonaceous remains.

INTRODUCTION

The only evidence of prehistoric life preserved in the geologic record consists of various types of body and trace fossils. **Fossils**, which are the remains or traces of prehistoric life, are preserved in rocks of the Earth's crust and make up the fossil record. The quality of the fossil record varies depending on the types of organisms living at a given time and the environment they lived in. Some groups have an excellent fossil record, while for others it is poor or even nonexistent. Organisms that have hard parts and live in areas where rapid sedimentation occurs have a better chance of being preserved. Therefore, the fossil record is biased toward marine organisms with preservable skeletons. Nevertheless, the fossil record provides us with an overview of the history of life.

PRECAMBRIAN LIFE

Prior to the mid-1950s, we had very little knowledge of fossils older than Paleozoic. Scientists had long assumed that the fossils so abundant in Cambrian sedimentary rocks must have had a long earlier history, but no record of these earlier organisms was known. Some enigmatic Precambrian fossils had been reported, but they were mostly dismissed as some type of inorganic features. In fact, the Precambrian was once referred to as the *Azoic*, meaning devoid of life.

In the early 1900s, Charles Walcott, who discovered the Burgess Shale fauna (see the Prologue), described layered, moundlike structures from the Precambrian of Ontario, Canada, that are now called **stromatolites**. Walcott proposed that they represented reefs constructed by algae, but paleontologists did not demonstrate that stromatolites are the products of organic activity until 1954. Studies of present-day stromatolites in such areas as Shark Bay, Australia, show that they originate when sediment grains are trapped on sticky mats of photosynthesizing cyanobacteria, commonly called blue-green algae.

The oldest known stromatolites are in 3.3- to 3.5-billion-year-old rocks in Australia. Indirect evidence for even more ancient life comes from 3.8-billion-year-old rocks in Greenland. These rocks contain small carbon spheres that may be of biologic origin, but the evidence is currently not conclusive.

The earliest life-forms were all varieties of single-celled bacteria that lacked a cell nucleus and apparently reproduced asexually, much as bacteria do today ($\triangleright$ Figure 19-2).

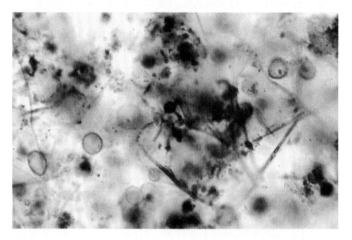

$\triangleright$ **FIGURE 19-2** Photomicrographs of spheroidal and filamentous microfossils from stromatolitic chert of the Gunflint Iron Formation, Ontario.

During this time in Earth history (3.8 to about 1.4 billion years ago), evolution was a comparatively slow process, and organic diversity was limited. About 1.4 billion years ago, more complex cells that had a distinct nucleus and were capable of sexual reproduction appeared in the fossil record. These organisms were still single celled, but once they evolved, organic diversity increased markedly. Multicelled algae are found in rocks at least 1 billion years old, and during the Late Precambrian, about 700 million years ago, the first multicelled animals evolved ($\triangleright$ Figure 19-3).

These multicelled animals are collectively referred to as the *Ediacaran fauna*, and their fossils are found on all continents except Antarctica. The type of animals comprising the Ediacaran fauna is currently the subject of debate. Some investigators are of the opinion that these oldest known animals include jellyfish and sea pens, segmented worms, and primitive arthropods. One wormlike fossil (Figure 19-3d) has even been cited as a possible ancestor of the trilobites that were so common during the Early Paleozoic. Other researchers disagree and think that these animals represent an early evolutionary development quite distinct from the ancestry of any present-day animals.

Regardless of which opinion is correct, the fact remains that complex, multicelled soft-bodied animals were not only present but distributed nearly worldwide by the Late Proterozoic. Their fossil record is poor because they lacked durable skeletons. Near the end of the Proterozoic, several

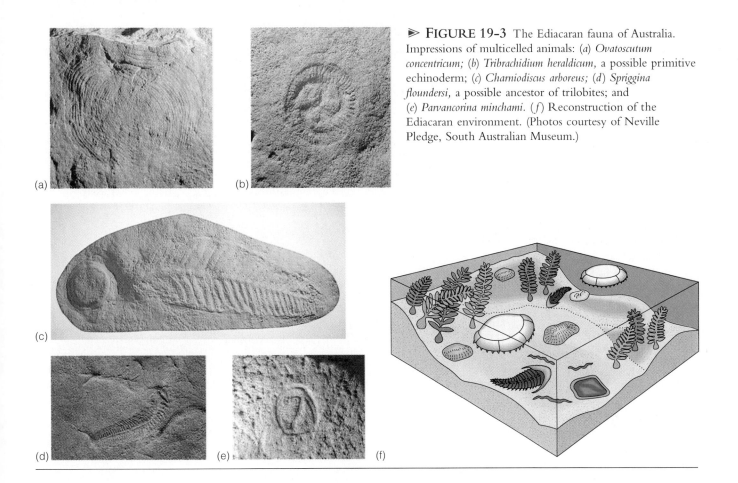

▷ **FIGURE 19-3** The Ediacaran fauna of Australia. Impressions of multicelled animals: (*a*) *Ovatoscutum concentricum;* (*b*) *Tribrachidium heraldicum,* a possible primitive echinoderm; (*c*) *Charniodiscus arboreus;* (*d*) *Spriggina floundersi,* a possible ancestor of trilobites; and (*e*) *Parvancorina minchami.* (*f*) Reconstruction of the Ediacaran environment. (Photos courtesy of Neville Pledge, South Australian Museum.)

(a) (b) (c) (d) (e) (f)

animals possessing skeletons did evolve. Evidence for their existence comes from minute scraps of shell-like material and spicules, presumably from sponges. Nevertheless, animals with durable skeletons of chitin (a complex organic substance), silica (SiO_2), and calcium carbonate ($CaCO_3$) were not abundant until the beginning of the Paleozoic.

🜂 PALEOZOIC LIFE

Long ago, geologists observed that the remains of animals with skeletons appeared rather abruptly in the fossil record at the beginning of the Paleozoic. Charles Darwin discussed this seemingly abrupt event and observed that without a convincing explanation, it was difficult to reconcile with his newly expounded evolutionary theory. Scientists now recognize that a number of multicelled animals existed during the late Precambrian and that some of these were surely the ancestors of the Paleozoic skeletonized animals.

The evolution of numerous animals with durable skeletons is sometimes described as an explosive development of new types of animals. In fact, their appearance in the fossil record was rapid only in the context of geologic time, occurring over millions of years during the Cambrian and on into the Ordovician periods (▷ Figure 19-4). Had

Cambrian			Ordovician	
Early	Middle	Late	Early	Middle
Arthropoda (Trilobites)				→
Brachiopoda				→
Echinodermata				→
Mollusca (Gastropoda)				→
Porifera				→
	Mollusca (Bivalvia)			→
	Mollusca (Cephalopoda)			→
		Protozoa		→
			Bryozoa	→
			Cnidaria (Tabulate corals)	→
				Cnidaria (Rugose corals) →

▷ **FIGURE 19-4** The first recorded occurrence of selected members of the major groups of marine invertebrates.

TABLE 19-1 The Major Invertebrate Groups and Their Stratigraphic Ranges

Phylum Protozoa	Cambrian-Recent	**Phylum Mollusca**	Cambrian-Recent
Class Sarcodina	Cambrian-Recent	Class Monoplacophora	Cambrian-Recent
Order Foraminifera	Cambrian-Recent	Class Gastropoda	Cambrian-Recent
Order Radiolaria	Cambrian-Recent	Class Bivalvia	Cambrian-Recent
		Class Cephalopoda	Cambrian-Recent
Phylum Porifera	Cambrian-Recent		
Class Demospongea	Cambrian-Recent	**Phylum Annelida**	Precambrian-Recent
Order Stromatoporoida	Cambrian-Oligocene		
		Phylum Arthropoda	Cambrian-Recent
Phylum Archaeocyatha	Cambrian	Class Trilobita	Cambrian-Permian
		Class Crustacea	Cambrian-Recent
Phylum Cnidaria	Cambrian-Recent	Class Insecta	Silurian-Recent
Class Anthozoa	Ordovician-Recent		
Order Tabulata	Ordovician-Permian	**Phylum Echinodermata**	Cambrian-Recent
Order Rugosa	Ordovician-Permian	Class Blastoidea	Ordovician-Permian
Order Scleractinia	Triassic-Recent	Class Crinoidea	Cambrian-Recent
		Class Echinoidea	Ordovician-Recent
Phylum Bryozoa	Ordovician-Recent	Class Asteroidea	Ordovician-Recent
Phylum Brachiopoda	Cambrian-Recent	**Phylum Hemichordata**	Cambrian-Recent
Class Inarticulata	Cambrian-Recent	Class Graptolithina	Cambrian-Mississippian
Class Articulata	Cambrian-Recent		

humans been present to observe this event, it would have seemed incredibly gradual.

Rather than focusing on the history of each invertebrate group (◉ Table 19-1), we will survey the evolution of the Paleozoic marine invertebrate communities through time, concentrating on their major features and the changes that occurred.

Cambrian Marine Invertebrate Community

Although almost all the major invertebrate groups appeared in the fossil record during the Cambrian Period, many were represented by only a few species. Though trace fossils are common and echinoderms diverse, trilobites, brachiopods, and archaeocyathids (bottom-dwelling organisms that constructed reeflike structures and lived only during the Cambrian) comprised the majority of Cambrian skeletonized life (▷ Figure 19-5). It is important to remember, however, that the fossil record is biased toward organisms with durable skeletons and that we generally know little about the soft-bodied organisms of that time (see the Prologue). At the end of the Cambrian Period, trilobites suffered mass extinctions, and even though they persisted until the end of the Paleozoic Era, their numbers were considerably diminished.

Ordovician Marine Invertebrate Community

A major transgression that began during the Middle Ordovician resulted in a widespread inundation of the craton. This vast shallow sea opened numerous new marine habitats that were soon filled by a variety of organisms such that the Ordovician is characterized by a dramatic increase in the

▷ **FIGURE 19-5** Reconstruction of a Cambrian marine community. Floating jellyfish, swimming arthropods, sponges, and scavenging trilobites are shown.

diversity of the total shelly fauna (Table 19-1). The end of the Ordovician, however, was a time of mass extinctions in the marine realm. More than 100 families of marine invertebrates did not survive into the Silurian.

Silurian and Devonian Marine Invertebrate Community

The mass extinction at the end of the Ordovician was followed by rediversification and recovery of many of the decimated groups. Brachiopods, bryozoans, gastropods, bivalves, corals, crinoids, and graptolites were just some of the groups that rediversified beginning during the Silurian. In fact, the Silurian and Devonian were times of major reef building in which organic reef-builders diversified in new ways, building massive reefs larger than any produced during the Cambrian or Ordovician (➤ Figure 19-6). Another mass extinction occurred near the end of the Devonian and resulted in a worldwide near-total collapse of the massive reef communities.

Carboniferous and Permian Marine Invertebrate Community

The Carboniferous invertebrate marine community responded to the Late Devonian extinctions in much the same way the Silurian invertebrate marine community responded to the Late Ordovician extinctions—that is, by renewed diversification. Large organic reefs like those existing earlier in the Paleozoic virtually disappeared, however, and were replaced by small patch reefs that flourished during the Late Paleozoic.

The Permian invertebrate marine faunas resembled those of the Carboniferous. They were not distributed as widely, though, due to the restricted size of the shallow seas on the cratons and the reduced shelf space along the continental margins (➤ Figure 19-7).

The Permian Marine Invertebrate Extinction Event

The greatest recorded mass extinction event to affect the marine invertebrate community occurred at the end of the Permian Period (➤ Figure 19-8). Before the Permian ended, roughly one-half of all marine invertebrate families and perhaps 90% of all marine invertebrate species became extinct. Among these extinct invertebrates were two groups of primitive corals, two groups of bryozoans, many brachiopods, and all remaining trilobites; several other groups either

➤ **FIGURE 19-7** A Permian patch-reef community from the Glass Mountains of west Texas. Shown are algae, brachiopods, cephalopods, and corals.

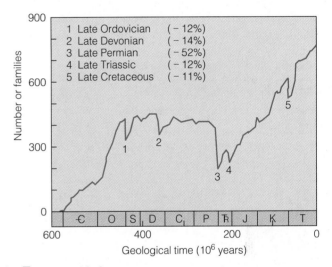

➤ **FIGURE 19-8** Phanerozoic diversity for marine invertebrate and vertebrate families. Note the three episodes of Paleozoic mass extinctions, with the greatest occurring at the end of the Permian Period.

➤ **FIGURE 19-6** Reconstruction of a Middle Devonian reef from the Great Lakes area. Shown are corals, cephalopods, trilobites, and brachiopods. (From the Field Museum, Chicago # Geo80821c.)

became extinct or were greatly reduced in number. On land, many amphibian and reptile families also died out.

What caused such a crisis for the marine invertebrates? Many hypotheses have been proposed, but no completely satisfactory answer has been found. Two currently discussed hypotheses are (1) a reduction of living area related to widespread regression of the seas and suturing of the continents when Pangaea formed and (2) decreased ocean salinity due to widespread arid climates. Continental convergence resulted in regression of the shallow seas from the cratons and reduction of the shallow-water shelf area surrounding each continent. Decreased ocean salinity would affect those organisms with narrow salinity tolerances, which includes most marine invertebrates.

The Permian mass extinctions were probably caused by a combination of many interrelated geologic and biologic factors. In any case, the surviving marine invertebrate faunas of the Early Triassic were of very low diversity and were widely distributed around the world.

Vertebrate Evolution

In addition to the numerous invertebrate groups, **vertebrates** (animals with a segmented vertebral column) also evolved and diversified during the Paleozoic. Remains of the most primitive vertebrates, the fishes, are found in Late Cambrian-aged marine rocks in Wyoming. These fish, referred to as *ostracoderms*, were jawless and had poorly developed fins and an external covering of bony armor (▷ Figure 19-9). Except for their distant relatives, the lamprey and slime hag, ostracoderms are now all extinct.

During the Silurian, two more groups of fish evolved. One of these groups, the *placoderms* (heavily armored jawed fish), included one of the largest marine predators that ever existed (Figure 19-9). The other group, the *acanthodians*,

were enigmatic fish characterized by large spines, scales covering much of the body, jaws, teeth, and reduced bony armor (Figure 19-9). Many scientists think the acanthodians included the probable ancestors of the present-day bony and cartilaginous fish groups (Figure 19-9).

Among the bony fishes, one group known as lobe-fins was particularly important because they included the ancestor of the amphibians, the first land-dwelling vertebrate animals (▷ Figure 19-10). In fact, the structural similarity between the group of lobe-finned fishes that gave rise to the amphibians and the earliest amphibians is striking and is one of the better documented transitions in the geologic record.

Although amphibians were living on land by the Devonian, they were not the first land-dwelling organisms. Plants made the transition to land during the Ordovician, and several invertebrates including insects, millipedes, spiders, and snails invaded the land before amphibians. All of these organisms encountered several problems during the transition from water to land. The most critical obstacles for animals were drying out, reproduction, the effects of gravity, and the extraction of oxygen from the atmosphere by lungs rather than from water by gills. These problems were partly solved by some of the lobe-finned fishes; they already had a

▷ **FIGURE 19-9** Recreation of a Devonian sea floor showing (*a*) an ostracoderm (*Hemicyclaspis*), (*b*) a placoderm (*Bothriolepis*), (*c*) an acanthodian (*Parexus*), and (*d*) a ray-finned fish (*Cheirolepis*).

➤ **FIGURE 19-10** A Late Devonian landscape in the eastern part of Greenland. Shown is *Ichthyostega*, an amphibian that grew to a length of about 1 m. The flora of the time was diverse, consisting of a variety of small and large seedless vascular plants.

backbone and limbs that could be used for support and walking on land and lungs to extract oxygen from the atmosphere.

Amphibians' ability to colonize the land was limited both because they had to return to water to lay their gelatinous eggs and because they never completely solved the problem

of drying out. In contrast, the reptiles evolved skin or scales to preserve their internal moisture and an egg in which the developing embryo is surrounded by a liquid-filled sac and provided with both a food and a waste sac. The evolution of such an egg allowed vertebrates to colonize all parts of the land because they no longer had to return to the water as part of their reproductive cycle. The oldest known reptiles appeared during the Mississippian Period. They were small, agile animals known as *stem reptiles* because they were the ancestors of all other reptiles.

One of the descendant groups of the stem reptiles was the *pelycosaurs*, which were the dominant reptiles by the Permian (➤ Figure 19-11). The pelycosaurs, also known as primitive mammal-like reptiles, were the first land-dwelling vertebrates to become diverse and widespread. Moreover, they were the Permian ancestors of the *therapsids*, or advanced mammal-like reptiles, which gave rise to mammals during the Triassic Period.

Plant Evolution

Plants encountered many of the same problems animals faced when they made the transition to land: drying out, the effects of gravity, and reproduction. Plants adapted by evolving a variety of structural features that allowed them to invade the land during the Ordovician and later periods. Most experts agree that the ancestors of land plants were green algae that first evolved in a marine environment, then moved into freshwater, and finally onto land.

The commonest and most widespread land plants are vascular plants, which have a tissue system of specialized cells for the movement of water and nutrients. Nonvascular plants, such as mosses and fungi, lack these specialized cells and are typically small and usually live in low, moist areas. Nonvascular plants were probably the first to make the transition to land, but their fossil record is poor.

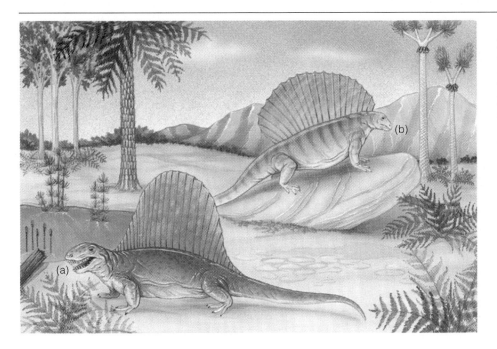

➤ **FIGURE 19-11** Most pelycosaurs have a characteristic sail on their back. One hypothesis explains the sail as a type of thermoregulatory device. Other hypotheses are that it was a type of sexual display or a device to make the reptile look more intimidating. Shown here are (*a*) the carnivore *Dimetrodon* and (*b*) the herbivore *Edaphosaurus*.

The earliest known, well-preserved fossils of vascular land plants are small, leafless, Y-shaped stems from the Middle Silurian of Wales and Ireland (▶ Figure 19-12). They are known as **seedless vascular plants** because they did not produce seeds. Although these plants lived on land, they never completely solved the problem of drying out and were thus restricted to moist areas. Even their living descendants, such as ferns, are usually found in moist areas. During the Pennsylvanian Period, the seedless vascular plants became very abundant and diverse because of the widespread coal-forming swamps that were ideally suited to their lifestyle (▶ Figure 19-13).

In addition to the evolution of diverse seedless vascular plants, another significant floral event occurred by the Late Devonian. The evolution of the seed at this time liberated vascular plants from their dependence on moist conditions and allowed them to spread over all parts of the land. The first to do so were the flowerless seed plants, or **gymnosperms**, which include the living cycads, conifers, and ginkgoes. While the seedless vascular plants dominated the flora of the Pennsylvanian coal-forming swamps, the gymnosperms made up an important element of the Late Paleozoic flora, particularly in the nonswampy areas.

MESOZOIC LIFE

The Mesozoic Era is commonly referred to as the "Age of Reptiles," alluding to the fact that reptiles were the most diverse and abundant land-dwelling vertebrates. Many people find the Mesozoic the most interesting time in the history of life because among the reptiles were dinosaurs, flying reptiles, and marine reptiles. In addition to the diversification of reptiles, other equally important events occurred. Mammals evolved from mammal-like reptiles during the Triassic, and birds evolved from reptiles, probably small carnivorous dinosaurs, during the Jurassic.

Invertebrates

Following the wave of extinctions at the end of the Paleozoic, the Mesozoic was a time during which marine invertebrates repopulated the seas. Among the mollusks, the clams, oysters, and snails became increasingly diverse and abundant, and the cephalopods were among the most important Mesozoic invertebrate groups. On the other hand, the brachiopods never completely recovered from their near extinction and have remained a minor invertebrate group ever since. In areas of warm, clear, shallow marine waters, corals again proliferated, but these corals were of a new and more familiar type (▶ Figure 19-14).

Single-celled animals known as foraminifera (Table 19-1) were also very important; they diversified tremendously during the Jurassic and Cretaceous periods. Floating or planktonic forms in particular became extremely common, but many of them became extinct at the end of the Mesozoic, and only a few types survived into the Cenozoic.

Dinosaurs

A common but erroneous perception of **dinosaurs** is that they were poorly adapted animals that had trouble surviving. True, they became extinct, but to consider this a failure is to ignore the fact that for more than 140 million years they were the dominant land vertebrates. During their existence, they diversified into numerous types and adapted to a wide variety of environments. Nor were dinosaurs the lethargic beasts often portrayed in various media. Recent evidence indicates that at least some may have been very active and possibly warm-blooded (see Perspective 19-1). It also appears that some species cared for their young long after hatching, a behavioral characteristic most often associated with birds and mammals. Although many dinosaurs were large, many were smaller than present-day elephants and one was no larger than a chicken. Eventually, the dinosaurs did die out, an event that then enabled mammals to become the dominant land vertebrates.

▶ FIGURE 19-12 The earliest known fertile land plant was *Cooksonia,* seen in this fossil from the Upper Silurian of South Wales. *Cooksonia* consisted of upright, branched stems terminating in sporangia (spore-producing structures). It also had a resistant cuticle and produced spores typical of a vascular plant. These plants probably lived in moist environments such as mud flats. This specimen is 1.49 cm long. (Photo courtesy of Dianne Edwards, University College, England.)

> **FIGURE 19-13** Reconstruction of a Pennsylvanian coal swamp with its characteristic vegetation. The amphibian is *Eogyrinus*.

> **FIGURE 19-14** Scleractinian corals evolved during the Triassic and proliferated in the warm, clear, shallow marine waters of the Mesozoic Era. Most living corals are scleractinians, represented here by the so-called staghorn coral. (Photo courtesy of Sue Monroe.)

Among the dinosaurs, two orders are recognized, the *saurischians* and *ornithischians* (> Figure 19-15). The saurischian dinosaurs had a lizard-like pelvis and are therefore referred to as lizard-hipped dinosaurs. Ornithischians had a bird-like pelvis; hence they are called bird-hipped dinosaurs.

Two groups of saurischian dinosaurs are recognized: *theropods* and *sauropods* (Figure 19-15). Theropods were carnivorous dinosaurs ranging in size from tiny *Compsognathus*, which was the size of a chicken, to *Tyrannosaurus*, the largest terrestrial carnivore known (> Figure 19-16).

Included among the sauropods were the giant herbivores (plant eaters) such as *Apatosaurus*, *Diplodocus*, and *Brachiosaurus* (Figure 19-15), the largest known land animals of any kind. Evidence from fossil trackways indicates that sauropods moved in herds. They depended on their size and herding behavior rather than speed as their primary protection from predators.

The great diversity of ornithischians is manifested by the fact that five distinct groups are recognized: *ornithopods, pachycephalosaurs, ankylosaurs, stegosaurs,* and *ceratopsians*

WARM BLOODED DINOSAURS?

All living reptiles are *ectotherms*, that is, cold-blooded animals whose body temperature varies in response to the outside temperature. *Endotherms*, warm-blooded animals such as birds and mammals, are capable of maintaining a rather constant body temperature regardless of the outside temperature. Some investigators think that dinosaurs, or at least some dinosaurs, were endotherms.

Proponents of dinosaur endothermy note that dinosaur bones are penetrated by numerous passageways that, when the animals were living, contained blood vessels. Bones of endotherms typically have this structure, but considerably fewer of these passageways are found in bones of ectotherms. Living crocodiles and turtles have this so-called endothermic bone structure, yet we know that they are ectotherms and not capable of maintaining a constant body temperature. It may be that bone structure is more related to body size and growth patterns than to endothermy.

Because endotherms have high metabolic rates, they must eat more than ectotherms of comparable size. Consequently, endothermic predators require large prey populations. They would therefore constitute a much smaller proportion of the total animal population than their prey. In contrast, ectothermic predators are much more numerous in relation to their prey population. Where data are sufficient to allow an estimate, dinosaur predators appear to have made up 3 to 5 percent of the total population. These figures are comparable to present-day mammalian populations. Still, a number of uncertainties about the composition of fossil communities make this argument for endothermy unconvincing to many paleontologists.

Large endotherms have a large brain in relation to body size. A relatively large brain is not necessary for endothermy, but endothermy does seem to be a prerequisite for having a large brain because a complex nervous system requires a rather constant body temperature. Some dinosaurs, particularly the small carnivores, did have a large brain in relation to their body, but many did not. That the small carnivorous ones had a large brain seems to be a good argument for endothermy, but other evidence is even more compelling. The relationship of birds to small carnivorous dinosaurs implies that these dinosaurs were endothermic or at least trending in that direction.

The large sauropods were probably not endothermic, but nevertheless may have been able to maintain their body temperatures within narrow limits as endotherms do. A large animal heats up and cools down slowly because it has a small surface area compared to its volume. With proportionately less surface area to allow heat loss, sauropods probably retained body heat more efficiently than smaller dinosaurs.

One further point on endothermy in dinosaurs is that the flying reptiles, the *pterosaurs*, evolved from the same ancestral reptile group as did the dinosaurs. At least one species of pterosaur had hair or hairlike feathers. This is interesting because an insulating covering of hair or feathers is known only in endotherms. Furthermore, the physiology of active flight requires endothermy. Such evidence indicates that perhaps both the dinosaurs' ancestors and the dinosaurs themselves were endothermic.

Obviously, considerable disagreement exists on dinosaur endothermy. In general, a fairly good case can be made that small, carnivorous dinosaurs and pterosaurs were endothermic, but for the others the question is still open.

(Figures 19-15 and 19-16). Ornithopods include the duck-billed dinosaurs, which had flattened bill-like mouths (Figure 19-15). Some species were characterized by head crests that may have functioned as resonating chambers to amplify bellowing. Some duck-billed dinosaurs practiced colonial nesting and care of the young. All ornithopods were herbivores and walked primarily on their hind feet.

The pachycephalosaurs constitute a most peculiar group of ornithischian dinosaurs. The most distinctive feature of these bipedal (upright walking) herbivores is the dome-shaped skull that resulted from thickening of the bones (Figure 19-15). Some scientists think that these domed skulls were used in intraspecific butting contests for dominance and mates.

Ankylosaurs were heavily armored herbivores, and some were quite large (Figure 19-16). Bony armor protected the back, flanks, and top of the head, and the tail ended in a large, bony clublike growth.

The stegosaurs were herbivores with bony spikes on the tail, which were undoubtedly used for defense, and body plates on the back (Figure 19-15). The exact arrangement of these plates is debated, but many paleontologists think they functioned as a device to absorb and dissipate heat.

A rather good fossil record indicates that large, Late Cretaceous ceratopsians such as *Triceratops* evolved from small, Early Cretaceous ancestors (Figure 19-15 and 19-16). The later ceratopsians were characterized by huge heads, a large bony frill over the top of the neck, and a large horn or horns on the skull. Fossil trackways indicate that these large herbivores moved in herds.

Although dinosaurs attract the most attention, several other Mesozoic reptiles were common. The flying reptiles

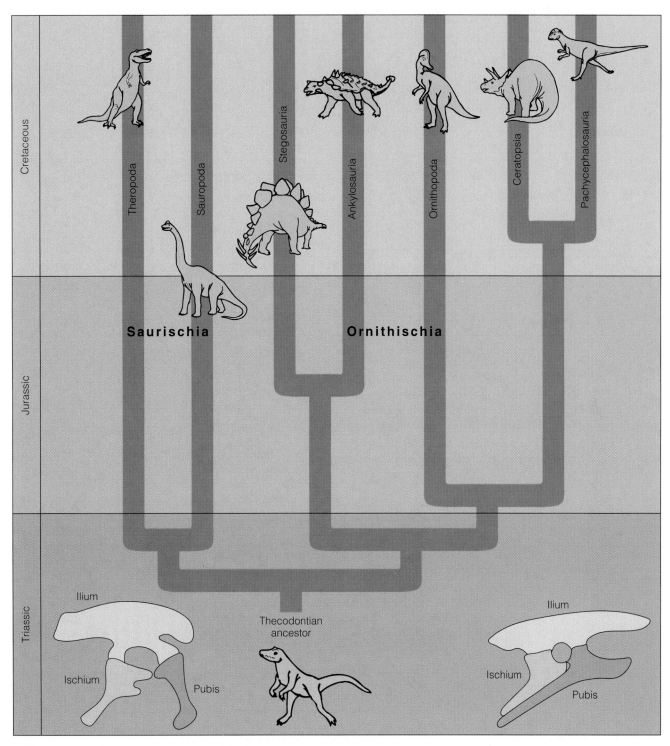

▷ **FIGURE 19-15** Origin of and inferred relationships among dinosaurs. Both ornithischian and saurischian dinosaurs evolved from thecodontians, but each may have had an independent origin from that group. The pelvis of each order of dinosaurs is shown.

and marine reptiles were contemporaries of dinosaurs as were crocodiles, turtles, lizards, and snakes. In addition, it appears that during the Jurassic Period, small carnivorous dinosaurs gave rise to the birds, which, although warm-blooded and feather-covered, retain a number of reptilian characteristics, such as the type of egg.

Flying Reptiles

Pterosaurs, the first vertebrate animals to fly, evolved during the Triassic from the same group of reptiles that the dinosaurs did, and were abundant until their extinction at the end of the Mesozoic (▷ Figure 19-17). Pterosaur flight

▷ **FIGURE 19-16** Scene from the Late Cretaceous showing the ankylosaur *Euoplocephalus* (foreground), the large theropod *Tyrannosaurus,* and the ceratopsian *Triceratops* (right).

adaptations include a wing membrane supported by an elongate fourth finger, light hollow bones, and development of those parts of the brain associated with muscular coordination and sight. Size varied considerably. Some early species ranged from sparrow to robin size, while one Cretaceous pterosaur from Texas had a wingspan of at least 12 m. At least one pterosaur had a hair- or feather-covered body and was a flier, strongly suggesting that it, and perhaps all pterosaurs, were endotherms.

Marine Reptiles

Ichthyosaurs are probably the most familiar of the Mesozoic marine reptiles (▷ Figure 19-18a). Most of these animals were about 3 m long, although one species was up to 15 m long, and were completely aquatic. Their numerous sharp teeth indicate that ichthyosaurs were fish eaters. Some fossils with young ichthyosaurs within the body cavity support the interpretation that female ichthyosaurs retained the eggs in their bodies and gave birth to live young.

(a)

(b)

➤ **FIGURE 19-17** (*a*) Long-tailed pterosaur from the Jurassic of Europe. *Pteranodon* (*b*) was a short-tailed Cretaceous pterosaur with a wingspan of more than 6 m.

A second group of Mesozoic marine reptiles, the **plesiosaurs**, occurred in two varieties: short necked and long necked (Figure 19-18b). Most plesiosaurs were between 3.6 and 6 m long, but one species from Antarctica measures 15 m. Long-necked plesiosaurs were heavy-bodied animals with mouthfuls of sharp teeth and limbs specialized into oarlike paddles. They probably rowed themselves through the water and may have used their long necks in snakelike fashion to capture fish. Plesiosaurs probably came ashore to lay their eggs.

Mammals

In a previous section, we briefly mentioned therapsids, or advanced mammal-like reptiles. During the Late Triassic, one group of therapsids gave rise to mammals. The transi-

➤ **FIGURE 19-18** Mesozoic marine reptiles. (*a*) Ichthyosaurs and (*b*) a long-necked plesiosaur.

(a)

(b)

➤ FIGURE 19-19 The oldest known placental mammals were members of the order Insectivora such as those in this scene from the Late Cretaceous. These animals probably fed on insects, worms, and grubs.

tion from mammal-like reptile to mammal is well documented by fossils and is so gradational that classification of some fossils as either reptile or mammal is difficult. Even though mammals appeared at the same time as dinosaurs, their diversity remained low, and all of them were small animals during the rest of the Mesozoic Era (➤ Figure 19-19).

Plants

Triassic and Jurassic land-plant communities, like those of the Late Paleozoic, were composed of seedless vascular plants and various flowerless seed plants. During the Cretaceous, these two groups were largely replaced by the flowering plants, or **angiosperms**. Since they first evolved, angiosperms have adapted to nearly every terrestrial habitat from mountains to deserts. Some have even adapted to shallow coastal waters. Another measure of their success is that they now account for about 96% of all vascular plant species.

The Cretaceous Mass Extinction Event

Another mass extinction occurred at the end of the Mesozoic Era. It was second in magnitude only to the extinctions at the end of the Paleozoic. Casualties of this extinction included dinosaurs, pterosaurs, marine reptiles, and several types of marine invertebrates. Among the latter were the ammonites, a specialized group of cephalopods that had been very abundant throughout the Mesozoic, a type of reef-building clam, and some species of planktonic foraminifera.

Some scientists think a large meteorite or comet hit the Earth at the end of the Mesozoic Era, setting in motion a chain of events that led to widespread extinctions. According to this idea, enough material was ejected into the atmosphere to block out sunlight for several months, resulting in the cessation of photosynthesis and collapse of the food chain. Certainly, an impact of this size would have had worldwide effects, and recently a possible impact crater was discovered centered on the north end of the Yucatán Peninsula of Mexico. Many scientists concede that the evidence for an impact of some sort is compelling, but question whether it caused the extinctions. Some scientists are of the opinion that a variety of organisms, including dinosaurs, were already on the decline and headed for extinction before the end of the Mesozoic and that a meteorite or comet impact may have simply hastened the process.

◉ CENOZOIC LIFE

The world's flora and fauna continued to evolve during the Cenozoic Era as more familiar types of plants and animals appeared. The flowering plants continued to diversify, but some gymnosperms, especially conifers, remained abundant, and seedless vascular plants still occupied many habitats. The Cenozoic marine environment was populated by those plants and animals that survived the Mesozoic extinction event. The planktonic foraminifera comprised a major component of the marine invertebrate community, but corals, clams, and snails also proliferated. Following the extinction of the ammonites at the end of the Mesozoic, only their relatives the nautiloids, squids, and octopuses were present.

Mammals

For more than 100 million years, mammals had coexisted with dinosaurs, yet their fossil record indicates that even during the Cretaceous Period only a limited number of varieties existed. The mass extinction eliminated the dinosaurs and their relatives, thereby creating numerous adaptive opportunities that were quickly exploited by mammals. In fact, the Cenozoic Era is called the "Age of Mammals."

Among the mammals, only *monotremes* lay eggs, whereas the *marsupials* and *placentals* give birth to live young. **Marsupials** are born in a very immature, almost embryonic condition and then undergo further development in the mother's pouch. In the **placentals**, a placenta nourishes the embryo, permitting the young to develop much more fully before birth. The fossil record of monotremes is so poor that their relationship to other mammals is not clear. The marsupials and placentals, on the other hand, diverged from a common ancestor during the Cretaceous and since then have had separate evolutionary histories.

The phenomenal success of placental mammals is related in part to their reproductive method. A measure of this success is that more than 90% of all mammals, fossil and living, are placental. The only long-term success for marsupials has been in the Australian region where most present-day species live. Marsupials were also common in South America during much of the Cenozoic before a land connection existed between that continent and North America. Once a land connection formed a few million years ago, many placentals migrated south, and all species of indigenous marsupials except opossums became extinct.

Although a major diversification of placental mammals began early in the Cenozoic, these mammals are considered archaic because they were primitive and included several varieties that were holdovers from the Mesozoic (➤ Figure 19-20). Most of these Early Cenozoic mammals had not yet become clearly differentiated from their ancestors, and the differences between herbivores and carnivores were slight. If we could somehow go back and visit this time, we would probably not recognize many of the animals. Some would be at least vaguely familiar, but the ancestors of horses, camels, whales, and elephants would bear little resemblance to their living descendants. By the Middle Cenozoic, all of the major groups of living placental mammals had evolved, and many of these would be easily recognized.

Numerous groups of mammals evolved during the Cenozoic, but some such as camels and horses and their relatives have excellent fossil records. Camels evolved from small, four-toed ancestors and were particularly abundant in North America where most of their evolutionary history is recorded. They became extinct in North America during the Pleistocene, but not before some species migrated to South America and Asia where they still survive. Horses and their living relatives, the rhinoceroses and tapirs, also evolved from small Early Cenozoic ancestors (➤ Figure 19-21). Horses and rhinoceroses were also common in North America, and like camels, they died out here but survived in the Old World.

Primates, the order of mammals to which humans belong, may have evolved by the Late Cretaceous, but they were undoubtedly present by the Early Cenozoic. The **hominids** (family Hominidae), the primate family that includes present-day humans and their extinct ancestors, have a fossil record extending back only about 4 million years

---

➤ **FIGURE 19-20** The archaic mammalian fauna of the Paleocene Epoch included such animals as *Ptilodus* (right foreground), insectivores (right background), *Protictis,* an early carnivore, (left background), and *Pantolambda*, which stood about 1 m tall.

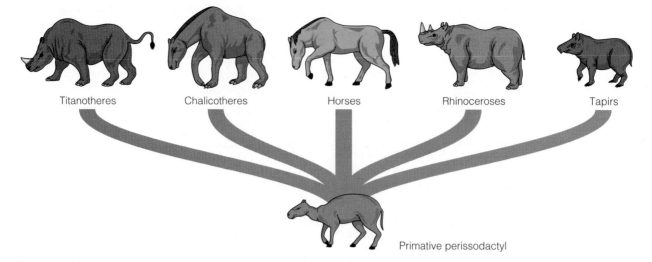

➤ **FIGURE 19-21** Evolution of perissodactyls (odd-toed mammals). Adaptive radiation and divergence from a common ancestor explain why perissodactyls share several characteristics yet differ markedly in appearance.

➤ **FIGURE 19-22** Re-creation of a Pliocene landscape showing *Australopithecus afarensis*, a species of early hominids, gathering and eating various fruits and seeds.

(► Figure 19-22). Several features distinguish them from other primates: hominids walk on two legs rather than four, they show a trend toward a large, complex brain, and they manufacture and use sophisticated tools.

One of the most remarkable aspects of the Cenozoic history of mammals is that so many very large species existed during the Pleistocene Epoch. In North America there were mastodons and mammoths, giant bison, huge ground sloths, giant camels, and beavers nearly 2 m tall at the shoulder. Kangaroos 3 m tall, wombats the size of rhinoceroses, leopard-sized marsupial lions, and large platypuses existed in Australia. In Europe and parts of Asia, cave bears, elephants, and the giant deer commonly called the Irish elk (► Figure 19-23) were abundant. In addition to mammals, giant birds up to 3.5 m tall and weighing 585 kg existed in New Zealand, Madagascar, and Australia, and giant vultures with a wingspan of 3.6 m are known from California.

Many smaller mammals were also present, many of which still exist. The major evolutionary trend in mammals, however, was toward large body size. Perhaps this was an adaptation to the cooler temperatures of the Pleistocene. Large animals have proportionately less surface area compared to their volume and therefore retain heat more effectively than do small animals.

About 10,000 years ago, almost all of the large terrestrial mammals of North America, South America, and Australia became extinct. Extinctions also occurred on the other continents, but they were of considerably lesser impact. This extinction was modest by comparison to earlier ones, but it was unusual in that it affected, with few exceptions, only large terrestrial mammals. The debate over the cause of this extinction continues between those who think that the large mammals could not adapt to the rapid climatic changes at the end of the Ice Age and those who think that these mammals were killed off by human hunters, a hypothesis known as *prehistoric overkill.*

CHAPTER SUMMARY

1. During much of the Precambrian, all known life-forms were single-celled varieties of bacteria. Multicelled algae are known from rocks at least 1 billion years old, and multicelled animals first appeared about 700 million years ago.

2. Marine invertebrate animals with durable skeletons appeared in abundance during the Cambrian and Ordovician periods and diversified throughout the rest of the Paleozoic.

3. The Cambrian invertebrate community was dominated by three major groups—trilobites, brachiopods, and archaeo-

cyathids. The trilobites suffered mass extinctions at the end of the Cambrian.

4. The Ordovician marine invertebrate community marked the beginning of dominance by the shelly fauna and the start of large-scale reef building. The end of the Ordovician Period was a time of major extinctions for many of the invertebrate phyla.

5. The Silurian and Devonian periods were times of diverse faunas dominated by reef-building animals, while the Carboniferous and Permian periods saw a great decline in invertebrate diversity and mass extinctions at the end of the Permian.

6. The fish are the earliest known vertebrates, appearing during the Late Cambrian. They have had a long and varied history, including jawless and jawed armored forms (ostracoderms and placoderms), cartilaginous forms, and bony forms. A group of lobe-finned fish gave rise to the amphibians.

7. The evolution of an egg that could be laid on land was the critical factor in the reptiles' ability to colonize all parts of the land. Pelycosaurs and therapsids were the dominant reptile groups during the Permian.

8. Plants had to overcome the same basic problems as the animals—namely, drying out, reproduction, and gravity—in making the transition from water to land.

9. The ancestor of terrestrial vascular plants was probably some type of green alga. The earliest seedless vascular plants were small, leafless stalks. From this simple beginning, plants evolved many of the major structural features characteristic of today's plants.

10. Among the marine invertebrates, survivors of the Permian extinction event diversified and gave rise to increasingly complex Mesozoic marine invertebrate communities.

11. Triassic and Jurassic land-plant communities were composed of seedless plants and gymnosperms. Angiosperms appeared during the Early Cretaceous, diversified rapidly, and soon became the dominant land plants.

12. Dinosaurs evolved during the Late Triassic, but were most abundant and diverse during the Jurassic and Cretaceous. Based on pelvic structure, two distinct orders of dinosaurs are recognized—saurischians (lizard-hipped) and ornithischians (bird-hipped).

13. Pterosaurs were the first flying vertebrate animals. At least one pterosaur species had hair or feathers, so it was very likely endothermic.

14. The fish-eating, porpoise-like ichthyosaurs were thoroughly adapted to an aquatic life. Female ichthyosaurs probably retained eggs within their bodies and gave birth to live young. Plesiosaurs were heavy-bodied marine reptiles that probably came ashore to lay eggs.

15. Birds probably evolved from small carnivorous dinosaurs during the Jurassic, while the earliest mammals evolved during the Late Triassic. Several types of Mesozoic mammals existed, but all were small, and their diversity was low.

16. Mesozoic mass extinctions account for the disappearance of dinosaurs, several other groups of reptiles, and a number of marine invertebrates at the end of the Cretaceous. One hypothesis holds that the extinctions were caused by the impact of a large meteorite or comet with the Earth.

17. Marine invertebrate groups that survived the Mesozoic extinctions continued to expand and diversify during the Cenozoic.

18. The Early Cenozoic mammalian fauna was composed of Mesozoic holdovers and a number of new groups. This was a time of diversification among mammals. Placental mammals owe much of their success to their method of reproduction.

19. The primates probably evolved during the Late Cretaceous. Hominids, the group to which humans and their ancestors belong, exhibit several traits that characterize and differentiate them from other mammal groups. These include walking on two legs rather than four, an increase in brain size, and the ability to manufacture and use sophisticated tools.

20. One of the most remarkable aspects of the Cenozoic history of mammals is that so many very large species existed during the Pleistocene Epoch. Perhaps this was an adaptation to the cooler temperatures that existed during this time.

IMPORTANT TERMS

angiosperm
dinosaur
fossil
gymnosperm

hominid
ichthyosaur
marsupial
placental

plesiosaur
primate
pterosaur

seedless vascular plant
stromatolite
vertebrate

1. The greatest recorded mass extinction event to affect the marine invertebrate community occurred at the end of which period?
   a. ____ Cambrian;
   b. ____ Ordovician;
   c. ____ Silurian;
   d. ____ Devonian;
   e. ____ Permian.
2. Jawless armored fish are:
   a. ____ ostracoderms;
   b. ____ placoderms;
   c. ____ cartilaginous;
   d. ____ bony;
   e. ____ lobe-finned.
3. Amphibians evolved from which fish group?
   a. ____ ostracoderms;
   b. ____ placoderms;
   c. ____ cartilaginous;
   d. ____ bony;
   e. ____ lobe-finned.
4. Which algal group was the probable ancestor of vascular plants?
   a. ____ red;
   b. ____ blue-green;
   c. ____ green;
   d. ____ brown;
   e. ____ yellow.
5. The most significant evolutionary change that allowed reptiles to colonize all parts of the land was:
   a. ____ endothermy;
   b. ____ origin of limbs capable of supporting the animals on land;
   c. ____ evolution of an egg that contained a food and waste sac and surrounded the embryo in a fluid-filled sac;
   d. ____ evolution of a watertight skin;
   e. ____ evolution of tear ducts.
6. Sauropod dinosaurs were:
   a. ____ carnivorous;
   b. ____ the largest dinosaurs;
   c. ____ descendants of therapsids;
   d. ____ particularly varied and abundant during the Early Triassic;
   e. ____ none of these.
7. An important Mesozoic event in the history of land plants was the:
   a. ____ extinction of gymnosperms;
   b. ____ origin of seedless vascular plants;
   c. ____ first appearance of angiosperms;
   d. ____ dominance of seedless vascular plants in coal-forming swamps;
   e. ____ all of these.
8. Which of the following are flying reptiles?
   a. ____ dinosaurs;
   b. ____ pterosaurs;
   c. ____ plesiosaurs;
   d. ____ ichthyosaurs;
   e. ____ pelycosaurs.
9. The large body size of Pleistocene mammals may have been an adaptation to:
   a. ____ increased predation;
   b. ____ more seasonal climates;
   c. ____ cooler temperatures;
   d. ____ higher elevations;
   e. ____ longer summers.
10. Which of the following is a hypothesis for Pleistocene extinctions?
    a. ____ meteorite impact;
    b. ____ prehistoric overkill;
    c. ____ reduced area of continental shelves;
    d. ____ freezing;
    e. ____ extensive volcanism.
11. Which of the following groups are mammals?
    a. ____ placentals;
    b. ____ marsupials;
    c. ____ monotremes;
    d. ____ answers (a) and (b);
    e. ____ answers (a), (b), and (c).
12. Describe the problems that had to be overcome before organisms could inhabit the land.
13. Why were the reptiles so much more successful at extending their habitat than the amphibians?
14. What are the names of the different groups of dinosaurs, and how are they differentiated from one another?
15. What is the evidence for endothermy in dinosaurs and pterosaurs?
16. Briefly summarize the evidence for the proposal that a meteorite impact caused Mesozoic mass extinctions.
17. How do placental mammals differ from marsupials and monotremes?
18. Briefly summarize the Cenozoic evolutionary history of mammals.

POINTS TO PONDER

1. Discuss the implications for humans if a meteorite the size that struck the Earth 66 million years ago should again hit the Earth.
2. Based on the history of life as preserved in the geologic record, can you make any predictions as to the future direction of life? What factors do you think will affect future evolutionary events?

ANSWERS TO MULTIPLE-CHOICE AND FILL-IN-THE-BLANK REVIEW QUESTIONS

CHAPTER 1
1. c; 2. a; 3. d; 4. a; 5. d; 6. c; 7. a; 8. c; 9. a; 10. e; 11. b.

CHAPTER 2
1. d; 2. a; 3. e; 4. e; 5. b; 6. c; 7. d; 8. b; 9. c; 10. a; 11. b; 12. c.

CHAPTER 3
1. b; 2. d; 3. b; 4. b; 5. a; 6. c; 7. a; 8. b; 9. c.

CHAPTER 4
1. b; 2. d; 3. a; 4. d; 5. d; 6. e; 7. b; 8. d; 9. a.

CHAPTER 5
1. a; 2. a; 3. e; 4. e; 5. a; 6. c; 7. c; 8. d; 9. a.

CHAPTER 6
1. b; 2. a; 3. b; 4. c; 5. d; 6. a; 7. d; 8. e; 9. b.

CHAPTER 7
1. a; 2. e; 3. d; 4. b; 5. c; 6. a; 7. c; 8. b; 9. b.

CHAPTER 8
1. c; 2. e; 3. a; 4. c; 5. a; 6. c; 7. c; 8. d; 9. b; 10. d; 11. a; 12. d.

CHAPTER 9
1. c; 2. d; 3. b; 4. d; 5. d; 6. b; 7. b; 8. c; 9. a.

CHAPTER 10
1. d; 2. d; 3. b; 4. c; 5. c; 6. c; 7. a; 8. a; 9. b.

CHAPTER 11
1. e; 2. e; 3. b. 4. a; 5. e; 6. a; 7. e.

CHAPTER 12
1. a; 2. c; 3. b; 4. c; 5. b; 6. c; 7. d; 8. c; 9. c.

CHAPTER 13
1. b; 2. d; 3. e; 4. d; 5. e; 6. b; 7. e; 8. e; 9. b.

CHAPTER 14
1. c; 2. a; 3. b; 4. e; 5. b; 6. b; 7. e; 8. b; 9. a.

CHAPTER 15
1. d; 2. c; 3. e; 4. d; 5. a; 6. e; 7. b; 8. a.

CHAPTER 16
1. d; 2. d; 3. b; 4. a; 5. c; 6. b; 7. a; 8. e; 9. b.

CHAPTER 17
1. c; 2. c; 3. a; 4. e; 5. d; 6. a; 7. d; 8. b; 9. c; 10. e; 11. b.

CHAPTER 18
1. c; 2. e; 3. d; 4. a; 5. b; 6. a; 7. a; 8. c; 9. b; 10. c; 11. a; 12. c.

CHAPTER 19
1. e; 2. a; 3. e; 4. c; 5. c; 6. b; 7. c; 8. b; 9. c; 10. b; 11. e.

PERIODIC TABLE
OF THE ELEMENTS

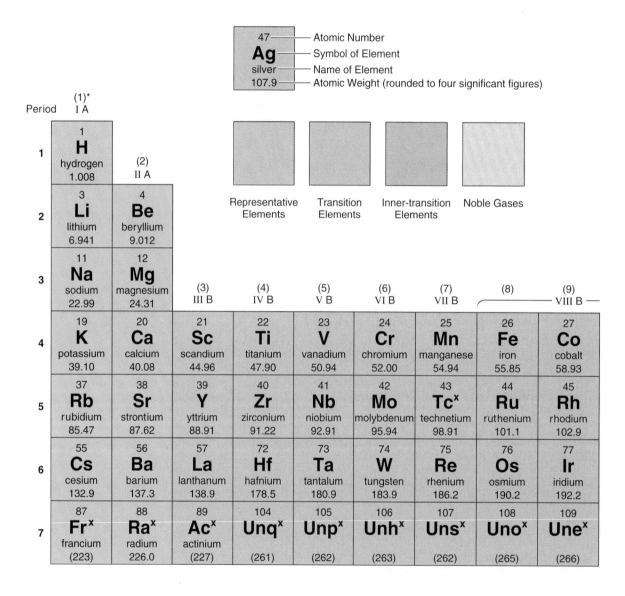

	Key	
47	Atomic Number
Ag	Symbol of Element
silver	Name of Element
107.9	Atomic Weight (rounded to four significant figures)

Representative Elements

Transition Elements

Inner-transition Elements

Noble Gases

Period	(1)* I A	(2) II A	(3) III B	(4) IV B	(5) V B	(6) VI B	(7) VII B	(8) VIII B	(9) VIII B
1	1 **H** hydrogen 1.008								
2	3 **Li** lithium 6.941	4 **Be** beryllium 9.012							
3	11 **Na** sodium 22.99	12 **Mg** magnesium 24.31							
4	19 **K** potassium 39.10	20 **Ca** calcium 40.08	21 **Sc** scandium 44.96	22 **Ti** titanium 47.90	23 **V** vanadium 50.94	24 **Cr** chromium 52.00	25 **Mn** manganese 54.94	26 **Fe** iron 55.85	27 **Co** cobalt 58.93
5	37 **Rb** rubidium 85.47	38 **Sr** strontium 87.62	39 **Y** yttrium 88.91	40 **Zr** zirconium 91.22	41 **Nb** niobium 92.91	42 **Mo** molybdenum 95.94	43 **Tc**^{x} technetium 98.91	44 **Ru** ruthenium 101.1	45 **Rh** rhodium 102.9
6	55 **Cs** cesium 132.9	56 **Ba** barium 137.3	57 **La** lanthanum 138.9	72 **Hf** hafnium 178.5	73 **Ta** tantalum 180.9	74 **W** tungsten 183.9	75 **Re** rhenium 186.2	76 **Os** osmium 190.2	77 **Ir** iridium 192.2
7	87 **Fr**^{x} francium (223)	88 **Ra**^{x} radium 226.0	89 **Ac**^{x} actinium (227)	104 **Unq**^{x} (261)	105 **Unp**^{x} (262)	106 **Unh**^{x} (263)	107 **Uns**^{x} (262)	108 **Uno**^{x} (265)	109 **Une**^{x} (266)

Lanthanides	58 **Ce** cerium 140.1	59 **Pr** praseodymium 140.9	60 **Nd** neodymium 144.2	61 **Pm**^{x} promethium (147)	62 **Sm** samarium 150.4

Actinides	90 **Th**^{x} thorium 232.0	91 **Pa**^{x} protactinium 231.0	92 **U**^{x} uranium 238.0	93 **Np**^{x} neptunium 237.0	94 **Pu**^{x} plutonium (244)

x: All isotopes are radioactive.

( ) Indicates mass number of isotope with longest known half-life.

* Number in ( ) heading each column represents the group designation recommended by the ACS Committee on Nomenclature.

			(13) III A	(14) IV A	(15) V A	(16) VI A	(17) VII A	(18) Noble Gases
								2 **He** helium 4.003
			5 **B** boron 10.81	6 **C** carbon 12.01	7 **N** nitrogen 14.01	8 **O** oxygen 16.00	9 **F** fluorine 19.00	10 **Ne** neon 20.18
(10)	(11) I B	(12) II B	13 **Al** aluminum 26.98	14 **Si** silicon 28.09	15 **P** phosphorus 30.97	16 **S** sulfur 32.06	17 **Cl** chlorine 35.45	18 **Ar** argon 39.95
28 **Ni** nickel 58.71	29 **Cu** copper 63.55	30 **Zn** zinc 65.37	31 **Ga** gallium 69.72	32 **Ge** germanium 72.59	33 **As** arsenic 74.92	34 **Se** selenium 78.96	35 **Br** bromine 79.90	36 **Kr** krypton 83.80
46 **Pd** palladium 106.4	47 **Ag** silver 107.9	48 **Cd** cadmium 112.4	49 **In** indium 114.8	50 **Sn** tin 118.7	51 **Sb** antimony 121.8	52 **Te** tellurium 127.6	53 **I** iodine 126.9	54 **Xe** xenon 131.3
78 **Pt** platinum 195.1	79 **Au** gold 197.0	80 **Hg** mercury 200.6	81 **Tl** thallium 204.4	82 **Pb** lead 207.2	83 **Bi** bismuth 209.0	84 **Po**^{x} polonium (210)	85 **At**^{x} astatine (210)	86 **Rn**^{x} radon (222)

63 **Eu** europium 152.0	64 **Gd** gadolinium 157.3	65 **Tb** terbium 158.9	66 **Dy** dysprosium 162.5	67 **Ho** holmium 164.9	68 **Er** erbium 167.3	69 **Tm** thulium 168.9	70 **Yb** ytterbium 173.0	71 **Lu** lutetium 175.0
95 **Am**^{x} americium (243)	96 **Cm**^{x} curium (247)	97 **Bk**^{x} berkelium (247)	98 **Cf**^{x} californium (251)	99 **Es**^{x} einsteinium (254)	100 **Fm**^{x} fermium (257)	101 **Md**^{x} mendelevium (258)	102 **No**^{x} nobelium (255)	103 **Lr**^{x} lawrencium (256)

GLOSSARY

A

aa—A lava flow with a surface of rough, angular blocks and fragments.

abrasion—The wearing and scraping of exposed rock surfaces by the impact of solid particles.

Absaroka sequence—A widespread sequence of Pennsylvanian and Permian sedimentary rocks bounded above and below by unconformities; deposited during a transgressive-regressive cycle of the Absaroka Sea.

absolute dating—The process of assigning ages in years before the present to geologic events. Various radioactive decay dating techniques yield absolute ages. See also *relative dating*.

abyssal plain—Vast, flat area on the sea floor adjacent to the continental rises of passive continental margins.

Acadian orogeny—A Devonian orogeny in the northern Appalachian mobile belt resulting from a collision of Baltica with Laurentia.

active continental margin—A continental margin characterized by volcanism and seismicity at the leading edge of a continental plate where oceanic lithosphere is subducted. Possesses a narrow continental shelf and a slope that descends directly into an oceanic trench. See also *passive continental margin*.

aftershock—An earthquake following a main shock resulting from adjustments along a fault. Aftershocks are common after a large earthquake, but most are smaller than the main shock.

alluvial fan—A cone-shaped alluvial deposit formed where a stream flows from mountains onto an adjacent lowland.

alluvium—A collective term for all solid materials transported and deposited by streams.

alpha decay—A type of radioactive decay involving the emission of a particle consisting of two protons and two neutrons from the nucleus of an atom; decreases the atomic number by two and the atomic mass number by four.

angiosperm—Any of the vascular plants having flowers and seeds; the flowering plants.

angular unconformity—An unconformity below which older strata dip at a different angle (usually steeper) than the overlying strata. See also *disconformity* and *nonconformity*.

anticline—An up-arched fold in which the oldest exposed rocks coincide with the fold axis, and all strata dip away from the axis.

aphanitic—An igneous rock texture in which individual mineral grains are too small to be seen without magnification; results from rapid cooling and generally indicates an extrusive origin.

aquiclude—Any material that prevents the movement of groundwater.

aquifer—A permeable layer that allows the movement of groundwater.

arête—A narrow, serrated ridge separating two glacial valleys or adjacent cirques.

artesian system—A confined groundwater system in which high hydrostatic pressure builds, causing water in the system to rise above the level of the aquifer.

ash—Pyroclastic material measuring less than 2 mm.

assimilation—A process in which magma changes composition by reacting with country rock.

asthenosphere—The part of the upper mantle over which lithospheric plates move.

atom—The smallest unit of matter that retains the characteristics of an element.

atomic mass number—The total number of protons and neutrons in the nucleus of an atom.

atomic number—The number of protons in the nucleus of an atom.

aureole—A zone surrounding a pluton in which contact metamorphism has occurred.

B

barchan dune—A crescent-shaped sand dune with the tips of the crescent pointing downwind.

barrier island—A long, narrow island composed of sand and separated from the mainland by a lagoon.

basal slip—A type of glacial movement in which a glacier slides over its underlying surface.

basalt plateau—A plateau built up by numerous lava flows from fissure eruptions.

base level—The lowest limit to which a stream can erode.

basin—The circular equivalent of a syncline. All strata in a basin dip toward a central point, and the youngest exposed rocks are in the center of the fold.

batholith—A discordant, irregularly shaped pluton composed chiefly of granitic rocks.

baymouth bar—A spit that has grown until it cuts off a bay from the open ocean or a lake.

beach—A deposit of unconsolidated sediment extending landward from low tide to a change in topography or where permanent vegetation begins.

bed (bedding)—Bed refers to an individual layer of rock, especially sedimentary rock, whereas bedding refers to the layered arrangement of rocks. See also *strata (stratification)*.

bed load—The part of a stream's sediment load that is transported along its bed; consists of sand and gravel.

beta decay—A type of radioactive decay during which a fast-moving electron is emitted from a neutron, thereby converting it to a proton; results in an increase of one atomic number, but does not change atomic mass number.

biochemical sedimentary rock—A sedimentary rock resulting from the chemical processes of organisms.

bonding—The process whereby atoms are joined to other atoms.

Bowen's reaction series—A mechanism accounting for the derivation of intermediate and felsic magmas from a mafic magma. It has a discontinuous branch of ferromagnesian minerals that change from one to another over specific temperature ranges and a continuous branch of plagioclase feldspars whose composition changes as the temperature decreases.

braided stream—A stream possessing an intricate network of dividing and rejoining channels. Braiding occurs when sand and gravel bars are deposited within channels.

breaker—A wave that oversteepens as it enters shallow water until its crest plunges forward.

butte—An isolated, steep-sided, pinnacle-like erosional feature found in arid and semiarid regions.

C

caldera—A large, steep-sided circular to oval volcanic depression usually formed by summit collapse resulting from partial draining of an underlying magma chamber.

carbon 14 dating technique—An absolute age dating method relying on determining the ratio of C^{14} to C^{12} in a sample; useful back to about 70,000 years ago; can be applied only to organic substances.

carbonate mineral— A mineral containing the negatively charged carbonate ion $(CO_3)^{-2}$, e.g., calcite $(CaCO_3)$.

carbonate rock— A sedimentary rock containing mostly carbonate minerals (e.g., limestone and dolostone).

cave—A naturally formed subsurface opening that is generally connected to the surface and is large enough for a person to enter.

cementation—The precipitation of minerals as binding material between and around sediment grains, thus converting sediment to sedimentary rock.

chemical sedimentary rock—Rock formed of minerals derived from the ions taken into solution during chemical weathering.

chemical weathering—The decomposition of rock materials by chemical alteration of parent material.

cinder cone—A small steep-sided volcano composed of pyroclastic materials that accumulated around a vent.

circum-Pacific belt—A zone of seismic and volcanic activity that nearly encircles the Pacific Ocean basin.

cirque—A steep-walled, bowl-shaped depression formed by erosion at the upper end of a glacial trough.

cleavage—The breaking or splitting of mineral crystals along planes of weakness.

columnar joint—A type of joint in some igneous rocks in which six-sided columns form as a result of cooling.

compaction—A method of lithification whereby the pressure exerted by the weight of overlying sediment reduces the amount of pore space and thus the volume of a deposit.

composite volcano—A volcano composed of pyroclastic layers, lava flows typically of intermediate composition, and mudflows; also called stratovolcano.

compound—A substance resulting from the bonding of two or more different elements, e.g., water (H_2O) and quartz (SiO_2).

compression—Stress resulting when rocks are squeezed by external forces directed toward one another.

concordant—Refers to plutons whose boundaries are parallel to the layering in the country rock. See also *discordant*.

cone of depression—The cone-shaped depression in the water table around a well; results from pumping water from an aquifer faster than it can be replenished.

contact metamorphism—Metamorphism of country rock adjacent to a pluton.

continental-continental plate boundary—A convergent plate boundary along which two continental lithospheric plates collide, e.g., the collision of India with Asia.

continental drift—The theory that the continents were once joined into a single landmass that broke apart with the various fragments moving with respect to one another.

continental glacier—A glacier covering at least 50,000 km² and unconfined by topography. Also called an ice sheet.

continental margin—The area separating the part of a continent above sea level from the deep-sea floor. Consists of a continental shelf and slope and, in some places, a rise.

convergent plate boundary—The boundary between two plates that are moving toward one another See also *continental-continental plate boundary, oceanic-continental plate boundary*, and *oceanic-oceanic plate boundary*.

core—The innermost part of the Earth below the mantle at about 2,900 km; divided into an outer liquid core and an inner solid core.

Coriolis effect—The deflection of winds to the right of their direction of motion (clockwise) in the Northern Hemisphere and to the left (counterclockwise) in the Southern Hemisphere due to the Earth's rotation.

correlation—The demonstration of time equivalency of rock units in different areas.

covalent bond—A bond formed by the sharing of electrons between atoms.

crater—A circular or oval depression at the summit of a volcano resulting from the extrusion of gases, pyroclastic materials, and lava.

craton—The relatively stable part of a continent; consists of a shield and a buried extension of a shield known as a platform; the ancient nucleus of a continent.

creep—A type of mass wasting involving slow downslope movement of soil or rock.

cross-bedding—Layers in sedimentary rocks that were deposited at an angle to the surface upon which they were accumulating.

crust—The outermost part of the Earth overlying the mantle. See also *continental crust and oceanic crust*.

crystal settling—The physical separation and concentration of minerals in the lower part of a magma chamber by crystallization and gravitational settling.

crystalline solid—A solid in which the constituent atoms are arranged in a regular, three-dimensional framework.

Curie point—The temperature at which iron-bearing minerals in a cooling magma attain their magnetism.

cyclothem—A vertical sequence of cyclically repeated sedimentary rocks resulting from alternating periods of marine and nonmarine deposition; commonly contain a coal bed.

D

daughter element—An element formed by the radioactive decay of another element, e.g., argon 40 is the daughter element of potassium 40. See also *parent element*.

debris flow—A mass wasting process; much like a mudflow but more viscous and containing larger particles.

deflation—The removal of loose surface sediment by the wind.

deformation—A general term referring to any change in shape or volume, or both, of rocks in response to stress. Deformation involves folding and fracturing.

delta—An alluvial deposit formed where a stream discharges into a lake or an ocean.

depositional environment—Any area where sediment is deposited such as on a stream's floodplain or on a beach.

desert—Any area that receives less than 25 cm of rain per year and has a high evaporation rate.

desert pavement—A surface mosaic of close-fitting gravel particles formed by the removal of sand-sized and smaller particles by the wind.

desertification—The expansion of deserts into formerly productive lands.

detrital sedimentary rock—Sedimentary rock consisting of the solid particles (detritus) of preexisting rocks, e.g., sandstone and conglomerate.

differential pressure—Pressure that is not applied equally to all sides of a rock body.

differential weathering—Weathering of rock at different rates, producing an uneven surface.

dike—A tabular or sheetlike discordant pluton.

dinosaur—Any of the Mesozoic reptiles belonging to the groups designated as ornithischians and saurischians.

dip—A measure of the maximum angular deviation of an inclined plane from horizontal.

dip-slip fault—A fault on which all movement is parallel with the dip of the fault plane. See also *normal fault* and *reverse fault*.

discharge—The volume of water in a stream moving past a particular point in a given period of time.

disconformity—An unconformity above and below which the strata are parallel. See also *angular unconformity* and *nonconformity*.

discordant—Refers to plutons with boundaries cutting across the layering in the country rock. See also *concordant*.

dissolved load—The part of a stream's load consisting of ions in solution.

divergent plate boundary—The boundary between two plates that are moving apart.

divide—A topographically high area that separates adjacent drainage basins.

dome—A circular equivalent of an anticline. All strata dip outward from a central point, and the oldest exposed rocks are in the center of the dome.

drainage basin—The surface area drained by a stream and its tributaries.

drainage pattern—The regional arrangement of stream channels in a drainage system.

dripstone—Various cave deposits resulting from the deposition of calcite.

drumlin—An elongate hill of till formed by the movement of a continental glacier.

dune—A mound or ridge of wind-deposited sand.

dynamic metamorphism—Metamorphism occurring in fault zones where rocks are subjected to high differential pressure.

E

earthflow—A type of mass wasting process; involves downslope flowage of water-saturated soil.

earthquake—Vibration of the Earth caused by the sudden release of energy, usually as a result of displacement of rocks along faults.

elastic rebound theory—A theory that explains the sudden release of energy when rocks are deformed by movement on a fault.

elastic deformation—A type of deformation in which the material returns to its original shape when stress is relaxed.

electron—A negatively charged particle of very little mass that orbits the nucleus of an atom.

electron capture—A type of radioactive decay in which a proton captures an electron and is thereby converted to a neutron; results in a loss of one atomic number, but no change in atomic mass number.

electron shell—Electrons orbit rapidly around the nuclei of atoms at specific distances known as electron shells.

element—A substance composed of all the same atoms; it cannot be changed into another element by ordinary chemical means.

emergent coast—A coast where the land has risen with respect to sea level.

end moraine—A pile of rubble deposited at the terminus of a glacier. See also *recessional moraine* and *terminal moraine*.

epicenter—The point on the Earth's surface directly above the focus of an earthquake.

erosion—The removal of weathered materials from their source area.

esker—A long, sinuous ridge of stratified drift formed by deposition by running water in a tunnel beneath stagnant ice.

evaporite—A sedimentary rock formed by inorganic chemical precipitation of minerals from solution e.g., rock salt and rock gypsum.

Exclusive Economic Zone—An area extending 371 km seaward from the coast of the United States and its territories in which the United States claims all sovereign rights.

exfoliation—The process whereby slabs of rock bounded by sheet joints slip or slide off the host rock.

exfoliation dome—A large rounded dome of rock resulting from the process of exfoliation.

F

fault—A fracture along which movement has occurred parallel to the fracture surface.

felsic magma—A type of magma containing more than 65% silica and considerable sodium, potassium, and aluminum, but little calcium, iron, and magnesium. See also *intermediate magma* and *mafic magma*.

ferromagnesian silicate—A silicate mineral containing iron and magnesium or both. See also *nonferromagnesian silicate*.

fetch—The distance the wind blows over a continuous water surface.

fiord—A glacial valley below sea level.

firn—Granular snow formed by the partial melting and refreezing of snow.

firn limit—The elevation to which the snowline recedes during a wastage season.

fissure eruption—An eruption in which lava or pyroclastic materials are emitted along a long, narrow fissure or group of fissures.

floodplain—A low-lying, relatively flat area adjacent to a stream that is partly or completely covered by water when the stream overflows its banks.

fluid activity—An agent of metamorphism in which water and carbon dioxide promote metamorphism by increasing the rate of chemical reactions.

focus—The place within the Earth where an earthquake originates and energy is released.

foliated texture—A texture of metamorphic rocks in which platy and elongate minerals are arranged in a parallel fashion.

footwall block—The block of rock that lies beneath a fault plane.

fossil—Remains or traces of prehistoric organisms preserved in rocks of the Earth's crust.

fracture—A break in rock resulting from intense applied pressure.

frost action—The mechanical weathering process that disaggregates rocks by repeated freezing and thawing of water in cracks and crevices.

frost heaving—The process whereby a mass of sediment or soil undergoes freezing, expansion, and actual lifting, followed by thawing, contraction, and lowering of the mass.

frost wedging—The opening and widening of cracks by the repeated freezing and thawing of water.

G

geologic time scale—A chart with the designation for the earliest interval of geologic time at the bottom, followed upward by designations for more recent intervals of time.

geology—The science concerned with the study of the Earth; includes studies of Earth materials (minerals and rocks), surface and internal processes, and Earth history.

geothermal energy—Energy that comes from the steam and hot water trapped within the Earth's crust.

geothermal gradient—The temperature increase with depth in the Earth; it averages about 25°C/km near the surface.

geyser—A hot spring that intermittently ejects hot water and steam.

glacial budget—The balance between accumulation and wastage in a glacier.

glacial drift—A collective term for all sediment deposited by glacial activity, including till deposited directly by glacial ice and outwash deposited by streams derived from melting ice.

glacial ice—Ice that has formed from firn.

glacier—A mass of ice on land that moves by plastic flow and basal slip.

Glossopteris flora—A Late Paleozoic association of plants found only on the Southern Hemisphere continents and India.

Gondwana—One of six large Paleozoic continents; composed mostly of present-day South America, Africa, Antarctica, Australia, and India.

graded bedding—A type of sedimentary bedding in which an individual bed shows a decrease in grain size from bottom to top.

graded stream—A stream possessing an equilibrium profile in which a delicate balance exists between gradient, discharge, flow velocity, channel characteristics, and sediment load so that neither significant erosion nor deposition occurs within the channel.

gradient—The slope over which a stream flows; usually expressed in m/km or ft/mi.

ground moraine—A deposit formed of sediment liberated from melting ice as a glacier's terminus retreats.

groundwater—The underground water stored in the pore spaces in rocks, sediment, or soil.

guide fossil—Any fossil that can be used to determine the relative geologic ages of rocks and to correlate rocks of the same age in different areas.

guyot—A flat-topped seamount of volcanic origin rising more than 1 km above the sea floor.

gymnosperm—The flowerless, seed-bearing land plants.

H

half-life—The time required for one-half of the original number of atoms of a radioactive element to decay to a stable daughter product, e.g., the half-life of potassium 40 is 1.3 billion years.

hanging valley—A tributary glacial valley whose floor is at a higher level than that of the main glacial valley.

hanging wall block—The block of rock that lies above a fault plane.

heat—An agent of metamorphism.

Hercynian-Alleghanian orogeny—Pennsylvanian to Permian orogeny in the Hercynian mobile belt of southern Europe and the Appalachian mobile belt from New York to Alabama.

hominid—Abbreviated form of Hominidae, the family to which humans belong. Bipedal primates such as *Australopithecus* and *Homo* are hominids.

horn—A steep-walled, pyramidal peak formed by the headward erosion of cirques.

hot spot—A localized zone of melting below the lithosphere.

hot spring—A spring in which the water temperature is warmer than the temperature of the human body (37°C).

humus—The material in soils derived by bacterial decay of organic matter.

hydraulic action—The power of moving water.

hydrologic cycle—The continuous recycling of water from the oceans, through the atmosphere, to the continents, and back to the oceans.

hydrolysis—The chemical reaction between the hydrogen (H^+) ions and hydroxyl (OH^-) ions of water and a mineral's ions.

hypothesis—A provisional explanation for observations; subject to continual testing and modification. If well supported by evidence, hypotheses are then generally called theories.

I

ichthyosaur—Any of the porpoise-like, Mesozoic marine reptiles.

igneous rock—Any rock formed by cooling and crystallization of magma or lava, or by the accumulation and consolidation of pyroclastic materials.

incised meander—A deep, meandering canyon cut into solid bedrock by a stream.

index mineral—A mineral that forms within a specific temperature and pressure range during metamorphism.

infiltration capacity—The maximum rate at which sediment or soil can absorb water.

inselberg—An isolated steep-sided erosional remnant rising above a desert plain.

intensity—The subjective measure of the kind of damage done by an earthquake as well as people's reaction to it.

intermediate magma—A magma having a silica content of 53 to 65% and an overall composition intermediate between felsic and mafic magmas. See also *felsic magma* and *mafic magma*.

intrusive igneous rock—See *plutonic rock*.

ion—An electrically charged atom produced by adding or removing electrons from its outermost electron shell.

ionic bond—A bond resulting from the attraction of positively and negatively charged ions.

isostatic rebound—The phenomenon in which unloading of the Earth's crust causes it to rise upward until equilibrium is again attained. See also *principle of isostasy*.

isotope—Two or more forms of an element having the same atomic number and the same chemical properties, but a different atomic mass number, e.g., carbon 12 and carbon 14.

J

joint—A fracture along which no movement has occurred parallel with the fracture surface.

K

Kame—Conical hill of stratified drift originally deposited in a depression on a glacier's surface.

karst topography—A topography with numerous caves, springs, sinkholes, solution valleys, and disappearing streams developed by groundwater erosion.

Kaskaskia sequence—A widespread sequence of Devonian and Mississippian sedimentary rocks bounded above and below by unconformities; deposited during a transgressive-regressive cycle of the Kaskaskia Sea.

L

laccolith—A concordant pluton with a mushroomlike geometry.

lahar—A mudflow consisting of volcanic materials such as ash.

Laramide orogeny—A Late Cretaceous to Early Cenozoic episode of deformation in the area of the present-day Rocky Mountains.

lateral moraine—The sediment deposited as a long ridge of till along the margin of a valley glacier.

laterite—A red soil rich in iron and aluminum or both that forms in the tropics by intense chemical weathering.

Laurasia—A Late Paleozoic Northern hemisphere continent consisting of present-day continents of North America, Greenland, Europe, and Asia.

lava—Magma at the Earth's surface; the molten rock material that flows from a volcano or a fissure.

lava dome—A bulbous, steep-sided mass of very viscous magma forced upward through a volcanic conduit.

lava flow—A stream of magma issuing from a volcano or fissure.

leaching—The dissolution or removal of soluble minerals from a soil or rock by percolating water.

lithification—The process of converting sediment into sedimentary rock.

lithosphere—The outer, rigid part of the Earth consisting of the upper mantle, oceanic crust, and continental crust.

lithostatic pressure—Pressure exerted on rock by the weight of overlying rock; it is applied equally in all directions.

loess—Windblown silt and clay deposits derived from deserts, Pleistocene glacial outwash, and floodplains of streams in semiarid regions.

longitudinal dune—A long ridge of sand aligned generally parallel to the direction of the prevailing wind.

longshore current—A current between the breaker zone and the beach flowing parallel to the shoreline and produced by wave refraction.

longshore drift—The movement of sediment along a shoreline by longshore currents.

low-velocity zone—The zone within the mantle between 100 and 250 km deep where the velocity of both P- and S-waves decreases markedly; it corresponds closely to the asthenosphere.

M

mafic magma—A silica-poor magma containing 45 to 52% silica and proportionately more calcium, iron, and magnesium than intermediate and felsic magmas. See also *felsic magma* and *intermediate magma*.

magma—Molten rock material generated within the Earth.

magma mixing—The process of mixing together magmas of different composition, thereby producing a modified version of the parent magmas.

magnetic anomaly—Any change, such as a change in average strength, of the Earth's magnetic field.

magnetic reversal—The phenomenon in which the north and south magnetic poles are completely reversed.

magnitude—The total amount of energy released by an earthquake at its source. See also *Richter Magnitude Scale*.

mantle—The thick layer between the Earth's crust and core.

mantle plume—A stationary column of magma originating deep within the mantle that slowly rises to the Earth's surface to form volcanoes or flood basalts.

marine terrace—A wave-cut platform now elevated above sea level.

marsupial—Any of the pouched mammals such as opossums, kangaroos, and wombats. At present, marsupials are common only in Australia.

mass wasting—The downslope movement of rock, sediment, or soil under the influence of gravity.

meandering stream—A stream possessing a single, sinuous channel with broadly looping curves.

mechanical weathering—The disaggregation of rock materials by physical forces yielding smaller pieces that retain the chemical composition of the parent material.

medial moraine—A moraine formed where lateral moraines of two valley glaciers merge.

Mediterranean belt—A zone of seismic and volcanic activity extending westerly from Indonesia through the Himalayas, across Iran and Turkey, and through the Mediterranean region of Europe.

mesa—A broad, flat-topped erosional remnant bounded on all sides by steep slopes and capped by resistant rock.

metamorphic rock—Any rock that has been altered by heat, pressure, or chemical fluids or a combination of these agents of metamorphism.

metamorphic zone—The region between lines of equal metamorphic intensity.

microplate—A small lithospheric plate that is clearly of different origin than rocks of the surrounding area.

Milankovitch theory—A theory that explains cyclic variations in climate and the onset of glacial episodes as a consequence of irregularities in the Earth's rotation and orbit.

mineral—A naturally occurring, inorganic, crystalline solid having characteristic physical properties and a narrowly defined chemical composition.

mobile belt—An elongated area of deformation as indicated by folds and faults; generally adjacent to a craton.

Modified Mercalli Intensity Scale—A scale having values ranging from I to XII that is used to characterize earthquake intensity based on damage.

Mohorovičić discontinuity (Moho)—The boundary between the Earth's crust and mantle.

monocline—A simple bend or flexure in otherwise horizontal or uniformly dipping rock layers.

mud crack—A sedimentary structure found in clay-rich sediment that has dried out. When drying occurs, the sediment shrinks and intersecting fractures form.

mudflow—A mass wasting process; a flow consisting of mostly clay- and silt-sized particles and more than 30% water.

N

native element—A mineral composed of a single element, e.g., gold and silver.

natural levee—A ridge of sandy alluvium deposited along the margins of a stream channel during floods.

nearshore sediment budget—The balance between additions and losses of sediment in the nearshore zone.

neutron—An electrically neutral particle found in the nucleus of an atom.

Nevadan orogeny—Late Jurassic to Cretaceous deformation that strongly affected the western part of North America.

nonconformity—An unconformity in which stratified rocks above an erosion surface overlie igneous or metamorphic rocks. See also *angular unconformity* and *disconformity*.

nonferromagnesian silicate—A silicate mineral that does not contain iron or magnesium. See also *ferromagnesian silicate*.

nonfoliated texture—A metamorphic texture in which there is no discernible preferred orientation of minerals.

normal fault—A dip-slip fault on which the hanging wall block has moved down relative to the footwall block. See also *reverse fault*.

nucleus—The central part of an atom consisting of one or more protons and neutrons.

nuée ardente—A mobile dense cloud of hot pyroclastic materials and gases ejected from a volcano.

O

oblique-slip fault—A fault having both dip-slip and strike-slip movement.

oceanic-continental plate boundary—A type of convergent plate boundary along which oceanic lithosphere and continental lithosphere collide; characterized by subduction of the oceanic plate beneath the continental plate and by volcanism and seismicity.

oceanic-oceanic plate boundary—A type of convergent plate boundary along which two oceanic lithospheric plates collide and one is subjected beneath the other; characterized by seismicity, volcanism, and the origin of a volcanic island arc.

oceanic ridge—A submarine mountain system found in all of the oceans; it is composed of volcanic rock (mostly basalt) and displays features produced by tension.

oceanic trench—A long, narrow depression in the sea floor where subduction occurs.

ooze—Deep-sea pelagic sediment composed mostly of shells of marine animals and plants.

orogeny—The process of forming mountains, especially by folding and thrust faulting; an episode of mountain building.

Ouachita orogeny—An orogeny involving deformation of the Ouachita mobile belt during the Pennsylvanian Period.

outwash plain—The sediment deposited by the meltwater discharging from the terminus of a continental glacier.

oxbow lake—A cutoff meander filled with water.

oxidation—The reaction of oxygen with other atoms to form oxides or, if water is present, hydroxides.

P

pahoehoe—A type of lava flow with a smooth ropy surface.

paleocurrent—The direction of an ancient current as indicated by sedimentary structures such as cross-bedding.

paleomagnetism—The study of remanent magnetism in rocks so that the intensity and direction of the Earth's past magnetic field can be determined.

Pangaea—The name proposed by Alfred Wegener for a supercontinent consisting of all the Earth's landmasses that existed at the end of the Paleozoic Era.

parabolic dune—A crescent-shaped dune in which the tips of the crescent point upwind.

parent element—An unstable element that changes by radioactive decay into a stable daughter element. See also *daughter element*.

parent material—The material that is mechanically and chemically weathered to yield sediment and soil.

passive continental margin—The trailing edge of a continental plate consisting of a broad continental shelf and a continental slope and rise, commonly with an abyssal plain adjacent to the rise. Passive continental margins lack volcanism and intense seismic activity. See also *active continental margin*.

pedalfer—A soil formed in humid regions with an organic-rich A horizon and aluminum-rich clays and iron oxides in horizon B.

pediment—An erosion surface of low relief gently sloping away from a mountain base.

pedocal—A soil of arid and semiarid regions with a thin A horizon and a calcium carbonate–rich B horizon.

pelagic clay—Generally brown or reddish deep-sea sediment composed of clay-sized particles derived from the continents and oceanic islands.

permeability—A material's capacity for transmitting fluids.

phaneritic—A coarse-grained texture in igneous rocks in which the minerals are easily visible without magnification; results from slow cooling and generally indicates an intrusive origin.

pillow lava—Bulbous masses of basalt resembling pillows formed when lava is rapidly chilled underwater.

placental—Any of the mammals that have a placenta to nourish the embryo; most living and fossils mammals are placentals.

plastic flow—The flow that occurs in response to pressure and causes permanent deformation.

plastic deformation—The result of stress in which a material cannot recover its original shape and retains the configuration produced by the stress such as folding of rocks.

plate—An individual piece of lithosphere that moves over the asthenosphere.

plate tectonic theory—The theory that large segments of the lithosphere move relative to one another.

playa—A dry lake bed found in deserts.

plesiosaur—A type of Mesozoic, marine reptile.

plunging fold—A fold with an inclined axis.

pluton—An intrusive igneous body that forms when magma cools and crystallizes within the Earth's crust, e.g., batholith and sill.

plutonic (intrusive igneous) rock—Igneous rock that crystallizes from magma intruded into or formed in place within the Earth's crust.

point bar—The sedimentary body deposited on the gently sloping side of a meander loop.

porosity—The percentage of a material's total volume that is pore space.

porphyritic—An igneous texture with mineral grains of markedly different sizes.

pressure release—A mechanical weathering process in which rocks formed deep within the Earth expand upon being exposed at the surface due to release of pressure.

primate—Any of the mammals belonging to the order Primates; characteristics include large brain, stereoscopic vision, and grasping hand.

principle of cross-cutting relationships— A principle used to determine the relative ages of events; holds that an igneous intrusion or fault must be younger than the rocks that it intrudes or cuts.

principle of fossil succession—A principle holding that fossils, and especially assemblages of fossils, succeed one another through time in a regular and determinable order.

principle of inclusions—A principle that holds that inclusions, or fragments, in a rock unit are older than the rock unit itself, e.g., granite fragments in a sandstone are older than the sandstone.

principle of isostasy—The theoretical concept of the Earth's crust "floating" on a denser underlying layer.

principle of lateral continuity—A principle that holds that sediment layers extend outward in all directions until they terminate.

principle of original horizontality—A principle that holds that sediment layers are deposited horizontally or very nearly so.

principle of superposition—A principle that holds that younger layers of strata are deposited on top of older strata.

principle of uniformitarianism—A principle that holds that we can interpret past events by understanding present-day processes; based on the assumption that natural laws have not changed through time.

proton—A positively charged particle found in the nucleus of an atom.

pterosaur—Any of the Mesozoic flying reptiles.

P-wave—A compressional, or push-pull, wave; the fastest seismic wave and one that can travel through solids, liquids, and gases; also known as a primary wave.

P-wave shadow zone—The area between 103° and 143° from an earthquake focus where little P-wave energy is recorded.

pyroclastic material—Fragmental material such as ash explosively ejected from a volcano.

pryoclastic (fragmental) texture—A fragmental texture found in igneous rocks composed of pyroclastic materials.

pyroclastic sheet deposit—Vast, sheet-like deposits of felsic pyroclastic materials erupted from fissures.

Q

quick clay—A clay that spontaneously liquefies and flows like water when disturbed.

R

radioactive decay—The spontaneous change of an atom to an atom of a different element.

rainshadow desert—A desert found on the leeward side of a mountain range; forms because moist marine air moving inland yields precipitation on the windward side of the mountain range and the air descending on the leeward side is much warmer and drier.

rapid mass movement—A type of mass movement involving a visible movement of material.

recessional moraine—A type of end moraine formed when a glacier's terminus retreats, then stabilizes and deposits till. See also *end moraine* and *terminal moraine*.

reef—A moundlike, wave-resistant structure composed of the skeletons of organisms.

reflection—The return to the source of some of a seismic wave's energy when it encounters a boundary separating materials of different density or elasticity within the Earth.

refraction—The change in direction and velocity of a seismic wave when it travels from one material into another of different density and elasticity.

regional metamorphism—Metamorphism that occurs over a large area resulting from high temperature and pressure, and the action of chemical fluids within the Earth's crust.

regolith—The layer of unconsolidated rock and mineral fragments and soil that covers much of the Earth's surface.

relative dating—The process of determining the age of an event relative to other events; involves placing geologic events in their correct chronological order, but involves no consideration of when the events occurred in terms of number of years ago. See also *absolute dating*.

reserve—The part of the resource base that can be extracted economically.

resource—A concentration of naturally occurring solid, liquid, or gaseous material in or on the Earth's crust in such form and amount that economic extraction of a commodity from the concentration is currently or potentially feasible.

reverse fault—A dip-slip fault in which the hanging wall block moves upward relative to the footwall block. See also *normal fault*.

Richter Magnitude Scale—An open-ended scale that measures the amount of energy released during an earthquake.

rill erosion—Erosion by running water that scours small channels in the ground.

rip current—A narrow surface current that flows out to sea through the breaker zone.

ripple mark—Wavelike (undulating) structure produced in granular sediment such as sand by unidirectional wind and water currents, or by oscillating wave currents.

rock—A solid aggregate of minerals of one or more kinds, as in granite, and limestone, or a consolidated aggregate of particles of other rocks, as in sandstone and conglomerate; although exceptions to this definition, coal and natural glass are considered rocks.

rock cycle—A sequence of processes through which Earth materials may pass as they are transformed from one rock type to another.

rockfall—A common type of extremely rapid mass wasting in which rocks fall through the air.

rock-forming mineral—A common mineral that comprises a significant portion of a rock.

rock glide—A type of rapid mass wasting in which rocks move downslope along a more or less planar surface.

rounding—The process by which the sharp corners and edges of sedimentary particles are abraded during transport.

runoff—The surface flow of streams.

S

salt crystal growth— A mechanical weathering process in which rocks are disaggregated by the growth of salt crystals in crevices and pores.

Sauk sequence—A widespread sequence of sedimentary rocks bounded above and below by unconformities; deposited during a latest Proterozoic to Early Ordovician transgressive-regressive cycle of the Sauk Sea.

scientific method—A logical, orderly approach that involves gathering data, formulating and testing hypotheses, and proposing theories.

sea-floor spreading—The theory that the sea floor moves away from spreading ridges and is eventually consumed at subduction zones.

seamount—A structure of volcanic origin rising more than 1 km above the sea floor.

sediment—Loose aggregate of solids derived from preexisting rocks, or solids precipitated from solution by inorganic chemical processes or extracted from solution by organisms.

sedimentary rock—Any rock composed of sediment, e.g., sandstone and limestone.

sedimentary structure—Any structure in sedimentary rock such as cross-bedding, mud cracks, and ripple marks that formed at the time of deposition or shortly thereafter.

seedless vascular plant—A type of land plant with vascular tissues for transport of fluids and nutrients throughout the plant; reproduces by spores rather than seeds, e.g., ferns and horsetail rushes.

seismology—The study of earthquakes.

Sevier orogeny—Cretaceous deformation that affected the continental shelf and slope areas of the Cordilleran mobile belt.

shear—The result of forces acting parallel to one another but in opposite directions; results in deformation by displacement of adjacent layers along closely spaced planes.

shear strength—The resisting forces helping to maintain slope stability.

sheet erosion—Erosion that is more or less evenly distributed over the surface and removes thin layers of soil.

sheet joint—A large fracture more or less parallel to a rock surface resulting from pressure released by expansion of the rock.

shield volcano—A large, dome-shaped volcano with a low rounded profile built up mostly of overlapping basalt lava flows (e.g., Mauna Loa and Kilauea on the island of Hawaii).

shoreline—The line of intersection between the sea or a lake and the land.

silica—A compound of silicon and oxygen atoms.

silica tetrahedron—The basic building block of all silicate minerals. It consists of one silicon atom and four oxygen atoms.

silicate—A mineral containing silica (e.g., quartz [SiO_2] and orthoclase [$KAlSi_3O_8$]).

sill—A tabular or sheetlike concordant pluton.

sinkhole—A depression in the ground that forms in karst regions by the solution of the underlying carbonate rocks or by the collapse of a cave roof.

slide—A type of mass movement involving movement of material along one or more surfaces of failure.

slow mass movement—Mass movement that advances at an imperceptible rate and is usually only detectable by its effects.

slump—A type of mass wasting that occurs along a curved surface of failure and results in the backward rotation of the slump block.

soil—Regolith consisting of weathered material, water, air, and humus that can support plants.

soil degradation—Any processes leading to a loss of soil productivity; may involve erosion, chemical pollution, and compaction.

soil horizon—A distinct soil layer that differs from other soil layers in texture, structure, composition, and color.

solifluction—A type of mass wasting involving the slow downslope movement of water-saturated surface materials.

solution—A reaction in which the ions of a substance become dissociated in a liquid, and the solid substance dissolves.

sorting—A term referring to the degree to which all particles in sediment or sedimentary rock are about the same size.

spheroidal weathering—A type of chemical weathering in which corners and sharp edges of angular rocks weather more rapidly than flat surfaces, thus yielding spherical shapes.

spit—A continuation of a beach forming a point of land that projects into a body of water, commonly a bay.

spring—A place where groundwater flows or seeps out of the ground. Springs occur where the water table intersects the ground surface.

stock—An irregularly-shaped discordant pluton with a surface area less than 100 km^2.

stoping—A process in which rising magma detaches and engulfs pieces of country rock.

strata (stratification)—Strata (singular *stratum*) refers to the layers in sedimentary rocks, whereas stratification refers to the layered aspect of sedimentary rocks. See also *bed (bedding)*.

stratified drift—Glacial drift displaying both sorting and stratification.

stream—Runoff confined to channels regardless of size.

stream terrace—An erosional remnant of a floodplain that formed when a stream was flowing at a higher level.

strike—The direction of a line formed by the intersection of a horizontal plane with an inclined plane, such as a rock layer.

strike-slip fault—A fault involving horizontal movement so that blocks on opposite sides of a fault plane slide sideways past one another.

stromatolite—A structure in sedimentary rocks, especially limestones, produced by entrapment of sediment grains on sticky mats of photosynthesizing bacteria.

subduction zone—A long, narrow zone at a convergent plate boundary where an oceanic plate descends beneath another plate, e.g., the subduction of the Nazca plate beneath the South American plate.

submergent coast—A coast along which sea level rises with respect to the land or the land subsides.

suspended load—The smallest particles carried by a stream, such as silt and clay, which are kept suspended by fluid turbulence.

s-wave—A shear wave that moves material perpendicular to the direction of travel, thereby producing shear stresses in the material it moves through; also known as a secondary wave.

s-wave shadow zone—Those areas more than 103° from an earthquake focus where no S-waves are recorded.

syncline—A down-arched fold in which the youngest exposed rocks coincide with the fold axis, and all strata dip inward toward the axis.

T

Taconic orogeny—An Ordovician orogeny that resulted in deformation of the Appalachian mobile belt.

talus—An accumulation of angular pieces of mechanically weathered rock at the base of a slope.

tension—A type of stress in which forces act in opposite directions but along the same line and tend to stretch an object.

terminal moraine—A type of end moraine; the outermost moraine marking the greatest extent of a glacier. See also *end moraine* and *recessional moraine*.

theory—An explanation for some natural phenomenon that has a large body of supporting evidence; to be considered scientific, a theory must be testable e.g., plate tectonic theory.

thermal convection cell—A type of circulation of material in the asthenosphere during which hot material rises, moves laterally, cools and sinks, and is reheated and reenters the cycle.

thermal expansion and contraction—A type of mechanical weathering in which the volume of rock changes in response to heating and cooling.

tide—The regular fluctuation in the sea's surface in response to the gravitational attraction of the Moon and Sun.

till—All sediment deposited directly by glacial ice.

Tippecanoe sequence—A widespread sequence of sedimentary rocks bounded above and below by unconformities; deposited during an Ordovician to Early Devonian transgressive-regressive cycle of the Tippecanoe Sea.

transform fault—A type of fault that changes one type of motion between plates into another type of motion.

transform plate boundary—Plate boundary along which plates slide past one another, and crust is neither produced nor destroyed; on land recognized as a strike-slip fault.

transport—The mechanism by which weathered material is moved from one place to another, commonly by running water, wind, or glaciers.

transverse dune—A long ridge of sand perpendicular to the prevailing wind direction.

tsunami—A destructive sea wave that is usually produced by an earthquake but can also be caused by submarine landslides or volcanic eruptions.

U

unconformity—An erosion surface separating younger strata from older rocks. See also *angular unconformity, disconformity*, and *nonconformity*.

U-shaped glacial trough—A valley with very steep or vertical walls and a broad, rather flat floor. Formed by the movement of a glacier through a stream valley.

V

valley glacier—A glacier confined to a mountain valley or to an interconnected system of mountain valleys.

valley train—A long, narrow deposit of stratified drift confined within a glacial valley.

velocity—A measure of the downstream distance water travels per unit of time.

ventifact—A stone whose surface has been polished, pitted, grooved, or faceted by wind abrasion.

vertebrate—Any animal having a segmented vertebral column; includes fish, amphibians, reptiles, mammals, and birds.

vesicle—A small hole or cavity formed by gas trapped in a cooling lava.

viscosity—A fluid's resistance to flow.

volcanic island arc—A curved chain of volcanic islands parallel to a deep-sea trench where oceanic lithosphere is subducted causing volcanism and the origin of volcanic islands.

volcanic neck—An erosional remnant of the material that solidified in a volcanic pipe.

volcanic pipe—The conduit connecting the crater of a volcano with an underlying magma chamber.

volcanic (extrusive igneous) rock—Igneous rock formed when magma is extruded onto the Earth's surface where it cools and crystallizes, or when pyroclastic materials become consolidated.

volcanism—The process whereby magma and its associated gases rise through the Earth's crust and are extruded onto the surface or into the atmosphere.

volcano—A conical mountain formed around a vent as a result of the eruption of lava and pyroclastic materials.

W

water table—The surface separating the zone of aeration from the underlying zone of saturation.

water well—A well made by digging or drilling into the zone of saturation.

wave—An undulation on a water surface.

wave base—A depth of about one-half wave length, where the diameter of the orbits of water particles in waves is essentially zero; the depth below which water is not affected by surface waves.

wave-cut platform—A beveled surface that slopes gently in a seaward direction; formed by erosion and landward retreat of a sea cliff.

wave refraction—The bending of waves so that they more nearly parallel the shoreline.

weathering—The physical breakdown and chemical alteration of rocks and minerals at or near the Earth's surface.

Z

zone of accumulation—In soil terminology, another name for horizon B where soluble minerals leached from horizon A accumulate as irregular masses. In glacial terminology, the part of a glacier where additions exceed losses and the glacier's surface is perennially covered by snow.

zone of aeration—The zone above the water table that contains both water and air within the pore spaces of the rock, sediment, or soil.

zone of saturation—The zone below the water table in which all pore spaces are filled with groundwater.

zone of wastage—The part of a glacier where losses from melting, sublimation, and calving of icebergs exceed the rate of accumulation.

MINERAL
IDENTIFICATION TABLES

			Hardness Specific		
Mineral	*Chemical Composition*	*Color*	*Gravity*	*Other Features*	*Comments*
Cassiterite	SnO_2	Brown to black	6½ 7.0	High specific gravity for a nonmetallic mineral	The main ore of tin. Most is concentrated in alluvial deposits because of its high specific gravity.
Chalcopyrite	$CuFeS_2$	Brassy yellow	3½–4 4.1–4.3	Usually massive; greenish-black streak; iridescent tarnish	The most common copper mineral and an important source of copper. Occurs mostly in hydrothermal rocks.
Galena	PbS	Lead gray	2½ 7.6	Cubic crystals; 3 cleavages at right angles	The ore of lead. Occurs mostly in hydrothermal rocks.
Graphite	C	Black	1–2 2.09–2.33	Greasy feel; writes on paper; 1 direction of cleavage	Used for pencil "leads" and as a dry lubricant. Mostly in metamorphic rocks.
Hematite	Fe_2O_3	Red brown	6 4.8–5.3	Usually granular or massive; reddish-brown streak	Most important ore of iron. Occurs as an accessory mineral in many rocks.
Magnetite	Fe_3O_4	Black	5½–6½ 5.2	Strong magnetism	An important ore of iron. Occurs as an accessory mineral in many rocks.
Pyrite	FeS_2	Brassy yellow	6½ 5.0	Cubic and octahedral crystals	The most common sulfide mineral. Occurs in some igneous and hydrothermal rocks, and in sedimentary rocks associated with coal.

Metallic Luster

Mineral	Chemical Composition	Color	Hardness Specific Gravity	Other Features	Comments
Anhydrite	$CaSO_4$	White, gray	3½ 2.9-3.0	Crystals with 2 cleavages; usually in granular masses	Found in limestones, evaporite deposits, and the cap rocks of salt domes. Used as a soil conditioner.
Apatite	$Ca_5(PO_4)_3F$	Blue, green, brown, yellow, white	5 3.1-3.2	6-sided crystals; in massive or granular masses	An accessory mineral in many rocks. The main constituent of bone and dentine. A source of phosphorous for fertilizer.
Augite	$Ca(Mg,Fe,Al)(Al,Si)_2O_6$	Black, dark green	6 3.25-3.55	Short 8-sided crystals; 2 cleavages; cleavages nearly at right angles	The most common pyroxene mineral. Found mostly in mafic igneous rocks.
Barite	$BaSO_4$	Colorless, white, gray	3 4.5	Tabular crystals; high specific gravity for a nonmetallic mineral	Commonly found with ores of a variety of metals, and in limestones and hot spring deposits. A source of barium.
Biotite (Mica)	$K(Mg,Fe)_3AlSi_3O_{10}(OH)_2$	Black, brown	2½ 2.9-3.4	1 cleavage direction; cleaves into thin sheets	Occurs in both felsic and mafic igneous rocks, in metamorphic rocks, and in clay-rich sedimentary rocks.
Calcite	$CaCO_3$	Colorless, white	3 2.71	3 cleavages at oblique angles; cleaves into rhombs; reacts with dilute hydrochloric acid	The most common carbonate mineral. The main component of limestones and marble. Also common in hydrothermal rocks.
Chlorite	$(Mg,Fe)_3(Si,Al)_4O_{10}$ $(Mg,Fe)_3(OH)_6$	Green	2 2.6-3.4	1 cleavage; occurs in scaly masses	Common in low-grade metamorphic rocks such as slate.

Mineral	Chemical Composition	Color	Hardness Specific Gravity	Other Features	Comments
Corundum	Al_2O_3	Gray, blue, pink, brown	9 4.0	6-sided crystals and great hardness are distinctive	An accessory mineral in some igneous and metamorphic rocks. Used as a gemstone and for abrasives.
Dolomite	$CaMg(CO_3)_2$	White, yellow, gray, pink	3½–4 2.85	Cleavage as in calcite; reacts with dilute hydrochloric acid when powdered	The main constituent of dolostones. Also found associated with calcite in some limestones and marble.
Fluorite	CaF_2	Colorless, purple, green, brown,	4 3.18	4 cleavage directions; cubic and octahedral crystals	Occurs mostly in hydrothermal rocks and in some limestones and dolostones. Used in the manufacture of steel and the preparation of hydrofluoric acid.
Garnet	$Fe_3Al_2(SiO_4)_3$	Dark red	7–7½ 4.32	12-sided crystals common; uneven fracture	Occurs mostly in gneiss and schist. Used as a semiprecious gemstone and for abrasives
Gypsum	$CaSO_4 \cdot 2H_2O$	Colorless, white	2 2.32	Elongate crystals; fibrous and earthy masses	The most common sulfate mineral. Found mostly in evaporite deposits. Used to manufacture plaster of Paris and cements.
Halite	$NaCl$	Colorless, white	3–4 2.2	3 cleavages at right angles; cleaves into cubes; cubic crystals; salty taste	Occurs in evaporite deposits. Used as a source of chlorine and in the manufacture of hydrochloric acid, many sodium compounds, and food seasoning.
Hornblende	$NaCa_2(Mg,Fe,Al)_5$ $(Si,Al)_8O_{22}(OH)_2$	Green, black	6 3.0–3.4	Elongate, 6-sided crysals; 2 cleavages intersecting at 56° and 124°	A common rock-forming amphibole mineral in igneous and metamorphic rocks.

Nonmetallic Luster

Mineral	Chemical Composition	Color	Hardness Specific Gravity	Other Features	Comments
Illite	$(Ca,Na,K)(Al,Fe^{3+}, Fe^{2+},Mg)_2 (Si,Al)_4O_{10}(OH)_2$	White, light gray, buff	1–2 2.6–2.9	Earthy masses; particles too small to observe properties	A clay mineral common in soils and clay-rich sedimentary rocks.
Kaolinite	$Al_2Si_4O_{10}(OH)_8$	White	2 2.6	Massive; earthy odor	A common clay mineral formed by chemical weathering of aluminum-rich silicates. The main ingredient of kaolin clay used for the manufacture of ceramics.
Muscovite (Mica)	$KAl_2Si_3O_{10}(OH)_2$	Colorless	2–2½ 2.7–2.9	1 direction of cleavage; cleaves into thin sheets	Common in felsic igneous rocks, metamorphic rocks, and some sedimentary rocks. Used as an insulator in electrical appliances.
Olivine	$(Fe,Mg)_2SiO_4$	Olive green	6½ 3.3–3.6	Small mineral grains in granular masses; conchoidal fracture	Common in mafic igneous rocks.
Plagioclase feldspars	Varies from $CaAl_2Si_2O_8$ to $NaAlSi_3O_8$	White, gray, brown	6 2.56	2 cleavages at right angles	Common in mafic, intermediate, and felsic igneous rocks and a variety of metamorphic rocks. Also in some arkoses.
Potassium feldspar — Microcline	$KAlSi_3O_8$	White, pink, green	6 2.56	2 cleavages at right angles	Common in felsic igneous rocks, some metamorphic rocks, and arkoses. Used in the manufacture of porcelain.
Potassium feldspar — Orthoclase	$KAlSi_3O_8$	White, pink			
Quartz	SiO_2	Colorless, white, gray, pink, green	7 2.67	6-sided crystals; no cleavage; conchoidal fracture	A common rock-forming mineral in all major rock groups and hydrothermal rocks. Also occurs in varieties known as chert, flint, agate, and chalcedony.

Nonmetallic Luster					
Mineral	*Chemical Composition*	*Color*	*Hardness Specific Gravity*	*Other Features*	*Comments*
Siderite	$FeCO_3$	Yellow, brown	4 3.8-4.0	3 cleavages at oblique angles; cleaves into rhombs	Found mostly in concretions and sedimentary rocks associated with coal.
Smectite	$(Al,Mg)_8(Si_4O_{10})_3(OH)_{10}$ $12{\cdot}H_2O$	Gray, buff, white	1-1.5 2.5	Earthy masses; particles too small to observe properties	A clay mineral with the unique property of swelling and contracting as it absorbs and releases water.
Sphalerite	ZnS	Yellow, brown, black	3½-4 4.0-4.1	6 cleavages; cleaves into dodecahedra	The most important ore of zinc. Commonly found with galena in hydrothermal rocks.
Talc	$Mg_3Si_4O_{10}(OH)_2$	White, green	1 2.82	1 cleavage direction; usually in compact masses	Formed by the alteration of magnesium silicates. Occurs mostly in metamorphic rocks. Used in ceramics, cosmetics, and as a filler in paints.
Topaz	$Al_2SiO_4(OH,F)$	Colorless, white, yellow, blue	8 3.5-3.6	High specific gravity; 1 cleavage direction	Found in pegmatites, granites, and hydrothermal rocks. An important gemstone.
Zircon	$SrSiO_4$	Brown, gray	7½ 3.9-4.7	4-sided, elongate crystals	Most common as an accessory in granitic rocks. An ore of zirconium, and used as a gemstone.

ADDITIONAL READINGS

CHAPTER 1

Dietrich, R. V. 1989. Rock music. *Earth Science* 42, no. 2: 24–25.

_____. 1990. Rocks depicted in painting and sculpture. *Rocks & Minerals* 65, no. 3: 224–36.

_____. 1991. How can I get others interested in rocks? *Rocks & Minerals* 67, no. 2: 124–28.

Dietrich, R. V., and B. J. Skinner. 1990. *Gems, granites, and gravels.* New York: Cambridge University Press.

Ernst, W. G. 1990. *The dynamic planet.* Irvington, N.Y.: Columbia University Press.

Francis, P., and S. Self. 1983. The eruption of Krakatau. *Scientific American* 249, no. 5: 172–87.

Hively, W. 1988. How much science does the public understand? *American Scientist* 76, no. 5: 439–44.

Mirsky, A. 1989. Geology in our everyday lives. *Journal of Geological Education* 37, no. 1: 9–12.

Officer, C., and J. Page. 1993. *Tales of the Earth.* New York: Oxford University Press.

CHAPTER 2

Bonatti, E. 1987. The rifting of continents. *Scientific American* 256, no. 3: 96–103.

Brimhall, G. 1991. The genesis of ores. *Scientific American* 264, no. 5: 84–91.

Condie, K. 1989. *Plate tectonics and crustal evolution.* 3d ed. New York: Pergamon.

Cromie, W. J. 1989. The roots of midplate volcanism. *Mosaic* 20, no. 4: 19–25.

Dalziel, I.W.D. 1995. Earth before Pangea. *Scientific American* 272, no. 1: 58–63.

Kearey, P., and F. J. Vine. 1990. *Global tectonics.* Palo Alto, Calif.: Blackwell Scientific.

Luhmann, J. G., J. B. Pollack, and L. Colin. 1994. The Pioneer mission to Venus. *Scientific American* 270, no. 4: 90–97.

Murphy, J. B., and R. D. Nance. 1992. Mountain belts and the supercontinent cycle. *Scientific American* 266, no. 4: 84–91.

Nance, R. D., T. R. Worsley, and J. B. Moody. 1988. The supercontinent cycle. *Scientific American* 259, no. 1: 72–79.

Parks, N. 1994. Exploring Loihi: The next Hawaiian Island. *Earth* 5, no. 5: 56–63.

Vink, G. E., W. J. Morgan, and P. R. Vogt. 1985. The Earth's hot spots. *Scientific American* 252, no. 4: 50–57.

CHAPTER 3

Berry, L. G., B. Mason, and R. V. Dietrich. 1983. *Mineralogy.* 2d ed. San Francisco: W. H. Freeman.

Blackburn, W. H., and W. H. Dennen. 1988. *Principles of mineralogy.* Dubuque, Iowa: William C. Brown.

Dietrich, R. V., and B. J. Skinner. 1979. *Rocks and rock minerals.* New York: Wiley.

_____. 1990. *Gems, granites, and gravels: Knowing and using rocks and minerals.* New York: Cambridge Univ. Press.

Klein, C., and C. S. Hurlbut Jr., 1985. *Manual of mineralogy* (after James D. Dana). 20th ed. New York: Wiley.

Pough, F. H. 1987. *A field guide to rocks and minerals.* 4th ed. Boston: Houghton Mifflin.

Vanders, I., and P. F. Kerr. 1967. *Mineral recognition.* New York: Wiley.

CHAPTER 4

Baker, D. S. 1983. *Igneous rocks.* Englewood Cliffs, N.J.: Prentice-Hall.

Best, M. G. 1982. *Igneous and metamorphic petrology.* San Francisco, Calif.: W. H. Freeman and Co.

Dietrich, R. V., and B. J. Skinner. 1979. *Rocks and rock minerals.* New York: Wiley.

Dietrich, R. V., and R. Wicander. 1983. *Minerals, rocks, and fossils.* New York: Wiley.

Ernst, W. G. 1969. *Earth materials.* Englewood Cliffs, N.J.: Prentice Hall.

Hall, A. 1987. *Igneous petrology.* Essex, England: Longman Scientific and Technical.

Hess, P. C. 1989. *Origins of igneous rocks.* Cambridge, Mass.: Harvard University Press.

McBirney, A. R. 1984. *Igneous petrology.* San Francisco, Calif.: Freeman, Cooper and Co.

MacKenzie, W. S., C. H. Donaldson, and C. Guilford. 1982. *Atlas of igneous rocks and their textures.* New York: Halsted Press.

Middlemost, E. A. K. 1985. *Magma and magmatic rocks.* London: Longman Group.

CHAPTER 5

Aylesworth, T. G., and V. Aylesworth. 1983. *The Mount St. Helens disaster: What we've learned.* New York: Franklin Watts.

Bullard, F. M. 1984. *Volcanoes of the Earth.* 2d ed. Austin, Tex.: University of Texas Press.

Coffin, M. F., and O. Eldholm. 1993. Large igneous provinces. *Scientific American* 269, no. 4: 42–49.

Decker, R. W., and Decker, B. B. 1991. *Mountains of fire: The nature of volcanoes.* New York: Cambridge University Press.

Erickson, J. 1988. *Volcanoes & earthquakes.* Blue Ridge Summit, Pa.: Tab Books.

Harris, S. L. 1976. *Fire and ice: The Cascade volcanoes.* Seattle, Wash.: The Mountaineers.

Lipman, P. W., and D. R. Mullineaux, eds. 1981. The 1980 eruptions of Mount St. Helens, Washington. *United States Geological Survey Professional Paper 1250.*

McClelland, L., T. Simkin, M. Summers, E. Nielsen, and T. C. Stein, eds. 1989. *Global volcanism 1975–1985.* Englewood Cliffs, N.J.: Prentice-Hall.

Rampino, M. R., S. Self, and R. B. Strothers. 1988. Volcanic winters. *Annual Review of Earth and Planetary Sciences* 16: 73–99.

Simkin, T., et al. 1981. *Volcanoes of the world: A regional gazetteer, and chronology of volcanism during the last 10,000 years.* Stroudsburg, Pa.: Hutchison Ross Publishing Co.

Tilling, R. I. 1987 *Eruptions of Mount St. Helens: Past, present, and future.* U.S. Geological Survey.

Tilling, R. I., C. Heliker, and T. L. Wright. 1987. *Eruptions of Hawaiian volcanoes: Past, present, and future.* U.S. Geological Survey.

Volcanoes and the Earth's interior. 1982. Readings from Scientific American. San Francisco, Calif.: W. H. Freeman and Co.

Wenkam, R. 1987. *The edge of fire: Volcano and earthquake country in western North America and Hawaii.* San Francisco, Calif.: Sierra Club Books.

Wolfe, G. W. 1992. The 1991 eruptions of Mount Pinatubo, Philippines. *Earthquakes and Volcanoes* 23, no. 1: 5–37.

Wright, T. L., and T. C. Pierson. 1992. Living with volcanoes: The U.S. Geological Survey's volcano hazards program. *U.S. Geological Survey Circular 1073.*

CHAPTER 6

Bear, F. E. 1986. *Earth: The stuff of life.* 2d revised ed. Norman, Okla.: University of Oklahoma Press.

Birkeland, P. W. 1984. *Soils and geomorphology.* New York: Oxford University Press.

Buol, S. W., F. D. Hole, and R. J. McCracken. 1980. *Soil genesis and classification.* Ames, Iowa: Iowa State University Press.

Carroll, D. 1970. *Rock weathering.* New York: Plenum Press.

Coughlin, R. C. 1984. *State and local regulations for reducing agricultural erosion.* American Planning Association, Planning Advisory Service Report No. 386.

Courtney, F. M., and S. T. Trudgill. 1984. *The soil: An introduction to soil study.* 2d ed. London: Arnold.

Gibbons, B. 1984. Do we treat our soil like dirt? *National Geographic* 166, no. 3: 350–89.

Loughnan, F. C. 1969. *Chemical weathering of the silicate minerals.* New York: Elsevier.

Ollier, C. 1969. *Weathering.* New York: Elsevier.

Parfit, M. 1989. The dust bowl. *Smithsonian* 20, no. 3: 44–54, 56–57.

World Resources 1992–93. 1992. A Report by the World Resources Institute. New York: Oxford University Press.

CHAPTER 7

Blatt, H., G. Middleton, and R. Murray. 1980. *Origin of sedimentary rocks.* New York: W. H. Freeman.

Boggs, S., Jr. 1987. *Principles of sedimentology and stratigraphy.* Columbus, Ohio: Merrill Publishing Co.

Collinson, J. D., and D. B. Thompson. 1982. *Sedimentary structures.* London: Allen & Unwin.

Fritz, W. J., and J. N. Moore. 1988. *Basics of physical stratigraphy and sedimentology.* New York: John Wiley & Sons.

LaPorte, L. F. 1979. *Ancient environments.* 2d ed. Englewood Cliffs, N.J.: Prentice-Hall.

Moody, R. 1986. *Fossils.* New York: Macmillan Publishing Co.

Selley, R. C. 1978. *Ancient sedimentary environments.* Ithaca, NY.: Cornell University Press.

_____. 1982. *An introduction to sedimentology.* 2d ed. New York: Academic Press.

Simpson, G. G. 1983. *Fossils and the history of life.* New York: Scientific American Books.

CHAPTER 8

Best, M. G. 1982. *Igneous and metamorphic petrology.* San Francisco: W. H. Freeman.

Bowes, D. R., ed. 1989. *The encyclopedia of igneous and metamorphic petrology.* New York: Van Nostrand Reinhold.

Gillen, C. 1982. *Metamorphic geology.* London: George Allen & Unwin.

Gunter, M. E. 1994. Asbestos as a metaphor for teaching risk perception. *Journal of Geological Education* 42, no. 1: 17–24.

Hyndman, D. W. 1985. *Petrology of igneous and metamorphic rocks.* 2d ed. New York: McGraw-Hill.

Margolis, S. V. 1989. Authenticating ancient marble sculpture. *Scientific American* 260, no. 6: 104–11.

Kokkou, A. 1993. *The Getty kouros colloquium.* Athens, Greece. Kapon Editions.

CHAPTER 9

Anderson, D. L., and A. M. Dziewonski. 1984. Seismic tomography. *Scientific American* 251, no. 4: 60–68.

Bolt, B. A. 1982. *Inside the Earth: Evidence from earthquakes.* San Francisco: W. H. Freeman.

_____. 1988. *Earthquakes.* New York: W. H. Freeman.

Bonatti, E. 1994. The Earth's mantle below the ocean. *Scientific American* 270, no. 3: 44–51.

Canby, T. Y. 1990. California earthquake—prelude to the big one? *National Geographic* 177, no. 5: 76–105.

Dawson, J. 1993. CAT scanning the Earth. *Earth 2*, no. 3: 36–41.

Fischman, J. 1992. Falling into the gap: A new theory shakes up earthquake predictions. *Discover* October 1992: 56–63.

Fowler, C. M. R. 1990. *The solid Earth.* New York: Cambridge Univ. Press.

Frohlich, C. 1989. Deep earthquakes. *Scientific American* 260, no. 1: 48–55.

Hanks, T. C. 1985. *National earthquake hazard reduction program: Scientific status.* U.S. Geological Survey Bulletin 1659.

Jeanloz, R. 1983. The Earth's core. *Scientific American 249*, no. 3: 56–65.

_____. 1990. The nature of the Earth's core. *Annual Review of Earth and Planetary Sciences,* 18: 357–86.

Jeanloz, R. and T. Lay. 1993. The core-mantle boundary. *Scientific American* 268, no. 5: 48–55.

Johnston, A. C., and L. R. Kanter. 1990. Earthquakes in stable continental crust. *Scientific American* 262, no. 3: 68–75.

McKenzie, D. P. 1983. The Earth's mantle. *Scientific American* 249, no. 3: 66–78.

Monastersky, R. 1994. Scrambled Earth: Researchers look deep to learn how the planet cools its heat. *Science News* 145, no. 15: 235–37.

Wesson, R. L., and R. E. Wallace. 1985. Predicting the next great earthquake in California. *Scientific American 252,* no. 2: 35–43

CHAPTER 10

Davis, G. H. 1984. *Structural geology of rocks and regions.* New York: Wiley.

Dennis, J. G. 1987. *Structural geology: An introduction.* Dubuque, Iowa. William C. Brown.

Hatcher, R. D., Jr. 1990. *Structural geology: Principles, concepts, and problems.* Columbus, Ohio: Merrill.

Howell, D. G. 1985. Terranes. *Scientific American* 253, no. 5: 116–25.

———. 1989. *Tectonics of suspect terranes: Mountain building and continental growth.* London: Chapman and Hall.

Jones, D. L., A. Cox, P. Coney, and M. Beck. 1982. The growth of western North America. *Scientific American* 247, no. 5: 70–84.

Lisle, R. J. 1988. *Geological structures and maps: A practical guide.* New York: Pergamon.

Miyashiro, A., K. Aki, and A. M. C. Segnor. 1982. *Orogeny.* New York: Wiley.

Molnar, P. 1986. The geologic history and structure of the Himalaya. *American Scientist* 74, no. 2: 144–54.

———. 1986. The structure of mountain ranges. *Scientific American* 255, no. 1: 70–79.

Spencer, E. W. 1988. *Introduction to the structure of the Earth.* New York: McGraw-Hill.

CHAPTER 11

Crozier, M. J. 1989. *Landslides: Causes, consequences, and environment.* Dover, N. H.: Croom Helm.

Fleming, R. W., and F. A. Taylor. 1980. *Estimating the cost of landslide damage in the United States.* U.S. Geological Survey Circular 832.

McPhee, J. 1989. *The control of nature.* New York: Farrar, Straus & Giroux.

Parks, N. 1993. The fragile volcano. *Earth* 6, no. 4: 42–49.

Small, R. J., and M. J. Clark. 1982. *Slopes and weathering.* New York: Cambridge Univ. Press.

Zaruba, Q., and V. Mencl. 1982. *Landslides and their control.* 2d ed. Amsterdam, The Netherlands: Elsevier.

CHAPTER 12

Baker, V. R. 1982. *The channels of Mars.* Austin, Tex.: Univ. of Texas Press.

Beven, K., and P. Carling, eds. 1989. *Floods.* New York: Wiley.

Frater, A., ed. 1984. *Great rivers of the world.* Boston: Little, Brown.

Knighton, D. 1984. *Fluvial forms and processes.* London: Edward Arnold.

Leopold, L. B., M. G. Wolman, and J. P. Miller. 1964. *Fluvial processes in geomorphology.* San Francisco: W. H. Freeman.

McPhee, J. 1989. *The control of nature.* New York: Farrar, Straus, & Giroux.

Morisawa, M. 1968. *Streams: Their dynamics and morphology.* New York: McGraw-Hill.

Petts, G., and I. Foster. 1985. *Rivers and landscape.* London: Edward Arnold.

Rachocki, A. 1981. *Alluvial fans.* New York: Wiley.

Schumm, S. A. 1977. *The fluvial system.* New York: Wiley.

CHAPTER 13

Courbon, P., C. Chabert, P. Bosted, and K. Lindslay. 1989. *Atlas of the great caves of the world.* St. Louis: Cave Books.

Dietrich, R. V. 1993. How are caves formed? *Rocks and Minerals* 68, no. 4: 264–68.

Dolan, R., and H. G. Goodell. 1986. Sinking cities. *American Scientist* 74, no. 1: 38–47.

Fetter, C. W. 1988. *Applied hydrogeology.* 2d ed. Columbus, Ohio: Merrill.

Grossman, D., and S. Schulman. 1994. Verdict at Yucca Mountain. *Earth* 3, no. 2: 54–63.

Jennings, J. N. 1983. Karst landforms. *American Scientist* 71, no. 6: 578–86.

Monastersky, R. 1988. The 10,000-year test. *Science News* 133: 139–41.

Price, M. 1985. *Introducing groundwater.* London: Allen & Unwin.

Rinehart, J. S. 1980. *Geysers and geothermal energy.* New York: Springer-Verlag.

Sloan, B., ed. 1977. *Caverns, caves, and caving.* New Brunswick, N.J.: Rutgers Univ. Press.

CHAPTER 14

Bell, M. 1994. Is our climate unstable? *Earth* 3, no. 1: 24–31.

Broecker, W. S., and G. H. Denton. 1990. What drives glacial cycles? *Scientific American* 262, no. 1: 49–56.

Carozzi, A. V. 1984. Glaciology and the ice age. *Journal of Geological Education* 32: 158–70.

Covey, C. 1984. The Earth's orbit and the ice ages. *Scientific American* 250, no. 2: 58–66.

Drewry, D. J. 1986. *Glacial geologic processes.* London: Edward Arnold.

Grove, J. M. 1988. *The Little Ice Age.* London: Methuen.

Imbrie, J., and K. P. Imbrie. 1979. *Ice ages: Solving the mystery.* Hillside, N.J.: Enslow Press.

John, B. S. 1977. *The ice age. Past and present.* London: Collins.

———. 1979. *The winters of the world.* London: David & Charles.

Kurten, B. 1988. *Before the Indians.* New York: Columbia Univ. Press.

Schneider, S. H. 1990. *Global warming: Are we entering the greenhouse century?* San Francisco: Sierra Club Books.

Sharp, R. P. 1988. *Living ice: Understanding glaciers and glaciation.* New York: Cambridge Univ. Press.

Williams, R. S., Jr. 1983. *Glaciers: Clues to future climate?* United States Geological Survey.

CHAPTER 15

Cooke, R., A. Warren, and A. Goudie. 1993. *Desert geomorphology.* London: UCL Press.

Dorn, R. I. 1991. Rock Varnish. *American Scientist* 79, no. 6: 542–53.

Ellis., W. S. 1987. Africa's Sahel: The stricken land. *National Geographic* 172, no. 2: 140–79.

Greeley, R., and J. Iversen. 1985. *Wind as a geologic process.* Cambridge, Mass.: Cambridge Univ. Press.

Walker, A. S. 1982. Deserts of China. *American Scientist* 70, no. 4: 366–76.

Waters, T. 1993. Dunes. *Earth* 2, no. 1: 44–51.

Wells, S. G., and D. R. Haragan. 1983. *Origin and evolution of deserts.* Albuquerque, N. Mex.: Univ. of New Mexico Press.

Whitney, M. A. 1985. Yardangs. *Journal of Geological Education* 33, no. 2: 93–96.

CHAPTER 16

Bird, E. C. F. 1984. *Coasts: An introduction to coastal geomorphology.* New York: Blackwell.

Bird, E. C. F., and M. L. Schwartz. 1985. *The world's coastline.* New York: Van Nostrand Reinhold.

Flanagan, R. 1993. Beaches on the brink. *Earth* 2, no. 6: 24–33.

Fox, W. T. 1983. *At the sea's edge.* Englewood Cliffs, N.J.: Prentice-Hall.

Garrett, C., and L. R. M. Maas. 1993. Tides and their effects. *Oceanus* 36, no. 1: 27–37.

Hecht, J. 1988. America in peril from the sea. *New Scientist* 118: 54–59.

Komar. P. D. 1976. *Beach processes and sedimentation.* Englewood Cliffs, N.J.: Prentice-Hall.

———. 1983. *CRC handbook of coastal processes and erosion.* Boca Raton, Fla.: CRC Press.

Pethick, J. 1984. *An introduction to coastal geomorphology.* London: Edward Arnold.

Schneider, S. H. 1990. *Global warming: Are we entering the greenhouse century?* San Francisco: Sierra Club Books.

Snead, R. 1982. *Coastal landforms and surface features.* Stroudsburg, Pa.: Hutchinson Ross.

Walden, D. 1990. Raising Galveston. *American Heritage of Invention & Technology* 5: 8–18.

Williams, S. J., K. Dodd, and K. K. Gohn. 1990. Coasts in crisis. *U.S. Geological Survey Circular 1075.*

CHAPTER 17

Berry, W. B. N. 1987. *Growth of a prehistoric time scale.* 2d ed. Palo Alto, Calif.: Blackwell Scientific.

Boslough, J. 1990. The enigma of time. *National Geographic* 177, no. 3: 109–32.

Geyh, M. A., and H. Schleicher. 1990. *Absolute age determination.* New York: Springer-Verlag.

Gould, S. J. 1987. *Time's arrow, time's cycle.* Cambridge, Mass.: Harvard Univ. Press.

Harland, W. B., R. L. Armstrong, A. V. Cox, L. E. Craig, A. G. Smith, and D. G. Smith. 1990. *A geologic time scale 1989.* New York: Cambridge Univ. Press.

Itano, W. M., and N. F. Ramsey. 1993. Accurate measurement of time. *Scientific American* 269, no. 1: 56–57.

Ramsey, N. F. 1988. Precise measurement of time. *American Scientist* 76, no. 1: 42–49.

CHAPTER 18

Bally, A. W., and A. R. Palmer, eds. 1989. *The geology of North America: An overview.* The Geology of North America. vol. A. Boulder, Colo.: Geological Society of America.

Condie, K. C. 1989. *Plate tectonics and crustal evolution.* 3d ed. New York: Pergamon Press.

Hatcher, R. D., Jr., W. A. Thomas, and G. W. Viele, eds. 1989. *The Appalachian-Ouachita orogen in the United States.* The Geology of North America. vol. F-2. Boulder, Colo.: Geological Society of America.

Jones, D. L., A. Cox, P. Coney, and M. Beck. 1982. The growth of western North America. *Scientific America* 247, no. 5: 70–84.

McKerrow, W. S., and C. R. Scotese, eds. 1990. *Palaeozoic palaeogeography and biogeography.* The Geological Society Memoir No. 12. London: Geological Society of London.

Morrison, R. B., ed. 1991. *Quaternary nonglacial geology: Conterminous U.S.* The geology of North America, vol. K-2. Boulder, Colo.: Geological Society of America.

Moullade, M., and A. E. M. Nairn, eds. 1991. *The Phanerozoic geology of the world I: The Palaeozoic.* New York: Elsevier Science Publishing.

Nisbet, E. G. 1987. *The young Earth: An introduction to Archean geology.* Boston: Allen & Unwin.

CHAPTER 19

Bakker, R. T. 1993. Bakker's field guide to Jurassic dinosaurs. *Earth* 2, no. 5: 33–43.

Colbert, E. H., and M. Morales. 1991. *Evolution of the vertebrates,* 4th ed. New York: Wiley.

Donovan, S. K., ed. 1989. *Mass extinctions.* New York: Columbia University Press.

Gore, R. 1993. Dinosaurs. *National Geographic* 183, no. 1: 2–53.

Gray, J., and W. Shear. 1992. Early life on land. *American Scientist* 80, no. 5: 444–56.

Lucas, S. G. 1994. *Dinosaurs: The textbook.* Dubuque, Iowa: Wm. C. Brown.

Schopf, J. W., ed. 1992. *Major events in the history of life.* Boston: James & Bartlett.

Thomas, B., and R. Spicer. 1987. *Evolution and palaeobiology of land plants.* Portland, Oreg.: Dioscorides Press.

Wicander, R., and J. S. Monroe. 1993. *Historical geology: Evolution of the Earth and life through time,* 2d ed. St. Paul, Minn.: West Publishing Co.

INDEX

Lassen Volcanic National Park, 76
Lateral continuity, principle of, 315, 318
Lateral erosion, 222, 223, 276
Lateral moraines, 262, 263
Laterites, 104–5
Laurasia, 20, 21
    collision with Gondwana, 339, 346
    formation during Silurian, 335
    during Mesozoic, 346
Laurentia, 335
    collision with Baltica, 335, 345
Lava, 58, *See also* Lava flows; Magma;
    Volcanism; Volcanoes
    fountains, 76, 80
    pillow, 30, 78, 79
    textures of, 59–60
Lava domes, 84, 87, 88
Lava flows, 58, 77–78, 81
    columnar joints in, 77–78, 173
    composite volcanoes and, 82, 83
    dating, use in, 328
    fissure eruptions and, 86
    on other planets and moons, 18
    paleomagnetism in, 23–24, 25
    rhyolite, 66
    shield volcanoes and, 80
    types, 77
    viscosity of, 59, 77
Leaching, 104
Lead, 36, 123, 143
    uranium-lead decay series/dating, 326,
    327
Levees, 208
    natural, 216
Lewis and Clark Caverns (Montana), 236
Lewis overthrust (Montana), 175, 353
Life history, 360–77
    Cenozoic, 374–77
    Cretaceous extinction, 374
    dinosaurs, 368–71
    flying reptiles, 371–72
    mammals, 373–74, 374–77
    marine invertebrates, 363–65, 368
    marine reptiles, 372–73
    Mesozoic, 368–74
    Paleozoic, 363–68
    Permian extinction, 365–66
    plants, 367–68, 374
    Precambrian, 362–63
    vertebrate evolution, 366–67
Lignite, 118, 123
Limbs (fold), 171
Limestone, 52, 339
    acid rain and, 100
    caves in, 99, 228, 231
    characteristics and types, 116, 117
    dissolution of, 234
    marble from, 13, 130, 139
    marine, 12
    mineral assemblages for, 136
    porosity and permeability, 230
    residual concentrations of minerals in, 107

Limonite, 101, 113
Lingula, 321, 322
Liquefaction, 157
Liquids, 41
    inability to transmit S-waves, 153, 161,
    183
Lisbon earthquake, 147
Lithification, 11, 13, 113, 114
Lithosphere, 2, 7, 27, 163
    defined, 8, 10
    oceanic vs. continental, 31–32
Lithostatic pressure, 132
Little Ice Age, 250, 264
Lobe-finned fish, 366–367
Loess, 103, 275, 276, 278
Loihi seamount/volcano (Hawaii), 35, 88,
    89
Loma Prieta (California) earthquake, 157,
    159
Longitudinal dunes, 275, 276
Longshore currents, 298, 299
Longshore drift, 299–300, 305
Long Valley (California) caldera, 86
Lovelock, James, 9
Low-velocity zone (in the mantle), 163
Luray Cavern (Virginia), 237
Luster (mineral), 49
Lyell, Charles, 315, 321
Lystrosaurus, 22–23

M

Mafic magma, 87
    basalt and gabbro and, 64–65
    in Bowen's reaction series, 61
    composition, 59
    contact metamorphism and, 134
    crystal settling, assimilation, and mixing,
    62–64
    at spreading ridges, 88, 89
    volcanic gases in, 76
Magma, 8, 10, 163. *See also* Felsic magma;
    Igneous rocks; Intermediate magma;
    Lava; Mafic magma; Volcanism
    assimilation and mixing, 63–64
    batholiths, emplacement of, and, 69–71
    Bowen's reaction series and, 61–62
    characteristics, 59
    contact metamorphism and, 134, 143
    cooling and texture of rocks, 59–61
    crystal settling, 62–63
    igneous rocks formed from, 12–13, 61
    magma chambers, craters, and calderas,
    80
    magnetism and, 23–24
    metamorphism and, 131, 143
    radiometric dating, 322–24, 326
    volcanic gases in, 76
Magnesium, 7, 8
    in dolostones, 116
    in ferromagnesian silicates, 48
Magnetic anomalies, 25, 26–27

Magnetic field
    earthquakes and, 160
    polar wandering and, 23–25
    source, 23
Magnetic reversals
    plate movement, calculating rate from, 34
    sea-floor spreading and, 25–27
Magnetite, 143
Magnitude (earthquake), 153, 155, 157
Malachite, 39
Mammals, 367, 368
    Cenozoic, 374–77
    Mesozoic, 373–74
Mammoth Cave (Kentucky), 99, 228, 231,
    236
Mancos Shale, 69
Manganese, 53, 107, 281
    nodules, 308
Manly, Lake, 281
Mantle, 162–63
    boundary with core, 162, 164
    boundary with crust, 163
    composition, density, and structure, 7, 8,
    163, 183
    convection cells in, 8, 26, 35, 36, 164
    geothermal gradient in, 163
    isostasy and, 182–83
    temperature of, 164
    ultramafic rocks in, 64
Mantle plumes, 34–35, 88
Maps
    metamorphic zones, 141, 142
    of sea floor, 291
    seismic risk maps, 158, 160
    strike and dip symbol on, 170
Marble, 13, 129, 131, 136, 143
    authenticating sculptures, 130–31
    dissolution of, 99, 100
    formation and characteristics, 139–40,
    141
Marine depositional environment, 113
Marine invertebrates, 363
    Cambrian, 364
    Carboniferous and Permian, 365
    Mesozoic, 368, 374
    Ordovician, 365
    Permian extinction event, 365–66
    Silurian and Devonian, 364–65
Marine terraces, 307, 308
Marrella splendens, 361
Mars, 18
Marsupials, 375
Mass spectrometer, 322
Mass wasting, 187–204, 211, 222, 223,
    256, 282
    defined, 188
    factors influencing, 188–92
    minimizing effects of, 200–204
    types, 192–200
Matter, 41
    bonding and compounds, 42–44
    elements and atoms, 41–42

CREDITS

CHAPTER 1

Opener:	Photo ©Krafft, K. Explorer le Monde de L'Image
1-1:	Krakatau 1883, by Tom Simpkin and Richard S. Fiske, Smithsonian Institution, 1983
1-2:	American Association of Petroleum Geologists/ IBM
1-3:	Collection of the New York Public Library Astor, Lenox, and Tilden Foundations
1-5:	Precision Graphics
1-6:	Victor Royer
1-7:	Victor Royer
1-8:	Carlyn Iverson
1-9:	Precision Graphics
1-10:	Carlyn Iverson
1-11:	Precision Graphics
1-13:	Carlyn Iverson
1-14:	Precision Graphics. From A.R. Palmer, "The Decade of North American Geology, 1983 Geologic Time Scale. *Geology* (Boulder, Colo.: Geological Society of America, 1983): 504. Reprinted by permission of the Geological Society of America.

CHAPTER 2

Opener:	Robert Caputo/Aurora
2-3:	Bildarchiv Preussischer Kulturbesitz
2-4:	Precision Graphics. From E. Bullard, J.E. Everett, and A.G. Smith, "The Fit of the Continents Around the Atlantic." *Philosophical Transactions of the Royal; Society of London* 258 (1965). Reproduced with permission of the Royal Society, J.E. Everett, and A.G. Smith.
2-5:	Precision Graphics
2-6a:	Precision Graphics
2-6b:	Photo courtesy of Scott Katz
2-7:	Precision Graphics. Modified from E.H. Colbert, *Wandering Lands and Animals* (1973): 72, Figure 31.
2-8:	Precision Graphics
2-9:	Precision Graphics. Reprinted with permission from Cox and Doell, "Review of Paleomagnetism," *GSA,* v. 71, 1960, p. 758 (Fig. 33), Geological Society of America.
2-10:	Precision Graphics. From A. Cox, "Geomagnetic Reversals." *Science* 163 (17 January 1969): 240, Figure 4. Copyright 1969 by the AAAS.
2-12:	Precision Graphics. From *The Bedrock Geology of the World.* Copyright 1984 by R.L. Larson and

W.C. Pitman III. Reprinted with permission by W.H. Freeman and Company.

2-13:	Precision Graphics
Perspective 2-1, Figure 1:	Carlyn Iverson
2-14:	Carlyn Iverson
2-15:	Precision Graphics
2-16:	Carlyn Iverson
2-17:	Carlyn Iverson
2-18:	Carlyn Iverson
2-19:	Carlyn Iverson
2-20:	Precision Graphics
2-21a:	Carlyn Iverson
2-21b:	Precision Graphics
2-22:	Precision Graphics

CHAPTER 3

Opener:	Photo courtesy of Jeff Scovil
3-2:	Precision Graphics
3-3:	Precision Graphics
3-4:	Precision Graphics
3-5:	Precision Graphics
3-6:	Precision Graphics
3-7:	Precision Graphics
3-8:	Precision Graphics. From R.V. Dietrich and Brian J. Skinner, *Gems, Granites, and Gravels: Knowing and Using Rocks and Minerals* (New York: Cambridge University Press, 1990): 39, Figure 3.4
3-9:	Precision Graphics
3-10:	Precision Graphics
Perspective 3-1, Figure 1:	©Gemological Institute of America. Photo by Shane McClure and Robert Kane.
3-14:	Precision Graphics
3-15:	Precision Graphics
Review Question 11:	Precision Graphics

CHAPTER 4

4-2:	Precision Graphics
4-5:	Precision Graphics
4-6:	Precision Graphics
4-7:	Precision Graphics
4-9:	Precision Graphics
4-10:	Precision Graphics. Modified from R.V. Dietrich, *Geology and Michigan: Fortynine Questions and Answers.* 1979.

4-15: Precision Graphics

Perspective 4-1,
Figure 1: Precision Graphics
4-17: Precision Graphics
4-18: Precision Graphics

CHAPTER 5

5-1: United States Department of the Interior, U.S. Geological Survey, David A. Johnston Cascades Volcano Observatory, Vancouver, Washington.
5-2: U.S. Geological Survey
5-3a: D.W. Peterson, U.S. Geological Survey
5-3b: J.B. Stokes, U.S. Geological Survey
5-5: Reproduced by permission of Marie Tharp, 1 Washington Ave., South Nyack, NY 10960.
5-6: University of Colorado
5-7: J.P. Lockwood, U.S. Geological Survey
5-8a-d: Precision Graphics. From Howel Williams, *Crater Lake: The Story of Its Origin* (Berkeley, Calif.: University of California Press): Illustrations from page 84. Copyright ©1941 Regents of the University of California, ©renewed 1969 Howel Williams.
5-9: Precision Graphics
5-10a: Precision Graphics
5-10b: Solarfilma/GeoScience Features

Perspective 5-1,
Figure 1: D.R. Crandell, U.S. Geological Survey
Perspective 5-1,
Figure 2: Keith Ronnholm
5-11a: Precision Graphics
5-11b: Lawrence R. Solkoski, consulting geologist, Vancouver, B.C. Canada.
5-12: Precision Graphics
5-13: I.C. Russell, U.S. Geological Survey
5-14: Precision Graphics. From R.I. Tilling, U.S. Geological Survey
5-15: Ward's Natural Science Establishment, Inc.
5-16: Precision Graphics. Modified from R.I. Tilling, C. Heliker, and T.L. Wright, *Eruptions of Hawaiian Volcanoes: Past, Present, and Future.* 1987. U.S. Geological Survey.
5-17: Carlyn Iverson
5-18: Precision Graphics
5-19: Carlyn Iverson

CHAPTER 6

Opener: Paul Johnson
6-1: Visuals Unlimited/©Frank Lembrecht
6-4: Precision Graphics. From A. Cox and R.R. Doell, "Review of Paleomagnetism." *GSA Bulletin* 71 (1960): 758, Figure 33.
6-5: University of Colorado
6-9: Precision Graphics
6-10: Bill Beatty/Visuals Unlimited
6-11: Precision Graphics
6-12 a-c: Precision Graphics
6-13a: John S. Shelton
6-13b: John D. Cunningham/Visuals Unlimted
6-14: Precision Graphics

6-15: Walt Anderson/Visuals Unlimited
6-16: Kansas State Historical Society
6-18: Science VU/Visuals Unlimited

CHAPTER 7

7-2: R.L. Elderkin, U.S. Geological Survey
7-4: Carlyn Iverson
7-5: Precision Graphics
7-12: Precision Graphics
7-18: Precision Graphics
Perspective 7-1,
Figure 1: Precision Graphics. Data from *Oil and Gas Journal,* v. 90, pp. 44-45 (Dec. 28, 1992).
Perspective 7-1,
Figure 2: Precision Graphics. From Robert S. Dietz and John C. Holden, "Reconstruction of Pangaea: Breakup and Dispersion of Continents, Permian to Present." *Journal of Geophysical Research* 75, no. 6 (10 September 1970): 4949, Figure 5. Copyright by the American Geophysical Union.

CHAPTER 8

8-1: Collection of the J. Paul Getty Museum, Malibu, California
8-2: Precision Graphics
8-3: Precision Graphics. From C. Gillen, *Metamorphic Geology* (London: Chapman & Hall, 1982): 24 and 73, Figures 2.3 and 4.4. Reprinted by permission of Chapman & Hall and C. Gillen.
Perspective 8-1,
Figure 1: Smithsonian Institution
8-5: Precision Graphics
8-7a: Precision Graphics
8-15: Precision Graphics. From H.L. James, *GSA Bulletin 66* (1955): 1454, Plate 1. Reprinted by permssion of the Geological Society of America.
8-16: Carlyn Iverson

CHAPTER 9

Opener: AP/Wide World Photos
9-1a: Precision Graphics
9-1b: Lee Stone/Sygma
9-1c: Ted Soqui/Sygma
9-1d: Don Boomer/Time
9-1e: Photo by Rick Rickman
9-2a: Precision Graphics
9-2b: U.S. Geological Survey
9-3: Precision Graphics
9-4: Precision Graphics
9-5: Precision Graphics. Data from National Oceanic and Atmospheric Administration.
9-6: J.K. Hillers, U.S. Geological Survey
9-7: Precision Graphics. From *Nuclear Explosions and Earthquakes: The Parted Veil* by B.A. Bolt. Copyright ©1976 by W.H. Freeman and Co. Reprinted by permission.
9-8a: Precision Graphics
9-8b: Precision Graphics. Data from C.F. Richter, *Elementary Seismology.* 1958. W.H. Freeman and Co.

13-11:	Precision Graphics
13-12:	Daniel W. Gotshall/Visuals Unlimited
13-13:	Precision Graphics. From J.B. Weeks et al., *U.S. Geological Survey Professional Paper 1400-A*. 1988.
13-14:	Precision Graphics
13-15:	Sarah Stone/Tony Stone Worldwide
13-16:	City of Long Beach Department of Oil Properties
13-17:	Precision Graphics
Perspective 13-1, Figure 1a:	Courtesy of United States Department of Energy
Perspective 13-1, Figure 1b:	Precision Graphics
Perspective 13-1, Figure 2:	Precision Graphics. Modified from *U.S. News & World Report* (18 March 1991): 72-73.
13-19:	VU/©1993 Hal Beral
13-20:	Precision Graphics

CHAPTER 14

Opener:	Photo courtesy of Michael J. Hambrey
14-1:	Reproduced courtesy of the Board of Directors of the Budapest Museum of Fine Arts
14-2:	Precision Graphics
14-3:	Precision Graphics
14-4:	Engineering Mechanics, Virginia Polytechnic Institute and State University
14-5:	Frank Awbrey/Visuals Unlimited
14-6:	Precision Graphics
14-7:	Precision Graphics
14-8:	National Park Service photograph by Ruth and Louis Kirk
14-12:	Precision Graphics
14-16:	Swiss National Tourist Office
14-17:	Precision Graphics
14-18ab:	Precision Graphics
14-18c:	Engineering Mechanics, Virginia Polytechnic Institute and State University
14-20:	Precision Graphics
14-22:	Precision Graphics
Perspective 14-1, Figure 1:	Precision Graphics
Perspective 14-1, Figure 2:	P. Weis, U.S. Geological Survey

CHAPTER 15

Opener:	Steve McCurry/Magmum
15-1:	Precision Graphics
15-2a:	Precision Graphics
15-2b:	Martin Miller
15-4:	Martin G. Miller/Visuals Unlimited
15-5:	Precision Graphics
15-6:	Precision Graphics
15-7a:	Precision Graphics
15-7b:	John S. Shelton
15-8a:	Precision Graphics
15-8b:	©1994 CNES. Provided by SPOT Image Corporation
15-9a:	Precision Graphics
15-9b:	Alan and Linda D. Mayo, GeoPhoto Publishing Company

15-10a:	Precision Graphics
15-11:	Precision Graphics. Based on F.K. Lutgens and E.J. Tarbuck, *The Atmosphere: An Introduction to Meteorology* (Englewood Cliffs, New Jersey: Prentice-Hall, 1979): 150, Figure 7.3.
15-12:	Precision Graphics
Perspective 15-1, Figure 1:	Precision Graphics. From C.B. Hunt and D.R. Mabey, U.S. Geological Survey Professional Paper 494A (1966): A5, Figure 2.
Perspective 15-1, Figure 2:	John D. Cunningham/Visuals Unlimited
15-14:	John S. Shelton
15-15:	Martin G. Miller/Visuals Unlimited
15-16:	Alan and Linda D. Mayo, GeoPhoto Publishing Company.
15-17a:	Precision Graphics

CHAPTER 16

16-1:	Precision Graphics
16-2:	Precision Graphics
16-3:	World Ocean Floor map by Bruce C. Heezen and Marie Tharp, 1977. Copyright ©Marie Tharp 1977. Reproduced by permission of Marie Tharp, 1 Washington Ave., South Nyack, NY 10960.
16-4:	Precision Graphics
16-5:	Precision Graphics
16-6a:	Precision Graphics
16-6b:	Carlyn Iverson
16-7a:	Precision Graphics
16-7b:	©Douglas Faulkner, Science Source/Photo Researchers
16-8:	Precision Graphics
16-9:	Precision Graphics
16-10:	John S. Shelton
16-11:	John S. Shelton
16-12b:	Michael Slear
16-13a:	Precision Graphics
16-13b:	John S. Shelton
16-14:	Precision Graphics
16-15:	NASA
Perspective 16-1, Figure 1:	Precision Graphics
Perspective 16-1, Figure 2:	Steve Starr/SABA
16-16:	Precision Graphics
16-17a:	Precision Graphics
16-18a:	Precision Graphics
16-18b:	Nick Harvey
16-18c:	Suzanne and Nick Geary, Tony Stone Worldwide.
16-19:	GEOPIC, Earth Satellite Corporation

CHAPTER 17

| 17-1: | Precision Graphics. From A.R. Palmer, "The Decade of North American Geology, 1983 Geologic Time Scale." *Geology* (Boulder, Colo.: Geological Society of America, 1983): 504. Reprinted by permission of the Geological Society of America. |

17-4:	Precision Graphics
17-5:	Precision Graphics. From *The Story of the Great Geologists* by Carroll Lane Fenton and Mildred Adams Fenton. Used by permission of Doubleday, a division of Bantam Doubleday Dell Publishing Group, Inc.
17-6:	Precision Graphics
17-7a:	Precision Graphics
17-8a:	Precision Graphics
17-9a:	Precision Graphics
17-10:	Precision Graphics. From *History of the Earth: An Introduction to Historical Geology*, second edition, by Bernhard Kummer. ©1970 by W.H. Freeman and Company; reprinted by permission.
17-11:	Precision Graphics
17-12:	Precision Graphics
17-13:	Precision Graphics. Data from S.M. Richardson and H.Y. McSween, Jr., *Geochemistry—Pathways and Processes* (Englewood Cliffs, NJ: Prentice-Hall, 1989).
Perspective 17-1, Figure 1:	Precision Graphics
Perspective 17-1, Figure 2:	Precision Graphics. Data from Environmental Protection Agency.
17-14:	Precision Graphics
17-15:	Precision Graphics
17-16:	Precision Graphics
17-17:	Precision Graphics
17-18:	Precision Graphics

CHAPTER 18

18-1:	Herb Orth, Life Magazine. ©Time Warner Inc.; painting by Chesley Bonestell, ©the estate of Chesley Bonestell
18-2:	Precision Graphics
18-3:	Precision Graphics
18-4:	Precision Graphics. Reproduced with permission from P.F. Hoffman, "United Plates of America, The Birth of a Craton: Early Proterozoic Assembly and Growth of Laurentia," *Annual Review of Earth and Planetary Sciences*, V. 16, p. 544, ©1988 by Annual Reviews, Inc.
18-5:	Carto-Graphics. Topography reprinted with permission from Bambach, Scotese, and Ziegler, "Before Pangaea: The Geographies of the Paleozoic World, *American Scientist*, v. 68, no. 1, 1980.
18-6:	Carto-Graphics. Topography reprinted with permission from Bambach, Scotese, and Ziegler, "Before Pangaea: The Geographies of the Paleozoic World, *American Scientist*, v. 68, no. 1, 1980.
18-7:	Precision Graphics. Reprinted with permission from L.L. SIOSS, "Sequences in the cratonic interior of North America," *GSA Bulletin,* v. 74, 1963, p. 110 (Fig. 6), Geological Society of America.
18-8:	Carto-Graphics
18-9:	Carto-Graphics
18-10a:	Precision Graphics
18-10c:	Precision Graphics
18-11:	Carto-Graphics

18-12:	Precision Graphics
18-13:	Carto-Graphics. Reprinted with permission from Dietz and Holden, "Reconstruction of Pangaea: Breakup and Dispersion of Continents, Permian to Present," *Journal of Geophysical Research,* 1970, v. 75, no. 26, pp. 4939-4956, ©1970 by the American Geophysical Union.
18-14:	Carto-Graphics
18-15:	Carto-Graphics
18-16:	Carto-Graphics
Perspective 18-1, Figure 1:	Stephen J. Kraseman/Photo Researchers
18-19:	Precision Graphics. Reprinted by permission of the Geological Society and A.M. Spencer, ed., *Mesozoic-Cenozoic Orogenic Belts* (Bath: Geological Society Publishing House, 1974).
18-20:	Precision Graphics
18-21:	Precision Graphics. Figure 175 (Page 386) from *Structural Geology of North America,* 2nd Edition by A.J. Eardley. Copyright ©1962 by A.J. Eardley. Copyright renewed. Reprinted by permission of HarperCollins, Publishers, Inc.
18-22:	Precision Graphics. Reprinted with permission from W.R. Dickinson, "Cenozoic Plate Tectonic Setting of the Cordilleran Region in the U.S.," in *Cenozoic Paleogeography of the Western U.S.,* Pacific Coast, Symposium 3, 1979, pp. 10-11 and pp 2,4.
18-23:	Precision Graphics

CHAPTER 19

Opener:	Smithsonian Institution. Transparency No. 86-13471A.
19-1a:	Smithsonian Institution
19-1b:	Smithsonian Institution
19-1c:	Smithsonian Institution
19-1d:	Smithsonian Institution
19-3a:	Photo courtesy of Nevill Pledge, South Australian Museum
19-3b:	Photo courtesy of Nevill Pledge, South Australian Museum
19-3c:	Photo courtesy of Nevill Pledge, South Australian Museum
19-3d:	Photo courtesy of Nevill Pledge, South Australian Museum
19-3e:	Photo courtesy of Nevill Pledge, South Australian Museum
19-3f:	Precision Graphics
19-4:	Precision Graphics
19-5:	Cernegie Museum of Natural History
19-6:	From the Field Museum, Chicago # Geo80821c
19-7:	©American Museum of Natural History, Trans. #K10269
19-8:	Precision Graphics. Reprinted with permission from Raup and Sekoski, "Mass Extinctions in the Marine Fossil Record," Science, v. 215, 1982, p. 1502, ©1982 by the AAAS.
19-9:	Carlyn Iverson
19-10:	Carlyn Iverson
19-11:	Carlyn Iverson
19-12:	Photo courtesy of Dianne Edwards, University College, England

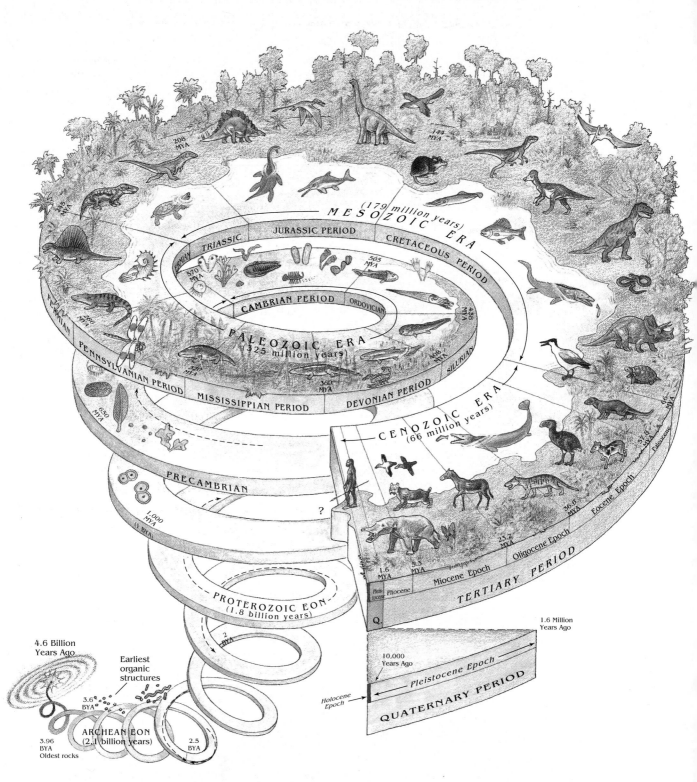

➤ Geologic Time Depicted in a Spiral History of the Earth